A.

TED.

Heinhold
Power Cables and their Application
Part 1

Authors	Section
Lothar Heinhold	1, 2.3–7.3, 12, 13, 15, 17, 28–30, 33–36.2, 37, 38
Karl-Heinz Ittmann	8–11.1
Dirk Rittinghaus	11.2
Alfred Roller	26
Friedrich Schaller	27
Otto-Ernst Schröter	14
Reimer Stubbe	31, 32
Herbert Sutter	36.3–36.6
Rudolf Wiedenmann	2.1, 2.2
Franz Winkler	16, 18–25

Translators

Wolf T. Bothmann and Fred E. Waspe

Power Cables and their Application

Part 1

Materials · Construction
Criteria for Selection
Project Planning
Laying and Installation · Accessories
Measuring and Testing

Editor: Lothar Heinhold

3rd revised edition, 1990

Siemens Aktiengesellschaft

Deutsche Bibliothek Cataloguing in Publication Data:

Power cables and their application / ed.: Lothar Heinhold. – Berlin ;
München : Siemens-Aktienges. [Abt. Verl.]
  Einheitssacht.: Kabel und Leitungen für Starkstrom <engl.>
NE: Heinhold, Lothar [Hrsg.]; EST
Pt. 1. Materials, construction, criteria for selection, project planning, laying and installation,
  accessories, measuring and testing. – 3., rev. ed. – 1990
  ISBN 3-8009-1535-9

Title of German original edition:
Kabel und Leitungen für Starkstrom, Teil 1
Herausgeber: Lothar Heinhold, Siemens Aktiengesellschaft 1987
ISBN 3-8009-1472-7

The following product names referred to in this book are registered trade marks of Siemens AG.

ARCOFLEX, CORDAFLEX, EVATHERM, FLEXIPREN, OZOFLEX, PLANOFLEX, PROTODUR,
PROTOFIRM, PROTOFLEX, PROTOFORM, PROTOLIN, PROTOLON, PROTOMONT,
PROTOTHEN, SIENOPYR, SIFLA, SIKABIT, SINOTHERM, SUPROMONT

ISBN 3-8009-1535-9

Published by Siemens Aktiengesellschaft, Berlin and Munich

Preface to the Third Edition

Progress in the fields of materials and production engineering has led further to new developments in the technology of power cables. On that account a complete revision of the book was necessary in order to reflect the present state of the technique. The update and current national and international specifications should also be covered. It is to be expected that on the one hand, the routine work will lead to further important changes in a few years, while on the other hand, particularly with regard to the medium-voltage cables, the domestic and world market will cause a selective influence on the enormous number of types available at present. For this reason, in the 4th edition, information is given, as far as possible, on the development trends which are emerging; furthermore the state of technique is described.

This edition, like previous ones, gives a review as comprehensive as possible on the application-oriented technology of power cables. In fields in which specialized literature is available, a brief general review is followed by corresponding information. In other fields, as for example in the section "Planning of Cable Installations", besides the data and calculation methods necessary for the solution of problems, an introduction to the fundamental is presented. Great stress is set on particular aspects of the technology of cable materials. This is a useful aid to everyone responsible for designing, installing and operating power cables.

In order to handle the book easily, although many topics are covered, the tables giving constructional data and electrical values as well as brief explanations and information are collected in a separate volume (Part 2).

We wish to thank all those who have participated in the preparation of this book for their efforts and advice given and for documents made available. Our special gratitude is due to Dirk Rittinghaus for his assistance with technical presentation, revisional work and proof reading.

Erlangen, February 1990

Lothar Heinhold

Observations on the German terms 'Kabel' and 'Leitungen' and the VDE Specifications

'Kabel' and 'Leitungen'

Power cables are used for the transmission of electrical energy or as control cables for the purpuses of measurement, control and monitoring in electric power installations. In German usage, a distinction is made traditionally between 'Kabel' and 'Leitungen'.

'Leitungen' (literally 'leads') are used, generally speaking, for wiring in equipment, in wiring installations and for connections to moving or mobile equipments and units. The term can thus be translated as 'insulated wires' or 'wiring' or 'flexible cables' or 'cords'.

'Kabel' (cables) are used principally for power transmission and distribution in electricity supply-authority systems, in industry and in mines etc.

With the use of modern insulating and sheathing materials the constructional differences between 'Kabel' and 'Leitungen' are in many cases no longer discernible.

The distinction is therefore observed purely in terms of the area of application, as described in DIN VDE 0298 Part 1 for power cables and Part 3 for wiring and flexible cables, and in the design specifications referred to therein, e.g. DIN VDE 0250 for wiring and flexible cables and DIN VDE 0271 for PVC insulated cables.

Further factors in the choice between 'Kabel' and 'Leitungen' are the equipment Specifications (e.g. DIN VDE 0700), the installation Specifications (e.g. DIN VDE 0100) or the operating stresses to be expected.

It can be taken as a rule of thumb that 'Leitungen' must not be laid in the ground, and that cables of flexible construction are classified as 'Leitungen', even if their rated voltage is higher than 0.6/1 kV – e.g. trailing cables. This apart, there are also types of 'Kabel' that are not intended for laying in the ground (e.g. halogen-free cables with improved performance in conditions of fire to DIN VDE 0266, or ship wiring cables to DIN VDE 0261).

In the present translation the terms 'cable' and 'power cable' have been used to include flexible and wiring cables where there is no risk of confusion.

VDE Specifications

From considerations of consistency in references and for greater clarity, the VDE Specifications applicable to power cables are generally quoted in accorcance with the new practice as 'DIN VDE ....'.

This applies equally to the older specifications which still retain the designation 'VDE ...' or 'DIN 57.../VDE ...' in their titles. Furthermore, since these specifications are of fundamental significance, the practice of quoting the date of publication has been dispensed with.

Contents

Constructional Elements of Insulated Cables

1 Conductors

The conductors in wiring cables and flexible cables consist nowadays of copper (Cu). The use of aluminium (Al), as well as copper, is also common in power cables. The cross-sectional area of the conductor is quoted basically not as the *geometrical* but as the *electrically effective* cross-sectional area, i.e. the cross-sectional area as determined by a resistance measurement.

In the international standard for copper, IEC 28 'International Standard of Resistance for Copper', the standard value for the resistivity at 20 °C is given as $\varrho_{20} = \frac{1}{58} = 0.017241 \; \Omega \, \mathrm{mm}^2/\mathrm{m}$. The temperature coefficient α_{20} at 20 °C for this copper is $\alpha_{20} = 3.93 \times 10^{-3}/\mathrm{K}$. This value increases or decreases approximately in proportion to the conductivity. Investigations have shown that the product of the temperature coefficient and the resistivity with different conductivities is nearly constant at $0.6776 \times 10^{-4} \; \Omega \, \mathrm{mm}^2/\mathrm{m} \, \mathrm{K}$.

Similar relationships exist for aluminium. In this case, IEC 111 'Resistivity of Commercial Hard Drawn Aluminium Conductor Wire' gives the resistivity at a temperature of 20 °C as $\varrho_{20} = 0.028264 \; \Omega \, \mathrm{mm}^2/\mathrm{m}$ and the temperature coefficient as $\alpha_{20} = 4.03 \times 10^{-3}/\mathrm{K}$. This coefficient is proportional to the degree of purity of the aluminium, and decreases with increasing impurity in the same way as the electrical conductivity. Here again, the product of resistivity and temperature coefficient remains approximately constant, in this case at $1.139 \times 10^{-4} \; \Omega \, \mathrm{mm}^2/\mathrm{m} \, \mathrm{K}$.

The temperature dependence of the resistivity is given in general by

$$\varrho_{\vartheta_2} = \varrho_{\vartheta_1}[1 + \alpha_{\vartheta_1}(\vartheta_2 - \vartheta_1)] \tag{1.0}$$

Thus:

for copper,

$$\varrho_\vartheta = \varrho_{20} + 0.68 \times 10^{-4}(\vartheta - 20) \; \Omega \, \mathrm{mm}^2/\mathrm{m} \tag{1.1}$$

for aluminium,

$$\varrho_\vartheta = \varrho_{20} + 1.1 \times 10^{-4}(\vartheta - 20) \; \Omega \, \mathrm{mm}^2/\mathrm{m} \tag{1.2}$$

with the temperature ϑ expressed in °C.

In the planning of cable installations, however, in view of the unavoidable uncertaincies in the given information, it is quite sufficient to calculate with the conventional temperature coefficients (see page 320):

for copper,

$$\alpha_{20} = 3.93 \times 10^{-3}/\mathrm{K}$$
$$\alpha_0 = 4.26 \times 10^{-3}/\mathrm{K} \tag{1.3}$$

$$\Theta_0 = \frac{1}{\alpha_0} = 234.5 \; \mathrm{K} \tag{1.4}$$

for aluminium,

$$\alpha_{20} = 4.03 \times 10^{-3}/\mathrm{K}$$
$$\alpha_0 = 4.38 \times 10^{-3}/\mathrm{K} \tag{1.5}$$

$$\Theta_0 = \frac{1}{\alpha_0} = 228 \; \mathrm{K} \tag{1.6}$$

In general,

$$\alpha_\vartheta = \frac{1}{\Theta_0 + \vartheta} \; 1/\mathrm{K}. \tag{1.7}$$

To convert a measured conductor resistance to the reference conditions of 20 °C and 1000 m length, the following expressions are applicable, according to IEC 228, 1966:

for copper,

$$R_{20} = R_\vartheta \frac{254.5}{234.5 + \vartheta} \times \frac{1000}{l} \; \Omega/\mathrm{km} \tag{1.8}$$

for aluminium,

$$R_{20} = R_\vartheta \frac{248}{248 + \vartheta} \times \frac{1000}{l} \; \Omega/\text{km} \qquad (1.9)$$

where

ϑ conductor temperature (°C)
R_ϑ measured conductor resistance at ϑ °C (Ω)
l length of cable (m)
R_{20} conductor resistance at 20 °C (Ω/km)

To permit the economical construction of cables with a small number of wire gauges, the conductor design has been slightly altered in accordance with IEC 228 (for details see IEC 228, 1966) and the resistance determined according to the expression

$$R_{20} = \frac{4A}{n \cdot \pi \cdot d^2} K_1 \, K_2 \, K_3 \; \Omega/\text{km} \qquad (1.10)$$

where

A resistivity at 20 °C
    for copper, $A = 17.241 \; \Omega\text{mm}^2/\text{km}$
    for aluminium, $A = 28.264 \; \Omega\text{mm}^2/\text{km}$
n number of wires in the conductor
d diameter of individual wires (mm)
K factor to allow for the effects of manufacturing processes:
    K_1 for wire diameter and surface treatment
    K_2 for conductor stranding
    K_3 for core stranding

Because of improved manufacturing techniques, particularly the compaction of stranded circular and sector-shaped conductors, the basic principles which had underlain the establishment of conductor resistances had lost something in validity, so that a revision of the existing IEC and VDE specifications became necessary. In particular the differences in the resistance values for solid and stranded conductors, and for single- and multi-core cables, in the former ranges were no longer applicable. It was thus possible in the 1978 edition of IEC 228 to achieve greater consistency of resistance value and a reduction in the number of conductor classifications from six to four. In 1980 this international agreement was incorporated into the standards for power cables, wires, and wiring cables and flexible cables (DIN VDE 0295). The new values are taken into account in the tables and planning sheets in the present book.

As well as plain aluminium conductors, the use has been tried in some countries of nickel-plated or tinned aluminium, and the so-called copper-clad alu-minium for conductors in wiring cables for fixed installations. These types, also mentioned in the new IEC specification, have not, however, been generally accepted so far.

The minimum number and the diameter of the wires and the resistance of the conductor are laid down in IEC 228 and DIN VDE 0295 (see also pages 455 to 457). Cables used abroad embody conductors in accordance with the respective national specifications, in the case that these differ from IEC.

If the conductors are insulated with a material which provokes an adverse chemical reaction with the copper, a metallic protective layer round the copper wire is necessary, e.g. of tin or some other barrier (see page 27).

1.1 Wiring Cables and Flexible Cables

Types of Conductor

For flexible and wiring cables in the Federal Republic of Germany, with few exceptions, circular copper conductors are used. These are aimed at two areas of application:

For Fixed Installation

The cables are subject to mechanical stresses due to bending only during installation. Accordingly, solid conductors are preferably used up to cross-sectional area of 10 mm^2 and stranded conductors above 10 mm^2.

For the Connection of Mobile Equipment

These cables, since they have to be flexible, embody fine-stranded conductors for all cross-sectional areas. Where a particularly high degree of flexibility is necessary, e.g. in the leads to welding-electrode holders,

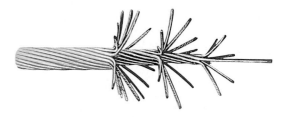

Fig. 1.1
Multiple stranded, circular flexible conductor

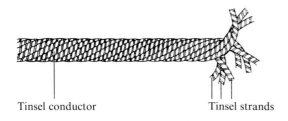

Tinsel conductor — Tinsel strands

Fig. 1.2 Tinsel conductor

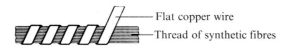

— Flat copper wire
— Thread of synthetic fibres

Fig. 1.3 Construction of tinsel strand

the conductor strands are made up of a number, appropriate to the cross-sectional area of the conductor, of extra fine substrands (multiple stranded, circular flexible conductors, Fig. 1.1). For very flexible connecting cords of very small cross-sectional area, e.g. 0.1 mm² for electric shavers, tinsel conductors (Figs. 1.2 and 1.3) are used.

Multiple stranded circular flexible conductors (Fig. 1.1) consist of strands whose individual wires are themselves stranded or bunched. The ability of the conductor to withstand mechanical stresses and its flexibility depend particularly on the stranding arrangement, as well as on the quality and diameter of the wires. The shorter the lay of the strands and substrands, the greater the flexibility and the ability to withstand bending. The strands may be laid in the same direction in all layers (uniform-lay stranding) or the direction may alternate from layer to layer (reversed-lay stranding). Conductors with uniform-lay stranding are preferred in flexible cables for hoists because of their better running behaviour when changing direction over rollers.

Tinsel conductors (Fig. 1.2) are made up of a number of tinsel threads stranded together. Each thread (Fig. 1.3) consists of a textile core with a helical wire strip (copper strip 0.2 to 0.3 mm wide and 0.01 to 0.02 mm thick).

1.2 Power Cables

Copper Conductors

Solid conductors are preferred up to 16 mm² cross-sectional area, stranded conductors for 25 mm² and above.

Given and adequate ability to withstand bending, the conductors should have a space factor which, together with the chosen conductor section, results in good utilization of the cross-sectional area of the cable. Accordingly, where possible, compacted circular conductors, or, if the cable construction permits, compacted sector-shaped conductors, are used. The space factor defines the percentage of the geometrical cross-sectional area of a conductor that is occupied by the individual wires. The construction of single-core cable and three-core separately-leaded (S.L.) cable requires the use of circular conductors.

Aluminium Conductors

DIN VDE 0295 permits the use of circular solid and stranded aluminium conductors from 25 mm² upwards and sector-shaped conductors from 50 mm² upwards.

Solid conductors are preferred in cables with polymer insulation and sector-shaped conductors in the range of cross-sectional areas from 50 to 185 mm². Single-core cables normally have stranded circular conductors: solid conductors are usually used only in laid-up single-core cables in cases of high thermal loading, because of the problems of thermal expansion (see page 292).

If cables with polymer insulation and aluminium protective (P) or neutral (PEN) conductors are laid in the ground or in an agressive atmosphere, in the event of damage to the sheath and the insulation these conductors may be open-circuited in the course of time through corrosion. The possibility of damage must therefore be taken into account, when such cables are installed, by the selection of appropriate protective measures.

Milliken Conductors

For high-power transmission with conductor cross-sectional areas of 1200 mm² or more, special measures are necessary to keep additional losses due to skin effect within tolerable limits. To this end, either the individual conductor strands are provided with an insulating layer (e.g. enamel) and so laid-up that

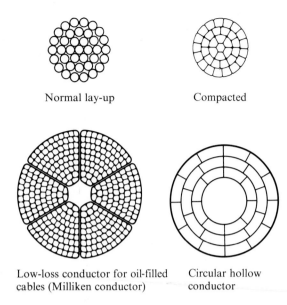

Normal lay-up

Compacted

Low-loss conductor for oil-filled cables (Milliken conductor)

Circular hollow conductor

Fig. 1.4
Construction of multi-core circular conductors

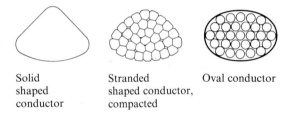

Solid
shaped
conductor

Stranded
shaped conductor,
compacted

Oval conductor

Fig. 1.5
Construction of sector shaped conductors

their position within the cross-section of the conductor changes periodically along the lenght of the conductor, or the conductors are made up of separate stranded, sector-shaped elements which are wrapped in conducting paper (Fig. 1.4). This latter type is also known as the milliken conductor.

Single-core oil-filled cables require a hollow conductor, while external-gas-pressure pipe type cables require oval conductors.

Superconductors

The most suitable conductor materials for superconducting cables are pure niobium and niobium-tin, whose critical temperatures are around 9.5 K and 18.4 K respectively. Since the current flows only in a very thin surface layer (0.1 µm), there is no need for the whole conductor to consist of this relatively expensive superconducting material. It is sufficient if a thin layer (10 to 100 µm) is deposited on a carrier material, e.g. high-purity copper or aluminium. The carrier metals must be so disposed that they are not traversed by the magnetic field of the conductor, and the generation of eddy-current losses is avoided (Fig. 1.6).

The development of superconducting cables is as yet in the early stages, although 110 kV cables capable of transmitting 2500 MVA have already been produced for experimental purposes.

Fig. 1.6
Model of a flexible superconducting cable core. Constructed of aluminium wires each with a thin coating of Niobium laid-up over a PE carrying tube. Above this an insulation of polymeric plastic foil is applied followed by the concentric return conductor and a profiled PE tape as protective layer

2 Insulation

For the insulation of wiring cables and flexible cables, synthetic materials and natural rubber are used, and for power cables, as well as these, impregnated paper. As a result of the development which has taken place in recent years, these materials can be produced with various electrical, thermal and mechanical properties according to their intended purpose. It is thus possible to manufacture cables for specific requirements and fields of application.

2.1 Polymers

A polymer is a macromolecule composed of a large number of basic units, the monomers. If the macromolecule is synthesized using only one kind of monomer, the product is a homopolymer. If the polymer chains are made up of two different types of monomer, the result is a copolymer, and of three different types a terpolymer.

Most of the important insulating materials are today produced synthetically. Only in the case of elastomers are partly natural products still of technical significance.

Technically important polymers are classified (Table 2.1) according to their physical properties as

▷ thermoplastics (plastomers),
▷ elastomers and
▷ thermosetting polymers (duromers).

The polymers principally used in cable engineering are listed in Table 2.2.

It is worth noting that materials which do not fit into this classification of thermoplastics, elastomerics and thermosetting materials are finding increasing application in cable engineering. These include the cross-linked polyolefines (e.g. cross-linked polyethylene), which behave as elastomers above the critical melting point, as manifested particularly in the heat-pressure characteristics at high temperatures (Fig. 2.1).

Also in this category are the so-called thermoplastic elastomers with their characteristic thermoplastic behaviour at processing temperatures and elastomeric characteristics at the temperatures at which they are used.

Table 2.1   Technically important polymers classified according to their physical properties

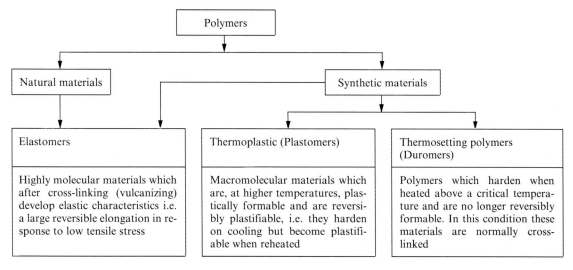

Table 2.2   Summary of the most important polymers used in the manufacture of cables

Thermoplastics (Plastomers)		Cross-linked Thermo-plastics	Thermoplastic Elastomers	Elastomers		Duroplastics (Duromers)	
Polyvinylchloride	PVC	Cross-linked Poly-ethylene XLPE	Blends of Polyfines and Natural Rubber	Natural Rubber	NR	Epoxy Resin	EP
Polyethylene	PE			Butyl Rubber (Isoprene Isobutylene Rubber)	IIR	Polyure-thane	
Ethylene Vinyl-Acetate Copolymer (VA ≤ 30%)	EVA		Three Block [2] Polymer Styrene-Alkylene-styrene			Resin	PUR
		Cross-linked Ethylene Copolymer		Styrene-Butadien Rubber	SBR		
Ethylene-Acrylate-Copolymer, e.g.:				Nitrile-Butadien Rubber	NBR		
Ethylene-Ethyl-Acrylate	EEA		Thermoplastic Polyurethane (PUR) and Polyester				
Ethylene-Butyl-Acrylate	EBA			Ethylene-Propylene Rubber	EPR [1]		
Polypropylene	PP						
Polyamide	PA			Ethylene-Propylene Diene Monomer Rubber	EPDM [1]		
Ethylene-Tetrafluoro-ethylene Copolymer	ETFE			Polychloroprene	CR		
Tetrafluoroethylene-Hexafluoropropylene-Copolymer (Fluorinated Ethylene Propylene)	FEP			Chlorsulphonyl Polyethylene	CSM		
				Chlorinated Poly-ethylene	CM		
				Silicone Rubber	SiK		
				Epichlorohydrin Rubber	ECO		
				Ethylene-Vinyl-Acetate-Copolymer (VA > 30%)	EVA		

[1] The general term for EPR and EPDM is EPR
[2] Block polymer: a copolymer whose chain is composed of alternating sequences of identical monomer units

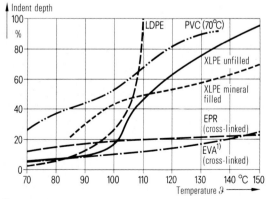

[1] VA content ≥ 30%

Fig. 2.1
Heat-pressure characteristics of polyolefines.

Heat-pressure test to DIN VDE 0472
Test sample: conductor 1.5 mm² with insulation 0.8 mm thick,
Test duration: 4 h

Determination of load using the formula:

$$F = 0.6 \cdot \sqrt{2 \cdot D \cdot \delta - \delta^2}$$

F Load in N
D Diameter of core in mm
δ Mean wall thickness of insulation in mm

2.1.1 Thermoplastics (Plastomers)

Thermoplastics are made up of linear or branched macromolecules, and unlike the elastomers and thermosetting polymers have reversible forming characteristics. The combination of properties of thermoplastics are determined by their structural and molecular arrangement. The thermoplastic polyethylene (PE) has the simplest structure of all plastics (Fig. 2.2).

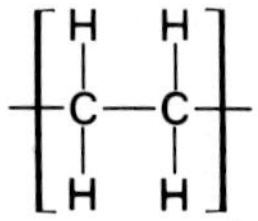

Fig. 2.2 Structural form of Polyethelene (PE)

In the so-called high-pressure polymerization of ethylene, chain molecules with lateral alkyl groups are formed by radical initiation (LDPE – *Low-D*ensity *PE*). Ionic polymerization at low pressure, on the other hand, leads to linear, very little branched chains (HDPE – *H*igh-*D*ensity *PE*). The less branched the chain molecules of a polyethylene are, the greater is its possible crystallinity. With increasing crystallinity, melting temperature, tensile strength, Young's modulus (stiffness), hardness and resistance to solvents increase, while impact strength, resistance to stress cracking and transparency decrease. Like all thermoplastics, the polyolefines – as in the case of e.g. polyethylene and polypropylene – also consist of a mixture of macromolecules of different sizes, and it is possible to control the mean molecular weight and the molecular weight distribution within certain limits through the choice of suitable polymerization conditions.

In the technical data sheets of the raw material manufacturers, instead of the mean molecular weight, the melt flow index[1] (for polyolefines) or the so-called K value (for polyvinyl chloride, PVC) is quoted (see page 18).

The mean molecular weight and the molecular weight distribution have a considerable effect on the mechanical properties. Thus, as a rule, tensile strength, elongation at tear and (notched) impact strength in-

crease with increasing chain length, as also the viscosity of the plasticized material. It should be borne in mind, however, that with increasing melting viscosity the material becomes more difficult to work.

The molecular chains (polyethylene, polyvinyl chloride) resulting from the synthesizing reactions, e.g. the polymerization of suitable monomers (ethylene, vinyl chloride) are formed by atomic forces (primary bonds). The cohesion of the molecular chains is due to secondary forces. In the polyolefines, for example, the dispersion or van der Waal forces predominate. In this case the forces of attraction between the molecules are unpolarized. In plastics with polarized groups, besides the dispersion forces, dipole orientation forces between the chains are also effective (e.g. in PVC). Strong forces of attraction between the chain molecules are also represented by the hydrogen bridges, as, for example, in polyamides, polyurethanes and fluoroplastics. With symmetrical structures the thermoplastics bonded by dispersion, dipole or hydrogen bonds tend towards crystallization. They are then hard and tough, and of high strength, and the softening range is small. To the extent that the macromolecular structure is asymmetrical (e.g. in PVC), the tendency to crystallization is reduced and the softening range extended.

Awareness of these relationships now makes it possible to manufacture plastics tailored to their application requirements. In addition to standard thermoplastic PVC and PE, thermoplastics and elastomers produced by specifically directed copolymerization of ethylenes with other copolymerable monomers have assumed significance in cable engineering.

Copolymers

The thermoplastic copolymers most frequently used in cable engineering are based on ethylene and are produced by copolymerization with vinyl acetate (EVA copolymer) or with alkyl acrylates (EEA and EBA copolymers). EVA copolymers with a vinyl acetate content up to 30% contain methylene units in crystalline formation and are therefore workable as thermoplastics. With a further increase in the vinyl acetate (VA) content the product becomes rubbery.

Polyethylenes and the ethylene copolymers, such as e.g. EVA, are of special significance in cable engineering because these thermoplastics can be cross-

---

[1] The melt flow index (MFI) is the quantity of material in g which under a fixed force is extruded through a given sized jet in a period of 10 minutes

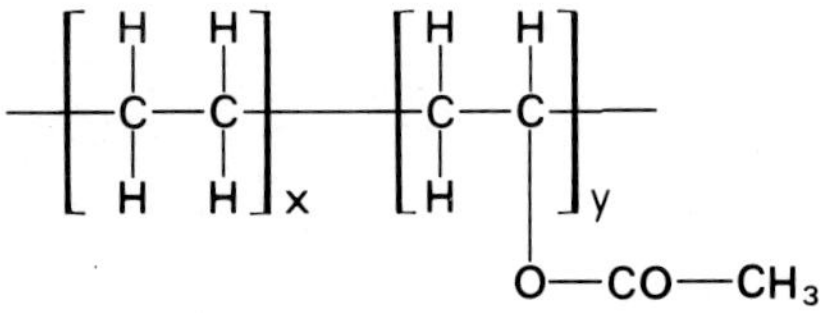

Fig. 2.3 Structural form of EVA

linked by means of organic peroxides or high-energy radiation. Cross-linking increases the thermomechanical stability – i.e., with a temperature loading beyond the crystallite melting point of the cross-linked thermoplastics the material no longer exhibits thermoplastic, but rather thermoelastic characteristics.

Fluoroplastics

Fluoroplastics are characterized by an outstanding combination of properties, such as good thermal stability, excellent electrical characteristics and high resistance to chemical attack and flame resistance. The best known fluoropolymers in cable engineering are the thermoplastically workable copolymers of ethylene and tetrafluoroethylene (ETFE) and of tetrafluoroethylene and hexafluoropropylene (FEP) (Fig. 2.4).

The various mechanical properties of the polymers (e.g. tensile strength, extension, elasticity and cold resistance), the various resistances to external influences (e.g. acids, alkalies, oil) and their electrical and thermal characteristics determine the areas of application of the cables in which they are used for insulation and sheathing.

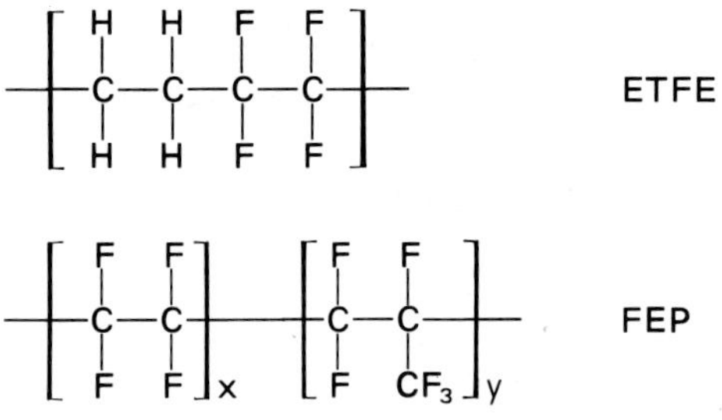

Fig. 2.4 Structural form of ETFE and FEP

Polyvinyl Chloride (PVC)

Among the insulating materials used for flexible and wiring cables, plastic compounds based on polyvinyl chloride (PVC) have assumed particular significance.

The starting material, the vinyl chloride, is nowadays produced mainly by the chlorination of ethylene (Fig. 2.5). It can be converted to polyvinylchloride by the emulsion (E-PVC), suspension (S-PVC) or mass polymerization (M-PVC) method.

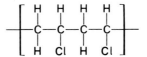

Fig. 2.5 Structural form of PVC

For insulating and sheathing mixtures in cable engineering, PVC obtained by the suspension method is usually used. These types of PVC, offered by the chemical industry as S-PVC, are distinguished by their grain structure and K value. The K value, according to Fikentscher (DIN 53726), characterizes the mean molecular weight of the PVC. The grain structure is significant from the point of view of the processing of the compound. For the manufacture of soft PVC compounds for the cable industry, an S-PVC with porous grain (plasticizer sorption) and a K value of about 70 has become generally accepted.

PVC and additives like plasticizers, mineral fillers, antioxidants, coulering pigment a.s.o. are prepared in a mixing and gelling process, under heat, to produce the working compound.

The compound, usually in granular form, is pressed onto the conductor as insulation, or onto the core as a sheath, by means of extruders.

Pure PVC resulting from polymerization is unsuitable for use as an insulating and sheathing material for flexible and wiring cables, because at its service temperature it is hard and brittle, and also thermally unstable. It is only through the incorporation of additives that the mechanical/thermal and electrical characteristics necessary in such materials, together with good processing properties, are obtained.

The most important additives are:

Plasticizers

The plasticizers normally used are esters of organic acids, such as DOP (Di-2-ethylhexylphthalate) or DIDP (Di-isodecylphthalate). Esters of azelain or sebacic acid are used for compounds with especially good cold resistance, while those for higher service temperatures contain trimellith acid esters or polyester plasticizers.

Stabilizers

These confer thermal and thermal oxidization stability on the PVC compound during processing and in service. Principally used as stabilizers are lead salts such as basic lead sulphate or lead phthalate. Antioxidants are necessary in addition, to prevent, for example, deterioration of the plasticizer through oxidation.

Fillers

These are used to obtain a specified combination of characteristics. In addition they contribute to reduce the cost. The most used fillers for PVC compounds are calcium carbonate and kaolin.

Lubricants

These improve the workability. Stearates are usually used.

PROTODUR Flexible and Wiring Cables

Cables with PVC insulation manufactured by Siemens are known by the trade name PROTODUR. They can be laid without special precautions in ambient temperatures above $-5\,°C$. If the cables are colder than this, they must be carefully warmed before installation. Flexible and wiring cables are generally of smaller diameter than power cables, and are therefore subject to lower stresses in installation, so that with careful handling they can be laid at lower temperatures. For countries such as Norway, Sweden or Finland, PVC compounds are available which afford the necessary bending capability down to low temperatures.

For installations with especially stringent requirements as to burning behaviour, compounds for cables have been developed which satisfy the bunched cable burning test, Test Category 3, of DIN VDE 0472, Part 804, lead to a lower emission of smoke and gas and do not release hydrogen chloride (see pages 79 and 125).

Polyethylene (PE)

Polyethylene is a macromolecular hydrocarbon with a structure similar to that of the paraffins (for the structural formula see page 17). This material, with its excellent dielectric properties, is used as an insulating material in power cable engineering in both non-cross-linked (thermoplastic PE) and cross-linked (XLPE) form. The power cables produced by Siemens with thermoplastic polyethylene insulation are known by the protected trade name PROTOTHEN-Y and those with cross-linked polyethelene insulation by the trade name PROTOTHEN-X. Of the wide range of types of polyethelene offered by the chemical industry, only specially prepared, purified and stabilized types are used in cable engineering.

Because both thermoplastic and cross-linked polyethelene are sensitive to ionization discharges, it is necessary for cables with rated voltages from $U_0/U = 3.5/6$ kV upwards to incorporate conducting layers over and under the insulation. The inner layer usually consists of a weakly conducting alkyl copolymer. Various methods were formerly used to provide the outer conducting layer:

▷ graphitizing or conducting lacquer or a conducting adhesive with weakly conducting tape applied to it;

▷ extruded conducting layers, which were either applied in a separate process or extruded in the same process with the insulation.

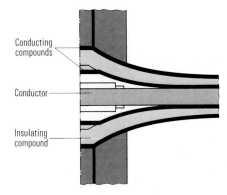

Fig. 2.6 Schematic arrangement of triple extrusion

According to the new specifications of DIN VDE 0273/..87, only outer conducting layers extruded with and bonded to the insulation are permitted.

The extruded conducting layers are very thin, and so firmly bonded to the insulation that they can be separated from it only with a scraper. In some countries conducting layers are used whose adhesion is somewhat lower, so that – if necessary after scoring with a tool – they can be stripped by hand (cables with strippable conducting layers). Because of the force required in the stripping operation, such layers are made somewhat thicker.

To ensure operational reliability in medium- and high-voltage power cables, it is particularly important, apart from using high-purity material and observing appropriate cleanliness in the manufacturing processes, that the insulation and the conducting layers should be free of bubbles, and that there should be good adhesion between the conducting layer and the insulation. According to DIN VDE 0273 this must be checked on every manufactured length by means of an ionization test.

In comparison with high polymers with polarized structures, such as PVC, high polymers with unpolarised structures, such as PE and XLPE, are characterized by outstanding electrical characteristics. They have, however, poor adhesion properties in relation to other materials, e.g. moulding compounds. This characteristic has to be taken into account in the design of accessories.

For the low-voltage range a polyethylene insulation compound has been successfully developed which bonds well to accessory materials and thus ensures the water-tightness of joints.

PROTOTHEN-Y

It is not usual to use thermoplastic polyethylene in power cables for $U_0/U = 0.6/1$ kV, because of the high conductor temperatures to be expected under short-circuit conditions. For higher rated voltages, while it offers advantages in comparison with PVC and paper insulation because of its satisfactory dielectric properties, it has declined in significance as power cable insulation, because of its poor heat/pressure characteristics (Fig. 2.1), in comparison with cross-linked polyethylene, and has been omitted from the new specification VDE DIN 0273/..87.

Cross-Linked Polyethylene (XLPE)

PROTOTHEN-X

The linear chain molecules of the polyethylene are knitted by the cross-linking into a three-dimensional network. There is thus obtained from the thermoplastic a material which at temperatures above the crystallite melting point exhibits elastomeric properties. By this means the dimensional stability under heat and the mechanical properties are improved. As a result, conductor temperatures up to 90 °C can be permitted in normal operation and up to 250 °C under short-circuit conditions.

There are three principal methods for cross-linking polyethylene insulation materials:

Cross-linking by Peroxides

Organic radical components, in particular specific organic peroxides, are incorporated. These decompose at temperatures above the extruding temperature, into highly reactive radicals. These radicals interlink the initially isolated polymer chains in the thermoplastic in such a way that a space network results (Fig. 2.7).

Formerly, polyethylene cable insulation was cross-linked mainly by 'continuous vulcanization in a steam tube', in the so-called CV[1] method (Fig. 2.8).

In this method the polyethylene, mixed with the peroxide as a cross-linking initiator, is pressed onto the conductor, by means of an extruder, at about 130 °C (below the temperature at which the peroxide decomposes). Following this, in the same process, the insulated core is passed through a tube, about 125 m long, filled with saturated steam at high pressure. At a pressure of 16 to 22 bar and a temperature of about 200 to 220 °C, the organic peroxide decomposes into reactive primary radicals, which effect the cross-linking. The cross-linking process is followed immediately by a cooling stage. This must similarly take place under pressure in tubes 25 to 50 m long, to prevent the formation of bubbles in the vulcanized material through the presence of gaseous products of the peroxide reaction.

An alternatives to this 'classical' cross-linking process, methods have been developed in which gases or liquids, e.g. silicone oil or molten salts (salt bath cross-linking) are used as media for the heat transfer.

---

[1] CV: continuous vulcanisation

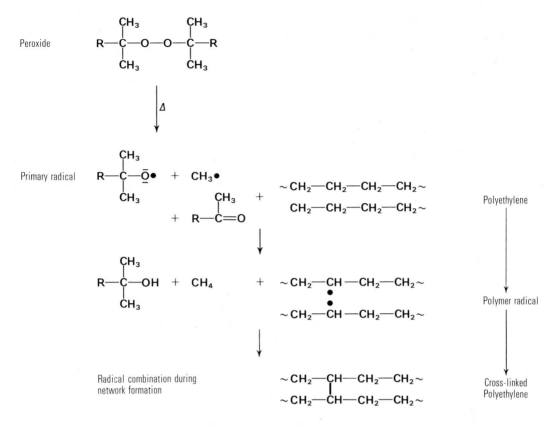

Fig. 2.7   Cross-linking of Polyethylene by organic peroxides

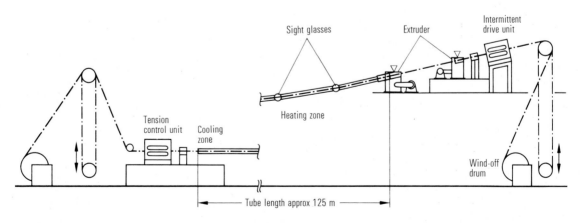

Fig. 2.8   Continuous cross-linking in a steam tube (CV process)

Compared to vulcanisation with steam, these methods permit cross-linking at higher temperatures and lower pressures.

Cross-linking by Electron Beams

The polymer chains are cross-linked directly by means of high-energy electron beams, without the necessity for the heating stage which is essential with peroxides. It will be clear from consideration of the reaction sequence in the cross-linking of polyethylene by electron irradiation, as illustrated in simplified form in Fig. 2.9, that in this case also gaseous reaction products are formed (mainly hydrogen).

Cross-linking by Siloxane Bridges

Polyolefines can also be cross-linked by means of siloxane bridges, whereby suitable alkoxysilanes are radically grafted into the polymer chains. In the presence of moisture and a condensation catalyst, hydrolysis takes place to form silanol groups, which then condense to the interlinking siloxane bonds (Fig. 2.10).

Because the grafted silane can contain up to three reactive alkoxy groups, this offers the possibility that bundled linking locations can be formed.

Although as regards the chemical structure of the cross-linking bridges the cross-linked polymer matrix appears to be quite different from those produced by the methods previously described, a combination of characteristics is nevertheless obtained which essentially corresponds to that of the cross-linked PE produced by the classical methods.

Like all polyolefines, cross-linked polyethylene is subject to a time and temperature-dependent oxidative decomposition, and it has to be protected against this by the addition of anti-oxidants, so that it can withstand continuous service at 90 °C over a long period of time (see page 27).

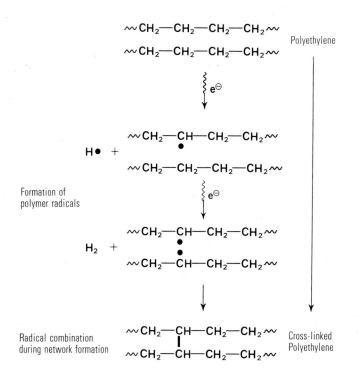

Formation of polymer radicals

Radical combination during network formation

Polyethylene

Cross-linked Polyethylene

Fig. 2.9
Cross-linking of Polyethylene by electron beams

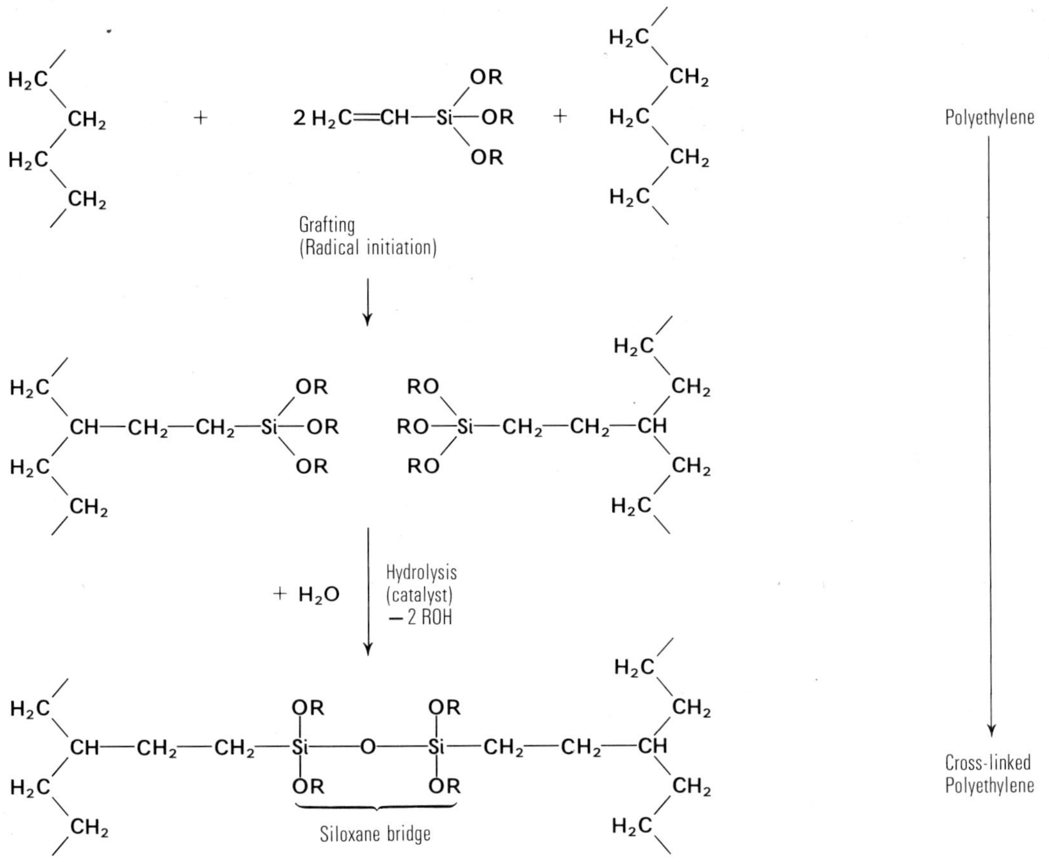

Fig. 2.10 Cross-linking of Polyethylene by Siloxane bridge method

2.1.2 Elastomers

In contrast to the thermoplastics, the molecule chains of elastomers form an extensive meshed network. This cross-linking, or vulcanization, gives rise to the elastic nature of the material: a large reversible extension in response to low tensile stress.

Elastomeric materials are used for insulation and for sheaths. They are applied mainly where the product has to be particularly flexible.

A wide range of elastomers is nowadays available to the cable industry. This makes possible the manufacture of compounds with specific properties, such as high abrasion and oil resistance, weather and heat resistance and flame resistance, combined with good overall electrical and mechanical characteristics.

The classical elastomeric material, natural rubber, has declined in significance in recent years. In its place, the synthetic elastomers, produced by the copolymerization of ethylene and propylene, are constantly finding new areas of application in cable engineering. These ethylene-propylene copolymers, known under the general term EPR, contain no double bonds, and cannot, therefore, be cross-linked by the vulcanization methods appropriate to the unsaturated rubbers (e.g. natural rubber, styrene butadiene rubber). On the other hand, because of the absence of double bonds in the main molecular chains, these elastomers have a significantly greater resistance to thermooxidative decomposition and to the effect of ultra-violet radiation, ozone and heat.

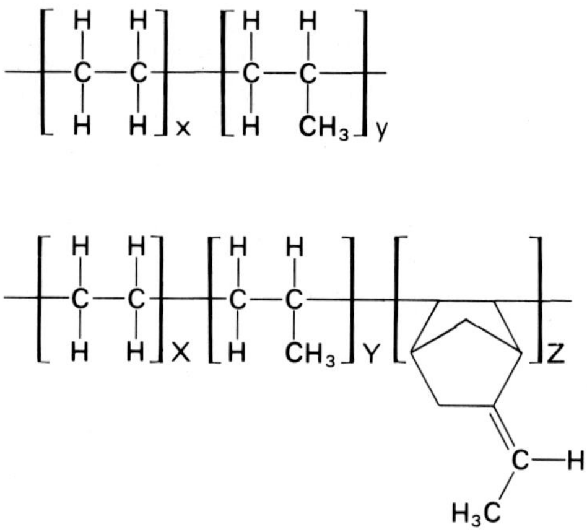

$$\left[\begin{array}{cc} H & H \\ | & | \\ -C - C - \\ | & | \\ H & H \end{array}\right]_x \left[\begin{array}{cc} H & H \\ | & | \\ C - C - \\ | & | \\ H & CH_3 \end{array}\right]_y$$

EPR

Fig. 2.11
Structural form of EPR and EPDM

$$\left[\begin{array}{cc} H & H \\ | & | \\ -C - C - \\ | & | \\ H & H \end{array}\right]_X \left[\begin{array}{cc} H & H \\ | & | \\ C - C - \\ | & | \\ H & CH_3 \end{array}\right]_Y \left[\quad\right]_Z$$

EPDM
with Ethylidiene
as tercomponent

$$\overset{}{\underset{H_3C}{C}} - H$$

With the incorporation of a diene[1], EPDM elastomers are obtained (Fig. 2.11), in which the double-bond active in cross-linking is arranged not in the main chain but in a side group.

Thermoplastic Elastomers (TPE)[2]

Technically interesting combinations of properties can be obtained through the admixture of thermoplastic olefines, e.g. polypropylene with ethylene propylene elastomers, or by the direct production of so-called polyolefine block polymers. Such copolymers of ethylene and propylene with a block structure consist of an EP elastomer phase with crystalline homopolymer end blocks, which represent the unstable reversible cross-linking centres. At temperatures above the crystallite melting point, these materials have thermoplastic properties; below the crystallite melting point they behave as elastomers. Polymers of this kind are therefore called thermoplastic elastomers (TPE)[2].

Another class of thermoplastic elastomers is represented by three-block polymers of styrene and ethyl-ene butylene blocks, which are so structured that ethylene butylene chains contain styrene units as end blocks. Polyesters and polyurethanes with TPE properties are also known.

Other types of elastomers used in cable engineering are polychloroprene, chlorosulphonated polyethylene and chlorinated polyethylene, which, because of their advantageous properties in relation to environmental influences, are preferably used as sheathing materials.

Conducting Rubber

Through the addition of conducting fillers, e.g. carbon black, natural rubber and synthetic elastomer compounds with a resistivity of from a few Ωcm up to several thousand Ωcm can be produced. Conducting rubber compounds are generally used in the monitoring of flexible and wiring cables in mines, and also for inner semiconducting layers and field limiting in synthetic elastomer insulated high-voltage cables (Ozonex principle).

Natural Rubber (NR)

Natural rubber is obtained in various countries in the equatorial belt from the rubber tree (hevea brasiliensis). This tree contains in the cambium cells under its bark a milky juice (latex), which flows out when the bark is cut. The rubber is obtained from this through coagulation with chemicals, electro coagu-

---

[1] The dienes used as tercomponents are special hydrate materials with double links which are non-conjugated
[2] In some publications for thermoplastic elastomers the abbreviation TPR is used as previously

lation or by other methods. The resulting rubber is supplied to the manufacturer in smoked form as 'smoked sheets' and in chemically bleached form as 'crepe'. Rubber is a hydrocarbon of high molecular weight with the monomer unit 1,4-polyisoprene. With the addition of vulcanization and aging-protection additives, specially selected fillers, and where appropriate by blending with synthetic elastomers, insulating compounds for cables and compounds for the sheathing of flexible and wiring cables can be manufactured.

Unlike synthetic elastomers, natural rubber has to be subjected to a so-called mastication process during manufacture, to make it receptive to the additives and to obtain the required plasticity in the compound. The significance of natural rubber in the cable industry has declined sharply in recent years in favour of synthetic elastomers.

Styrene Butadiene Rubber (SBR)

SBR is a copolymer of styrene and butadiene, referred to as either a hot or a cold polymer according to the method of manufacture. Cold polymers, with the normal styrene content of 24% (by weight) are characterized in comparison with the so-called hot-rubber types by higher tensile and tear strength and better working characteristics; they are therefore preferred as admixtures used in the production of SBR-NR compounds. SBR and SBR-NR mixtures are suitable for use as insulation in low-voltage flexible and wiring cables for operating temperatures up to 60 °C.

Nitrile Butadiene Rubber (NBR)

Through the copolymerization of acryl nitrile with butadiene, elastomers are obtained which are distinguished in comparison with the SBR types by high oil resistance and good weather resistance. For this reason they are preferably used for sheathing compounds.

By mixing with PVC, NBR-PVC blends are produced which have better flame resistance.

Butyl Rubber (IIR)

Butyl rubber is a copolymer of isobutylene and isoprene. To permit vulcanization, an unsaturated component of 1.5 to 4.5% (by weight) is introduced. The lower the isoprene content, the less is the extent to which the rubber ages under heat; on the other

hand, a lower isoprene content lowers the rate of vulcanization and makes the product less elastic.

The relatively small number of double bonds makes butyl rubber less susceptible to the effects of oxygen and ozone. The main advantages are very low water absorption and low gas permeability. The good heat resistance permits operating temperatures up to 90 °C with suitable compound structures. The mechanical properties can be improved by the addition of special active fillers; plasticizers, for example, improve the elastic properties, particularly at low temperatures. Since EPR and EPDM synthetic elastomers have become available, butyl rubber ist used only in special cases.

Ethylene-Propylene Rubber (EPR)

EPR is used as a general designation for the two sub-types

> ethylene-propylene rubber (EPR) and
> ethylene-propylene terpolymer rubber (EPDM).

Ethylene-propylene rubber (EPR) is a copolymer of low density without C-C double bonds, i.e. it is a completely saturated polymer which, like polyethylene, can only be cross-linked radically. The further development of this saturated rubber to EPDM (Fig. 2.11) through the incorporation of dienes with lateral double bonds permits a conventional sulphur vulcanization as well as radical cross-linking, e.g. with peroxides.

There is little difference between cross-linked EPR and EPDM as regards mechanical and electrical properties. Peroxide – i.e. radical – cross-linking, however, gives better long-term heat resistance and better heat/pressure characteristics than sulphur vulcanization.

Outstanding characteristics of these elastomers are resistance to ozone, oxygen and ionization, good flexibility at low temperatures and high resistance to weather and light. Because of their good dielectric properties EPR and EPDM, depending on the structure of the compound, are suitable for insulation at voltages up to 100 kV, with a maximum permissible service temperature between 80 and 90 °C. Such insulating materials will withstand temperatures up to 250 °C without damage under short-circuit conditions.

Cables with ethylene propylene rubber (EPR) manufactured by Siemens are known by the protected trade name PROTOLON.

Blending EPR with PE enables the mechanical strength and hardness to be increased significantly ('hard grade'). The insulating materials so produced closely resemble the elasticized polyethylenes in their combination of characteristics, i.e. they exhibit, as well as the improved mechanical characteristics, improved electrical characteristics, similar to those of polyethylene. They are known by the abbreviation HEPR.

Silicone Rubber (SiR)

Silicone rubber is produced by the polycondensation of hydrolyzed dimethyldichlorosilane and methylphenyldichlorosilane. The macromolecules in this case consist not of carbon chains, as in most other polymers, but of silicon-oxygen chains (Fig. 2.12), which is the reason for the remarkably high heat resistance.

During processing, fillers are added to the silicone rubber, together with organic peroxides for the purpose of cross-linking (vulcanization). The end product is characterized by a high heat resistance. Because of the excellent insulation properties and the practically unvarying flexibility over the temperature range from -50 to $+180\,°C$, flexible and wiring cables insulated with silicone rubber can be used continuously at conductor temperatures up to $180\,°C$ (up to $250\,°C$ for short periods).

The silicone-rubber-based SINOTHERM compounds manufactured by Siemens have outstanding electrical characteristics with good resistance to ozone. They are insensitive to moisture and exhibit good weather resistance. They are thus suitable for both insulation and sheathing. Another preferred area of application is that of accessories.

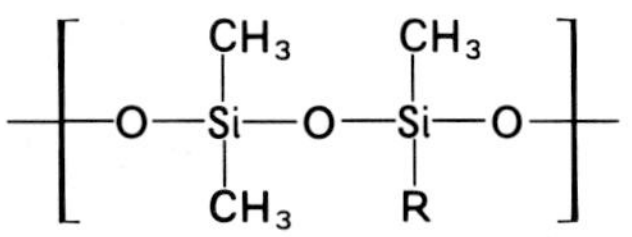

Ethylene Vinyl Acetate (EVA)

EVA is a copolymer which is used either as a thermoplastic (vinyl acetate content $\leq 30\%$) or, with suitable cross-linking, as an elastomer (see Fig. 2.3 for structural formula). The properties of an EVA copolymer are in the main determined by the ratio of vinyl to acetate content. Cross-linked EVA elastomers are characterised by good heat resistance and permit conductor temperatures up to $120\,°C$. They also exhibit excellent resistance to aging in hot air and superheated steam, together with very satisfactory heat/pressure characteristics (see page 16), particularly at high temperatures. In addition, EVA compounds have outstanding resistance to ozone and oxygen, weather resistance and colour stability. The dynamic freezing point is in the region of -20 to $-30\,°C$. The application of EVA as an insulating material is limited by its electrical characteristics to the low-voltage range. The compounds are used for heat-resistant non-sheathed cables and flexible cords, and for heating cables.

2.1.3 Thermosetting Polymers (Duromers)

Unlike the elastomers, thermosetting polymers are usually closely cross-linked and in general have better wear resistance and dimensional stability than thermoplastics and elastomers.

The application of the thermosetting polymers as insulating materials is limited to the use of epoxide and polyurethane resins for the filling of cable accessories. Filling resins based on epoxides are preferably converted to the thermosetting state by heat-curing. Polyurethane resin materials, on the other hand, harden at room temperature, and so offer advantages in application techniques. Both types of resin are notable for outstanding adhesive strength, particularly to metals.

A suitable choice of resins and hardeners enables a well balanced combination of thermomechanical and electrical properties to be obtained, together with good chemical resistance.

$$\left[O-\underset{\underset{CH_3}{|}}{\overset{\overset{CH_3}{|}}{Si}}-O-\underset{\underset{R}{|}}{\overset{\overset{CH_3}{|}}{Si}}-O \right]$$

$(R = CH_3 \text{ or } C_6H_5)$

Fig. 2.12 Structural form of SiK

2.2 Chemical Aging of Polymers

By the term 'aging' is understood the change in the properties of a material with time. Polymers are subject in use to chemical changes which have an adverse effect on their mechanical and electrical characteristics.

The chemical aging processes are accelerated with increasing temperature. It is therefore necessary to protect polymers which are exposed to high temperatures by means of stabilizers, in order to ensure an adequate service life for the products made from them.

Insulation and, particularly, sheaths that are exposed to direct sunlight (UV radiation) must be further protected against this effect by the addition of so-called light stabilizers to the compounds. Carbon black has proved to be an excellent light stabilizer, especially for polyolefines; an addition of 2 to 3% (by weight) of finely divided carbon black, well distributed, affords an effective protection.

In the presence of atmospheric oxygen, the chemical aging of many polymers, e.g. the polyolefines, arises from oxidation processes which are provoked, or accelerated, by heat and light. For the purpose of stabilization, anti-oxidants are added to the polymers in a proportion, normally, of 0.1 to 0.5% (by weight).

Aging reactions, especially oxidation, are accelerated catalytically by the presence of some metals. This is particularly marked in the case of contact between copper and polyolefines. In this situation anti-oxidants offer no appreciable protection.

In practice, to avoid direct contact between polyolefine insulation and copper conductors, separators are often introduced, e.g. plastic films with sufficient stability in contact with copper, or tinned copper conductors are used. In medium and high-voltage cable the field-limiting conducting layers act as separators. Conducting compounds are protected against the effect of direct contact with copper by their high carbon black content.

Where there is direct contact between polyolefine insulating materials and copper conductors, metal deactivators are added to the insulating compounds to counteract the catalytic effect. It has been possible to demonstrate by practical aging tests that by this means a service life can be achieved which is comparable with that obtained in compounds not subject to contact with copper.

Evaluation of the aging properties of polymer cable materials is based on two quantities: the *service life,* which denotes the period of time for which a material remains serviceable, and the *temperature limit,* up to which a material can be used subject to given boundary conditions. These two quantities are interrelated, so that an increase in the temperature limit results in a reduction in the service life.

In determining the corresponding pair of values for the service life and the temperature limit, the changes in the significant characteristics which are necessary for the function of the material must be examined as functions of temperature and time up to point where an end criterion is reached. The choice of characteristics and of the end criterion determine the results.

In cable engineering the tear strength is in most cases adopted as the essential characteristic, and as an end criterion, on practical reasons, the attainment of a particular elongation value, e.g. $\varepsilon_R = 50\%$ (residual elongation).

DIN VDE 0304 contains guidelines for the determination of the thermal stability of electrical insulating materials, the revised edition of December 1980 being a reproduction of IEC 216 (1974).

According to DIN VDE 0304, under the designation 'temperature index' (TI), a temperature limit for a service life of 20 000 h is stipulated. TI/164, for example, signifies that the material for which it is quoted remains serviceable for 20 000 h under any thermal stress at temperatures up to 164 °C.

To obtain the temperature index (TI), experimentally determined pairs of values for a range of temperatures – the test temperature T and the service life t_E – are plotted on a graph with time on a logarithmic horizontal axis and the reciprocal of the absolute temperature on the vertical axis. A straight line is drawn through the plotted points and by extrapolation gives the required temperature index. In the graphs of Figs. 2.13 to 2.16 the temperature values on the horizontal axis have been converted to degrees Celsius (°C) for ease of reading.

This method is based on the assumption that the rate at which the aging process takes place can be expressed in terms of the Arrhenius equation

$$t_E = A^* \cdot e^{(E/RT)} \qquad (2.1)$$

which on a log/linear plot leads to a linear relationship between the two plotted quantities:

$$\log t_E = A + B\,\frac{1}{T} \qquad (2.2)$$

where

t_E	service life, representing the time taken to reach the end criterion
E	activation energy
R	universal gas constant
T	test temperature, representing the temperature limit in K
A^*, A, B	constants
e	base of Naperian logarithms

The graphical representation in the diagram according to Arrhenius permits the rapid comparison of life values with stipulated service lives. Extrapolation far beyond the range of the measured experimental values, however, calls for a certain caution.

Figs. 2.13 to 2.16 show graphically the temperature relationships of the service lives of the cable materials in general use. The range of the graph corresponds to the measurement range in each case. The examples chosen are based on compounds formulated by Siemens.

Fig. 2.13 shows the temperature dependence of the service life of two PVC products with different thermal stresses. Curve 1 represents the compounds normally used for a maximum conductor temperature of 70 °C, Curve 2 those for a maximum conductor temperature of 90 °C.

Both curves apply to insulating coverings with an end criterion $\varepsilon_R = 50\%$.

The curves show for the 70 °C compound $TI = 88$ °C, and for the 90 °C compound $TI = 103$ °C.

Curve 2 represents at the same time the maximum degree of PVC stabilization attainable at the present time, from which it follows that the so-called 105 °C compounds have a reduced service life.

In Fig. 2.14 service life curves are plotted for two XLPE products.

Curve 1 shows the resistance to aging of a conventional peroxide cross-linked compound for medium-

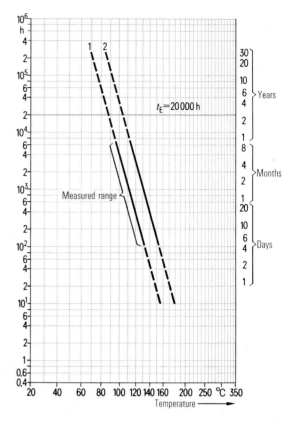

Curve 1 for 70 °C ⎱ Highest permissible operating
Curve 2 for 90 °C ⎰ temperature of conductor

Fig. 2.13
Service life of representative PVC insulating compounds

voltage cables. The measurements were made on model cores with 1.5 mm round solid conductors, 0.65 mm conducting layers and 1.0 mm XLPE insulation; the criterion was a wrapping test. The temperature index (*TI*) was 105 °C.

Curve 2 represents the resistance to aging of a 0.6/1 kV compound with a metal deactivator. The measurements were made on insulating coverings in direct contact with copper, the end criterion was a residual elongation $\varepsilon_R = 100\%$ and the temperature index (*TI*) was 114 °C.

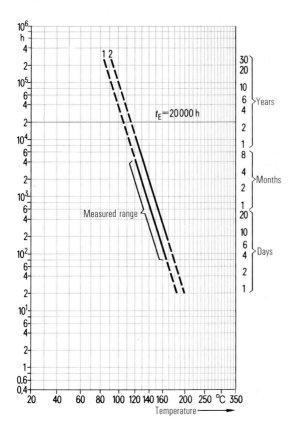

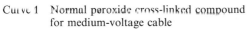

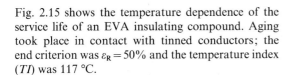

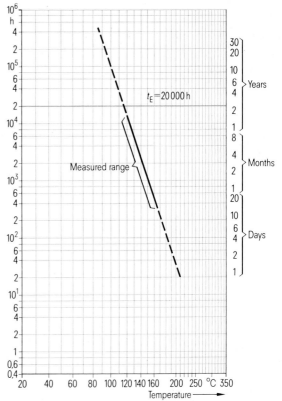

Curve 1   Normal peroxide cross-linked compound
for medium-voltage cable

Curve 2   Compound with metal deactivator for low-
voltage cable with copper conductor

Fig. 2.14
Service life of XLPE insulation compounds

Fig. 2.15
Service life of EVA insulation compounds

Fig. 2.15 shows the temperature dependence of the
service life of an EVA insulating compound. Aging
took place in contact with tinned conductors; the
end criterion was $\varepsilon_R = 50\%$ and the temperature index
(*TI*) was 117 °C.

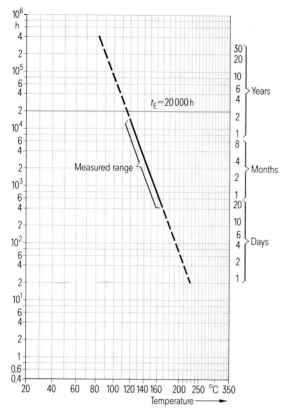

Fig. 2.16
Service life of EPR insulation compounds

Fig. 2.16 shows the service life characteristic for an EPR insulating compound for 0.6/1 kV cables. The values were obtained from insulating coverings in direct contact with copper (10% elongated conductors); the end criterion was $\varepsilon_R = 100\%$ and the temperature index (*TI*) was 113 °C.

2.3 The Influence of Moisture on Polyolefine Insulating Materials

Practical experience and long-term tests on model cables have shown that water has an adverse effect on polyolefine insulating materials such as PE and XLPE. Using appropriate dyeing techniques, it is possible to observe tree-like structures in such materials subjected to electrical stresses. These originate from practically unavoidable microscopically small fault locations and run in the direction of the electric field. This phenomenon, known as 'water treeing' (WT) is quite distinct from 'electrical' tree formation (electrical treeing, ET) caused, for example, by ionization.

The mechanism whereby WT structures arise has not so far been clearly explained. Because the WT growth is influenced by many factors besides water and electric fields, and these processes take a long time, investigation is very difficult and time-consuming. Ignoring the finer points, WT structures can be divided into two groups (Fig. 2.17):

▷ 'bow-tie trees' in the interior of the insulation
▷ 'vented trees' originating from the boundaries of the insulation.

Because of the low concentration of moisture in the interior of the insulation, the growth of 'bow-tie trees' is slowed down, so that they usually remain small (Figs. 2.18a and 2.18b). The serviceability of cables is therefore only rarely impaired by 'bow-tie trees'.

'Vented trees' (Figs. 2.19a to 2.19f) require more critical assessment. These can extend right through the insulation if sufficient water is available. In this way the electrical stability of the cable is gradually reduced, until a breakdown of the cable is initiated

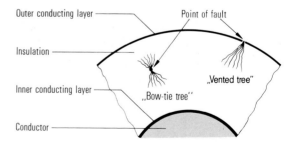

Fig. 2.17
Diagramatic example of WT structures

Fig. 2.18a
"Bow-tie tress" in a cable with XLPE insulation
(magnification 1:100)

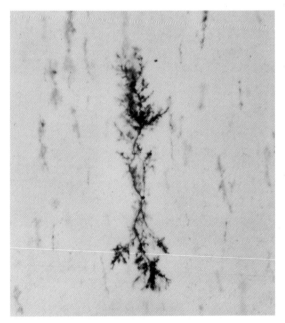

Fig. 2.18b
PE cable from the early days of PE technique with
"Bow-tie trees" of high density (magnification 1:160)

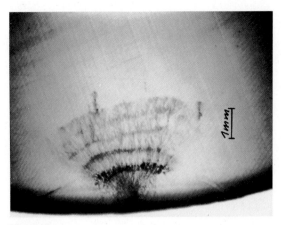

Fig. 2.19a
"Vented tree", growing from the outer graphited
conducting layer of a PE cable

Fig. 2.19b
PE cable from the early days of PE technique.
"Vented tree", growing from the outer graphited
conducting layer

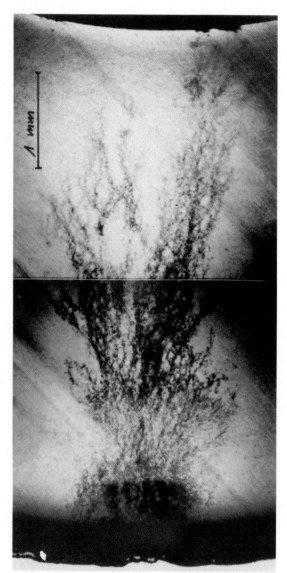

Fig. 2.19d
"Vented tree" (length approx. 700 µm) on the inner
extruded conducting layer of a 20 kV XLPE cable
after several years operation with water inside the
cable (magnification 1:135)

Fig. 2.19c
PE cable from the early days of PE technique.
"Vented tree" growing from the outer graphited
conducting layer. (The picture was constructed from
two photographs)

Fig. 2.19e
"Vented tree" (length approx. 50 µm) at the outer extruded conducting layer of a 20 kV XLPE cable after
several years of operation (magnification: 1:135)

200 μm

Fig. 2.19f
Structure change from WT to ET at the top of a
"Vented tree", XLPE cable after 6000h "Water treeing test" with 5 kV/mm and water in the conductor
and following short-time stressing with approximately
nine times operational field strength. (The picture was
constructed from two photographs)

by the conversion of the water tree into an electrical
tree (Fig. 2.19f).

Experiments on cables that had been in service for
about eight years, in which, as a result of lack of
care in installation, water had penetrated to the conductors and the screen regions, have confirmed the
results of accelerated laboratory tests on the deterioration to be expected in the electrical strength of insulation. In this connection, Fig. 2.20 shows, by means
of Weibull statistics (a method of evaluation specially
developed for the physics of breakdown mecha-
nisms), the determined residual strength as dependent
on the nature of the applied voltage, which by linear
regression of the measured values plotted in the probability diagram is established as 63.2% of the rated
value. It can be seen from this illustration that water
in the conductor has a particularly unfavourable effect on the insulation [2.1].

This knowledge has given rise to the following measures for the construction, manufacture and installation of cables with PE or XLPE insulation:

a) minimization of fault locations in insulation and
   at the boundaries of the conducting layers, i.e.:

   ▷ optimization of the purity of the insulating
     and conducting-layer materials and the clean-
     liness of the manufacturing process;

   ▷ extruded conducting layers to be preferred.

b) reduction of water content and prevention of ingress of moisture, i.e.:

   ▷ prevention of ingress of water into the conduc-
     tors and the screen region in manufacture,
     storage, transport and installation and in ser-
     vice (e.g. through subsequent damage to the
     sheath),

   ▷ use of mechanically resistant outer sheath, e.g.
     of PE,

   ▷ provision of lengthwise water-tight screen re-
     gion to limit the ingress of water in the event
     of damage to the sheath,

   ▷ in high-voltage cable, for $\geq 36/60$ kV, the use
     of a laminated aluminium sheath and a length-
     wise water-tight screen region.

In addition to this, intensive development is in progress to increase the resistance of XLPE insulating
compounds to WT[1] by means of additives [2.3].

[1] In the literature also referred to a "Water tree retardent compound"
    (WTR compound)

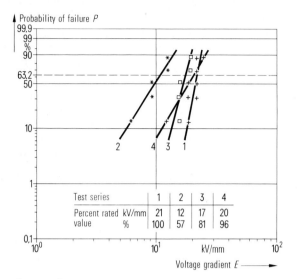

Test series		1	2	3	4
Percent rated	kV/mm	21	12	17	20
value	%	100	57	81	96

a) a.c. voltage

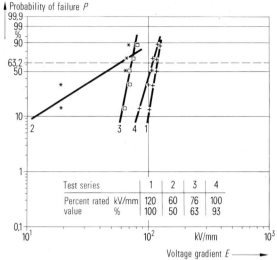

Test series		1	2	3	4
Percent rated	kV/mm	120	60	76	100
value	%	100	50	63	93

b) Impulse voltage

1 New condition, dry, not prestressed

2 Water in the conductor and below the sheath of the cable after eight years in operation

3 Water under the sheath of the cable after seven years in operation

4 Water at the undamaged sheath after nine years in operation

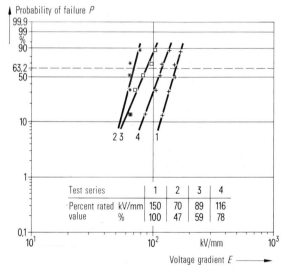

Test series		1	2	3	4
Percent rated	kV/mm	150	70	89	116
value	%	100	47	59	78

c) d.c. voltage

Fig. 2.20
Breakdown strength of 10 kV PE cables.
Weibull-Distribution:
Probability of failure P relative to mean voltage gradient E (break-down voltage divided by thickness of insulation)

2.4 Impregnated Paper

Impregnated paper was used for conductor insulation at the end of the last century. It made possible the manufacture of cables for higher voltages. Because of its good dielectric properties, paper insulation is still indispensible for cables used at up to the highest operating voltages customary today. In the low- and medium-voltage ranges, however – up to 30 kV – it has been replaced by polymers to an ever-increasing extent in recent decades. This development is now being extended increasingly also to the high-voltage range.

Cable paper consists of the purest possible long-stapled cellulose, obtained from northern timbers. It is also known as sodium cellulose paper, from the process by which it is prepared. The conductors are insulated with this special high-quality paper to the thickness required for the rated voltage. In the case of the higher-voltage cables, it is advantageous to provide conducting papers on the conductor and a screen of metallized paper on the core, whose thickness, up to a certain limit, according to DIN VDE 0225, is counted as part of the insulation thickness.

The single- or multi-core cable assembly, according to the cable construction, is dried in an impregnating tank and then impregnated with a degassed and dried impregnating medium ('impregnating compound') appropriate to the intended purpose of the cable. Paper-insulated cables are divided according to the method of impregnation into mass-impregnated cables and oil-filled cables. Cables whose insulation is filled after installation with nitrogen under pressure are known as internal gas pressure cable.

Low-voltage, medium-voltage and external gas pressure cables are impregnated with high-viscosity polybutene compounds, which have very low dielectric losses and excellent aging characteristics, from low to very high operating temperatures, in comparison with the oil-resin compounds produced from natural or synthetic hydrocarbon resins.

The viscosity of the impregnating compound is chosen in such a way that small differences in level do not cause the compound to migrate.

For special cables to be installed on steep slopes, 'non-draining cables' and internal gas pressure cables, special compounds are used, known as non-draining or nd compounds. These consist of polybutene modified by the addition of selected microcrystalline waxes; they shrink only slightly in the transition from the liquid to the semi-solid state, and are prevented from flowing in the permissible service temperature range by the microwax structure.

In addition, these non-draining compounds – like all polybutene compounds – have outstanding dielectric properties, even after long periods of service.

For extra-high-voltage low-pressure oil-filled cable, a low-viscosity gas-absorbing impregnant is used. This may be a mineral oil rich in aromatic compounds, a naphtha-based mineral oil with alkyl benzene additives or an alkyl benzene, ensuring good gas absorption in an electric field at all service temperatures, especially in regard to hydrogen. Other characteristics of these impregnants are adequate resistance to oxidation and little tendency to swell in the presence of the sealing materials used.

2.5 Literature Referred to in Section 2

[2.1] Kammel, G.; Sunderhauf, H: Längswasserdichte Kunststoffkabel (Lengthwise watertight cables), Elektrotechn. Z. (1982) No. 4, pp. 173–176
[2.2] Kalkner, W; Müller, U; Peschke, E.F.; Henkel, H.J.; Olshausen, R.v.: Water treeing in PE and XLPE insulated medium and high-voltage cables, Elektr.-Wirtsch. 81 (1982) No. 26, pp. 911–922
[2.3] Peschke, E.; Wiedenmann, R.: Ein neues VPE-Mittelspannungskabel mit Water-tree-retardierender (WTR-)Isolierung (A new XLPE medium-voltage cable with water-tree-retardant (WTR) insulation), Elektr.-Wirtsch. 86 (1987) No. 6, pp. 226–227

Experimental installation for the investigation
of the influence of water on polymer insulating materials
in medium- and high-voltage cables

3 Protective Sheaths

A distinction is made in the DIN-VDE specifications between sheaths and protective coverings or outer coverings of thermoplastics or elastomers.

Protective coverings and outer coverings serve as corrosion protection over a metal sheath or as light mechanical protection for flexible and wiring cables, whereas sheaths are dimensioned for greater mechanical stresses.

Since the properties of these components are similar except for dimensions, only the collective term 'sheath' is used in relation to cables in the following section.

3.1 Thermoplastic Sheaths

Polyvinyl Chloride (PVC)

PVC-based compounds are used predominantly as a sheathing material for power cables and for flexible and wiring cables because of the many advantages they offer.

The thermoplastic sheath is extruded onto the cable core assembly in a process providing a seamless cover. Cables with PVC outer sheaths have a clean, smooth surface.

The PVC compounds combine high tensile strength and elongation, pressure stability even in high-temperature regions, resistance to practically all chemicals in soils and most chemicals encountered in chemical plants, and especially flame resistance and resistance to aging. The sheaths used in PROTODUR cables are characterized by their comparative hardness, toughness and adequate pliability from the point of view of bending at low temperatures (see page 18).

The PVC outer sheath proved over many years for power and wiring cables in fixed installations is also used, in a suitably softer form, for flexible cables. Light and medium PVC-sheathed flexible cords have been introduced satisfactorily for household equipment because of their clear and durable colours and smooth surfaces. PVC-sheathed flexible cords are not suitable for use at low temperatures, in the open air or in heating appliances (e.g. smoothing irons), in which the cable can come into contact with hot parts; elastomer-sheathed cords should be used in these cases.

Polyethylene (PE)

Practical experience in supply-authority systems has shown that in many cases medium-voltage cables laid in the ground are subjected to considerably higher mechanical stresses than was originally assumed. Because of the danger presented to cables by the penetration of moisture, an undamaged impervious outer sheath has a decisive effect on the life expectancy of PE and XLPE insulation (see page 30). A mechanically resistant PE sheath is therefore increasingly preferred, especially for medium- and high-voltage cables with XLPE insulation. A PE sheath is recommended in the new specification DIN VDE 0273/..87 for XLPE cables laid in the ground.

The disadvantages of these materials, such as flammability, greater difficulty of handling in installation, inferior adhesion to the materials normally used in accesories and greater longitudinal shrinkage, are accepted in view of their greater hardness and abrasion resistance. From considerations of resistance to UV radiation and environmental stress cracking, only black PE sheaths are permitted. The most significant factor in the choice of the base polymers is the temperature to be expected in normal service. The screen temperature to be expected under fault conditions (see page 286) should be allowed for by suitable constructional measures.

Particularly advantageous is the combination of a PE sheath with the measures described in Section 7.3 for the sealing of the screen region of the cable against the ingress of moisture.

In connection with the leading-in or laying of cables in interior locations, it must be remembered that PE sheaths are not flame retardant. Where necessary, appropriate fire protection measures should be adopted at the site, e.g. spraying or painting the cable with a flame retardant protective coating.

Polyamide (PA) and Polyurethane (PUR)

Polyamides are polycondensation products with linear chain structures made up of dicarbon acids and diamines or aminocarbon acids. Polyurethanes are polyaddition products with a chain-formation to spatial structure of di-isocyanates or polyisocyanates and dialcohols or polyalcoholes respectively. Flexible and wiring cables subjected to particularly high mechanical stresses or to chemical influences, e.g. from benzene or agressive, mostly aromatic oils (e.g. coaltar oils), are provided with a protective layer of polyamide or polyurethane over the sheath or the insulation. These two materials are distinguished mainly by outstanding mechanical properties and good resistance to oils, fats, ketones, esters and chlorinated hydrocarbons. Polyamide protective coverings are applied to, among others, flexible and wiring cables for use in mineral oil extraction and in aircraft.

Polyamides are not suitable for use as insulating materials, on account of their poor dielectric characteristics, but because of their high abrasion resistance and toughness, together with their good resistance to organic solvents and fuels they are used as sheathing materials for special flexible and wiring cables.

Polyurethane sheaths have high impact resistance, high flexibility at low temperatures and good abrasion resistance.

Polypropylene (PP)

Polypropylene is of less importance because of its brittleness at low temperature and its special sensitivity to thermo-oxidative deterioration, especially when in contact with copper, can only be employed under limited conditions.

3.2 Elastomer Sheaths

In the Federal Republic of Germany, apart from use in ships wiring and halogen-free cables with improved properties under fire conditions, elastomer sheaths are only used for wiring and flexible cables. Because natural rubber has only limitted resistance

to weathering, chemicals and heat Siemens have developed special synthetic elastomer compounds for use as an outer sheath material.

Polychloroprene (PCP)

A polymer of 2-chlorine-butadien shows a good resistance against the influences of light, oxygen and ozone and a very good resistance against cold, heat and flames. Its excellent resistance against chemicals, which is very high for a elastomer deserves special mention.

It has, therefore, particular advantages for use as a basic material for sheathing compounds.

The cables and flexible cables manufactured by Siemens with a sheath based on polychloroprene are known under the trade mark PROTOFIRM. The mechanical strength of the vulcanized compound is very high, therefore, these cables have an increased service life under mechanical stresses of any kind. PROTOFIRM sheaths also offer advantages where good resistance to weathering, flame retardance and a certain ammount of resistance to oil is required, furthermore where a elastomer is preferred to PVC compounds because of its higher flexibility, resistance to abrasion and tear extension.

These sheaths, therefore, are particularly suited for flexible cables in underground mining applications and locations with fire hazard.

Chlorosulphonyl Polyethylene (CSM)

CSM is produced by chlorosulphonation of polyethylene. The partly crystalline polyethylene is in this process transferred into an amorphous elastomer. The cross-linking can be established by using either radical or conventional special sulphur compounds.

CSM is available as an industrial product under the trade mark HYPALON (Manufacturer: Dupont de Nemours International S.A.). Both, properties and the range of application correspond to those of polychloroprene (PCP), however, CSM has improved properties as regards colour fastness and resistance to heat.

Chlorinated Polyethylene (CM)

CM is a new sheath compound with characteristics virtually identical with those of HYPALON but with reduced flexibility at low temperatures. When blended with other elastomers special compounds can be produced, e.g. by the use of EPR or EPDM flexibility at low temperature is improved, or, by the use of nitril-butadien rubber the oil and fuel resistance is improved.

Nitril-Butadien Rubber (NBR)

Cables which are to be permanently immersed in oil are provided with an oil resistant sheath based on nitril rubber (NBR). Nitril rubber is predominantly used blended with PVC and is known for its good resistance to oil. This resistance to oil is based on the polarity of the nitril rubber molecule. Nitril rubber is therefore highly resistive to non-polarised oils and solvents but does swell considerably in highly polarised solvents.

3.3 Sheathing Materials for Special Purposes

Siemens have developed cables and flexible cables under the trade mark SIENOPYR FRNC (*F*lame-*R*etardent, *N*on *C*orrosive) which have particularly important characteristics in the event of fire, namely:

▷ reduced support of combustion even when bunched

▷ fumes do not contain corrosive substances

▷ greatly reduced smoke development

▷ retention of insutation

achieved by the use of special sheathing materials.

The base materials are olefincopolymere such as EVA or EEA. To achieve flame-retardant qualities, these materials being normally combustible, certain hydrate containing mineral fillers are used. Therefore to satisfy the above requirements all other additives such as antiaging agents are halogen-free.

3.4 Metal Sheath

Lead Sheath

Insulation materials, sensitive to humidity, e.g. impregnated paper are protected by a metal sheath. Since the beginning of cable manufacturing lead, which is easy to handle, has been the proven material for this purpose. Lead covered PVC-sheathed cables

NYBUY and PVC-insulated cables with lead sheath are used for filling stations as well as locations with fire and explosion hazard; these cables also have, as protection against corrosion and mechanical damage, an outer sheath of PVC.

For the lead sheath a cable lead Kb-Pb to DIN 17640 is used which is sufficiently resistant against the vibration which are normally present. The base material for this cable lead is pig lead Pb 99.94 to DIN 1719. To avoid a coarse grain structure this is blended with 0.03 to 0.05% (by weight) of copper (Table 3.1).

Table 3.1 Cable lead to DIN 17640.
A base metal of Pb 99.94 to DIN 1719 with an additional 0.001% Mg must be used

Desig-nation	Cable lead	Tellurium lead
Abbrevi-ation	Kb-Pb	Kb-Pb Te 0.04
Used for	Weak alloy cable sheaths; base metal for manufacture of alloy cable sheaths	Cable sheaths which are subjected to a high degree of vibration

Components in % (by weight)

Cu	0.03–0.05	0.03 to 0.05 [1]
Sb	–	–
Sn	[2]	–
Te	–	0.0350
Pb	remainder to 100%	remainder to 100%

Maximum amount of additives in %

Ag	0.001
As	0.001
Bi	0.050
Fe	0.001
Mg	0.001
Sb	0.005
Sn	0.005
Zn	0.001

[1] This Cu addition can be omitted by agreement between lead smelters and cable manufacturers

[2] This material Kb Pb may also have an Sn content of up to 0.05%

Table 3.2 Features of Lead and Aluminium

Features		Cable lead Kb-Pb to DIN 17640	Lead alloys to DIN 17640	Aluminium for cable sheaths
Density	g/cm³	11.34		2.7
Tensile strength	N/mm²	13 to 18	16 to 26	55 to 65
Elongation	%	40 to 50	25 to 35	25 to 35
Brinell Hardness to DIN 50351	HB 5/31, 25/30	4 to 5	5.0 to 6.5	–
	HB 2.5/31, 25/30	–	–	16 to 18
Melting point	°C	327		658
Specific resistance at 20 °C	$\Omega\,m$	$21.4 \cdot 10^{-8}$		$2.84 \cdot 10^{-8}$
Thermal conductivity	$\dfrac{W}{K\,m}$	34.8		218
Specific heat capacity	$\dfrac{J}{m^3\,K}$	$1.45 \cdot 10^6$		$2.5 \cdot 10^6$

For cables which are to be subjected to heavier vibration, e.g. cables for installation on bridges, railway cables or aerial cables, Siemens preferably use a lead-tellurium-alloy to DIN 17640 (Kb-Pb Te 0.04). The basis for this alloy is pig lead Pb 99.94 to DIN 1719, to which at least 0.035% of tellurium is added.

The main characteristics of lead and aluminium for use as cable sheaths are shown in Table 3.2.

Aluminium Sheath

In the 1940's Siemens AG were the first manufacturer to succeed in pressing aluminium, with its high stability and good conductivity, around a core assembly, which previously could only be done with lead. After having proven their worth, cables with aluminium sheath were included at first in VDE 0286/10.56 "Specifications for Metal-Sheathed Power Cables on Trial" since 1964 for aluminium-sheathed cables, DIN VDE 0255 "Specification for Cables with Mass-Impregnated Paper Insulation and Metal-Sheath in Power Plants" applies.

A reliable corrosion protection ensures that the aluminium sheath is not threatened even under unfavourable conditions. Aluminium-sheathed cables are installed in the same manner as paper lead cables. The smooth soft aluminium sheath allows sufficiently small bending radii (see page 400).

The good electrical conductivity of aluminium makes it possible to use the sheath as neutral conductor (PEN) in three-phase systems with earthed neutral point (three-phase four-wire systems).

To ensure these cables correspond to the normal $3^1/_2$ core paper-insulated cables, the thickness of the aluminium sheath is dimensioned such that the conductivity of the sheath has a value equal to or greater than that of the corresponding standardized neutral conductor.

The good electrical conductivity of the aluminium sheath ensures a good screening factor; the interference with control cables and communication cables is therefore lower than that of lead-sheathed cables (see page 352).

Normally cables with aluminium sheath do not have to be armoured, due to the mechanical stability of aluminium. This is of particular importance in the case of single-core cables where with lead-sheathed types the mechanical protection could only be achieved by a relatively expensive non-magnetic armour.

Aluminium is not susceptible to vibration and does not tend to re-crystallize even at higher ambient temperatures. These facts make aluminium-sheathed cables particularly useful for installations where subjection to heavy vibration is to be expected, e.g. on bridges, alongside railway tracks etc.

4 Protection against Corrosion

Metal-sheathed as well as armoured power cables must be provided with protection against corrosion.

4.1 Cable with Lead Sheath

Fibrous Materials in Bituminous Compounds

Unarmoured Cables

The protection against corrosion consists of several layers of bituminized paper and one layer of pre-impregnated jute, with intermediate coatings of neutral bituminous coatings (Asphalt). The outer surface is white-washed to prevent sticking of the cables whilst on the drum. Nowadays this type of protection against corrosion is only rarely used. It is increasingly preferred to use a plastic sheath bonded to the lead sheath by a suitable compound.

Armoured Cables

Belted cables are provided with a protective inner covering over the lead sheath consisting of several layers of bituminized fibrous material with intermediate layers of bituminous compound. This protective inner covering serves also as bedding for the armour. The armour is protected against corrosion by a layer of jute in bituminous compound. The outer surface is white-washed. This type of protection against corrosion is sufficient under normal conditions.

If there is a danger of heavy chemical or electrolytic corrosion at least one layer of elastomer type or polymeric foil has to be provided in addition to the tapes of fibrous material, unless the cable is protected by an outer polymeric sheath.

Separate Lead-Sheathed (S.L.) Cables

Separate lead-sheathed (S.L.) cables have, over each lead sheath, a bitumen layer and a layer of polymeric tape which is followed by a further layer of bitumen compound and one layer of bituminized fibre tape. Over the touch protected and laid-up cores an inner protective sheath of paper, or textile tape, and (or),

depending upon the requirements, a bedding of impregnated jute (NEKEBA) is applied.

If for practical reasons, a polymeric sheath was not selected a layer of polymeric foil must be laid over the armour if there is danger of chemical or electrolytic corrosion (NEKEBEA).

Polymeric Outer Sheath

A reliable corrosion protection is provided by a sheath of PVC which is chemically stable and flame retardent. Cables which are to be subjected to abnormal stresses, either in operation or during installation a PE-sheath or a reinforced PVC sheath to DIN VDE 0225 can be provided.

The PE outer sheaths of medium- or high-tension cables are always coloured black (see page 395).

Table 4.1 shows the colour of outer sheaths to DIN VDE 0206.

Table 4.1 Colour of outer sheath

Cable type	Rated voltage U_0/U in kV	Sheath colour
PE outer sheath		
Medium and high-voltage cable	$>0.6/1$	Black
PVC outer sheath		
Low-voltage cable	0.6/1	Black
Low-voltage cable for mining applications below ground	0.6/1	Yellow
Medium- and high-voltage cable	$>0.6/1$	Red

4.2 Aluminium-Sheathed Cables

Whilst being exceedingly durable when installed in free air, aluminium has to be protected by a water and ion-resistant anti-corrosive covering, if the cable is to be installed in the ground. In order to achieve a high degree of safety and mechanical strength a multi-layer corrosion protection is required.

According to DIN VDE 0255 Type A 5 it consists of a plastic foil applied overlapped and bonded to the aluminium sheath and to the outer PVC sheath by means of bitumen compound.

Special tests show that the corrosion protection adheres well to the aluminium sheath and that in the event of a locally limited damage to the cable eventual corrosion on the outside of the aluminium sheath is practically limited to the exposed area.

5 Armour

The armour protects the cable against mechanical stresses. In the case of polymeric cables for rated voltages above $U_0/U = 0.6/1$ kV it normally serves also as an electrical screening.

Paper-insulated lead-sheathed cables are normally armoured with two compounded steel tapes, each applied in an open helix in such a manner that the second tape covers the gap left by the first.

High-voltage cables with polymer insulation having a metallic copper screen as well as low-voltage cables with PVC or XLPE insulation and aluminium-sheathed cables do not require to be armoured if they are sufficiently protected against damage and not subjected to tensile stresses. The permissible pull during laying of cables is shown on page 406.

An armour of flat-steel wires may also serve as a screen in multi-core cables with polymer insulation not having a screen of copper. This design is common for PVC cables for 3.5/6 kV, where a screen around each core separately is not required, and also for cables to be installed in networks where double earth-faults or earth-faults in earthed neutral systems render an armour of steel wires advantageous in its function as a common metallic screen (see also page 281).

Cables which are to be subjected to higher mechanical stresses (especially tensile stress) must be armoured with galvanized steel wires. The right profile (e.g. flat, round or "Z"-wire), dimensions and strength of the wires has to be chosen according to the size and application of the cable, e.g. as river cable, submarine cable or shaft cable (see pages 129 and 130). A steel tape helix prevents bird-cageing of the wires.

Single-core cables in single- or three-phase a.c. systems are not armoured as a rule, in order to avoid additional losses. An armour of non-magnetic material, however, has to be provided wherever mechanical damage or higher tensile stresses are to be expected during or after laying of the cable. Occasionally mass-impregnated or oil-filled cables are manufactured with an open armour of steel wires, instead of a non-magnetic armour, for reasons of economics.

6 Concentric Conductors

The concentric conductors in low-voltage cables such as NYCY, 2XCY, NYCWY and 2XCWY are used as PE or PEN conductors (see page 397) and at the same time form touch protection. According to the VDE specifications these must be of copper. The cross-sectional area included in the type designation, however, relates only to the material used for the phase (main) conductors.

In a cable with copper conductors, for example, NYCWY 3×95 SM/50 0,6/1 kV, the value of direct current resistance of the concentric copper conductor, to comply with the above regulation must not be greater than the maximum value of that of a copper conductor of 50 mm^2. Similarly in a cable with aluminium conductors NAYCWY 3×95 SM/95 0.6/1 kV the value of d.c. resistance of the concentric copper conductor, to comply with the above regulation, must not be greater than the maximum value of that of an aluminium conductor of 95 mm^2.

The concentric conductor comprises either a helically applied layer of copper wires or a wave form layer of copper wires (CEANDER-cable) e.g. NYCWY or 2XCWY. In addition a copper tape is applied helically to interconnect the wires (transverse helical tape). In the Federal Republic of Germany aluminium is not permitted for use as a concentric conductor.

Concentric conductors are arranged under the outer polymer sheath to ensure they are protected against corrosion. If armour is arranged above the concentric conductor a separation sheath (impervious extruded sheath) of PVC must be applied between them. (Type reference designations for concentric conductor see page 101.)

7 Electrical Screening

Electrical screening is necessary only for cables with $U_0 > 0.6/1$ kV and fulfils the following functions:

▷ Potential grading and limiting of the electrical field

▷ Conduction of charge and discharge currents

▷ Touch protection

To satisfy these functions the screening normally comprises a combination of conducting layers with metalic elements. One differentiates between cables with non-radial characteristic fields (e.g. belted cables) and radial field cables. The radial characteristics of lines of field between conductor and screen is achieved by placing a conducting layer, a metal screen or a metal sheath over each individual core. Insulation is stressed only perpendicular to the wall thickness. In cables with laminated dielectric (paper insulation) this is the direction of the highest electrical withstand. Interstices of the cores in these cables remain field-free (see page 97).

7.1 Conducting Layers[1]

The magnitude of electric stress and the degree of sensitivity of the insulation material against partial discharge govern the type of screening of the insulation with conducting layers (Table 7.1).

Cable with Paper Insulation

The "inner conducting layer" consists of several layers of semi-conducting paper (also known as carbon black paper). This is often referred to as conductor smoothing because it is used to smooth local peaks in the electric field which could otherwise occur, e.g. because of irregularities in the surface of stranded conductors.

The "outer conducting layer" normaly consists of metallized paper, also known as Höchstädter Folie

(H-foil), if necessary in combination with conducting paper. It can also consist of a combination of aluminium tape with conducting paper tapes.

Cable with PVC Insulation

The "inner conducting layer" consists of a PVC compound having a high carbon-black content. This is normally applied together with the insulation in a single production process so that both layers are bonded firmly without gaps or cavities.

For the "outer conducting layer" elastic conducting adhesives with a cover of conducting tapes (textile or carbon black paper) is a preferred method.

Cable with PE or XLPE Insulation

Because of the higher sensitivity of PE and XLPE insulation to partial discharge the reliable well adhe-

Table 7.1
Arrangement of conducting layers above and below the cable insulation

Type of insulation	Conducting layers required	
	below the insulation (inner conducting layer over the conductor)	above the insulation (outer conducting layer)
	rated voltages exceeding U_0/U kV	kV
Impregnated paper belted cable[1] radial field cable	8.7/10 8.7/15	– 3.6/6
PVC EPR	6/10 6/10	6/10 6/10
PE XLPE	3.6/6 3.6/6	3.6/6 3.6/6

[1] Paper-insulated cable with non-radial field: permissible only for rated voltages $U_0/U \le 8.7/10$ kV. In Germany belted cables are normaly used for voltages $U_0/U \le 6/10$ kV (see page 148)

---

[1] Remark:
  In this book the simpler term 'conducting layer' has been used instead of 'semiconducting layer' chosen in the respective IEC standards

sive gap and cavity-free bonding to conducting layers is of greatest significance for the life expectancy of the cable. DIN VDE 0273 requires proof of non-partial discharge for each individual cable length for a voltage range up to $2 U_0$ and with a measuring sensitivity of < 5 pC.

The "inner conducting layer" normally consists of a polymer compound which is made conductive by adding carbon black and is, together with the insulation, applied to the cable in a single manufacturing process and in the case of XLPE cable, cross-linked with the insulation.

The "outer conducting layer" is formed by the insulation and a layer of conductive polymer compound being simultaneously applied to the cable and in the case of XLPE cable, cross-linked. This from a technical viewpoint is the most favourable solution where the conducting layer is firmly bonded to the insulation and requires a special tool to remove it during cable installation. In another variant this layer can be removed by hand after piercing with a tool.

In the Federal Republic of Germany it was previous practice to use cables in which the outer conducting layer consisted of graphite rubbed on the outer surface of the core with a conducting tape applied over it. This graphite required a special solvent to remove it during installation. The new regulation DIN VDE 0273/..87 does no more include this variant.

Cable with EPR Insulation

EPR is less sensitive to partial discharge in comparison with PE and XPLE but here also inner and outer conducting layers of polymer compound must be provided and firmly bonded to the insulation.

7.2 Metallic Components of Electrical Screening

The resistance of conducting layers is sufficient to control the very small partial charge and discharge currents over small distances, e. g. over the circumference of the core. For the transmission of these currents in the longitudinal direction of the cable towards the earthed point, additional elements having a substantially lower specific resistance are necessary.

This function is performed by metallic screens which are in electrical contact with the conducting layers. Depending on cable type these are:

▷ A layer of copper wires with a helix of copper tape or tapes *above the laid-up* individually screened *cores* (transverse helical tape) which may be each screened with conducting layers *or*;

▷ A layer of copper wires with a helix of copper tape or tapes or a layer of copper tape *over each individual core* which may be each screened with conducting layers; or they are

▷ Metal sheaths (e. g. paper-insulated cables) *above each individual core* or above the *laid-up cores* which may be each individually screened by metalized paper; or

▷ Steel-wire armouring (e. g. in cables with polymer sheath) *over the laid-up* individually screened *cores*; each screened by conducting layers

Other materials (e. g. aluminium) are not acceptable, particularly in Germany.

In contrast to the rules for concentric conductors, for copper screens it is not the electrical effective cross section which is the important factor because when considering earth fault or short-circuit stresses (see page 287) the geometric cross section is the more significant (Table 7.2). Screens are arranged below the outer polymer sheath to provide protection against corrosion.

If armouring is provided above the copper screen this must always be separated from the screen by an impervious separation sheath of PVC.

Table 7.2
Minimum cross-sectional areas of screens to VDE (geometric cross section)

Nominal cross-sectional area of main conductor mm^2	Nominal cross-sectional area of screen mm^2
25	16
35	16
50	16
70	16
95	16
120	16
150	25[1]
185	25[1]
240	25[2]
300	25
400	35

[1] For cables laid in earth a cross section of 16 mm^2 is permitted
[2] For single-core cables laid in earth a cross section of 16 mm^3 is permitted

Type designations of screens see page 101; details for current carrying capacity of screens in the event of earth fault, double earth fault and earth short circuit see page 281.

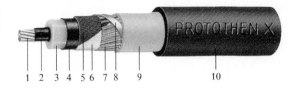

7.3 Longitudinally Water Proof Screens

If extreme conditions are to be considered (e. g. submarine cables, runs with great height differential, damage to outer sheath) additional measures can be taken. To avoid water penetrating the cable through damage to the outer sheath which could, in the area of the screen, spread over a large distance it is prudent to use cables which are protected against water penetration in the screen areas. To achieve this protection there are several constructional possibilities, e.g. in the screen area absorbent powders or tapes can be added which swell in the event of moisture ingres so that all cavities and gaps are filled and the longitudinal spread of moisture is limited.

In a construction developed by Siemens the screening wires are embedded in unvulcanised rubber. The gap sealing between screen wires and the extruded outer conducting layer is achieved by a bolster of low-conducting moisture swelling fibres or a combination of low-conducting crepe paper with a tape of non-conducting moisture swelling fibres. This bolster also ensures electrical contact of the screen wires with the outer conducting layer above the insulation. Outer mechanical protection is provided in each case by a tough PE sheath (Fig. 7.1).

This type of construction has significant advantages over the simple longitudinally water proof variant (swelling tape or swelling powder) and even though marginaly more costly has received good market acceptance. The special advantages of the inner covering of unvulcanized rubber are;

▷ good adhesion to the PE sheath which limits the unavoidable shrinking of PE sheath to a negligible degree;

▷ protection of the other component parts of the cable when, in the event of short-circuit or double earth short-circuit, the screen can attain a relatively high temperature;

▷ additional barrier against ingress of moisture from minor damage to the sheath when in such an event the moisture is prevented from reaching the inner core and the longitudinal sealing (swelling of tapes) may not be initiated.

1 Conductor
2 Inner-conducting layer
3 XLPE insulation
4 Outer extruded conducting layer
5 Semi-conducting crepe paper
6 Swelling tape
7 Copper wire screen
8 Helix of copper tape
9 Inner covering of unvulcanized rubber
10 PE outer sheath

Fig. 7.1
Single-core cable with XLPE insulation, longitudinally water proof screens and PE sheath Type NA2XS(F) 2Y 1 × 150 RM/25 6/10 kV

If transverse sealing of the sheath against diffusion of moisture is required as e.g. with high-voltage cables with rated voltage $U_0/U \geq 36/60$ kV, an aluminium tape, plastic coated on one side only, is applied in a longitudal direction between the PE sheath and the copper screen. This is closely bonded to the PE sheath at the overlapping area (Al peth-sheath). The area surrounding the screen is filled with swelling powder. A further possibility which is particularly suitable for submarine cable is a metal sheath (e.g. Pb, Al) which normally makes a copper screen unnecessary.

The power-supply cable to a mobile container crane is subjected to frequent reeling and unreeling and also to high-mechanical stresses.
PROTOLON trailing cables which are service-free offer safety in operation and long service life even under such extreme conditions

Insulated Wires and Flexible Cables

8 Types of Wires and Cables

8.1 National and International Standards

Insulated wires and flexible cables for electric power installations must be capable of withstanding the stresses experienced during both installation and in operation. In a typical normal plant, containing fixed cable runs and with provision for the connection of mobile loads, this can be best ensured by using cables which comply with the relevant national or international standards, not only with regard to construction and testing but also to the parameters and limitations for the type of application. For special applications only cables which comply, in their construction and characteristics, as close as possible to VDE or IEC specifications should be used.

8.1.1 VDE Specifications

The main VDE specifications governing construction, testing and application of flexible cables are:

DIN VDE 0207	Insulating and sheath compounds for cables and flexible cords
DIN VDE 0250	Cables, wires and flexible cords for power installation
DIN VDE 0281	PVC cables, wires and flexible cords for power installation
DIN VDE 0282	Rubber cables, wires and flexible cords for power installation
DIN VDE 0289	Definitions for cables, wires and flexible cords for power installation
DIN VDE 0293	Identification of cores in cables and flexible cords used in power installations with nominal voltages up to 1 000 V
DIN VDE 0295	Conductors of cables, wires and flexible cords for power installation
DIN VDE 0298 Parts 3 and 4	Application of cables, wires and flexible cords in power installations
DIN VDE 0472	Testing of cables, wires and flexible cords

When the VDE Approval Organization verifies that a flexible cable complies with the relevant VDE specifications it authorises the use of a black-red printed identification thread. A second identification thread is used as a manufacturers mark which shows for the products of Siemens AG the colours green-white-red-white. As an alternative to the identification threads or in addition to these the mark ◁VDE▷ and the manufacturers trade mark may be printed or embossed on the cable or sheath. In special cases the cables are marked with a word trade mark or by a protected trade mark such as, in the case of Siemens, a coloured line over the full length of the sheath. At the present time there are no VDE specifications to cover the cables shown in Sections 8.2.1 to 8.4 however, in producing these cables, the safety technical requirements laid down in VDE are adhered to such that all types of construction comply with the principles of these rules.

Information for the selection of cables is given on page 55.

8.1.2 Harmonized Standards [1]

It is the task of the European Committee for Electrotechnical Standardisation (CENELEC) to remove technical barriers to trade between member countries where differing standards, national regulations or approval proceedures exist. Within the committee each country is represented by its national delegates (representatives of consumers, manufacturers and standards organisations) who prepare a basic harmonization document which, after a period for public comment, is used as a basis for a final harmonized document which is then issued and brought into force. The relevant national committees are then obliged to accept the contents of these documents without deviation or addition and introduce them into their relevant national standards system.

---

[1] See also: "Heinhold, L.; Retzlaff, E.; Warner, A., Harmonisierung der Starkstromkabel und -leitungen". Booklet 31, VDE-Verlag GmbH

CENELEC harmonized documents for flexible cables are:

HD 21   Polyvinyl chloride-insulated cables of rated voltages up to and including 450/750 V – Part 1 up to Part 5

HD 22   Rubber-insulated cables of rated voltages up to and including 450/750 V – Part 1 up to Part 4.

These documents together with the associated amendments are aimed to achieve world-wide approval of the relevant IEC standards (see page 55).

In the Federal Republic of Germany they are published and in force as:

DIN VDE 0281   PVC cables, wires and flexible cords for power installation

DIN VDE 0282   Rubber cables, wires and flexible cords for power installation

The national standards DIN VDE 0250 for types of construction, which are replaced by the above harmonized standards, have meanwhile been withdrawn.

Marking

The harmonized standards relate firstly to the most commonly used cables such as insulated wires and flexible cords. For these a special marking was agreed containing the letters ◁HAR▷ or alternatively harmonization thread coloured black-red-yellow.

This marking together with VDE mark, authorized by the approval organization, and the manufacturers mark is shown on the insulation or sheath, hence products of Siemens AG are marked e. g.

SIEMENS   ◁VDE▷   ◁HAR▷

If identification threads are used the nationality of the approval organization can be determined from the differing lengths of colours on the thread (Table 8.1).

The marking is approved by the CENELEC member countries in accepting the HAR approval proceedure. The use of wires, cables and cords marked in this manner is accepted by these countries without further approval:

Belgium (B)   Italy (I)
Federal Republic of Germany (D) Netherlands (NL)
Denmark (DK)   Norway (N)
France (F)   Austria (A)
United Kingdom (UK)   Spain (E)
Ireland (IRL)   Sweden (S)

The countries Finland, Portugal and Switzerland recognise the harmonized standards but their use in these countries requires individual approval.

Type Designation

In order to avoid confusion due to language a new common system of type designations has been agreed. Initially this system will be used only for harmonized cables and approved supplementary types. This consists of three parts (Table 8.2).

The first part identifies the regulations to which the cable has been manufactured and the rated voltage. The letter "H" indicates that the cable in all respects complies with the harmonized standard. A letter 'A' is used to indicate that the cable complies basically with the harmonized standard but is only approved for use in a specific country (approved national supplementary type).

The rated voltage is expressed by two a.c. voltages U_0/U where:

U_0 is the r.m.s. value between any insulated conductor and earth and

U is the r.m.s. value between any two phase conductors in a multic-core cable or in a system of single core cables.

The second part contains the abbreviations for component parts. The third part contains information on the number of cores and rated cross-section as well as indication when a protective conductor (green-yellow) is included. For harmonized flexible cables the presence of a green-yellow core is no longer indicated by the letter "I" or "O" which previously was used as a suffix to the type designation.

Table 8.1 Approval authorities and harmonization marking

Country and approval authority	Harmonization marking either printed or embossed	Harmonization marking by black red yellow identification threads (colour length in cm)		
		black	red	yellow
Belgium Comite Electrotechnique Belge (CEBEC)	CEBEC ◁HAR▷	1	3	1
Federal Republic of Germany and West Berlin Verband Deutscher Elektrotechniker (VDE) e.V., Prüfstelle	◁VDE▷ ◁HAR▷	3	1	1
Denmark Danmarks Elektriske Materielkontroll (DEMKO)	◁DEMKO▷ ◁HAR▷	3	1	3
France Union Technique de l'Electricite (UTE)	USE ◁HAR▷	3	3	1
Ireland Institute for Industrial Research and Standards (IIRS)	◁IIRS▷ ◁HAR▷	3	3	5
Italy Istituto del Marchio Qualita (IMQ)	IEMMEQU ◁HAR▷	1	3	5
Netherlands N.V. tot Keuring van Elektrotechnische Materialien (KEMA)	KEMA-KEUR ◁HAR▷	1	3	3
Norway Norges Elektriske Materiellkontroll (NEMKO)	NEMKO ◁HAR▷	1	1	7
Austria Österreichischer Verband für Elektrotechnik (ÖVE)	◁ÖVE▷ ◁HAR▷	3	1	5
Sweden Svenska Elektriska Materielkontrollanstalter (SEMKO)	SEMKO ◁HAR▷	1	1	5
Spain Asociación Electrotéchica y Electrónica Española (AEE)	◇UNE◇ ◁HAR▷	3	1	7
United Kingdom British Approvals Service for Electric Cables	BASEC ◁HAR▷	1	1	3

Table 8.2 System for cable designation for flexible cables to harmonized standards

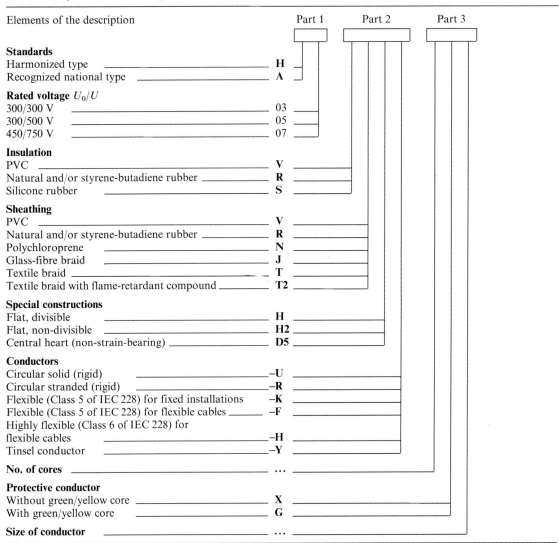

Elements of the description		Part 1	Part 2	Part 3

Standards
Harmonized type ———————————— **H**
Recognized national type ——————————— **A**

Rated voltage U_0/U
300/300 V ———————————————— 03
300/500 V ——————————————— 05
450/750 V ——————————————— 07

Insulation
PVC ————————————————— **V**
Natural and/or styrene-butadiene rubber ———— **R**
Silicone rubber ————————————— **S**

Sheathing
PVC ————————————————— **V**
Natural and/or styrene-butadiene rubber ———— **R**
Polychloroprene ————————————— **N**
Glass-fibre braid ————————————— **J**
Textile braid ——————————————— **T**
Textile braid with flame-retardant compound ———— **T2**

Special constructions
Flat, divisible ——————————————— **H**
Flat, non-divisible ————————————— **H2**
Central heart (non-strain-bearing) ——————— **D5**

Conductors
Circular solid (rigid) ——————————— **–U**
Circular stranded (rigid) ——————————— **–R**
Flexible (Class 5 of IEC 228) for fixed installations **–K**
Flexible (Class 5 of IEC 228) for flexible cables ——— **–F**
Highly flexible (Class 6 of IEC 228) for
flexible cables ——————————————— **–H**
Tinsel conductor ————————————— **–Y**

No. of cores ——————————————— ...

Protective conductor
Without green/yellow core ———————— **X**
With green/yellow core ———————————— **G**

Size of conductor ——————————— ...

Examples of type designations

Single-core non-sheathed cable for general purposes with rigid solid conductor **H07V-U 1.5 sw**

Heavy polychloroprene-sheathed flexible cable, three-core, 2.5 mm^2, with green/yellow core **H07RN-F 3G2.5**

Light PVC-sheathed circular cord, twin-core, 0.75 mm^2, without green/yellow core **H03VV-F 2X0.75**

Table 8.3 Summary of cables to harmonized standards

Cables to DIN VDE 0281	Type abbreviation	Rated voltage U_0/U V	No. of cores	Nominal cross-sectional area mm²	Superseded types to VDE 0250
Single-core non-sheathed cables for internal wiring – with solid conductur – with flexible conductor	H05V–U H05V–K	300/500	1	0.5 to 1	NYFA, NYA NYFAF, NYAF
Single-core non-sheathed cables for general purposes – with rigid solid conductur – with rigid stranded conductor – with flexible conductor	H07V–U H07V–R H07V–K	450/750	1 1 1	1.5 to 10 6 to 400 1.5 to 240	NYA NYA NYAF
Flat tinsel cords	H03VH–Y	300/300	2	~0.1	NLYZ
Flat non-sheathed cords	H03VH–H	300/300	2	0.5 and 0.75	NYZ
Light PVC-sheathed cords – circular – flat	H03VV–F H03VVH2–F	300/300	2 to 4 2	0.5 and 0.75 0.5 and 0.75	NYLHY rd NYLHY fl
Ordinary PVC-sheathed cords – circular – flat	H05VV–F H05VVH2–F	300/500 300/500	2 to 5 7 2	0.75 to 4 1 to 2.5 0.75	NYMHY rd NYMHY rd NYMHY fl
Flat PVC-sheathed flexible cables for lifts and similar application	H05VVH6–F H07VVH6–F	300/500 450/750	3 to 24 3 to 24	0.75 and 1 1.5 to 16	NYFLY NYFLY
Cables to DIN VDE 0282					
Heat-resistant silicone insulated cables	H05SJ–K	300/500	1	0.5 to 16	N2GAFU
Braided cords	H03RT–F	300/300	2 and 3	0.75 to 1.5	NSA
Ordinary tough-rubber-sheathed cords	H05RR–F	300/500	2 to 5 3 and 4	0.75 to 2.5 4 and 6	NLH, NMH
Ordinary polychloroprene-sheathed cords	H05RN–F	300/500	1 2 and 3 4	0.75 and 1 0.75 and 1 0.75	NMHöu NMHöu NMHöu
Heavy polychloroprene-sheathed flexible cables	H07RN–F	450/750	1 2 and 5 3 and 4	1.5 to 500 1 to 25 1 to 300	NMH, NMHöu and NSHöu
Rubber-insulated lift cables for normal use – braided cables	H05RT2D5–F H07RT2D5–F	300/500 450/750	4 to 24 4 to 24	0.75 1	NFLG NFLG
– armoured cables	H05RND5–F H07RND5–F	300/500 450/750	4 to 24 4 to 24	0.75 1	NFLGC NFLGC

Table 8.4
Comparison of flexible cables to harmonized standards DIN VDE 0281 and DIN VDE 0282 with IEC

Type of cable	Cables to DIN VDE 0281	Superseded types to DIN VDE 0250	Comparable construction to IEC 227
Single-core non-sheathed cables for internal wiring – with solid conductor – with flexible conductor	H05V–U H05V–K	NYFA, NYA NYFAF, NYAF	227 IEC 05 227 IEC 06
Single-core non-sheathed cables for general purposes – with rigid solid conductor – with rigid stranded conductor – with flexible conductor	H07V–U H07V–R H07V–K	NYA…e NYA…m NYAF	227 IEC 01 227 IEC 01 227 IEC 02
Flat tinsel cords	H03VH–Y	NLYZ	227 IEC 41
Flat non-sheathed cords	H03VH–H	NYZ	227 IEC 42
Light PVC-sheathed cords – circular – flat	H03VV–F H03VVH2–F	NYLHY rd NYLHY fl	227 IEC 52
Ordinary PVC-sheathed cords – circular – flat	H05VV–F H05VVH2–F	NYMHY NYMHY fl	227 IEC 53
Flat PVC-sheathed flexible cables for lift and similar application	H05VVH6–F H07VVH6–F	NYFLY NYFLY	– –
	Cables to DIN VDE 0282		Comparable construction to IEC 245
Heat-resistant silicone-insulated cables	H05SJ–K	N2GAFU	245 IEC 03
Braided cords	H03RT–F	NSA	245 IEC 51
Ordinary tough-rubber-sheathed cords	H05RR–F	NLH NMH	245 IEC 53 245 IEC 53
Ordinary polychloroprene-sheathed flexible cables	H05RN–F	NMHöu	245 IEC 57
Heavy polychloroprene-sheathed flexible cables	H07RN–F	NMHöu NSHöu	245 IEC 65 245 IEC 66
Rubber insulated lift cables for normal use – braided cables	H05RT2D5–F H07RT2D5–F	NFLG NFLG	– –
– armoured cables	H05RND5–F H07RND5–F	NFLGC NFLGC	– –

8.1.3 National Types

Apart from those cables shown in Table 8.3 a number of national types exist, namely:

1. The so called "approved national types" which are split into two groups,

   a) approved national types which are an addition to the approved harmonized cable type, e.g. in respect of number of cores or conductor cross-section (they correspond fully in all other aspects with the harmonized standard),

   b) approved national types which deviate from the harmonized types;

2. national types which have not yet been embraced by the harmonization procedure, e.g. flexible cables with a rated voltage ≥ 1 kV as well as multi-core cables for fixed wiring for which harmonization is not yet finalised.

All tables mentioned in 1 and 2 are however covered by VDE specifications, carry the VDE marking and are only approved for use in the Federal Republic of Germany.

8.1.4 IEC Standards

The following IEC publications are current in respect of cables:

IEC 227:
> "Polyvinylchloride-insulated flexible cables and cords with circular conductors and a rated voltage not exceeding 750 V"

IEC 245:
> "Rubber-insulated flexible cables and cords with circular conductors and a rated voltage not exceeding 750 V".

The cables to IEC correspond in construction with the types listed in Table 8.4 to DIN VDE 0281 and DIN VDE 0282. They differ in some cases in dimensions and test requirements.

In the IEC standards cables are identified by two numbers preceded by the abbreviated title of the relevant IEC standard. The first number designates the basic class of cable, the second the specific type within the basic class. The class separation of 'medium' and 'light' does not comply with the classifications in DIN VDE.

0 Non-sheathed cables for fixed wiring
  01 Single-core non-sheathed cable with rigid conductor for general purposes
  02 Single-core non-sheathed cable with flexible conductor for general purposes
  03 Heat-resistant silicone-insulated cable for a conductor temperature of maximum 180 °C
  05 Single-core non-sheathed cable with solid conductor for internal wiring for a conductor temperature of 70 °C
  06 Single-core non-sheathed cable with flexible conductor for internal wiring for a conductor temperature of 70 °C
  07 Single-core non-sheathed cable with solid conductor for internal wiring for a conductor temperature of 105 °C
  08 Single-core non-sheathed cable with flexible conductor for internal wiring for a conductor temperature of 105 °C

1 Sheathed cables for fixed wiring.
  10 Light polyvinyl chloride-sheated cable

4 Non-sheathed flexible cables for light duty.
  41 Flat tinsel cord
  42 Flat non-sheathed cord
  43 Cord for decorative chains

5 Flexible cords for normal duty.
  51 Braided cord
  52 Light polyvinyl chloride-sheathed cord
  53 Ordinary polyvinyl chloride or tough-rubber-sheathed cord
  57 Ordinary polychloroprene or other equivalent synthetic elastomer-sheathed cord

6 Flexible cables for heavy duty.
  66 Heavy polychloroprene or other equivalent synthetic elastomer-sheathed flexible cable

7 Sheated flexible cables for special duty.
  70 Braided lift cable
  71 Flat polyvinyl chloride-sheathed lift cables and cables for flexible connections
  74 Though-rubber-sheathed lift cable
  75 Polychloroprene or other equivalent synthetic elastomer-sheathed lift cable

8 Flexible cables for special application
  81 Tough-rubber-sheathed arc welding electrode cable
  82 Polychloroprene or other equivalent synthetic elastomer-sheathed arc welding electrode cable

8.2 Selection of Flexible Cables

When selecting the type required the relevant VDE specifications and the VDE specifications for the erection of power installations and also the special regulations issued by electricity supply authorities or others (factory inspectorate, mines and quarries inspectorate) where applicable, must be observed. In countries other than Germany, in addition to the harmonized types, cables made to the VDE specifications may be used provided their characteristics meet the requirements for the function and the relevant regulations for that country. Tables 8.5 and 8.6 show the most commonly used types and areas of application.

8.2.1 Cables for Fixed Installation

Table 8.5   Cables for fixed installation

Type	Type designation	Rated voltage U_0/U	Construction	Remarks	Standard
Single-core non-sheathed cables for internal wiring	H05V-U	300/500 V	1 2	To facilitate large-scale internal wiring, additional colours and two-colour combinations are allowed. Green and yellow may only by used individually, however, if it is permitted by the applicable safety requirements.	DIN VDE 0281
	H05V-K		1 3 1 PROTODUR insulation 2 Copper conductor, solid 3 Copper conductor, flexible		
Single-core non-sheathed cables for general purposes	H07V-U	450/750 V	1 2	H07V-K is flexible, because of its finely stranded conductor, and so offers advantages for installation in conduit in confined spaces or for connections to moving parts – e.g. hinged control panels. As equipotential bonding conductors, these wires can be laid directly on, in or under plaster, or on racks etc.	DIN VDE 0281
	H07V-R		1 3		
	H07V-K		1 4 1 PROTODUR insulation 2 Copper conductor, solid 3 Copper conductor, stranded rigid 4 Copper conductor, flexible		
Single-core Polychloro-prene-sheathed cable for special purposes	NSGAFÖU	1.8/3 kV	1 2 3 1 Polychloroprene sheath 2 PROTOLON insulation 3 Copper conductor flexible	Cables with a nominal voltage U_0/U of at least 1.8/3 kV are considered to be short-circuit and earth-fault-proof in switchboards and distribution boards rated at up to 1 000 V.	DIN VDE 0250
SIFLA flat building wires	NYIF	220/380 V	1 2 3 1 2 3 1 Rubber sheath 2 PROTODUR insulation 3 Copper conductor, solid	SIFLA building are indispensible when slot chasing is not possible in buildings of pre-stressed or poured concrete, or on light building boards.	DIN VDE 0250
Light PVC-sheathed cable	NYM	300/500 V	1 2 3 4 1 PRODODUR sheath 2 Extruded filler 3 PROTODUR insulation 4 Copper conductor, solid or stranded	For applications with more stringent requirements – e.g. in agricultural installations, dairies, cheese-making plants, laundries, industrial and administrative buildings.	DIN VDE 0250

Conductors		Standard colours of insulation or sheath	Applications			
Num-ber	Cross-sectional area mm^2		In dry locations	In damp and wet locations, and outdoors	In operating areas and store rooms subject to fire hazard	In areas with explosion hazard
1 1	0.5 to 1	Green-yellow Black Light blue Brown Violet Grey White Red	For internal wiring of equipment and protected installation in and on luminaires. Also for installation in conduit on and under plaster, but only for signalling systems.	Not permitted	Not permitted	Not permitted
1 1	1.5 to 400 1.5 to 240	Green-yellow Black Light blue Brown Violet Grey White Red	In conduit on or under plaster (only in plastic conduit in bathrooms and shower compartments in dwellings and hotels) and for open installation on insulators over plaster beyond arm's reach. In equipment, switchboards and distribution boards and in or on luminaires with a rated voltage of up to 1000 V a.c. or 750 V d.c. to earth. For use in rail vehicles the d.c. operating voltage may be up to 900 V to earth.	Open installation on insulators beyond arm's reach, but not outdoors	Installation in plastic conduit on and under plaster	In switchboards and distribution boards to DIN VDE 0165
1	1.5 to 300	Black	for traction vehicles and busses to DIN VDE 0115 as well as in dry rooms			
2 3 4 5	1.5 to 4 1.5 to 4 1.5 and 2.5 1.5 and 2.5	Natural	In or under plaster, including installations in bathrooms and shower compartments in dwellings and hotels. Without plaster covering in cavities of ceilings and walls of non-flammable materials. Not permitted in wooden houses or buildings used for agricultural purposes, or in adjacent sections of buildings not separated from them by fire-proof walls.	Not permitted	Not permitted	Not permitted
1 2 3 4 5	1.5 to 16 1.5 to 10 1.5 to 10 1.5 to 35 1.5 to 16	Grey	On, in and under plaster	On, in and under plaster	On, in and under plaster to DIN VDE 0100	On, in and under plaster, depending on special chemical and thermal factors (see DIN VDE 0165)

▷

57

Table 8.5 Cables for fixed installation (continued)

Type	Type designation	Rated voltage U_0/U	Construction	Remarks	Standard
Lead covered PVC-sheathed cable	NYBUY	300/500 V	1 PROTODUR outer sheath 2 Lead sheath 3 Extruded filler 4 PROTODUR insulation 5 Copper conductor, solid or stranded	This type is preferred for use where high safety is demanded – e.g. in chemical works, heavy in-dustry and mining installations.	DIN VDE 0250
PVC-sheathed metalclad cable	NHYRUZY	300/500 V	1 PROTODUR outer sheath 2 Folded metal (zinc) sheath 3 Extruded filler 4 Sheath wire 5 PROTODUR insulation 6 Copper conductor, solid or stranded	Used in place of light PVC-sheathed cables for fixed wiring where fixings on plaster are more widely spaced. NHYRUZY has an elastic rubber filler and a tinned-copper sheath wire (1.5 or 2.5 mm^2 with one, two or three strands) under the folded-metal sheath. The sheath wire must not be used as an earthing or protec-tive conductor.	DIN VDE 0250
Neon Lighting Cables	NYLY	4/8 kV	SIEMENS 4/8kV◁VDE▷	Application according to DIN VDE 0128.	DIN VDE 0250
	NYLRZY	4/8 kV	SIEMENS 4/8kV◁VDE▷ 1 PROTODUR sheath 2 Folded metal (zinc) sheath 3 Discharge wire 4 PROTODUR insulation 5 Copper conductor, flexible	The discharge wire consists of tinned copper strands of 0.3 mm dia. and has a cross-sectional area of 1.5 mm^2.	
SINOTHERM-insulated heat-resistant cables	SIA SIAF	300/500 V	1 Silicone rubber insulation 2 Copper conductor, solid 3 Copper conductor, flexible	Maximum operating conductor temperature 180 °C. For use in high-ambient temper-atures – e.g. in heating appli-ances, high-power luminaires, foundries and boiler rooms. Exposure to superheated steam and flue gases should be avoided.	Construction closely relates to DIN VDE 0250
Heat-resistant silicone-insulated-cable	H05SJ-K A05SJ-K	300/500 V	1 Glass-fibre braid 2 Silicone rubber insulation 3 Copper conductor, flexible		DIN VDE 0282

Conductors		Standard colours of insulation or sheath	Applications			
Number	Cross-sectional area mm^2		In dry locations	In damp and wet locations, and outdoors	In operating areas and store rooms subject to fire hazard	In areas with explosion hazard
2 3 4 5	1.5 to 35 1.5 to 35 1.5 to 35 1.5 to 6	Grey	On, in and under plaster, but not in bath-rooms and shower compartments in dwell-ings and hotels.	On, in and under plaster	On, in and under plaster permitted	On, in and under plaster, depending on special chemical and thermal factors (see DIN VDE 0165)
2 3 4 5	1.5 to 25 1.5 to 25 1.5 to 25 1.5 to 6	Grey	On, in and under plaster and in rooms con-taining high-frequency equipment, but not in bathrooms and shower compartments in dwellings and hotels.	On, in and under plaster, in rooms containing high-frequency equipment, but not outdoors	On, in and under plaster permitted and to DIN VDE 0298 in rooms con-taining high-frequency equipment	Not permitted
1	1.5	Yellow	Only in ventilated steel pipes to DIN 49020 or in equivalent conduits on and under plaster, also in metallic neon signs and reliefs as well as cable conduits of metal.		Not permissible	Not permissible
			On and under plaster	On and under plaster	Not permissible	Not permissible
1	1.5 to 10	Brown	In protected installations in equipment and in or on luminaires	Not permitted	Not permitted	Not permitted
1	0.75 to 120					
1	0.75 to 95	White	In conduit on and under plaster and in or on luminaires	Not permitted	Installation in plastic conduit on and under plaster	In switchboards and distribution boards to DIN VDE 0165

Table 8.5 Cables for fixed installation (continued)

Type	Type designation	Rated voltage U_0/U	Construction	Remarks	Standard
Heat-resistant synthetic elastomer-insulated cables	N4GA	450/750 V		Maximum operating conductor temperature 120 °C. For wiring subject to high-mechanical stresses.	DIN VDE 0250
	N4GAF				

1 Synthetic elastomer insulation
2 Copper conductor, solid or stranded, tinned
3 Copper conductor, flexible, tinned

Conductors		Standard colours of insulation	Applications			
Num-ber	Cross-sectional area mm^2		In dry locations	In damp and wet locations, and outdoors	In operating areas and store rooms subject to fire hazard	In areas with explosion hazard
1	0.5 to 95 1.5 and 2.5 1.5 and 2.5 1.5 and 2.5	Black Green-Yellow Blue Brown	In conduit on and under plaster, in or on luminaires and in protected installations in equipment.	Not permitted	Installation in plastic conduit on and under plaster	In switchboards and distribution boards to DIN VDE 0165
1	0.5 to 95	Black				

8.2.2 Flexible Cables

Table 8.6 Flexible cables

Type	Type designation	Rated voltage U_0/U	Construction	Remarks	Standards
Flat tinsel cord	H03VH-Y	300/300 V	1 PROTODUR insulation 2 Tinsel conductor	To avoid overloads, these cords may be used for permanent connection to appliances, or in conjunction with appliance connectors, only if the current does not exceed 1 A. Not suitable for cooking or heating appliances.	DIN VDE 0281
Flat non-sheathed cord	H03VH-H	300/300 V	1 PROTODUR insulation 2 Copper conductor, highly flexible	Not suitable for connecting cooking and heating appliances.	DIN VDE 0281
Light PVC-sheathed cord	H03VV-F H03VVH2-F	300/300 V	1 PROTODUR sheath 2 PROTODUR insulation 3 Copper conductor, flexible	Not suitable for connecting cooking and heating appliances. As well as the round type, there is also a flat version H03VVH2-F, twin-core, 0.5 and 0.75 mm².	DIN VDE 0281
Ordinary PVC-sheathed cord	H05VV-F H05VVH2-F	300/500 V	1 PROTODUR sheath 2 PROTODUR insulation 3 Copper conductor, flexible	Permitted for connecting cooking and heating appliances only if there is no possibility of contact between the cable and hot parts of the appliance or other sources of heat. As well as the round type, there is also a flat version H05VVH2-F, twin-core, 0.75 mm².	DIN VDE 0281

Conductors		Standard colours of insulation or sheath	Applications	
Number	Cross-sectional area mm^2		Location	Permissible stress
2	0.1	Black White Grey	In dry locations – e.g. in homes and offices	For connecting extremely light hand appliances – e.g. electric shavers. The current loading must not exceed 1 A and the length must not exceed 2 m.
2	0.5 and 0.75	Black White Brown	In dry locations – e.g. in homes, kitchens and offices	For light electrical equipment with very low mechanical stresses – e.g. radios, table lamps etc.
2 3 4	0.5 and 0.75 0.5 and 0.75 0.5 and 0.75	Black White	In dry locations – e.g. in homes, kitchens and offices. Not in industrial or agricultural premises.	For light electrical equipment with low mechanical stresses – e.g. office machines, table lamps, kitchen appliances etc.
2 3 4 5 7	0.75 to 2.5 0.75 to 2.5 0.75 to 2.5 0.75 to 2.5 1 to 2.5	Black White	In dry locations; for domestic and cooking appliances also in damp and wet locations. Not in industrial or agricultural premises, but permitted in tailors shops and similar premises.	For connecting electrical appliances with medium mechanical stresses – e.g. washing machines, spin driers, refrigerators etc. The cables may be installed permanently – e.g. in furniture, decorative panelling, screens etc.

▷

Table 8.6 Flexible cables (continued)

Type	Type designation	Rated voltage U_0/U	Construction	Remarks	Standards
Braided cord	H03RT-F	300/300 V	1 Braid 2 Textile filler 3 Synthetic elastomer insulation 4 Copper conductor, flexible, tinned	The braid consists of polished rayon yarn. Further development led to the design of rubber-sheathed cables suitable to withstand the high-mechanical and thermal stresses when used with domestic irons (see Section FLEXO-cables).	DIN VDE 0282
Ordinary tough rubber-sheathed cord	H05RR-F	300/500 V	1 Natural rubber sheath 2 Synthetic elastomer insulation 3 Copper conductor, flexible, tinned	These cables are not suitable for continuous use outdoors.	DIN VDE 0282
Ordinary polychloroprene-sheathed cord	H05RN-F	300/500 V	1 Polychloroprene sheath 2 Synthetic elastomer insulation 3 Copper conductor, flexible, tinned	For use where there is a possibility of exposure to fats and oils.	DIN VDE 0282
Heavy polychloroprene-sheathed flexible cables	H07RN-F	450/750 V	1 Polychloroprene outher sheath 2 Polychloroprene inner sheath 3 Synthetic elastomer insulation 4 Copper conductor, flexible, tinned 5 Textile filler	Permitted for permanent protected installation in conduit or in equipment, and for rotor connecting cables for motors with rated voltages of up to 1 000 V a.c. or 750 V d.c. to earth. In rail vehicles the d.c. operating voltage may be up to 900 V to earth. The design of the OZOFLEX-H07RN-F highly flexible cable is similar to that of the H07RN-F, but with a short lay, textile filler and extremely finely stranded conductors.	DIN VDE 0282
PROTO-FIRM-sheathed cable	NSSHÖU	0.6/1 kV	1 PROTOFIRM (polychloroprene) outer sheath 2 Polychloroprene inner sheath 3 Synthetic elastomer insulation 4 Copper conductor, flexible, tinned	This type of cable is suitable for forced guiding and reeling only to a limited extent. For this kind of use, CORDAFLEX cables (NSHTÖU) are recommended (see page 72).	DIN VDE 0252

Conductors		Standard colours of insulation or sheath	Applications	
Number	Cross-sectional area mm^2		Location	Permissible stress
2 3	0.75 to 1.5 0.75 to 1.5	Blue-white	In dry locations – e.g. in homes, kitchens and offices. Not in industrial or agricultural premises, but permitted in tailors' shops and similar premises.	For light electrical appliances with low mechanical stresses – e.g. electric blankets
2 3 4 5	0.75 to 2.5 0.75 to 6 0.75 to 6 0.75 to 2.5	Black	In dry locations – e.g. in homes, kitchens and offices. Not in industrial or agricultural premises, but permitted in tailors' shops and similar premises.	For connecting electrical appliances with low mechanical stresses – e.g. vacuum cleaners, irons, kitchen equipment, soldering irons etc. These cables may also be installed permanently – e.g. in furniture, decorative panelling, screens etc.
1 2 3 4 5	0.75 and 1.0 0.75 and 1 0.75 and 1 0.75 0.75	Black	In dry, damp and wet locations and outdoors	For connecting electrical appliances and tools with low-mechanical stresses – e.g. deep-fat friers, kitchen equipment, soldering irons, hedge clippers etc. These cables may also be installed permanently – e.g. in cavities in prefabricated building sections.
1 2 3 4 5 7 to 18 19 to 36	1.5 to 500 1 to 25 1 to 300 1 to 300 1 to 25 1.5 to 4 1.5 and 2.5	Black	In dry, damp and wet locations and outdoors. In agricultural operating areas and those subject to fire hazard. In operating areas and storerooms to DIN VDE 0165 subject to explosion hazard.	For connecting electrical appliances and tools, including industrial equipment with medium-mechanical stresses – e.g. large water heaters, hot-plates, power drills, circular saws and mobile motors or machines on building sites. For permanent installation – e.g. in temporary buildings – and for direct installation in components of hoisting equipment, machines etc.
3 4 5	1 and 1.5 1 and 1.5 1 and 1.5			Where cables are subject to kinking and twisting – e.g. for hand grinders and power drills.
1 2 3 4 5 7 to 36	2.5 to 400 1.5 to 185 1.5 to 185 1.5 to 185 1.5 to 70 1.5 to 4	Yellow	In dry, damp and wet locations and outdoors. In agricultural operating areas and those subject to fire hazard. In operating areas and storerooms to DIN VDE 0165 subject to explosion hazard.	For heavy equipment and tools with high-mechanical stresses on building sites, in industry, in quarries and in open-cast and underground mines.

$\triangleright$

Table 8.6 Flexible cables (continued)

Type	Type designation	Rated voltage U_0/U	Construction	Remarks	Standards
HYDRO-FIRM cables	TGK TGKT TGW TGFLW	450/750 V	 1 Polychloroprene sheath 2 Synthetic elastomer insulation 3 Copper conductor flexible 4 Insulation cross-linked polyolefine	Suitable for operation in water at depths up to 500 m. The ability to operate continuously in water has been proved by tests. Design and dimensions are as for H07RN-F. Cables to meet special requirements on request.	Construction closely relates to DIN VDE 0282
Heat-resistant flexible cable	4GMH4G	300/500 V	 1 Heat-resistant synthetic elastomer sheath 2 Heat-resistant synthetic elastomer insulation 3 Copper conductor, flexible, tinned	Maximum operating conductor temperature 120 °C. These cables also remain flexible at low temperatures – down to about − 30 °C.	Construction closely relates to DIN VDE 0282
Heat-resistant silicone-insulated and sheathed cable	N2GMH2G	300/500 V	 1 Silicone-rubber sheath 2 Silicone-rubber insulation 3 Copper conductor, flexible	Maximum operating conductor temperature 180 °C. Exposure to superheated steam and flue gases is harmful. If air is excluded at temperatures above 100 °C the mechanical properties of the silicone rubber are impaired.	DIN VDE 0250
ARCO-FLEX welding cable	NSLFFÖU	200 V	 1 PROTOFIRM (polychloroprene) sheath 2 Separator 3 Copper conductor, flexible	These cables have a PROTOFIRM sheath which is oil-resistant, flame retardant and resistant to abrasion and indentation. Maximum operating conductor temperature 80 °C.	DIN VDE 0250
FLEXI-PREN welding cable (hand held)	NSLFFÖU	200 V	 1 PROTOFIRM (polychloroprene) sheath 2 Separator 3 Copper conductor, highly flexible	FLEXIPREN hand-welding cables have an extremely finely-stranded conductor, with thinner strands than are required by VDE 0250. This makes them exceptionally flexible.	DIN VDE 0250

Conductors		Standard colours of insulation or sheath	Applications	
Number	Cross-sectional area mm^2		Location	Permissible stress
1 2 3 4 5	1.5 to 500 1 to 25 1 to 300 1 to 300 1 to 25	Blue	In water, in dry, damp and wet locations and outdoors	For connecting electrical equipment with medium-mechanical stresses, especially equipment which operates continuously in water – e.g. submerged pumps and underwater floodlights. TGK for water temperatures up to 40 °C TGKT for continuous immersion in drinking water up to 40 °C TGW and TGFLW for water temperatures up to 60 °C.
3 4	1 to 70 1 to 70			
3 4 5 7	0.75 to 2.5 0.75 to 2.5 0.75 to 2.5 0.75 to 2.5	Grey	In dry, damp and wet locations and outdoors	For connecting cooking and heating appliances with medium-mechanical stresses and increased ambient temperatures – e.g. cookers, electric storage heaters etc.
2 3 4 5	0.75 to 4 0.75 to 4 0.75 to 4 0.75 to 4	Brown	In dry, damp and wet locations and outdoors	For low-mechanical stresses and high-ambient temperatures.
1	16 to 185	Black	In dry, damp and wet locations and outdoors	For very high-mechanical stresses as machine and hand-welding cables.
1	25 to 70	Black	In dry, damp and wet locations and outdoors	Highly-flexible hand-welding cable for very high-mechanical stresses.

▷

Table 8.6 Flexible cables (continued)

Type	Type designation	Rated voltage U_0/U	Construction	Remarks	Standards
PVC control cable	SYSL	300/500 V	 1 PROTODUR sheath 2 Separator (≥ 12 core) 3 PROTODUR insulation 4 Copper conductor flexible	For fixed installation but not with free movement and forced guidance on rollers or reeling duty.	Construction closely related to DIN VDE 0250
PROTO-FLEX PVC-control cable	NYSLYÖ	300/500 V	 1 PROTODUR sheath 2 Textile layer 3 PROTODUR insulation 4 Copper conductor, flexible	For arrangements affording freedom of movement; not for forced guiding over rollers or operation on reels. The outer sheath is substantially unaffected by moisture, oils, fats and chemicals.	DIN VDE 0250
PROTO-FLEX screened PVC-control cable	NYSLYCYÖ	300/500 V	 1 PROTODUR sheath 2 Tinned-copper braid 3 PROTODUR inner sheath 4 PROTODUR insulation 5 Copper conductor, flexible	These cables meet the requirements for "Electrical equipment of industrial machines" in accordance with IEC 204 and DIN VDE 0113. The screening braid of the NYSLYCYÖ, by virtue of its design, has a very low coupling of 250 Ω/km.	
Lift control cable	YSLTK-JZ	300/500 V	 1 PROTODUR sheath 2 Textile braid 3 Separator 4 PROTODUR insulation 5 Copper conductor, flexible 6 Sheathed strain bearing element	Insulation and sheath are made from a cold-resistant PVC compound (flexible down to $-20\,°C$) Cables with up to 18 cores have a textile strain bearing element; with 24 or more cores the strain bearing element is a steel rope. This will support the maximum suspension length with a factor of safety of five. The manufacturer's installation instructions must be adhered to.	Construction closely relates to DIN VDE 0250

Conductors		Standard colours of sheath	Applications	
Number	Cross-sectional area mm^2		Location	Permissible stress
2 to 60	0.5 to 6	Grey	In dry damp and wet locations	For control equipment, production lines and machine tools as connecting and inter-connecting cables with medium-mechanical stresses.
3 to 60	0.5 to 2.5	Grey	In dry, damp and wet locations	For control equipment, production lines and machine tools as connecting and inter-connecting cables with medium-mechanical stresses.
3 to 25	0.5 to 2.5	Grey	In dry, damp and wet locations	For control rooms, production lines and data-processing equipment with medium-mechanical stresses, where interference suppression is required.
7 12 18 24 30	1 1 1 1 1	Black	In dry, damp and wet location	A self-supporting flexible control cable with medium-mechanical stress – e.g. for lifts and conveyor systems; suspension lengths up to 50 m, cage velocity up to 1.5 m/s.

Table 8.6 Flexible cables (continued)

Type	Type designation	Rated voltage U_0/U	Construction	Remarks	Standards
Lift flexible control cables	YSLYTK-JZ YSLYCYTK-JZ	300/500 V	1 PROTODUR sheath 2 Textile braid 3 PROTODUR inner sheath 4 Separator 5 Textile filler 6 PROTODUR insulation 7 Copper conductor, flexible 8 Sheathed strain bearing element	Insulation and sheath are made from a cold-resistant compound (flexible down to −30 °C). The strain bearing element is a steel rope with reduced twisting which will support the maximum suspension length with a factor of safety of five. In equipment for which interference suppression to VDE 0875 is required, type YSLYCYTK-JZ must be used. The manufacturer's installation instructions must be adhered to.	Construction closely related to DIN VDE 0250
Flat PVC-sheathed flexible cables for lifts and similar application	H05VVH6-F	300/500 V	1 PROTODUR sheath 2 PROTODUR insulation 3 Copper conductor, flexible	Flat PVC-insulated cables are not intended for use outdoors.	DIN VDE 0281
	H07VVH6-F	500/700 V			
PLANO-FLEX flat flexible cables	NGFLGÖU	300/500 V	1 Polychloroprene sheath 2 Synthetic elastomer insulation 3 Copper conductor, highly flexible	A cold-resistant chloroprene rubber is used for the sheath, enabling the cables to remain sufficiently flexible down to −35 °C. The insulation consists of an ozone and weather-resistant PROTOLON synthetic-elastomer. Maximum operating conductor temperature 90 °C.	DIN VDE 0250

Conductors		Standard colours of sheath	Applications	
Number	Cross-sectional area mm^2		Location	Permissible stress
28 + 2 individually-screened communication cores	1 0.5 for the communication cores	Black	In dry, damp and wet location	Self-supporting flexible control cables with medium-mechanical stress – e.g. for lifts and conveyor systems; suspension lengths up to 150 m, cage velocity up to 10 m/s.
3 to 24	0.75 and 1	Black	In dry, damp and wet locations	Flexible power and control cables with medium-mechanical stresses and sharp bending in one plane in operation – e.g. in hoisting equipment, transport systems, machine tools etc., as power supply and control cable.
3 to 24	1.5 to 16			
3 to 24 3 to 8 3 to 7 3 and 4	1 to 2.5 1 to 4 1 to 35 1 to 95	Black	In dry, damp and wet locations and outdoors	

Table 8.6   Flexible cables (continued)

Type	Type designation	Rated voltage U_0/U	Construction	Remarks	Standards
CORDA-FLEX cable	NSHTÖU	0.6/1 kV	CORDAFLEX 1 2 3 4 1 Polychloroprene outer sheath 2 Polychloroprene inner sheath 3 Synthetic elastomer insulation 4 Copper conductor, flexible, tinned	Maximum operating conductor temperature 90 °C. Sheath of polychloroprene, flexible down to −30 °C.	DIN VDE 0250
CORDA-FLEX(K) cable	NSHTÖU	0.6/1 kV	CORDAFLEX 1 2 3 4 5 6 1 Polychloroprene outer sheath 2 Supporting braid 3 Polychloroprene inner sheath 4 Textile layer 5 Synthetic elastomer insulation 6 Copper conductor, flexible, tinned	Maximum operating conductor temperature 90 °C. Sheath of polychloroprene, flexible down to −35 °C.	DIN VDE 0250
CORDA-FLEX (SM) cable	NSHTÖU	0.6/1 kV	CORDAFLEX 1 2 3 4 5 6 1 Polychloroprene outer sheath 2 Supporting braid 3 Polychloroprene inner sheath 4 Synthetic elastomer insulation 5 Foil 6 Copper conductor, flexible, tinned	Maximum operating conductor temperature 90 °C. Sheath of polychloroprene, flexible down to −20 °C	DIN VDE 0250

Conductors		Standard colours of sheath	Applications	
Number	Cross-sectional area mm^2		Location	Permissible stress
5 to 24 4 to 24 4	1.5 2.5 4 to 50	Black	In dry, damp and wet locations and outdoors	For high-mechanical stresses on reels without guide rollers for apparatus with realing speedy up to 60 m/minute.
5 to 30 4	1.5 and 2.5 2.5 to 120	Black	In dry, damp and wet locations and outdoors	For high-mechanical stresses preferably for forced guiding – e.g. reels or guide rollers, for high acceleration and travel speeds in hoists, transportation and coveyor equipment. For travel speeds up to 120 m/minute.
4 to 30 4	1.5 and 2.5 10 to 50	Yellow	In dry, damp and wet locations and outdoors	For very high-dynamic stresses as e.g. operation of electro-hydraulic grab cranes, crane lifting magnets etc. as well as mobile cable carriers. For travel speeds up to 120 m/minute.

8.2.3 FLEXO Cords

FLEXO cords comprise elastomer- or PVC-sheathed cables having either vulcanized or moulded, non-separable connectors, such as plugs connectors or appliance plugs, factory attached at on or both ends. If cables and connectors are of elastomers these are vulcanized in the press. With PVC sheathed cables the connectors are also of PVC injection moulded to form one unit. Frequently plastic plugs are also moulded on to elastomer-sheathed cables. Prefabricated FLEXO cables save time and cost during installation, have a high degree of electrical safety and offer the user practical advantages.

Due to the additional requirements for connecting cables for heating applicances, especially domestic irons, a cable was developed which will withstand the high bending stresses and temperature. This cord is available under the trade name THERMOSTABIL and is designed closely to the type of construction O5RR and fully complies with the harmonized standard.

The pin support bridge is of moulded Duroplast and supports other contacts such as the protective conductor contact. It is not permitted to use thermoplastic materials for the manufacture of the pin support bridge. All component parts of the connector are firmly embedded and secured on all sides in either the elastomer or the plastic during moulding and therefore are electrically insulated and mechanically protected.

The conductors are either soldered or welded to the contact pieces. The attachment of the cord to the support bridge is formed by a tapered sleeve which prevents sharp bending and improves resistance to kinking. The arrangement of cord entry into the connector can be either central (Figs. 8.1 or 8.2) or angled (Fig. 8.3). Cables with central cable entry are suitable for appliances where the plug is frequently disconnected and connected. Connectors with angled cable entry are generally more suitable where space is limitted.

The relevant standards lay down the profile and dimensions for the pin end of the connector. However the overal shape of the body of the support bridge is left to the designer providing that all test requirements are fulfilled. A Siemens design, generally favoured by users, is protected under the trade name PROTOFORM.

Connectors of different constructions could also be manufactured, but because of the high cost of mould-

Fig. 8.1
European flat plug up to 2.5 A to DIN 49 464-F/CEE 7 sheet XVI

Fig. 8.2
SCHUKO plug for 10 A to DIN 49 441-R2/CEE 7 sheet VII with two protective contact systems, mainly for use in Belgium and France

Fig. 8.3
SCHUKO plug for 10 A to DIN 49 441-R1/CEE 7 sheet IV

ing and tooling this is only practical where large quantities are required.

The free cable end, used for the fixed connection, can be finished by the customer as desired and can be stripped, fitted with boot-lace sleeves, spade connectors or others.

As part of an elaborate quality control system all features relevant to safety are tested both during and immediately after manufacture. Safety features related to personel safety are routine tested.

Furthermore, during approval type testing the relevant national approval authorities carry out extensive electrical and mechanical tests.

The bridge piece of connectors is marked showing the permissible rated current and voltage, the manufacturer identity, a VDE mark if applicable together with the mark of the relevant national approval authority.

For factory made connection cords, special approvals are necessary for individual countries even though the cord itself complies in almost all cases with the harmonized standard. The approvals are based on the requirements of DIN/VDE regulations as well as CEE publications 7 and 22 respectively.

8.3 Flexible Cables for Mining and Industry

The high-mechanical stresses met in mining and heavy industry require tough cables with a type of construction suited to the relevant application.

These cables must have a particularly strong outer sheath. For elastomer-sheathed cables Siemens have developed the impact resistant, tear-abrasion resistant PROTOFIRM sheath (see page 38). Sheath colours see Table 8.7.

Table 8.7
Sheath colours to DIN VDE 0206
and DIN VDE 0118

Cable for	Sheath colour
Rated voltage $U_0/U \leq 0.6/1$ kV	Yellow
Rated voltage $U_0/U > 0.6/1$ kV	Red
Intrinsically safe equipment	Blue

Polychloroprene-Sheathed Cables for Heavy-Mechanical Stresses

Polychloroprene-sheathed cables for heavy-mechanical stresses for rated voltages up to 1000 V are manufactured by Siemens under the trade name PROTOMONT (Figs. 8.4 and 8.5). When selecting cables for mining the regulations DIN VDE 0118 and DIN VDE 0168 as well as any special regulations of the relevant mining authority must be observed. For industrial applications DIN VDE 0100 is similarly relevant.

Heavy PROTOMONT Polychloroprene-Sheathed Cables NSSHÖU

These cables are used for the connection of motors, fixed and movable heavy apparatus as well as industrial tools. This type of construction also replaces type NSHÖU, which was not covered by the harmonized standard, for cables subjected to high-mechanical stresses having cross-sectional areas up to 6 mm^2 and up to 5 cores. In mining below ground where cables are subjected to gas a construction type must be used which has a concentric protective conductor surrounding either all main phase conductors or is equaly divided arranged around each individual phase conductor. The later type is preferred in the mining industry (Fig. 8.5). The interstices of the cores may also be used to incorporate pilot cores.

1   2   3   4   4   4   5

1 PROTOFIRM outer sheath
2 Polychloroprene inner sheath
3 Textile layer
4 Numbered PROTOLON-insulated cores
5 Copper conductors, flexible, tinned

Fig. 8.4
Heavy PROTOMONT polychloroprene-sheathed cables NSSHÖU 19×2.5 0.6/1 kV

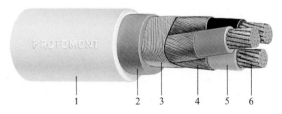

1   2   3   4   5   6

1 PROTOFIRM outer sheath
2 Polychloroprene inner sheath
3 Textile layer
4 Layer of tinned copper wires
5 PROTOLON insulation
6 Copper conductors, flexible, tinned

Fig. 8.5
Heavy PROTOMONT polychloroprene-sheathed cable NSSHÖU $3 \times 95 + 50/3$ E 0.6/1 kV

75

Cables for Coal-Cutters for Operation below Ground

Moving coal-cutters in mining below ground are connected with coal-cutter cable (Fig. 8.6). These cables are subjected to the heaviest mechanical stresses.

Depending on the type of cable arrangement: free dragging, forced guiding by protective cable drag chains or forced guiding with a cable roller system, three different cable constructions were developed particularly for the supply of power to coal-cutters.

The drag cable must be capable of withstanding the operational pulling forces with a high factor of safety. For this purpose a steel-copper braid is embedded in the outer sheath which performs additionally the function of concentric protective conductor.

In the drag chain system cables are guided without significant tensile stress. The cutter cable in this ar-

rangement, however, must be particularly flexible to allow easier installation and to achieve free movement of the chain. To achieve this the concentric protective conductor consisting of steel-copper braid is arranged between the inner and outer sheaths.

A later development for forced guiding of coal-cutter cable is the protected installation in an enclosed duct via a moving roller system. For this purpose a specially flexible range of coal-cutter cables, having a braid of strands of polymeric yarn embedded in the outer sheath, were developed. All cutter cables have control and monitoring cores laid into the interstices between the cores. For mechanical reasons these control and monitoring cores are constructed in concentric form. The monitoring conductor is electrically connected to the conductive rubber layers which surround the cores of the phase conductors. This connection is used together with a monitoring device to detect damage to the cable and initiate disconnection.

With increasing cross-sectional areas the problem of both mechanical handling and service life increase and because of this Siemens do not manufacture PROTOMONT coal-cutter cables larger than 150 mm^2 for $U_0/U = 0.6/1$ kV.

Polychloroprene-Sheathed Cables for Hoists

In interconnecting shafts between coal seams of different levels, hoists may be installed.

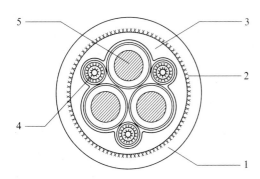

1 PROTOFIRM outer sheath
2 Armour (protective conductor) of steel-copper braid
3 Inner sheath
4 Control monitoring conductor:
control conductor 1.5 mm^2 flexible tinned copper conductor under a PROTOLON insulation coloured blue and monitoring conductor 1.5 mm^2 tinned concentric copper conductor and conducting layer
5 Phase conductor:
copper conductor, flexible tinned under a coloured PROTOLON insulation and conducting layer

Fig. 8.6
PROTOMONT coal-cutter cable NSSHCGEÖU
3 × 50/25 KON + 3 × (1.5 ST KON/1.5 ÜL KON)Z
0.6/1 kV

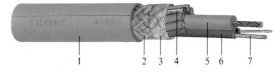

1 PROTOFIRM sheath
2 Wide mesh textile braid
3 Textile braid
4 Screened communication cores
5 PROTOLON insulation
6 Sheathed strain bearing element
7 Copper conductors, flexible, tinned.

Fig. 8.7
PROTOMONT rubber-sheathed cables for hoists.
NTMTWÖU 8 × 1.5 ST + 2 × 1 FM(C) 0.6/1 kV

For the control of hoist cages, a blue suspension cable NTMTWÖU is used.

A torsion-free centrally arranged steel rope allows free hanging of this cable up to 200 m. For communication purposes, two of the ten control cores are screened.

Polychloroprene-Sheathed Cables for Coal Face Lighting

For the connection of flameproof light fittings at the coal face NSSHÖU cables are used in which the individual core screening is used as a protective conductor. Where the coal face lighting installation incorporates telecommunications, then the coal face lighting cable must also contain a screened communications pair (Fig. 8.8).

Both cable types are connected to monitoring apparatus. In the case of NSSHÖU cables one phase conductor is used as a conductor in the monitoring system. In cables incorporating telecommunications one monitoring core is provided sheathed with conductive rubber which in turn is connected to the conductive rubber layers over the phase conductors and the control cores.

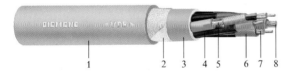

1 PROTOFIRM outer sheath
2 Textile layer
3 Polychloroprene sheath
4 Conducting rubber sheath
5 Protective conductor
6 PROTOLON insulation
7 Communication cores
8 Copper conductor, flexible, tinned

Fig. 8.8
PROTOMONT-polychloroprene sheathed cable for coal face lighting
NSSHKCGEFMÖU
$3 \times 6 + 2 \times 2.5 \text{ ST} + 2 \times 0.5 \text{ FM} + 2.5 \text{ ÜL } 0.6/1 \text{ kV}$

Mining Cables above 0.6/1 kV

Siemens provide mining cables for rated voltages $U_0/U \geq 0.6/1$ kV under the trade name SUPROMONT cable (Fig. 8.9).

1 PROTODUR outer sheath
2 Steel wire braid
3 Textile layer
4 Monitoring conductor
5 Conducting layer
6 Extruded filler
7 Protective conductor
8 Control core
9 PROTODUR insulation
10 Copper conductor, flexible

Fig. 8.9
SUPROMONT cable NYHSSYCY
$3 \times 35 + 3 \times 16/3 \text{ E} + 3 \times 2.5 \text{ ST} + \text{ÜL } 3.6/6 \text{ kV}$

SUPROMONT cables of either thermoplastic or elastomeric construction are used to bring the high tension network of 6 kV or 10 kV direct to the load centres. These cables are used, for example, for the incoming feed to transformers in mining work faces below ground and in tunnel operations for roads or underground rail systems. This avoids having low-voltage cables with large cross-sectional areas. For reasons of safety these cables are provided with a protective conductor, a monitoring conductor and steel-wire braid as armouring below the outer sheath. In addition control cores are incorporated in the interstices of cores. SUPROMONT cables are supplied, mostly, in lengths of either 100 m or 200 m with factory-fitted end terminations. The form of termination being arranged to suit the type of connecting or interconnecting boxes.

Trailing Cables and their Terminations

Trailing cables are used to transmit large amounts of energy in the voltage range of 1 to 35 kV and are subjected to high-mechanical stresses. These cables are used on large mobile machinery such as excavators, dredgers coal face equipment and hasting gear in the form of drum wound or trailed power supplies.

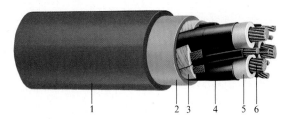

1 PROTOFIRM outer sheath
2 Polychloroprene sheath
3 Extruded filler
4 Conducting rubber
5 PROTOLON insulation
6 Copper conductor, flexible

Fig. 8.10
PROTOLON trailing cable
NTSCGEWÖU 3 × 25 + 3 × 25/3 6/10 kV

Table 8.8
Current-carrying capacities of PROTOLON trailing cables at 30 °C ambient

Nominal cross-sectional area mm²	Current-carrying capacities in A rated voltage	
	up to 10 kV	above 10 kV
2.5	32	–
4	43	–
6	56	–
10	78	–
16	104	110
25	138	146
35	171	181
50	213	226
70	263	279
95	317	336
120	370	391
150	425	450
185	485	514

For cables laid on the ground etc. a correction factor of 0.95 has to be applied

Table 8.9
Correction factors for ambient air temperatures other than 30 °C. To be applied to the current-carrying capacities shown in Table 8.8

Ambient temperature °C	25	30	35	40
Correction factor	1.05	1.0	0.95	0.89

PROTOLON trailing cables (Fig. 8.10) are constructed in line with DIN VDE 0250. For the current-carrying capacities to DIN VDE 0250 of 3 loaded conductors in free air at 30 °C the values shown in Table 8.8 apply. For ambient temperatures other than 30 °C these values require to be adjusted using the factors given in Table 8.9.

The cores for high-voltage cables from 6 kV upwards are constructed to the OZONEX principle (see page 24) developed by Siemens and which has been proved over several decades i.e. to avoid harmful partial discharge conducting rubber layers are placed over the conductors and above the PROTOLON insulation. The earth conductor, sheathed in conducting rubber, is divided and laid into the interstices between the cores.

In high-tension cables the conducting rubber layer also acts as touch protection. It is therefore necessary for the conductivity to be such that the resistance between the protective conductor and any point on the outer conducting layer not to exceed 500 Ω.

For particularly high stresses and travel speeds exceeding 60 m/min PROTOLON cables are fitted with an additional textile braid incorporated in the outer sheath. The textile braid increases mechanical strength and provides torsion protection.

Trailing cables must be provided with terminations to protect against ingress of moisture. With rated voltages greater than 6 kV the termination also provides an electrical function. The individual termination constructions are dependant on operating and installation conditions.

For motor connection boxes, transformer stations and gate-end boxes and similar equipment PROTOLON indoor terminations are used.

For rated voltages up to 10 kV a simple dividing box termination is sufficient (Fig. 8.11). From 15 kV upwards a dividing box with core sleeves is required (Fig. 8.12). Where space is severely limited a smaller divider with core sleeves over the cable tails is available (Fig. 8.13).

For use on outdoor trailing cables up to 35 kV a vulcanised water shed termination is available which may be directly connected to overhead supply wires (Fig. 8.14). These are mainly used in Electricity Board networks during network alterations, for the supply to floating dredgers, or open cast mining, or

Fig. 8.11
Dividing box termination for trailing cables NTSWÖU and NTSCGEWÖU with rated voltages from 1 to 10 kV

Fig. 8.12
Dividing box termination with stress cone for trailing cables NTSCGEWÖU with rated voltages from 15 to 35 kV

Fig. 8.13
Low-space dividing box with stress cone for trailing cables NTSCGEWÖU with rated voltages from 6 to 10 kV

Fig. 8.14
Vulcanized outdoor water shed termination with clamp on terminal

for the power supply to building sites. This outdoor termination is designed to withstand the stresses to be expected during frequent rearrangement of cables.

8.4 Halogen-Free SIENOPYR Wiring and Flexible Cables with Improved Performance in the Event of Fire

Experience gained from a number of large fires has shown that, particularly in buildings with a high density of installed cables and wires, e. g. hospitals, hotels etc., considerable consequential damage can be caused when the cable insulation is PVC based. In such conditions during the combustion of PVC materials chlorine and hydrogen is released which in the presence of moisture combine to form the highly corrosive hydrochloric acid. The consequential damage caused by this is often more extensive than the primary damage. In addition such materials in the event of fire lead to such a strong smoke development that rescue work and fire fighting is significantly hampered.

To reduce the risk, especially in buildings with a high concentration of people and/or high value contents, Siemens has developed halogen-free insulation materials having special profile characteristics to suit their applications and employ these on the most important basic types of cables and wires. These new products bear the trade name SIENOPYR and fulfil the general requirements for cables and wires in respect of electrical mechanical and chemical parameters and in addition have the following special characteristics:

▷ very little support of combustion

▷ no corrosive combustion gasses from halogens

▷ much reduced smoke density

The testing relating to combustion characteristics of cables and wires are laid down in DIN VDE 0472 for

the combustion characteristics   in Part 804
the corrosiveness of combustion gasses in Part 813
the smoke density   in preparation

For further details see also page 125 "Halogen-Free Cables with Improved Performance in the Event of Fire."

The preferred areas of application for SIENOPYR cables are in installations having increased safety requirements, e. g. hospitals, high-rise buildings, theatres, industrial buildings, power stations, hotels, schools, department stores, electronic data processing plants and the transport industry.

For the application of these cables the rules given in DIN VDE 0298 Part 3 "Application of cables, wires and flexible cords for power installations. General rules for cables", must be observed. In particular the data for the relevant basic types of cables, from which the SIENOPYR types were derived, must be noted. In addition also the relevent installation and apparatus standards as well as standards and directives of the relevant authorities or institutions must be observed.

In line with current market requirements the following cable types are readily available:

Light sheathed SIENOPYR cables,
Heat resistant non-sheathed single-core cables
Single-core synthetic elastomer-sheathed SIENOPYR cables for special purposes
Synthetic elastomer-sheathed flexible SIENOPYR(X) cables.

Apart from the above special types are manufactured e.g. control cables.

Light-Sheathed SIENOPYR Cables

As installation cables in buildings with high density of people and/or valuable contents for fixed installation above, on as well as in and under plaster SIENOPYR sheathed cables type NHXMH 300/500 V are recommended. These cannot only be used in dry but also in humid and wet rooms. The cable corresponds to DIN VDE 0250 Part 214. It is based, as regards dimensions and basic characteristics on the NYM type of construction and is designed for the same maximum conductor operating temperature of 70 °C.

1 Sheath of non-cross-linked polyolefine compound
2 Core insulation of non-cross-linked special compound
3 Insulation of cross-linked polyolefine compound
4 Copper conductor, solid or stranded

Fig. 8.15
Light sheathed SIENOPYR cable NHXMH 300/500 V

Heat-Resistant Non-Sheathed Single-Core SIENOPYR Cables

For switchgear and distribution boards in dry rooms as wiring cables, with increased performance in the event of fire, cables 450/750 V with solid conductor (N)HX4GA or with flexible conductor (N)HX4GAF are used. These cables are also suitable for internal wiring of apparatus having rated voltages up to 1 000 V a.c. or 750 V d.c. to earth. The maximum conductor operating temperature is 120 °C. These insulated wires also remain flexible at low temperatures and can be used down to −30 °C.

The construction complies with the regulations for heat-resistant synthetic elastomer-insulated cables (N4GA respectively N4GAF) DIN VDE 0250 Part 501.

1 Insulation of synthetic elastomer based on Ethylene-Vinylacetate-Copolymer
2 Copper conductor, solid, tinned
3 Copper conductor, flexible, tinned

Fig. 8.16
Heat-resistant non-sheathed single-core SIENOPYR cable. (N)HX4GA, (N)HX4GAF 450/750 V

Single-Core Synthetic Elastomer-Sheated SIENOPYR Cable for Special Purposes

These cables for 1.8/3 kV can be used for fixed installations in traction vehicles and buses to DIN VDE 0115 Section 2 as well as in dry rooms. DIN VDE 0100 permits these to be used as short-circuit fault proof and earth fault proof connections. The maximum conductor operating temperature is 90 °C. The sheath is oil resistant to DIN VDE 0472 Part 803 test type A.

The construction is in line with DIN VDE 0250 Part 602 for special rubber-insulated cables NSGA-FÖU.

1 Sheath of cross-linked synthetic elastomer based on Ethylene-Vinylacetate-Copolymer
2 Insulation of cross-linked synthetic elastomer based on Ethylene-Propylene-Rubber
3 Copper conductor, flexible, tinned

Fig. 8.17
Single-core synthetic elastomer-sheathed SIENOPYR cable for special purposes
(N)HXSGAFHXÖ 1.8/3 kV

SIENOPYR(X) Synthetic Elastomer-Sheathed Flexible Cable (N)HXSHXÖ

For flexible connection cables and interconnecting cables in buildings with high concentration of people and/or valuable contents with medium-high mechanical stress synthetic elastomer-sheathed SIENOPYR(X) cables are used, type (N)HXSHXÖ (Fig. 8.18). These can be used in dry, damp and wet rooms as well as outdoors. They may also be used in fixed installations.

The cables are constructed closely to DIN VDE 0250 Part 812 (NSSHÖU). The maximum conductor operating temperature is 90 °C. The sheath is oil resistant to DIN VDE 0472 Part 803, test type A. Furthermore the cable is KMV[1] fault resistant which means they are also suited to meet the special conditions of application in nuclear power stations.

---

[1] KMV = Loss of coolant medium

1 Outer sheath of cross-linked synthetic elastomer based on Ethylene-Vynilacetate-Copolymer
2 Inner sheath of cross-linked synthetic elastomer based on Ethylene-Vynilacetate-Copolymer
3 Insulation of cross-linked synthetic elastomer based on Ethylene-Propylene-Rubber
4 Copper conductor, flexible, tinned

Fig. 8.18
SIENOPYR(X) Synthetic elastomer-sheathed flexible cable (N)HXSHXÖ

9 Core Identification of Cables

The identification of cores for insulated cables has been agreed internationally and is incorporated in DIN VDE 0293 (Table 9.1).

If flexible cables have one core with a smaller cross section then this core must be marked green-yellow in cables with protective conductor or blue in cables without protective conductor.

Table 9.1 Core identification

N° of Cores	Cables with green/yellow core (marked 'J' to DIN VDE 0250 respectively 'G' to DIN VDE 0281/0282)	Cables without green/yellow core (marked '0' to DIN VDE 0250 respectively 'X' to DIN VDE 0281/0282)
Cables for fixed installation		
1	green/yellow	black (other colours [1])
2	green/yellow, black [2]	brown, blue
3	green/yellow, black, blue	black, blue, brown
4	green/yellow, black, blue, brown	black, blue, brown, black
5	green/yellow, black, blue, brown, black	black, blue, brown, black, black
6 and over	green/yellow, additional cores black and numbered	black and numbered
Flexible cables		
1	–	black [3]
2	–	brown, blue
3	green/yellow, brown, blue	black, blue, brown [4]
4	green/yellow, black, blue, brown	black, blue, brown, black [4]
5	green/yellow, black, blue, brown, black	black, blue, brown, black, black [4]
6 and over	green/yellow, additional cores black and numbered	black and numbered

[1] The individual colours green or yellow or any other colour combination except green/yellow is not permitted. Cables for wiring of apparatus and Partially Type Tested Factory-Built Assembles may however be marked green or yellow as well as with dual colours.
[2] This 2-core variant is to DIN VDE 0100 Part 540 and is only permissible for cross sections equal or greater than 10 mm^2 copper
[3] The core colour for illumination and lighting is brown
[4] 3 to 5-core cables without green/yellow core are not yet harmonized

82

It is important that:

The core marked green/yellow must be used exclusively for protective conductor (PE or PEN). This must not be used for any other purpose.

The core marked blue is used for neutral conductor (N). This core can be used as required (i.e. also as phase conductor) but not as protective conductor (PE) or combined neutral and protective conductor (PEN).

If power supply cables are used in telecomunication installations to VDE 0800 the green/yellow core must also be used exclusively as conductor with protective function.

10 Definition of Locations to DIN VDE 0100

The definition of locations in accordance with the following categories often requires exact knowledge of the conditions at site as well as of the operating conditions. If, for example, in a location high humidity occurs only at one definite place but the rest of the location is dry because of proper ventilation, the whole location need not be classified as damp.

Electrical Operating Areas

are rooms or locations used essentially for the operation of electrical equipment and generally accessible to qualified personnel only.

These include, for example, switch houses, control rooms, distribution installations in separate rooms, separate electrical test departments and laboratories, machine rooms in power stations and the like where the machines are attended solely by qualified persons.

Closed Electrical Operating Areas

are rooms or locations used exclusively for the operation of electrical equipment and kept locked. The locks may be opened by authorized persons only and only qualified persons are permitted to enter these areas.

These include, for example, closed switchboards and distribution boards, transformer cubicles, switchgear cubicles, distribution boards in sheet-steel housings or in other forms of enclosure, pole-mounted substations.

Dry Locations

are rooms or locations in which condensation of moisture does not usually occur or in which the air is not saturated with moisture. These include, for example, dwelling rooms (also hotel rooms), offices; they may also include: business premises, sales rooms, lofts, staircases and cellars with heating and ventilation.

Kitchens and bathrooms in dwellings and hotels are considered as dry rooms as regards the wiring installation, as moisture is present in them only temporarily.

Humid and Wet Areas and Locations

are areas in which electrical equipment must be at least drip protected (IP X1 to DIN 40050).

In areas and locations in which water jets are used for cleaning but where the electrical equipment is not normaly directly subjected to water jets the equipment must be at least splash protected (IP X4 to DIN 40050).

In areas and locations in which water jets are used and where the electrical equipment is directly subjected to the water jets than the equipment must have a sufficient type of protection or suitable additional protection which does not impair the proper operation of that equipment.

Note:

> Protection class IP X5 to DIN 40050 does protect the equipment against cleaning with high pressure water jets e.g. hosing down or high pressure cleaning.

Locations with Fire Hazard

are rooms or locations or sections in rooms or in the open where, owing to local conditions and the nature of the work, there is a risk that easily ignitable materials in dangerous quantities may come so close to electrical apparatus that high temperatures on this apparatus or arcs may cause fire.

These may include working, drying, storage rooms or sections of such rooms as well as locations of this nature outdoors, e.g. paper, textile and wood-working factories, hay, straw, jute, flax stores.

When classifying rooms as locations with fire hazard the relevant regulations must be observed.

Easily ignitable is applied to combustible solid materials which when exposed to the flame of a match for 10 s continue to burn or to glow after the source of ignition has been removed.

These may include hay, straw, straw dust, wood shavings, loose excelsior, magnesium turnings, brushwood, loose paper, cotton waste or cellulose fibre.

Mobile Buildings

are constructions which are suitable for and designed for repeated errection and dismantling such as fair ground equipment e.g. round-abouts, slides, areana stands, sales kiosks, tents, also buildings for travelling exhibitions, apparatus for artistic displays in the air and similar. Wagons which can be largely modified and used operationaly in a fixed location (e.g. Showman wagons) are also classified as mobile buildings.

Areas with Explosion Hazard

are areas in which because of local and operational conditions an explosive atmosphere can occur in a dangerous quantity (explosion hazard).

An explosive atmosphere is a mixture of combustible gasses, vapours, mist or dust with air, which contains the usual additional substances (e.g. moisture), under atmospheric conditions such that after ignition a burning reaction extends unaided.

Operational equipment for areas with explosion hazard are selected in respect of zones, temperature class and the explosion group of the combustible material; see DIN VDE 0165.

Installations on Building Sites

comprise the electrical equipment for carrying out structural work above or below ground on building sites as well as with structural steelwork. Building sites also include constructional work and parts of such which is extended, modified, put into service or demolished.

Places at which merely handlamps, soldering irons, welding equipments, electric tools to DIN VDE 0740, e.g. drilling machines, grinding wheels, polishers and other appliances used individually are not regarded as building sites.

Agricultural Operating Areas

are rooms, locations or areas which are used for agriculture or similar purposes e.g. horticulture.

Note:

In agricultural areas because of special ambient conditions e.g. ingress of moisture, dust, highly chemically agressive vapours acids or salts, to which the electrical equipment is subjected there is an increased risk of accident both to persons and to farm animals (large animals). Large farm animals include horse, cattle, sheep and pigs. In addition due to the presence of easily ignitable materials an increased danger of fire may be present. Further dangers are present in rooms for intensive farming and also for small animals e.g. failure of life sustaining systems.

Depending on the form of danger existing in the agriculture operating area, in addition DIN VDE 0100 Parts 720 and 737, covering humid and wet areas and rooms as well as outdoor installations, must be observed.

11 Application and Installation of Cables

Cables must be selected and applied in line with rules laid down in DIN VDE 0298 and the standards referred to therein. They must be installed only by using suitable installation fixing materials. They must be protected against mechanical, thermal and chemical damage by location or protective means.

Mechanical protection which is generally accepted include:

▷ Insulated wires in conduit which bear the identification mark A to DIN VDE 0605,

▷ Sheathed cables,

▷ Power cables,

▷ Installation in or under plaster,

▷ Installation in cavities,

▷ Installation in trunking to DIN VDE 0604.

At locations which are particularly hazardous additional protection must be provided e.g. by polymeric or steel conduit or cladding which is securely fixed. Such dangerous locations can exist for example at or near floor level.

Cable in or under plaster must be installed vertically or horizontally parallel to the room corners. In ceilings the may be installed using the most direct route (see also DIN VDE 18015). When selecting the type of cable fixing elements the cable construction and shape must be considered. Where expansion gaps are bridged the cable must be installed such that the expected movement does not cause damage to the electrical equipment.

Apart from the cable outside diameter and the cable construction also the type of installation and operation will affect the smallest allowable bending radius.

The relevant values are laid down in DIN VDE 0298 Part 3 (see Table 11.1).

Metal sheathing as well as any uninsulated sheath wires must not be used as operational current carrying conductors nor must they be used for neutral or protective conductor.

It is not permitted to bury flexible insulated power cables in the ground.

Flat underplaster cable must only be installed by the use of such means and processes that do not deform or damage the cable. Approved means of fixing are for instance use of pads of gypsum plaster or shaped cable clips of plastic or insulation coated metal, with adhesive or nailing with special fixings with insulated washers.

Flat underplaster cable can be installed under plasterboard sheets only when the boards are secured to the wall with pads of gypsum plaster.

Light PVC-sheathed cables may be installed in brickwork and cement. Direct embedding in concrete is not permitted where the concrete is tamped, vibrated or rammed.

Light PVC-sheathed cables NYM and lead covered PVC-sheathed cables NYBUY may however, be installed in underground conduit providing the cable remains accessible and replacable and the conduit is mechanically strong, protected against ingress of liquids and ventilated. This type of installation should be the exception and only used for short runs.

Where *trailing drum and drag cables* are used on mobile reclaimers, mining machinery and conveyors with accessories as well as conveying equipment above ground and open-cast mining or similar operations DIN VDE 0168 must be observed.

In *special cases*, e.g. with flexible cable for fast movements and forced guiding over rollers specially constructed cables, e.g. CORDAFLEX polychloroprene-sheathed cable NSHTÖU must be used.

Cables for mobile loads must be relieved of pull or push stresses at their connecting ends. The protective conductor core must be left longer than the other current carrying cores, such that failure of the strain relieving device the protective conductor core is only subjected to stress after the current carrying cores. Sleeves or rounding off of the cable entry point are used to protect the power cables against sharp bends at this point.

Table 11.1 Minimum permissible bending radii

Cable type	Rated voltage up to 0.6/1 kV				Rated voltage above 0.6/1 kV
Cables for permanent installation	Outer diameter *d* of the cable or thickness *d* of flat cable				
	up to 10 mm	over 10 to 25 mm	over 25 mm		
Permanent installation	4 *d*	4 *d*	4 *d*		6 *d*
Formed bend	1 *d*	2 *d*	3 *d*		4 *d*
Flexible cables	Outer diameter *d* of the cables or thickness *d* of flat cable				
	up to 8 mm	over 8 to 12 mm	over 12 to 20 mm	over 20 mm	
Fixed installation	3 *d*	3 *d*	4 *d*	4 *d*	6 *d*
Free moving	3 *d*	4 *d*	5 *d*	5 *d*	10 *d*
At cable entry	3 *d*	4 *d*	5 *d*	5 *d*	10 *d*
Forced guiding [1] e.g. drum operation	5 *d*	5 *d*	5 *d*	6 *d*	12 *d*
cable wagon operation	3 *d*	4 *d*	5 *d*	5 *d*	10 *d*
drag chain operation	4 *d*	4 *d*	5 *d*	5 *d*	10 *d*
roller guides	7.5 *d*	7.5 *d*	7.5 *d*	7.5 *d*	15 *d*

[1] The type of construction must be checked to ensure suitability for this type of operation

The application and installation of *neon lighting cables* is covered in DIN VDE 0128. In addition to this the following must also be observed:

Where neon lighting cables enter enclosures they must be covered with sleeving or enter through a gland. In outdoor installations the cable entry must be sealed by sheaths or covers of insulating material to prevent surface creepage.

In respect of cable constructions which are used for several different types of application it may be advisable to discus the application with the manufacturer.

11.1 Rated Voltage, Operating Voltage

The definitions of rated voltage and operating voltage of wiring cables and flexible cables is given in DIN VDE 0298 Part 3.

Rated Voltage

The rated voltage of an insulated cable is the voltage on which the construction and testing of the cable, in respect of electrical characteristics, is based. The rated voltage is expressed by two a.c. voltage values U_0/U expressed in V:

U_0 rms value between one-phase conductor and earth (non insulated surrounding),

U rms value between two-phase conductors of a multi-core cable or a system of single-core cables.

In a system with a.c. voltage the rated voltage of the cable must be at least equal to the rated voltage of the system to which it is connected. This condition applies to the value U_0 as well as to the value U.

In a system with d.c. voltage the rated system voltage must not exceed 1.5 times the rated a.c. voltage of the cable.

Table 11.2
Typical rated voltages of various cable types

Rated voltage U_0/U	Cable type
300/300 V	Tinsel cords and flat non-sheathed cords
220/380 V	Flat building wires
300/500 V	Light PVC-sheathed cables
450/750 V	Heavy polychloroprene-sheathed flexible cables
0.6/ 1 kV	PROTOFIRM-sheathed cables
1.8/ 3 kV	Single-core polychloroprene-sheathed cables for special purposes
4 / 8 kV	Neon lighting cables
6 /10 kV	
8.7/15 kV	
12 /20 kV	
14 /25 kV	Trailing cables
18 /30 kV	
20 /35 kV	

Operating Voltage

The operating voltage is the voltage between conductors or between conductors and earth present in a power installation under healthy stable conditions.

Cables with Rated Voltage
$U_0/U \leq 0.6/1\ kV$

These cables are suitable for application on 3-phase, single-phase and d.c. installations where the maximum permanently permissible operating voltage does not exceed the rated voltage of the cable by more than

10% for wiring cables and flexible cables with rated voltage

$$U_0/U \leq 450/750\ \text{V}$$

20% for wiring cables and flexible cables with rated voltage

$$U_0/U = 0.6/1\ \text{kV}$$

Wiring Cables and Flexible Cables with Rated Voltage
$U_0/U > 0.6/1\ kV$

These cables are suitable for application in 3-phase and single-phase installations having a maximum operating voltage not exceeding 20% above the rated voltage of the cable.

The cable may be used:

a) in 3-phase and single-phase installations where the star point is effectively earthed;

b) in 3-phase and single-phase installations where the star point is not effectively earthed providing that any individual earth fault is not sustained longer than 8 hours and the total of all earth fault times per year does not exceed 125 hours. If this situation can not be ensured than, to ensure a service life of the cable, a cable having a higher rated voltage should be selected.

Cables for Direct Current Installations

For cables in d.c. installations the permanent permissible operating d.c. voltage between conductors must not exceed 1.5 times the rated a.c. voltage of the cable. In 2-wire earthed d.c. installations this value must be multiplied by a factor of 0.5.

11.2 Selection of Conductor Cross-Sectional Area

General

The temperature rise, respectively current-carrying capacity, of a cable is dependent upon the type of construction, the characteristics of the materials used and also operating conditions. In order to achieve a safe design and a full service life of a cable the conductor cross-sectional area must be chosen such that the requirement

current-carrying capacity $I_z >$ loading I_b

for the conditions of

▷ normal operation and
▷ short circuit

are satisfied. This will ensure that no part of the cable at any point in time is heated above the rated maximum permissible operating temperature respectively short-circuit temperature.

Current-Carrying Capacity in Normal Operation

The design of installation projects is simplified by using established data collected over several decades in respect of *current-carrying capacity* under practical applications which has now been incorporated in various regulations governing apparatus and installation. For electrical installations in buildings the standards for electrical installation of buildings DIN VDE 0100 apply for power installations up to 1000 V. In this standard, up to the present day, in Part 523 the types of installation were divided into three groups:

Group 1: insulated conductors and single-core cables in a conduit
Group 2: multi-core cables for fixed installation
Group 3: single-core cables for fixed installations and power cables.

The associated values for current-carrying capacity had been determined originally for rubber insulated cables.

By todays standards, these groupings of types of installation appear extremely rough but have, however, proved to be adequate for the methods practised at that time. Meanwhile, however, different installation practises have developed and modern, more sophisticated materials have become available. These developments have necessitated the evolution of more detailed project planning.

In February 1988, the specification DIN VDE 0298 Part 4 "Recommended values for current-carrying capacity for sheathed and non-sheathed cables for fixed wirings, flexible cables and cords", was published. This publication contains comprehensive and detailed information on the relevant terms and regulations required to determine the cross-sectional area of conductors for normal operation and for short-circuit conditions. Firstly the precise operating conditions on which this data was based were defined and specified. Based on these *reference operating conditions,* which take into acount the *type of operation* as well as *installation* and *ambient conditions,* tabulated data was prepared of *rated values* of current-carrying capacity I_r (rated value). To cater for conditions which deviate from the agreed operating conditions, conversion factors were prepared. The relationship:

$$I_z = I_r \Pi f$$

applies where Πf is the product of all conversion factors which are applicable.

As a basis for type of operation, *continuous operation* was selected, which is operation at constant current for a duration sufficient for the cable to reach thermal equilibrium but otherwise not limited in time.

Short-time and *intermittend operation* e.g. for the starting currents of motors or the operation of crane installations are described in Section 18.6

The *specified ambient temperature* for all applications is 30 °C and it is required that the room is sufficiently large and ventilated such that the ambient temperature is not noticeably increased by the cable losses.

The *installation conditions,* in comparison to the previously used groups 1 to 3, are more precisely defined and enlarged. One differentiates now between:

Method of Installation type A:
Installation in walls having low thermal conductivity,

Method of installation type B1:
Installation of single-core insulated cables in conduit or duct on or in a wall,

Method of installation type B2:
Installation of multi-core insulated cables in conduit or duct on or in a wall,

Method of installation type C:
Installation of cable directly on or in a wall/under plaster,

Method of installation type E:
Installation of cable in free air.

Table 11.3
Current-carrying capacity. Cables for fixed installation. Method of installation A, B1, B2 and C

1	2	3	4	5	6	7	8	9
Insulating material	PVC							
Type designation [1]	NYM, NYBUY, NHYRUZY, NYIF, NYIFY, H07V-U, H07V-R, H07V-K, NHXMH [8]							
Maximum permissible operating temperature	70 °C							
No. of loaded conductors	2	3	2	3	2	3	2	3
Method of installation	A		B1		B2		C	
	In thermally insulated walls		On or in walls or under plaster in conduit or trunking				direct installation	
	Insulated conductors in conduit [2] [5]		Insulated conductors in conduit on the wall [3]		Multi-core cable in conduit on the wall or on the floor		Multi-core cable on the wall or on the floor [4]	
	Multi-core cable in conduit [5]		Insulated conductors in trunking on the wall		Multi-core cable in trunking on the wall or on the floor		Single-core sheathed cable on the wall or on the floor	
	Multi-core cable in the wall		Insulated conductors, single-core sheathed cables, multi-core cable in conduit in masonry [6]				Multi-core cable, flat webbed building wires in the wall or under plaster [7]	
Copper conductor Nominal cross-sectional area in mm²	Current-carrying capacity in A							
1.5	15.5	13	17.5	15.5	15.5	14	19.5	17.5
2.5	19.5	18	24	21	21	19	26	24
4	26	24	32	28	28	26	35	32
6	34	31	41	36	37	33	46	41
10	46	42	57	50	50	46	63	57
16	61	56	76	68	68	61	85	76
25	80	73	101	89	90	77	112	96
35	99	89	125	111	110	95	138	119
50	119	108	151	134	–	–	–	–
70	151	136	192	171	–	–	–	–
95	182	164	232	207	–	–	–	–
120	210	188	269	239	–	–	–	–

Table 11.4

Current-carrying capacity. Cables for fixed installation. Method of installation in free air

1	2	3
Insulating material	PVC	
Type designation [1]	NYM, NYMZ, NYMT, NYBUY, NHYRUZY, NHXMH [2], NYHSSYCY [3]	
Maximum permissible operating temperature	70 °C	
No. of loaded conductors	2	3
Method of installation	E	E
	≥0,3d	≥0,3d
Copper conductor Nominal-cross sectional area in mm²	Current-carrying capacity in A	
1.5	20	18.5
2.5	27	25
4	37	34
6	48	43
10	66	60
16	89	80
25	118	101
35	145	126
50 [4]	–	153
70 [4]	–	196
95 [4]	–	238

[1] Type designation and further details in Section 8.1
[2] Not included in DIN VDE 0298 Part 4. Insulation of cross-linked polyolefine compound
[3] Rated voltage 3.6/6 kV
[4] Not included in DIN VDE 0298 Part 4

For each of these *reference operating conditions* the recommended rated values of current-carrying capacity I_r are shown in Tables 11.3 and 11.4. The headings of the tables include diagramatic representations of the installations for ease of understanding which, together with the footnotes, provide a detailed description. The current-carrying capacities apply for:

▷ two-core cables with two conductors loaded as well as for two loaded single-core insulated conductors or two loaded single-core sheathed cables given in Table 11.3 columns 2, 4, 6 and 8 as well as Table 11.4 columns 2 and for

▷ three-core cables with three conductors loaded as well as for three loaded single-core insulated conductors or three loaded single-core sheathed cables given in Table 11.3 columns 3, 5, 7 and 9 as well as Table 11.4 column 3.

◁ [1] Type designation and further details in section 8.1
[2] Also applies to insulated conductors in conduit in enclosed floor trench
[3] Also applies to insulated conductors in conduit in ventilated floor trench
[4] Also applies to multi-core cable in open or ventilated trench
[5] Also applies to insulated conductors, single-core sheathed cable, multi-core cable in ducting in the floor
[6] Also applies to insulated conductors in conduit in the ceiling
[7] Also applies to multi-core cable in the ceiling
[8] Not included in DIN VDE 0298 Part 4. Insulation of cross-linked polyolefine compound

Where, for example, in a multi-core cable all conductors are not loaded at the same time, the value of current-carrying capacity is possibly greater than that given in Tables 11.3 and 11.4. The relevant values depend on the type of construction of the cable and the installation conditions such that common conversion factors cannot be prepared. For reasons of safety it is recommended to consider only the number of loaded conductors, disregarding the total number of cores, when allocating a value of current-carrying capacity from Tables 11.3 and 11.4. Only in this way can overloading be safely avoided.

The current-carrying capacity at ambient temperatures other than 30 °C can be established using the conversion factors given in Table 1.2.10 in part 2 of this work.

The current-carrying capacity quantities given in Tables 11.3 and 11.4 apply with the proviso that only either one multi-core cable or two respectively three insulated conductors, alternatively sheathed single-core cable, are installed. If several cables are arranged next to one another, above one another or adjacent to or above power cables then the carrying capacity is reduced corresponding to the hindered heat dissipation, respectively the additional heat generated. Conversion factors which cater for this *grouping* of cables for fixed installation (possibly with power cables), are given in Tables 1.2.11 and 1.2.12 in Section 1.2.2 of part 2 of this work.

Also included in part 2 will be found informations on flexible cables and methods of installation which could not be included in part 1 for reasons of space. Conversion factors are also incorporated for *heat resistant cables,* for *multi-core cables with more than five cores* and for *cables operating on drums.*

Current-Carrying Capacity under Short Circuit

The thermally permissible short-circuit current I_{thz} is determined from the *rated short-time current density* J_{thr} from Table 1.2.16 in Section 1.2.3 of part 2 of this work from:

$$I_{thz} = I_{thr} \sqrt{t_{kr}/t_k}$$
$$I_{thr} = J_{thr} S_n$$

where

I_{thz} Thermally permissible short-circuit current-carrying capacity

I_{thr} Thermally effective short-circuit current

t_{kr} Rated short-circuit duration ($t_{kr} = 1$ s)

t_k Short-circuit duration in s

J_{thr} Rated short-circuit current density

S_n Nominal cross-sectional area of conductor

Determination of Voltage Drop

Especially in low-voltage networks, apart from the current-carrying capacity, the conductor cross-sectional area must also be considered in respect of voltage drop Δu to ensure this does not exceed the permissible value. For the calculation of this, comprehensive aids for design are included in Section 1.2.4 of part 2 of this work.

Protection against Excessive Temperature Rise

It is possible to heat cables above the permissible limit by an operational overcurrent as well as by short circuit-current. Protective devices must be incorporated, therefore, for protection against overcurrent as listed in e.g. DIN VDE 0636, DIN VDE 0641 and DIN VDE 0660.

The co-ordination of these overcurrent protection devices to the conductor cross-sectional area is made by reference to DIN VDE 0100 Part 430. Details for this are also included in Section 1.2.5 in part 2 of this work.

Power Cables

12 National and International Standards

12.1 VDE Specifications

In respect of construction, testing and application of power cables, the relevant valid editions of the following VDE specifications and DIN standards apply:

DIN VDE 0206 Recommendations on colours for polymeric sheaths and coverings with polymeric and rubber insulation for cables and flexible cords

DIN VDE 0207 Insulating and sheathing compounds for cables and flexible cords

DIN VDE 0255 Regulations for mass-impregnated paper-insulated metal-sheathed power cables (except external gas-pressure and oil-filled cables)

DIN VDE 0256 Low-pressure oil-filled cables and their accessories for nominal voltages up to U_0/U 230/400 kV

DIN VDE 0257 External-gas-pressure pipe type cables and their accessories for alternating voltages up to 275 kV

DIN VDE 0258 Internal-gas-pressure cables and accessories for alternating voltages up to 275 kV

DIN VDE 0265 Cables with plastic insulation and lead sheath for power installation

DIN VDE 0266 Halogen-free cables with improved characteristics in the case of fire; nominal voltages: U_0/U 0.6/1 kV

DIN VDE 0271 PVC-insulated cables with nominal voltages up to and including 6/10 kV

DIN VDE 0272 Cross-linked polyethylene-insulated cables; nominal voltage: U_0/U 0.6/1 kV

DIN VDE 0273 Cross-linked polyethylene-insulated cables, nominal voltages: U_0/U 6/10, 12/20 and 18/30 kV

DIN VDE 0274 Cross-linked polyethylene-insulated conductors for overhead transmission lines, nominal voltage: U_0/U 0.6/1 kV

DIN VDE 0289 Definitions for cables, wires and flexible cords for power installation

DIN VDE 0293 Identification of cores in cables and flexible cords used in power installations with nominal voltages up to 1000 V

DIN VDE 0295 Conductors of cables, wires and flexible cords for power installation

DIN VDE 0298 Application of cables, wires and flexible cords in power installations

    Part 1: General for cables with rated voltages U_0/U up to 18/30 kV

    Part 2: Recommended values for current-carrying capacity of cables for fixed installation with rated voltages U_0/U up to 18/30 kV

DIN VDE 0299 Calculation method on the basis of fictitious diameters for determination of dimensions of protective coverings for cables and flexible cords for power installations.

    Part 1: Power cables

DIN VDE 0304 Thermal properties of electrical insulating materials

    Part 21: General procedures for the determination of thermal endurance properties, temperature indices and thermal endurance profiles

    Part 22: List of materials and available tests

DIN VDE 0472 Testing of cables, wires and flexible cords

DIN VDE 0100 Erection of power installations with rated voltages up to 1000 V

DIN VDE 0101 Erection of power installations with rated voltages above 1 kV

DIN VDE 0103 Mechanical and thermal short-circuit strength of electrical power installations

DIN VDE 0105 Operation of power installations

DIN VDE 0111 Insulation co-ordination to equipment for three-phase a.c. systems above 1 kV

DIN VDE 0115 Rail-borne and trackless vehicles

DIN VDE 0118 Specifications for the erection of electrical installations in mines below ground

DIN VDE 0168 Specification for the erection and operation of electrical installations in open-cast mines, quarries and similar works

DIN VDE 0211 Planning and design of overhead power lines with rated voltages up to 1000 V

DIN VDE 0228 Proceedings in the case of interference on telecommunication installations by electric power installations

DIN 17640 Lead and lead alloys for cable sheaths

DIN 89150 Cables and flexible cords for installation on ships; survey, current ratings, overcurrent protection, direction for laying

DIN 89158 Power cables with copper braid; type MGCG

DIN 89159 Communication cables, type FMGCG; nominal cross-sectional area 0.5 mm^2 and 0.75 mm^2

DIN 89160 Power cables without copper braid; type MGG

12.2 Standards of Other Countries

Cables manufactured to standards of other countries e.g. British standards (BS), French standards (NF), Italian standards (CEI), Swedish standards (SEN), comply normaly in their basic construction with cables manufactured to VDE standards but deviate in dimensions and test requirements. Where required Siemens AG can also supply cables to meet these or other standards. Certain types are already approved by the relevant standards institutes.

12.3 IEC and CENELEC Standards

The international commissions IEC and CENELEC have the responsibility to unify the varying standards which exist within the E.E.C. At the present time it cannot be forseen by what date full harmonization will be achieved. In deference to IEC, CENELEC does not normally prepare independant specifications. The "final harmonized documents" are issued by CENELEC and after a short introductory period, enforced to be incorporated in the national specifications of the relevant standards institutions without any change of content. In certain circumstances CENELEC also issues European Norms (EN norms) which must be accepted by the member countries unchanged in form and content.

The following IEC publications are relevant to power cables:

IEC 28 (1925) Revised ed.
  International standard of resistance for copper

IEC 55
  Paper-insulated metal-sheathed cables for rated voltages up to 18/30 kV (with copper or aluminium conductors and excluding gas-pressure and oil-filled cables)

  55-1 (1978) Fourth ed.
  Part 1: Tests

  55-2 (1981) First ed.
  Part 2: General and construction requirements

IEC 92
  Electrical installation in ships
  92-3 (1965) Second ed.
  Part 3: Cables (construction, testing and installations)

IEC 111 (1983) Second ed.
Resistivity of commercial hard-drawn aluminium electrical conductor wire

IEC 141
Tests on oil-filled and gas-pressure cables and their accessories

141-1 (1976) Second ed.
Part 1:
Oil-filled, paper-insulated, metal-sheathed cables and accessories for alternating voltages up to and including 400 kV

141-2 (1963) First ed.
Part 2:
Internal gas-pressure cables and accessories for alternating voltages up to 275 kV

141-3 (1963) First ed.
Part 3:
External gas-pressure (gas compression) cables and accessories for alternating voltages up to 275 kV

141-4 (1980) First ed.
Part 4:
Oil-impregnated paper-insulated high pressure oil-filled pipe-type cables and their accessories for alternating voltages up to and including 400 kV

IEC 167 (1964) First ed.
Methods of test for the determination of the insulation resistance of solid insulating materials

IEC 183 (1984) Second ed.
Guide to the selection of high-voltage cables

IEC 216
Guide for the determination of thermal endurance properties of electrical insulating materials

261-1 (1974) Second ed.
Part 1:
General procedures for the determination of thermal endurance properties, temperature indices and thermal endurance profiles
(s. DIN IEC 216 Teil 1 / VDE 0304 Teil 21)

216-2 (1974) Second ed.
Part 2: List of materials and available tests
(s. DIN IEC 216 Teil 1 / VDE 0304 Teil 21)

IEC 228 (1978) Second ed.
Conductors of insulated cables

IEC 229 (1982) Second ed.
Tests on cable oversheaths which have a special protective function and are applied by extrusion

IEC 230 (1966) First ed.
Impulse tests on cables and their accessories

IEC 287 (1982) Second ed.
Calculation of the continuous current rating of cables (100% load factor)

IEC 331 (1970) First ed.
Fire-resisting characteristics of electric cables

IEC 332
Test on electric cables under fire conditions

332-1 (1979) Second ed.
Part 1: Test on a single vertical insulated wire or cable

332-3 (1982) First ed.
Part 3: Tests on bunched wires or cables

IEC 502 (1983) Third ed.
Extruded solid dielectric insulated power cables for rated voltages from 1 kV up to 30 kV

IEC 540 (1982) Second ed.
Test methods for insulations and sheaths of electric cables and cords (elastomeric and thermoplastic compounds)

IEC 754-1 (1982) First ed.
Test on gases evolved during combustion of electric cables.
Part 1: Determination of the amount of halogen acid gas evolved during the combustion of polymeric materials taken from cables

IEC 811
Common test methods for insulating and sheathing materials of electric cables
811-1 Part 1
Methods for general application

811-1-1 (1985) First ed.
Section One:
– Measurement of thickness and overall dimensions
– Tests for determining the mechanical properties

811-1-2 (1985) First ed.
Section Two:
– Thermal ageing methods

811-1-3 (1985) First ed.
Section Three:
– Method for determining the density
– Water absorption test
– Shrinkage test

811-1-4 (1985) First ed.
Section Four:
– Test at low temperature

811-2 Part 2
Methods specific to elastomeric compounds

811-2-1 (1986) First ed.
Section One:
– Ozone resistance test
– Hot set test
– Mineral oil immersion test

811-3 Part 3
Methods specific to PVC compounds

811-3-1 (1985) First ed.
Section One:
– Pressure test at high temperature
– Test for resistance to cracking

811-3-2 (1985) First ed.
Section Two:
– Loss of mass test
– Thermal stability test

811-4 Part 4
Methods specific to polyethylene and polypropylene compounds

811-4-1 (1985) First ed.
Section One:
– Resistance to environmental stress cracking
– Wrapping test after thermal ageing in air
– Measurement of the melt flow index
– Carbon black and/or mineral content measurement in PE

IEC 840 (1980) First ed.
Tests for power cables with extruded insulation for rated voltages above 30 kV ($U_m = 36$ kV) up to 150 kV ($U_m = 170$ kV)

13 Types of Construction of Low- and High-Voltage Cables

13.1 General

When designing cables it is necessary to take account of both ambient conditions and the electrical stresses which may occur. Whereas the ambient conditions are important when selecting the right type of protective covering and armour, the electrical stresses are the decisive factor for, amongst others, the thickness of the insulation and the right type of screen. A distinction is made between cables having a non-radial electric field (e.g. belted cables) and radial field cables.

Multi-core cables with non-radial field have only one screen above the laid-up cores. They are, as can be seen from Table 7.1 (see page 45) depending on the material used for insulation only permissible up to cable rated voltage of maximum $U_0/U = 8.7/10$ kV. Paper insulated cables with non-radial field are also known as *belted cables*, as above the laid-up cores an additional thin layer of insulation is added, namely the belt insulation.

High-tension cables manufactured to VDE are normally used for networks which have a non-earthed star point. The insulation "conductor/metal sheath" is therefore dimensioned such that the cable can also, in the event of an earth fault, remain in operation for several hours without incurring any damage (see page 147).

To meet the requirements of standards effective in other countries for networks with non-earthed star point (e.g. networks with earth-fault compensation or insulated star point) belted cables with increased belt insulation are often used.

Radial field cables have a screen above the insulation of each individual phase conductor which directs the field lines in a radial formation. Radial field cables include all single-core cables having a concentric conductor and screen as well as multi-core cables having a conducting layer above the insulation of each individual core; the interstices are field-free.

The metallic components of the screen (see page 46) can be arranged also either above each individual core or above the laid-up cores.

In the Federal Republic of Germany, with cables having PVC insulation on rated voltages of $U_0/U = 6/10$ kV, normally multi-core cables are used. For medium-voltage cables with XLPE insulation however single-core types are mostly preferred. Three laid up single-core XLPE cables have advantages where installation space is at a premium.

The service life of medium-voltage cables with XLPE insulation is influenced by ingress of moisture (see page 30). Therefore further development has the target to limit and localise the ingress of water when the cable sheath is damaged. Types of construction with PE sheath (see page 37) and longitudinal water-tight screens are therefore of increased significance.

Dimensions and test specifications of cables as well as regulations for their installation are laid down in national VDE specifications and international IEC standards (see page 93). Apart from these a large number of other types of constructions exist for special applications (see page 124).

The cables must be selected depending on rated voltage, the requirements in operation and also economic considerations. Table 13.1 shows the various basic types of construction of cables. On pages 102 to 123 the cable types together with terminations and jointing methods commonly used in Germany are illustrated. For the selection of cables the chapter "Planning of cable installations" (page 141) should be observed.

Table 13.1   Basic constructions of cables

Cables with insulation of PVC or XLPE

Diagram of electric field	Rated voltages U_0/U kV	Type of construction		Examples of type designation

Non-radial field cables

Diagram of electric field	Rated voltages U_0/U kV	Type of construction		Examples of type designation
[1]	0.6/1	Cables without metal covering		NYY, NAYY 2XY [2], NA2XY
	0.6/1	Multi-core cables – with concentric conductor		NYCY, 2XCY [2] NYCWY, NAYCWY 2XCWY [2], NA2XCWY
	up to 3.6/6	– with flat steel wire armour		NYFY, NAYFY

Radial field cables

Diagram of electric field	Rated voltages U_0/U kV	Type of construction		Examples of type designation
	0.6/1	Single-core cables – with concentric conductor		NYCY, NAYCY 2XCY [2], NA2XCY
	3.6/6	– with copper screen		NYSY, NAYSY
	for XLPE from 3.6/6	Single-core cables with copper screen	Partial discharge-free construction by use of a conducting layer between conductor and insulation (inner conducting layer) as well as an outer conducting layer firmly bonded to the insulation	N2XS2Y, NA2XS2Y
	for PVC 6/10			NYSY, NAYSY
	for XLPE from 3.6/6	Three-core cables with copper screen above each individual core		N2XSE2Y, NA2XSE2Y
	for PVC 6/10			NYSEY, NAYSEY
	for XLPE from 3.6/6	Three-core cables with a conducting inner layer above the laid-up cores and a concentric copper screen or flat steel wire armour		N2XS2Y, NA2XS2Y N2XF2Y, NA2XF2Y
	for PVC 6/10			NYFY, NAYFY

[1] The field of a cable installed in air without metal sheath as e.g. NYY or NA2XY is not controlled
[2] Generally as DIN VDE 0272 as currently low-voltage cables with copper conductors and XLPE insulation are not yet included in DIN VDE 0272

Cable with paper insulation and metal sheath [3] (mass impregnated cable)

Rated voltages U_0/U kV	Type of construction		Examples of type designation
up to 6/10	Belted cables: The laid-up cores are surrounded by a common insulation – belt insulation – followed by the metal sheath		NKLEY NAKLEY NKBA NAKBA
up to 18/30	Single-core cables: Conductor with insulation, metal sheath and protective layers	For voltages $U_0/U = 3.6/6$ kV and above a layer of metallised paper is included above the insulation, to ensure any cavities in the dielectric remain field free and hence corona-free. Cavities may occur below the metal sheath due to changes in volume of the impregnating mass caused through variations in load.	NKLEY NAKLEY NKY NAKY
6/10 to 18/30	Separate lead-sheathed (S.L.) cables: Three single-core cables each with lead sheath and corrosion protection, laid-up and provided with the necessary overal protective layers. S.L. cables for 11.6/20 kV and 17.3/30 kV are commonly used in Germany.		NEKEBA NAEKEBA
6/10 to 18/30	H-Cable: Paper-insulated cores covered by H-foil are laid-up and are surrounded by textile webbing which contains aluminium wires interwoven. A common metal sheath and protective layer is then added overal. This type of construction is now used in Germany only for special applications.		NHKRGY NAHKRGY

[3] Cables with lead sheath (paper-lead cable) and cables with aluminium sheath

13.2 Type designation

Cables are designated with:

▷ Abbreviated description of the cable design and its component materials

▷ Number of cores by nominal cross-sectional area of conductor in mm²

▷ Indications of shape type of conductor

▷ Where applicable nominal cross-sectional area of screen or concentric conductor in mm²

▷ Rated voltage in kV (see Section 17).

The description of type of construction is derived by adding a combination of letters after the first letter 'N' building the order of construction outwards from the conductor. The letter 'N' indicates "Norm type" and designates cable types which comply with the VDE specifications mentioned in Sections 12 and 13.

The following are not indicated:

▷ Copper conductor
▷ Insulation of impregnated paper (core, belt)
▷ Inner and outer conducting layers in cables with polymer insulation
▷ Inner coverings
▷ Fillers of the interstices
▷ Inner beddings of fibreous materials.

**Additional Symbols for Cables
with Improved Characteristics in the Case of Fire**

HX	Insulation of cross-linked Halogen-free polymer compound
HX	Sheath of cross-linked Halogen-free polymer compound
H	Sheath of non-cross-linked Halogen-free polymer compound
FE	Insulation retention (symbol appears after the designation of conductor)

Symbols for Ships Cable

M	Power supply ships cable to HNA standard
G	Insulation of EPR
G	Sheath of CR
C	Screen of copper braid

Symbols for Conductor

RE	Solid circular conductor
RM	Stranded circular conductor
SE	Solid sector shaped conductor
SM	Stranded sector shaped conductor

RF	Flexible circular conductor
RM/V	Stranded circular conductor compacted by either squeezing through rollers or the use of shaped wires (for thermally stable cables with paper insulation)
H	Circular hollow conductor, the diameter of the oil channel given in mm precedes the letter 'H' e.g. RM/V 14H
OM	Stranded conductor of oval cross-section

The rated cross-sectional area of copper screens is given after an oblique sign '/' located after the symbols for the phase conductors
e.g. NYSEY 3 × 95 RM/16 6/10 kV.

The rated cross-sectional area of the concentric conductor is also indicated following a '/' sign after the symbols for the phase conductors
e.g. NYCWY 3 × 95 SM/50 0.6/1 kV.

Further Commonly Used Symbols

YV	Reinforced PVC sheath
2YV	Reinforced PE sheath
O	Open armour (FO or RO)
AA	Double outer protective layer of fibrous material
Te	Lead sheath of lead Tellur alloy
sv	Special impregnation for cable with paper insulation for steeply sloping cable runs (sv = nd: non-draining compound see page 35)

Table 13.2 Summary of the main letters used for the type designation of cable

Construction element	Paper-insulated cables DIN VDE 0255	Polymer-insulated cables DIN VDE 0265, 0271, 0272, 0273	High- and extra high-voltage cables DIN VDE 0256, 0257, 0258
Norm type	N	N	N
Conductor			
of copper	no letter	no letter	no letter
of aluminium	A	A	A
Insulation			
Paper with mass impregnation	no letter	–	–
Paper oil impregnated	–	–	Ö
– with high-pressure oil cables in steel pipe	–	–	ÖI
Paper with mass impregnation			
– for external gas pressurised cable	–	–	P
– for internal gas pressurised cable	–	–	I
PVC, polyvinylchloride	–	Y	–
PE, polyethylene	–	2Y	–
XLPE, cross-linked polyethylene	–	2X	–
Concentric copper conductor			
with longitudinal layer	–	C	–
with wave form lay	–	CW	–
Copper screen			
– for single-core cables or for multi-core cables with common metallic screen	–	S	–
– metallic screen on each core in multi-core cables	–	SE	–
Screening in multi-core cables with paper insulation and common metal sheath (H-cable)			
– single-core screening with metallised paper (Höchstädter Folie)	H	–	–
Metal sheath			
of lead			
– for single-core cables and multi-core cables with common sheath	K	K	K
– for three-core cables with corrosion protection on each sheath	EK	–	
– non-magnetic pressure protection bandage on the lead sheath	–	–	UD
of aluminium			
smooth	KL	–	KL
– corrugated	–	–	KLD
Pipe type cable			
– non laid-up cores	–	–	u
– laid-up cores	–	–	v
Thermoplastic sheath and inner protective covering			
– PVC sheath	Y	Y	Y
– PE sheath	–	2Y	2Y
– lapped bedding with additional layer of plastic tape	E	–	E
Armour			
– steel tape	B	–	–
– flat steel wire	F	F	F
– round steel wire	R	R	R
– spiral binder tape	G	G	G
– skid wire (non-magnetic)	–	–	GL
– steel tube	–	–	ST
Outer protection			
– fibreous material (jute) in compound	A	–	A
– PVC sheath	Y	Y	Y
– PE sheath	–	2Y	2Y
Cable with $U_0/U = 0.6/1$ kV without concentric conductor			
– with green/yellow core	-J	-J	–
– without green/yellow core	-O	-O	–

13.3 Selection of Cables and Accessories

Table 13.3 Cables and associate accessories

Construction	Designation, standards	Preferred application	Limited application
N Y **Y** 1 PROTODUR-insulation (PVC) 2 PROTODUR-sheath (PVC) 3 Cu-conductor 4 Tape or extruded filter	NYY NAYY DIN VDE 0271 IEC 502	Power cable: Indoors, cable trunking, outdoors and buried in the ground, for power stations, industry and switchgear as well as for urban supply networks, if mechanical damage is unlikely	It may be necessary to observe relevant local regulations when installing in countries other than Germany.
N Y **Y** 1 PROTODUR-insulation (PVC) 2 PROTODUR-sheath (PVC) 3 Cu-conductor 4 Tape or extruded filter	NYY DIN VDE 0271 IEC 502	Control cable: as for power cables	
N A **Y** **CW** **Y** 1 Al-conductor 2 PROTO-DUR-insulation (PVC) 3 Concentric, protective or PEN conductor (Cu wires and transverse helical tape) 4 PROTO-DUR-sheath (PVC) 5 Extruded filter	NYCWY[1] NAYCWY[1] NYCY[2] DIN VDE 0271 IEC 502	For installation in the ground, indoors, cable trunking and outdoors if subsequent mechanical damage is likely. For urban networks, household feeders and street lighting. The concentric conductor in wave form is not cut at branch points.	Where high-mechanical stresses may occur during installation and operation. The concentric conductor should not be considered as armour.
N Y **F** **G** **Y** 1 PROTO-DUR-insulation (PVC) 2 Flat steel wire armour 3 Steel tape binder helix 4 PROTO-DUR-sheath (PVC) 5 Cu-conductor 6 Filler 7 Lapped inner covering	NYFGY NAYFGY DIN VDE 0271 IEC 502	For installation in the ground, indoors, cable trunking and outdoors if increased mechanical protection is required or where high-pulling stresses may occur during installation or operation	

[1] Concentric conductor in wave formation   [2] Concentric conductor helically applied

$$U_0/U = 0.6/1 \text{ kV } (U_m = 1.2 \text{ kV})$$

Indoor sealing ends (examples)	Outdoor sealing ends (examples)	Cable joints (examples)

Shrink-on straight joint

4-core cable
with fanned tails

PROTOLIN-cable end
PEA

PROTOLIN-straight joint

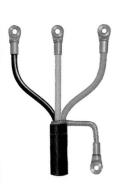

PROTOLIN-branch Y-joint

3-core cable
with concentric conductor
and parallel tails

T-joint HM

3-core cable
with flat steel-wire armour and
fanned tails

Shrink-on cable end

PROTOLIN-transition joint for cable with polymer
insulation to mass-impregnated cable

Table 13.3 Cables and associated accessories (continued)

Construction	Designation, standards	Prefered application	Limited application
N A 1 Al-conductor **2X** 2 PROTOTHEN-X-insulation (XLPE) **Y** 3 PROTODUR-sheath (PVC) 4 Extruded filler	2XY [1] NA2XY DIN VDE 0272 IEC 502	Power cables for urban networks, for installation in the ground, cable trunking, in- and outdoors. Cable with copper conductor also for power stations, industry and switchgear.	It may be necessary to observe relevant local regulations when installing in countries other than Germany.
2X 1 PROTOTHEN-X-insulation (XLPE) **Y** 2 PROTODUR-sheath (XLPE) 3 Cu-conductor 4 Tape or filler	2XY [1] IEC 502	Control cable: as for power cable	
N A 1 Al-conductor **2X** 2 PROTO-THEN-X-insulation (XLPE) **CW** 3 Concentric, protective or (PEN) conductor (Cu wires and transverse helical tape) **Y** 4 PROTO-DUR-sheath (PVC) 5 Extruded filler	2XCWY [1] NA2XCWY DIN VDE 0272 IEC 502	For urban networks with concentric conductor of wave formation which is not cut at branch points. For installation in the ground, indoors, cable trunking and outdoors where subsequent mechanical damage is likely. For street lighting and household feeders in urban networks.	Where high-mechanical stresses may occur during installation and operation. The concentric conductor should not be considered as armour.
2X 1 PROTOTHEN-X-insulation (XLPE) **F** 2 Flat steel wire armour **Y** 3 PROTODUR-sheath (PVC) 4 Cu-conductor 5 Filler 6 Lapped inner covering	2XFY [1] IEC 502	For installation in the ground, indoors, cable trunking and outdoors where higher-mechanical protection is required or where high pulling stresses may occur during installation or operation.	

[1] Cable for $U_0/U = 0.6/1$ kV with XLPE insulation and copper conductors as well as armoured cables are not yet included in DIN VDE 0272

$U_0/U = 0.6/1 \text{ kV } (U_\mathrm{m} = 1.2 \text{ kV})$

Indoor sealing ends (examples)	Outdoor sealing end (examples)	Cable joints (examples)

Indoor sealing ends (examples)

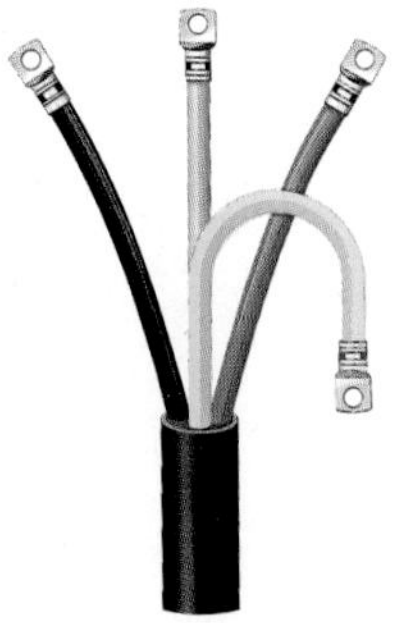

4-core cable
with fanned tails

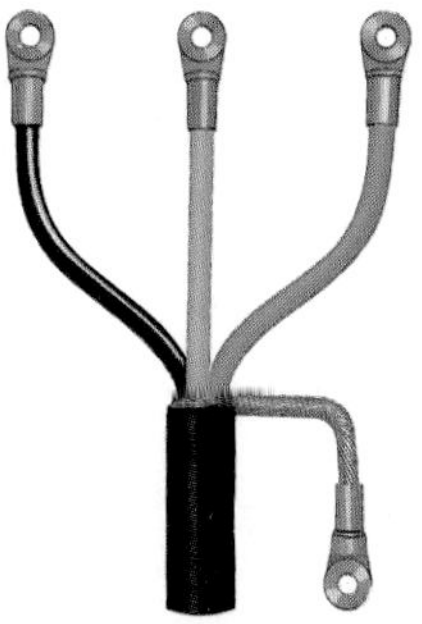

3-core cable
with concentric conductor
and parallel tails

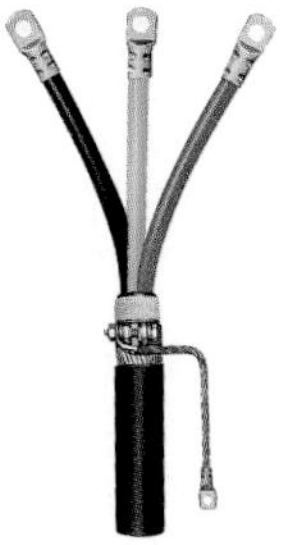

3-core cable
with flat steel wire armour and
fanned tails

Outdoor sealing end (examples)

Shrink-on cable
end

Cable joints (examples)

Shrink-on straight joint

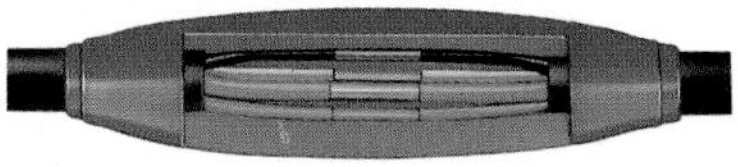

PROTOLIN-straight joint

PROTOLIN-branch Y-joint

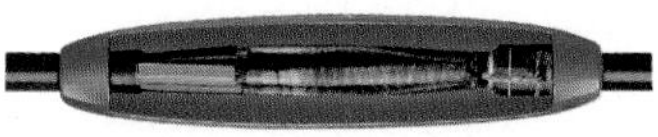

PROTOLIN-transition joint for cable with polymer
insulation to mass impregnated cable

$\triangleright$

105

Table 13.3 Cables and associated accessories (continued)

Construction	Designation, standards	Preferred application	Limited application
N A **KL** **E** **Y** 1 Al-conductor 2 Aluminium-sheath 3 Mass embedded plastic tape 4 PROTODUR-sheath (PVC) 5 Insulation (impregnated paper) 6 Filler 7 Belt insulation (impregnated paper)	NKLEY NAKLEY DIN VDE 0255	Cable previously used for urban networks, aluminium sheath used as neutral conductor (N) respectively as PEN conductor	Not used indoors, not suitable for areas liable to subsidence
N K **B** **A** 1 Lead sheath 2 Steel-tape armour 3 Jute serving 4 Conductor 6 Filler 8 Mass impregnated paper 5 Insulation (impregnated paper) 7 Belt insulation (impregnated paper)	NKBA NAKBA DIN VDE 0255	Cable previously used for urban networks where additional-earthing through the lead sheath was required.	Indoor and in cable trunking only with flame retardant outer cover or after removal of the jute serving. Where there is danger of corrosion additional corrosion protection is required e.g. PVC outer sheath (designation NKBY, NAKBY), where additional earthing via lead sheath is not necessary.

$$U_0/U = 0.6/1 \text{ kV } (U_m = 1.2 \text{ kV})$$

Indoor sealing end (examples)	Outdoor sealing end (examples)	Cable joints (examples)
Cylindrical sealing end	Pole mounted sealing end	T-joint The connection of the neutral conductors in cables with aluminium sheath is made by slitting helically and opening up the sheath Straight joint without lead inner casing

Table 13.3 Cables and associated accessories (continued)

Construction			Designation, standards	Preferred application	Limited application
N Y 1 PROTODUR- insulation (PVC)	F 2 Flat steel wire armour	Y 3 PROTODUR- sheath (PVC)	NYFY NAYFY	Indoor, cable trunking, out-doors and in ground for power stations, industry and switchgear	
4 Cu-Insulation	5 Lapped inner covering		DIN VDE 0271		
N Y 1 PROTODUR- insulation (PVC)	S 2 Cu-screen	Y 3 PROTODUR- sheath (PVC)	NYSY NAYSY	Because of small bending radii indoors in confined spaces, for power stations and switch-gear as well as in stations, as underground cable, because of its light weight favoured in situations where installation is difficult e.g. steep slopes	When selecting screen cross-sections the earth fault respec-tively double earth fault within the network must be consid-ered. When installing single-core cables in air adequate fix-ing must be provided because of dynamic effect of short-cir-cuit currents (see page 297).
4 Cu- conductor	5 Inner and outer conducting layer	6 Conducting 7 Tape tape	DIN VDE 0271		

$$U_0/U = 3.6/6 \text{ kV } (U_m = 7.2 \text{ kV})$$

Indoor sealing ends (examples)	Outdoor sealing ends (examples)	Cable joints (examples)
PROTOLIN-sealing end PEB	Encased sealing end FF 10 with porcelain insulators	PROTOLIN-straight joint
Push-on sealing end IAES 10	Push-on sealing end FAE 10 Sealing end FEP with porcelain insulator	Straight joint WP

▷

Table 13.3 Cables and associated accessories (continued)

Construction	Designation, standards	Preferred application	Limited application
N A **K** **B** **A** 1 Al-conductor 2 Lead sheath 3 Steel-tape armour 4 Jute serving 1 2 3 4 5 6 7 8 5 Insulation (impregnated paper) 7 Belt insulation (impregnated paper) 6 Filler 8 Mass impregnated paper	NKBA NAKBA DIN VDE 0255	In the ground, if no particular stresses are present	Indoors and in cable trunking only with flame retardant outer sheath, if need be outer serving must be removed; where differences in level occur (e.g. steep slopes) cables with polymer insulation must be used
N A **K** **B** **Y** 1 Al-conductor 2 Lead sheath 3 Steel tape armour 4 PROTODUR-sheath (PVC) 1 2 3 4 5 6 7 8 5 Insulation (impregnated paper) 7 Belt insulation (impregnated paper) 6 Filler 8 Mass impregnated paper	NKBY NAKBY DIN VDE 0255	In ground if increased corrosion protection is rquired; also suitable indoors	Where differences in level occur (e.g. steep slopes) cables with polymer insulation must be used
N K **R** **A** 1 Lead sheath 2 Armour of round steel wire 3 Jute serving 1 2 3 4 5 6 7 8 4 Cu-conductor 6 Filler 8 Mass-impregnated paper 5 Insulation (impregnated paper) 7 Belt insulation (impregnated paper)	NKFA NAKFA NKRA NAKRA DIN VDE 0255	With flat steel-wire- F and round steel-wire armour R, if particular mechanical stresses are to be expected e.g. pulling. With double outer jute serving as river or sea cable	Indoors and in cable trunking only with flame retardant outer sheath with steel spiral binder tape

$$U_0/U = 3.6/6 \text{ kV } (U_m = 7.2 \text{ kV}); \quad U_0/U = 6/10 \text{ kV } (U_m = 12 \text{ kV})$$

Indoor sealing end (examples)	Outdoor sealing end (examples)	Cable joint (examples)

Straight joint (special with lead inner casing) |
| Sealing end IKM | Encased sealing end FF 10 with porcelain insulators | |

Table 13.3 Cables and associated accessories (continued)

Construction	Designation, standards	Preferred application	Limited application

N Y   S   Y
1 PROTODUR- 2 Cu-screen   3 PROTODUR-
  insulation   sheat
  (PVC)   (PVC)
 	
4 Cu-conductor   5 Inner and outer   6 Conducting 7 Tape
                 conducting layer   tapes
	NYSY
	NAYSY
	
	DIN VDE 0271
	IEC 502
		For power stations and switchgear as well as stations because of small bending radii in confined spaces indoors. As underground because of light weight where installation conditions are difficult (e.g. steep slopes)	When selecting screen cross-sections earth-faulth respectively double earth-fault conditions of the network must be considered.
			When installing single-core cables in air adequate fixing must be provided because of dynamic effect of short-circuit currents (see page 297).

N A   Y   SE   Y
1 Al-   2 PROTODUR- 3 Cu-screen 4 PROTODUR-
  conductor   insulation   above each   sheath
            (PVC)   individual   (PVC)
                            core
 	
5 Inner and outer   6 Conducting tape   7 Extruded filler
  conducting layer
	NYSEY
	NAYSEY
	
	DIN VDE 0271
	IEC 502
		Indoors, cable trunking, outdoors and in ground; for power stations, industry and switchgear	When selecting screen cross-sections the earth fault respectively double earth fault conditions of the network must be considered

N Y   F   Y
1 PROTODUR-   2 Flat steel   3 PROTODUR-
  insulation (PVC)   wire armour   sheath (PVC)
 	
4 Cu-conductor   5 Inner and outer   6 Extruded
                 conducting layer   conducting filler
	NYFY
	NAYFY
	
	NYRY
	NAYRY
	
	DIN VDE 0271
	IEC 502
		With flat steel-wire armour F or armour of round steel-wire R in difficult installation or operating conditions	

$$U_0/U = 6/10 \text{ kV } (U_m = 12 \text{ kV})$$

Indoor sealing ends (examples)	Outdoor sealing ends (examples)	Cable joints (examples)

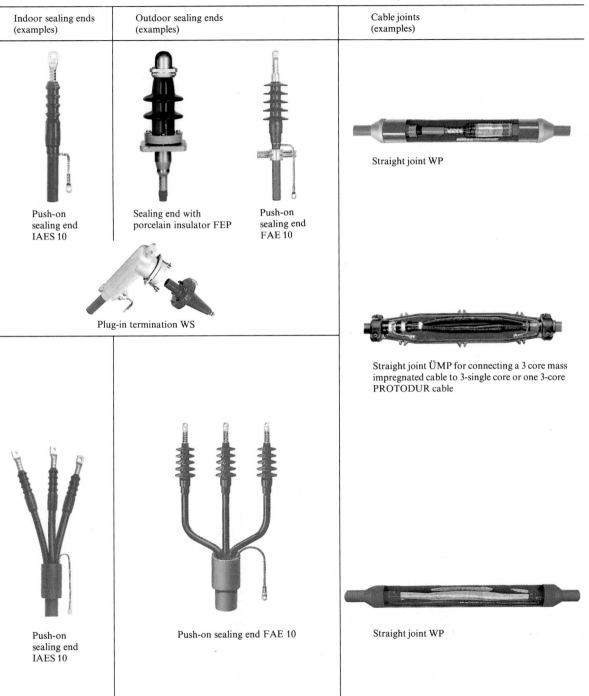

Push-on sealing end IAES 10

Sealing end with porcelain insulator FEP

Push-on sealing end FAE 10

Plug-in termination WS

Push-on sealing end IAES 10

Push-on sealing end FAE 10

Straight joint WP

Straight joint ÜMP for connecting a 3 core mass impregnated cable to 3-single core or one 3-core PROTODUR cable

Straight joint WP

Table 13.3   Cables and associated accessories (continued)

Construction	Designation, standards	Preferred application	Limited application
N 2X S Y 1 PROTO- THEN-X- insulation (XLPE)　　2 Cu-screen　　3 PROTODUR- sheath (PVC) 4 Cu-　　　5 Inner and outer　6 Conducting　7 Tape conductor　　conducting layer　tape	N2XSY NA2XSY DIN VDE 0273 IEC 502	These cables with PVC sheath previously used in urban networks are in Germany increasingly superseded by mechanically superior type with PE sheath.	In ground if because of mechanical stresses damage to the PVC sheath is likely. When installing single-core cables in air adequate fixing must be provided because of the dynamic effect of short-circuit currents (see page 297). For the selection of screen cross sections, the earth-fault respectively double earth-fault conditions of the network must be considered.
N A 2X S 2Y 1 Al- Conductor　2 PROTO- THEN-X- insulation (XLPE)　3 Cu-screen　4 PROTO- THEN- sheath (PE) 5 Inner and outer　　6 Conducting tape　7 Tape conducting layer	N2XS2Y NA2XS2Y DIN VDE 0273 IEC 502	In ground for urban networks because of extremely low electrical losses. To case installation 3 cables can be layed up and supplied on a single drum.	If after mechanical damage ingress of water is likely cable having longitudinal water tightness in the screen area has advantages. When used indoors it must be observed that the PE sheath is not flame retardant. When installing single-core cables in air adequate fixing must be provided because of dynamic effect of short-circuit currents (see page 297). For the selection of screen cross-sections, earth-fault respectively double earth-fault conditions of the network must be considered.
N A 2X S(F) 2Y 1 Al- conductor　2 PROTO- THEN-X- insulation　3 Copper screen longitudinally water tight　4 PROTO- THEN- sheath (PE) 5 Inner and outer　6 Gap sealing　　7 Extruded filler conducting layer　(conducting tape with swelling tape)	N2XS(F)2Y NA2XS(F)2Y DIN VDE 0273 IEC 502	In unfavourable installation conditions especially if after mechanical damage ingress of water in longitudinal direction must be avoided, longitudinally water tight cables with extruded filler and gap sealing in the screen area offer advantages.	In cable trunking and indoors it must be noted the PE sheath is not flame retardant. When installing single-core cables in air adequate fixing must be provided because of dynamic effect of short-circuit currents (see page 297). For the selection of screen cross section the earth-fault respectively double earth-fault conditions of the network must be considered.

$$U_0/U = 6/10 \text{ kV } (U_m = 12 \text{ kV})$$

Indoor sealing ends (Examples)	Outdoor sealing ends (Examples)	Cable joints (Examples)

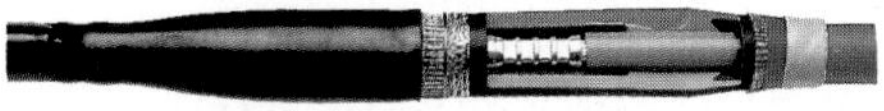

Push-on straight joint AMS

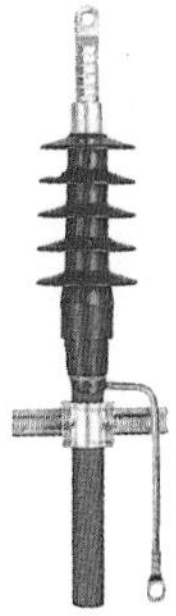

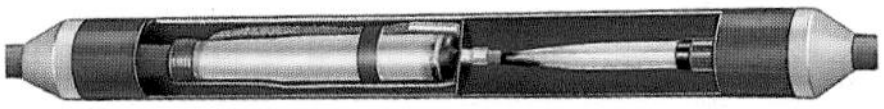

Push-on sealing end
FAE

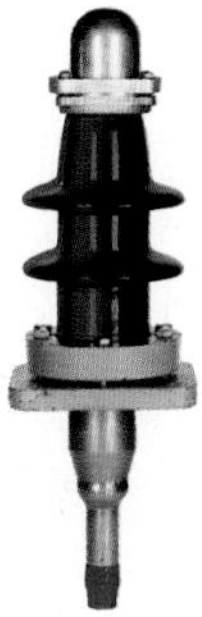

Straight joint WP

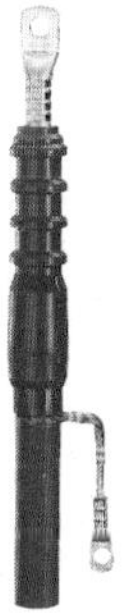

Push-on sealing
end IAES

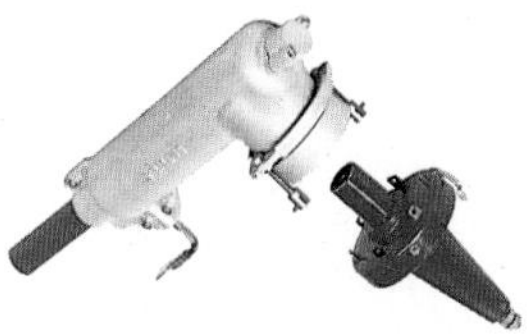

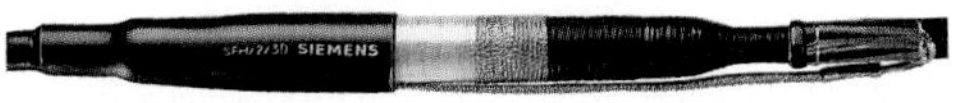

Straight joint WPS with shrink-on sleeve

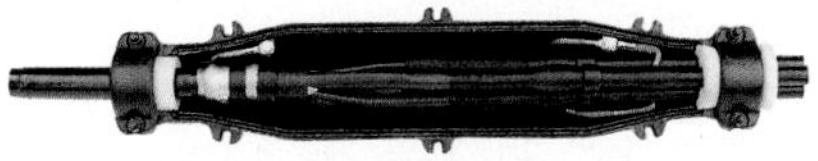

Sealing end with
porcelain insulator FEP

Straight transition joint ÜMP for connecting
a 3-core mass impregnated cable
to 3-single-core XLPE cables

Plug-in termination WS

115

Table 13.3 Cables and associated accessories (continued)

Construction	Designation, standards	Preferred application	Limited application

N A   2X   S   2Y
1 Al-   2 PROTO-   3 Cu-screen   4 PROTO-
conductor   THEN-X-   THEN-
        insulation   sheath
        (XLPE)   (PE)
 	
5 Inner and outer   6 Conducting tape   7 Tape
   conducting layer
	N2XS2Y
	NA2XS2Y
	
	DIN VDE 0273
	IEC 502
		In ground for urban networks because of extremely low dielectric losses.
		Cables with PVC sheath previously used in urban networks are in Germany increasingly superseded by the mechanical superior type with PE sheath.
			Larger termination space than for single-core cable is required.
			For installation in trunking and indoors it must be noted that the PE sheath is not flame retardant.
			Multi-core cables with polymer insulation are not available in longitudinal water tight form.

N 2X   SE   2Y
1 PROTOTHEN-X-   2 Copper-screen   3 PROTOTHEN-
   insulation   over each   sheath
   (XLPE)   indivudal core   (PE)
 	
4 Cu-   5 Inner and outer   6 Conducting   7 Extruded
   conductor   conducting layer   tape   filler
	N2XSE2Y
	NA2XSE2Y
	
	DIN VDE 0273
	IEC 502		

N 2X   SE   Y   F   Y
1 PROTO-   2 Copper-   3 PROTO-   4 Flat   5 PROTO-
   THEN-X-   screen   DUR-   steel-wire   DUR-
   insulation   over each   sheath   armour   sheath
   (XLPE)   individual   (PVC)   (PVC)
           core
 	
6 Cu-   7 Inner and   8 Conducting   9 Extruded
   conductor   outer   tape   filler
            conducting
            layer   10 Tape
	N2XSEYFY
	NA2XSEYFY
	
	N2XSEYRY
	NA2XSEYRY
	
	DIN VDE 0273
	IEC 502
		With flat steel-wire armour F or armour of round steel-wire R for underground mining.
		With PE sheath instead of PVC sheath also for installation in ground where installation and operating conditions are difficult.
			The previously used cables with PVC sheath are in Germany increasingly superseded by superior types with PE sheath.

$$U_0/U = 6/10 \text{ kV } (U_\text{m} = 12 \text{ kV})$$

Indoor sealing ends (examples)	Outdoor sealing ends (examples)	Cable joints (examples)

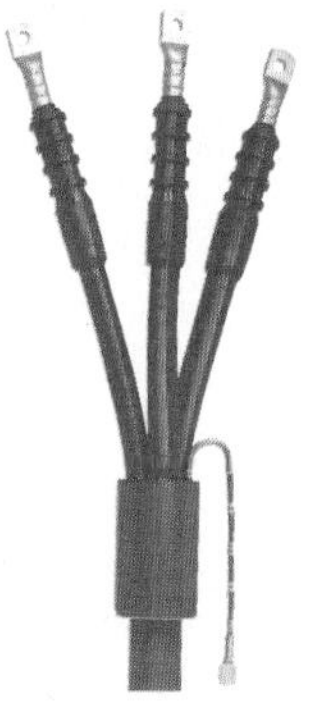

Straight joint WP

Push-on sealing end
IAES 10

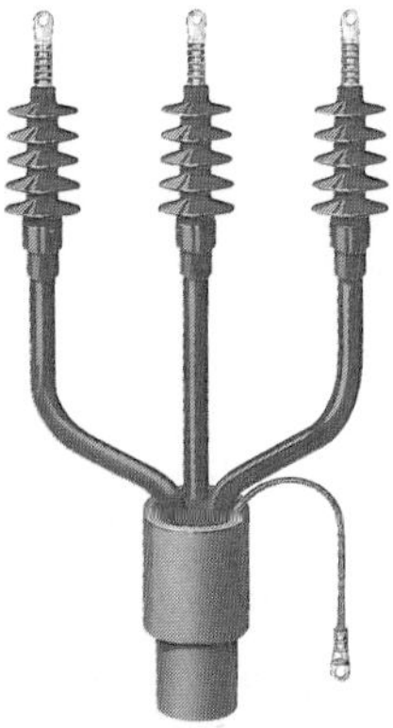

Push-on sealing end
FAE 10

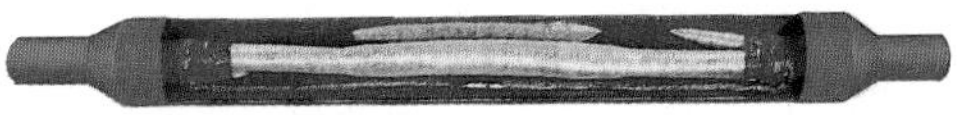

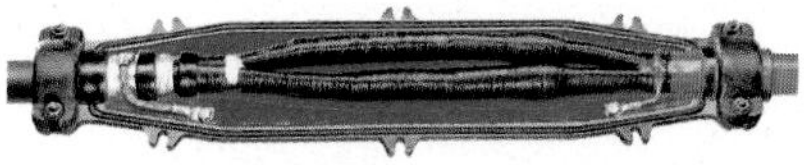

Straight transition joint ÜMP for connecting
a 3-core mass impregnated cable
to a 3-core XLPE cable

Table 13.3 Cables and associated accessories (continued)

Construction	Designation, standards	Preferred application	Limited application
N EK **E** **B** **A** 1 Individually 2 Plastic 3 Steel-tape 4 Jute screened lead- tape armour serving sheathed core 5 Cu- 6 Conducting 7 Insulation 8 Layers of mass- conduc- paper (impregnated paper) impregnated tor fibrous material	NEKEBA NAEKEBA DIN VDE 0255	In ground if no particular stresses are present	Indoors and in cable trunking only with flame retardant outer sheath alternatively with outer sheath removed; where there is danger of corrosion additional corrosion protection is required; where differences in level occur (e.g. steep slopes) cables with polymer insulation must be used.

N A   **EK**   **E**   **B**   **Y**
1 Al-   2 Individually   3 Plastic   4 Steel-tape   5 PROTO-
con-   sheathed lead-   tape   armour   DUR-
ductor   sheathed core   sheath
                                           (PVC)
 	
6 Conducting   7 Insulation   8 Layers of mass-
paper   (impregnated paper)   impregnated
                                        fibrous materials
	NEKEBY
	NAEKEBY
	
	DIN VDE 0255
		In ground increased corrosion protection is required; also suitable for indoors	Where differences in level occur (e.g. steep slopes) cables with polymer insulation must be used

N H   **K**   **F**   **A**
1 Screening of   2 Lead   3 Flat steel   4 Jute
metallised   sheath   wire armour   serving
paper
 	
5 Cu-conductor   7 Insulation   8 Filler
                       (impregnated
6 Conducting paper   paper)   9 Layers of mass-
                                       impregnated
                                       fibrous materials
	NHKFA
	NAHKFA
	
	NHKRA
	NAHKRA
	
	NHKBA
	NAHKBA
	
	DIN VDE 0255
		With flat steel-wire armour F or amour of round steel-wires R as river or sea cable; in ground where particularly high-mechanical stresses can be expected; to provide increased protection against corrosion a red PROTODUR sheath (PVC) replaced the jute serving (Designation: NHKRY respectively NHKFY)	H-cables with steel-tape armour are rarely used, S.L.-cables are preferred; where differences in level occur (e.g. steep slopes) cables with polymer insulation must be used

$$U_0/U \geq 12/20 \text{ kV } (U_m \geq 24 \text{ kV})$$

Indoor sealing ends (examples)	Outdoor sealing ends (examples)	Cable joints (examples)

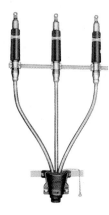

EoD with transparent
cast-resin insulators

Straight joint with individual lead inner casing

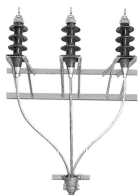

FEL with porcelain insulators

EoD with transparent
cast-resin insulators
with increased short-circuit
withstand

Straight joint with steel inner casing
for connecting H-cables to
S.L.-cables

Table 13.3 Cables and associated accessories (continued)

Construction	Designation, standards	Preferred application	Limited application
N A KL E Y 1 Al-conductor 2 Aluminium-sheath 3 Plastic tape mass embedded 4 PROTODUR-reinforced sheath (PVC) 5 Conducting paper 6 Insulation (impregnated paper) 7 Screening (conducting paper and Al foil)	NKLEY NAKLEY DIN VDE 0255	Cables previously used for urban networks; now being superseded by XLPE cables.	Not suitable for mechanical stresses and areas subject to subsidence; where differences in level occur (e.g. steep slopes) cables with polymer insulation must be used.
N 2X S Y 1 PROTOTHEN-X-insulation (XLPE) 2 Cu-screen 3 PROTODUR-sheath (PVC) 4 Cu-conductor 5 Inner and outer conducting layer 6 Conducting tape 7 Tape	N2XSY NA2XSY DIN VDE 0273 IEC 502	These cables with PVC sheath previously used in urban networks are in Germany increasingly superseded by mechanically superior type with PE sheath.	In ground if because of mechanical stresses damage to the PVC sheath is likely. When installing single-core cables in air adequate fixing must be provided because of the dynamic effect of short-circuit currents (see page 297). For the selection of screen cross sections, the earth-fault respectively double earth-fault conditions of the network must be considered.
N A 2X S 2Y 1 Al-conductor 2 PROTOTHEN-X-insulation (XLPE) 3 Cu-screen 4 PROTOTHEN-sheath (PE) 5 Inner and outer conducting layer 6 Conducting tape 7 Tape	N2XS2Y NA2XS2Y DIN VDE 0273 IEC 502	In ground for urban networks because of extremely low dielectric losses. To ease installation 3 cables can be layed up and supplied on a single drum.	If after mechanical damage ingress of water is likely cable having longitudinal water tightness in the screen area has advantages. When used indoors it must be observed that the PE sheath is not flame retardant. When installing single-core cables in air adequate fixing must be provided because of the dynamic effect of short-circuit currents (see page 297).

$$U_0/U \geq 12/20 \text{ kV } (U_m \geq 24 \text{ kV})$$

Indoor sealing ends (examples)	Outdoor sealing ends (examples)	Cable joints (examples)
EoD with transparent cast-resin insulator		Brass straight joint
JAE JAES Push-on sealing ends	Push-on sealing end FAE	Push-on straight joint AMS Straight joint WP
	Plug-in termination WS	Straight transition joint SM-WP for connecting a paper insulated S.L.-cable to 3 single-core XLPE cables

▷

Table 13.3 Cables and associated accessories (continued)

Construction	Designation, standards	Preferred application	Limited application
N A **2X** **S(F)** **2Y** 1 Al-conductor 2 PROTOTHEN-X-insulation (XLPE) 3 Copper screen (longitudinal water tight) 4 PROTOTHEN-sheath (PE) 5 Inner and outer conducting layer 6 Gap sealing (conducting tape with swelling tape) 7 Extruded filler	N2XS(F)2Y NA2XS(F)2Y DIN VDE 0273 IEC 502	In unfavourable installation conditions especially if, after mechanical damage ingress of water in longitudinal direction must be avoided longitudinally watertight cables with extruded filling compound and gap sealing in the screen area offer advantage.	In cable trunking and indoors it must be noted that the PE sheath is not flame retardant. When installing single-core cables in air adequate fixing must be provided because of dynamic effect of short-circuit currents (see page 297). For the selection of screen cross-section the earth-fault respectively double earth-fault conditions of the network must be considered.
N 2X **SE** **Y** 1 PROTOTHEN-X-insulation (XLPE) 2 Cu-screen over each individual core 3 PROTODUR-sheath (PVC) 4 Cu-conductor 5 Inner and outer conducting layer 6 Conducting tape 7 Extruded filler	N2XSEY NA2XSEY DIN VDE 0273 IEC 502	Types used in countries where 3-core cables are required	These cables with PVC sheath previously used in urban networks are in Germany increasingly superseded by mechanically superior types with PE sheath.
N 2X **SE** **Y** **F** **Y** 1 PROTOTHEN-X-insulation (XLPE) 2 Cu-screen over each individual core 3 PROTODUR-sheath (PVC) 4 Flat steel-wire armour 5 PROTODUR-sheath (PVC) 6 Cu-conductor 7 Inner and outer conducting layer 8 Conducting tape 9 Tape 10 Extruded filler	N2XSEYFY NA2XSEYFY N2XSEYRY NA2XSEYRY DIN VDE 0273 ICE 502	Types used in other countries with flat steel-wire armour F or armour of steel round-wire R where difficult installation and operating conditions exist. Preferred with PE sheath instead of PVC sheath for laying in ground.	

$U_0/U > 6/10$ kV ($U_m > 12$ kV)

Indoor sealing ends (examples)	Outdoor sealing ends (examples)	Cable joints (examples)

Straight joint WP

Push-on sealing end
FAE with core spreading

Push-on sealing end
IAES with core spreading

Straight transition joint SM-WP to
connect paper insulated S.L.
with 3-core XLPE cables

Sealing end FEP
with porcelain insulators
and core spreading

14 Power Cables for Special Applications

14.1 Cable with Elastomer Insulation

These cables have been superseded in Germany by cables with PVC insulation. Only on shipboard installations are cables with elastomer insulation still used to any great extent.

In a few countries low- and medium-voltage cables with an insulation of Ethylene-Propylene-Rubber (EPR) have major significance. The operating characteristics of these cables comply as to their permissible conductor operating temperature of 90 °C in normal operation and 250 °C in the event of short-circuit with those of XLPE cables. However it must be noted that in the medium-voltage range EPR offers a higher resistance to partial discharge but on the other hand has slightly higher dielectric losses.

tions in engine rooms under tropical conditions or when used outdoors in winter conditions.

The core designation for shipboard cables is shown in Table 14.1 and this differs from DIN VDE 0293.

Table 14.1 Core identification of shipboard cables

No. of cores	Colour
1	Light grey
2	Light grey-black
3	Light grey-black-red
4	Light grey-black-red-blue
5	Light grey-black-red-blue-black
6 and above	Light grey, each with a number designation in black commencing from the centre with number 1

14.2 Shipboard Power Cable

In cables to IEC 92 for merchant marine vessels elastomer insulation materials and normally also elastomer sheathing materials are used. PVC sheaths are also permitted in various combinations but, because of their thermoplastic characteristics and because of the installation conditions prevailing on shipboard, are not recommended.

14.2.1 Construction and Characteristics

In Germany shipboard cables are manufactured, complying with the relevant DIN norms, with EPR insulation and CR sheath. This insulation is suitable for the normal conditions of operating temperatures in ship building of up to 85 °C. The PROTOLON compounds used by Siemens have a particularly high resistance to high temperature and have a long service life under the influence of ozone and partial discharges. The PROTOFIRM sheath is highly resistant to notch impact and tearing as well as being oil resistant and flame retardant. It maintains its elasticity at both high and low temperatures, e.g. in installa-

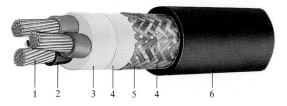

a) MGCG 0.6/1 kV with screen

b) MGG 0.6/1 kV without screen

1 Tinned stranded conductors
2 EPR insulation (PROTOLON)
3 Inner covering
4 Separating foil
5 Copper wire braid
6 Polychloroprene-sheath (PROTOFIRM)

Fig. 14.1 Shipboard power cables MGCG and MGG

14.2.2 Application and Installation

Cables of the type MGCG (Fig. 14.1a) can be installed in line with DIN 89150 and close to IEC Publication 92-352 as a permanent installation in any room and on open decks.

Cable MGG (Fig. 14.1b) should only be installed below the upper metal deck.

Due to the good electrical screening of the type MGCG, having a copper wire braid, radio interference and the disturbance of the operation of electronic equipment is reduced. The copper wire braid also acts as mechanical protection and in the event of a fault provides touch protection. For this purpose both ends must be securely earthed by screws.

The load capacities of shipboard cables is laid down in IEC 92-201 (values: see part 2).

Basically the cables type MGCG to DIN 89158 and MGG to DIN 89160 are recognized by the following classification bodies. However for the individual manufacturer approval may be required [1].

▷ American Bureau of Shipping (ABS)
▷ Bureau Veritas (BV)
▷ Det Norske Veritas (DNY)
▷ Germanischer Lloyd (GL)
▷ Lloyds Register of Shipping (LRS)
▷ Polski Rejestr Statkow (PRS)
▷ USSR Register of Shipping (RSU)

14.3 Halogen-Free Cables[2] with Improved Characteristics in the Case of Fire

The standard low-voltage cables (0.6/1 kV) for indoor installations (Section 13) and the standardized installation cables (Section 8) are flame retardant in the sense of IEC 332-1 and the identical regulation DIN VDE 0472 Part 804, type test B. These require a single test on the cable using a gas burner. PVC-insulated cables and flexible wires as well as the cable commonly used in some countries with XLPE insulation and a suitable sheath satisfy IEC test requirements. These cables have proved themselves over many years in service and will continue to be used in the future mainly in industrial installations.

For buildings or plant with a high concentration of either persons or high-value contents however very often more stringent safety requirements apply. To meet these requirements cables with improved performance in the event of fire are available.

14.3.1 Testing Performance under Conditions of Fire

The following characteristics of cables are relevant under conditions of fire:

▷ Spread of fire
▷ Corrosivity of combustion gases
▷ Smoke density
▷ Insulation retention during fire

Spread of Fire

Practical experience has shown that a single PVC-insulated and PVC-sheathed cable when installed in the normally vertical plane does not aid the spread of fire, however this does not apply when cables are bunched such as installation in parallel or in bundles (Fig. 14.2). An arrangement for comparative testing was developed to determine the burning characteristics of bunched cables (DIN VDE 0472 Part 804 type test c). For this test the cables are fixed to a ladder rack side by side in a vertical chimney, at the bottom of which a gas burner subjects the cables to flame for a given period of 20 minutes.

After turning off the burner the flames must not continue to spread to the upper end of the test arrangement. Cables which pass this test can be classed as

---

[1] Depending on classification authority this could mean either approval or acceptance

[2] See page 79. "Halogen-Free SIENOPYR Wiring and Flexible Cables with Improved Performance in the Event of Fire"

Fig. 14.2
Burning PVC during combustion testing

Fig. 14.3
Burning SIENOPYR-FRNC cable during combustion
testing

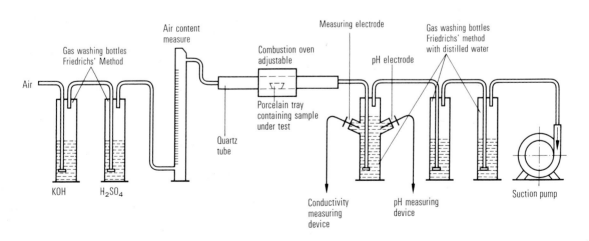

Fig. 14.4
Test arrangement to assess corrosivity of combustion gasses

resistant to the spread of fire (Fig. 14.3); type designation in SIENOPYR cables: FR (flame retardant). Similar tests are also included in other national test specifications for cables. The current international concept is laid down in IEC report 332-3 and an IEC Standard is under preparation.

As an aid for the selection of materials for cables with improved performance in fire oxygen index testing is also used. In this test a sample of the relevant material is combusted and it is recorded what degree of oxygen content, in a nitrogen/oxygen mixture, is necesssary to sustain combustion. The higher the recorded oxygen index OI[1] the more resistant the material is to supporting combustion under normal conditions a sure assessment of the characteristics and performance in fire for the complete cable can however only be made by testing in the chimney rig. The oxygen index test however is most useful for material quality assurance testing.

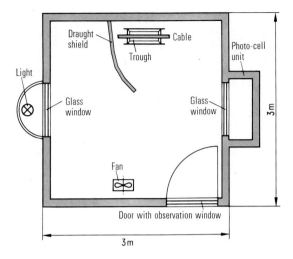

Fig. 14.5 Testing for smoke density (cube test)

Corrosivity of Combustion Gases

Halogens contained in cable materials such as chlorine in PVC, but also fluorine and bromine in other polymer insulation materials, raise their ignition temperatures and hence their resistance to the spread of fire along the cable. If however the cable insulation is burnt for example by external heat input corrosive gases of combustion are formed which, together with humidity from the air or with water from fire extinguishers, form acids (e.g. HCL). This acid can lead to corrosion damage to electrotechnical equipment and to parts of buildings. A test method was therefore devised to measure the amount of corrosive gas produced during combustion (Fig. 14.4). In the test proceedure laid down in DIN VDE 0472 Part 813 material samples are combusted and the gases produced passed through water while measurements are made of electrical conductivity and pH value of the water.

Cables containing only materials which comply with prescribed values of conductivity and pH level are classed as halogen-free and non-corrosive in the event of fire (designation in SIENOPYR cables: NC, non-corrosive).

Smoke Density

If cables with insulation or sheaths containing chlorine are combusted dense black smoke is produced. This smoke hampers fire fighting and also evacuation of any premises used by the public. To assess cables using special materials with lesser smoke density in the case of fire IEC TC 20 recommends an optical test procedure in an enclosed cube (Fig. 14.5).

The FRNC cables described on page 128 have, under comparable conditions of fire, a ten times lower smoke density than PVC-insulated and sheathed · cables of similar construction.

Insulation Retention under Conditions of Fire

In certain installations it is also required under conditions of fire for the cable to remain functional for a certain period of time i.e. to continue to supply electrical energy. To ensure this characteristic is satisfied cables are subjected to a test of insulation retention under conditions of fire (type test) (Fig. 14.6). In the type test laid down in DIN VDE 0472 Part 814 a single cable in horizontal position is subjected to flame from a long gas burner. A voltage of 380 V is applied to the core of the cable (rated voltage 0.6/1 kV) via a fuse and the test result is satisfactory if the fuse remains intact during the test.

---

[1] In international standards LOI

Fig. 14.6
Arrangement testing of insulation retention under conditions of fire

14.3.2 Construction and Characteristics

To improve the characteristics of cables in fire there are two courses of action. Firstly one can increase the halogen content of the materials or one can add components which together with halogen have a synergistic effect. By this means cable can be manufactured having a very good characteristic in regard to the limitation of spreading of fire. Corrosive and toxic gases are however developed during combustion. By special compounding it is possible to reduce, at least, the high smoke emission of the halogen containing materials. Such cables as FR-PVC-cable (flame-retardant low smoke) are used in some European countries e. g. for installation in power stations.

Another possibility for the improvement of the characteristics in fire is to prevent the formation of toxic or corrosive gases; here pure polyolefines are used as materials for insulation and sheaths. These materials are relatively easily flammable. An improvement of the characteristics of such polymers is possible by the use of special compounds (see page 39 "Sheathing Materials for Special Purposes"). One possibility for this is the addition of aluminiumoxyhydrate. In the case of fire water is than released which vapourises. This endotherm process leads to quicker extinguishing of the fire.

**Cables with Improved Characteristics
in Fire to DIN VDE 0266**

SIENOPYR-FRNC-cable to DIN VDE 0266 type NHXHX 0.6/1 kV (Fig. 14.7) have an insulation of

elastomer, halogen-free, material compound HI1, base EPR to DIN VDE 0207 Part 23.

The laid-up cores are surround by a core cover of halogen-free elastomer compound with low flammability. The sheath consists of halogen-free, low flammability elastomer compound HM1 to DIN VDE 0207 Part 24 (base EVA). If necessary a concentric conductor is arranged below the sheath (type NHXCHX).

Cables to DIN VDE 0266 normally have insulation retention (see page 127) of more than 20 minutes (designation FE). For this the conductors of smaller cross sections require additional silicone covers.

Fig. 14.7
SIENOPYR-FRNC-cable NHXHX 4 × 1.5 FE 0.6/1 kV with insulation retention

Fig. 14.8
SIENOPYR-FRNC-cable (N)2XH 4 × 1.5 0.6/1 kV without insulation retention

Cable with Improved Characteristics in the Event of Fire According to VDE Register Nr. 11 099 109/110

SIENOPYR-FRNC-cable Type (N)2XH (Fig. 14.8) have an insulation of special cross-linked polyethylene compound 2XI1 to DIN VDE 0207/Part 22 with the laid-up cores surrounded by a cover of halogen free, low flammability thermoplastic filler. The sheath consists of a low flammability thermoplastic compound with increased heat resistance HM2 to DIN VDE 0207 (base EVA). If required a concentric conductor can be arranged below the sheath (type (N)2XCH).

14.3.3 Laying and Installation

Cables with improved characteristics in the event of fire are used in indoor installations and outdoors in a similar manner to NYY cables. They can be arranged on racks or fixed to walls and ceilings in either vertical or horizontal directions by means of cable clips. The bending radius of the cable is 12 D (single-core 15 D). Installation directly in the ground is not recommended for these cables. When terminating these cables both inner and outer sheaths should be cut at the same point. When the insulation has been removed a silicone covering which may be included must also be removed. In addition the instructions for installation and transportation given in DIN VDE 0298 Part 1 must be observed. For jointing either flame retardant cast-resin joints or flame retardant shrink-on sleeves may be used. If insulation retention is required for the joints special measures must be taken.

Cables to DIN VDE 0266 (type NHXHX and NHXCHX) are designed for a maximum conductor operating temperature of 70 °C with a maximum permissible short-circuit temperature of 160 °C. For cables (type (N)2XH and (N)2XCH) with VDE Register Nr. 11099 109/110 the maximum permissible conductor operating temperature is 90 °C and the maximum permissible short-circuit temperature is 250 °C. The current carrying capacity must be taken from the relevant tables in DIN VDE 0298 Part 2 (see Section 18).

The cables with special characteristics in the event of fire are employed where special measures, for the protection of high-value equipment or persons, must be taken. In these circumstances apart from the reduced spread of fire, in some cases priority is given to the characteristics as regards corrosiveness of fumes and smoke density whereas in other cases insulation retention may be of prime importance.

14.4 Cables for Mine Shafts and Galleries

In mining installations below ground cables with polymer insulation are commonly used as mine shaft and gallery cables. These cables for rated voltages 0.6/1 kV always contain a protective conductor and most of them are armoured.

Construction and Characteristics

Cables used in mining applications normally have copper conductors. For plant with rated voltages up to 10 kV PVC insulation to DIN VDE 0118 is used. In areas subject to mining gas, however, only installations with rated voltages up to 6 kV are permitted. Installations having a rated voltage of 10 kV and with cable having XLPE insulation have been approved by Oberbergamt (OBA) (mining authority) Nordrhein-Westfalen. The protective conductor is incorporated in the following types:

▷ as separate insulated conductor marked green/yellow e. g. as in NYFGY-J 4 × 50 SM 0.6/1 kV

▷ concentric conductor above inner covering e. g. as in NYCYRGY 3 × 50 SM/25 3.6/6 kV

▷ equally split, concentric conductor over individual cores e. g. as in NYCEYRGY 3 × 50 RM/25 6/10 kV.

Fig. 14.9
Gallery cable NYCYFGY 3 × 120 SM/70 3.6/6 kV

Fig. 14.10
Mine shaft cable NYCYRGY 3 × 120 SM/70 3.6/6 kV

In order to avoid metallic contact between construction elements having different electrolytic potentials a protective extruded covering of PVC (separation sheath) is always included between a concentric conductor and the round steel-wire R or the flat steel-wire armour F. In addition, a protective PVC sheath is included above the armour. This sheath is coloured yellow in underground mining cables with rated voltage of 0.6/1 kV. For both construction and testing of these cables generally DIN VDE 0271 applies for PVC insulated cables and DIN VDE 0273 for XLPE insulated cables. In addition, where no other national standard exists, IEC 502 is relevant.

Installation and Fixing

In mining installations with slopes up to 50° the cable must be supported at distances of no more than 5 meters and must be hung with suitable sag between supports. This applies for armoured as well as unarmoured gallery cables.

In mining locations where the slopes exceed 50° cables must be tear resistant with a degree of safety factor 3. For the calculation of tear resistance of cables the armour is the deciding factor. These cables must, after hanging in, be fixed by clamps at distances not exceeding 6 meters. Where longer distances between supports is unavoidable the armour must withstand the tensile stress with a factor of safety of 5.

Accessories for mining applications below ground must, in Germany, apart from VDE approval, also have the special approval of the Oberbergämter (mining inspectorate).

14.5 River and Sea Cables

Cables used for crossing rivers or for laying in sea water, e.g. for the connection of island networks to the mainland, for the supply of energy to off-shore plant or for the operation of lighthouses or navigational aids, are usually fitted with a substantial armour to withstand the high-mechanical stresses during installation and when in operation.

Type of Construction

River and sea cables normally have polymer or paper insulation. Polymer-insulated cables have the advantage of low weight. Normally low-voltage cables are

PVC insulated and for medium voltage XLPE is more commonly used. For this application a cable with both longitudinal and radial water tightness is generally selected (see page 47). Cables having polymer insulation are particularly suited especially if a combination of communication cores as well as optical fibre cores are required. With paper lead cables the lead sheath must be thicker than those in DIN VDE 0255 to cater for the mechanical stresses to be expected. If the cable is to be subjected continuous vibration, e.g. by heavy surf, the lead sheath is then strengthened by alloy additions as protection against metal fatigue fractures. The armour of river and sea cables consists of tinned steel wires which are, depending on local conditions, either of round or flat profile. The shape and thickness are dependent upon the expected tensile stresses and any prevailing danger (by punting poles, anchors, movement of sea bed etc.). In particularly unfavourable conditions a double armour may be necessary. To provide safe protection against corrosion of the armour the outer serving comprises two layers of compounded special jute or a polymer sheath (PVC or PE).

Laying and Installation

The method of laying sea cables is dependent upon local conditions and on the type of equipment available (loading facilities and type of ship). The delivery length is often only limited by transportation capacity. If necessary individual lengths are spliced to achieve the total required length. Should it be necessary to joint cables at sea, cable joints can be provided which are also safe in operation under difficult installation conditions. River cables and cables laid in tidal inshore waters are often laid in trenches cut

Fig. 14.11
XLPE-insulated sea cable with round steel-wire armour for 12/20 kV

Fig. 14.12
Installation operation at sea; running off from cable stack over rollers to laying head at stern of vessel

in the bed using a scavenging keel with high pressure water jets.

For river crossings standard cables could also be used if laid in thermoplastic piping or laid in trenches cut in the bed.

14.6 Airport Cables

Airport cables supply energy to airport lighting apparatus.

Construction and Characteristics

The cables are single core and have a copper conductor of 6 mm^2 and a concentric conductor of 2.5, 4 or 6 mm^2 cross-section. Insulation and outer sheath are of PVC. The construction is in line with IEC 502 and DIN VDE 0271. Preferred rated voltages are 1/2 and 5/10 kV. Other than this standard construction which is used in Germany there are constructions having EPR insulation and CR sheath e.g. to the American standards of the Federal Aviation Administration (FAA L 824).

Fig. 14.13
Airport cable FLYCY 1 × 6 RE/2.5 1/2 kV

Application and Installation

Single-core airport cables supply airport lighting equipments connected in serries. Normally single-phase transformers are used to feed special lights. For this duty joints or plug connectors are used to connect the equipment via flexible tails of NSGAF.

14.7 Cable with Polymer Insulation and Lead Sheath

Cables with polymer insulation and lead sheath are used in Germany for special operating conditions whereas, in the main, PVC-insulated cables are used for rated voltages of 0.6/1 kV. In special circumstances lead-sheathed cables are also used for higher rated voltages (e. g. 3.6/6 kV) or with a different insulation of e. g. XLPE.

Construction and Characteristics

Cable with polymer insulation and lead sheath for 0.6/1 kV rated voltage are governed by DIN VDE 0265. These are constructed as multi-core cable and contain either solid or stranded copper conductors. The lead sheath is arranged over the inner covering. Below this an additional tinned conductor may be arranged as sheath wire. Above the lead sheath a PVC sheath is provided.

Application and Installation

Cables with polymer insulation and lead sheath are applied especially where there is danger of effect from

Fig. 14.14
Cable with PVC insulation and lead sheath
NYKY 0.6/1 kV

solvents and fuel oils. The lead sheath must not be used as neutral conductor (N). If a sheath wire is incorporated it can be used to earth the lead sheath e.g. in explosion proof installations (DIN VDE 0165). Because of the good coupling resistance of the closed lead sheath these cables have advantages where electromagnetic compatibility (EMC) is important.

14.8 Insulated Overhead Line Cables

Insulated overhead cables are not strictly power cables. Based on their application and construction they have become classified as overhead line cables. Because the same insulation materials are used as for power cables these overhead line cables are covered by the VDE regulations for cable.

Construction and Characteristics

Insulated overhead line cables for medium voltage are not standardized. The construction is largely in line with cable to DIN VDE 0273. For the construction and testing of overhead line cable for 0.6/1 kV rated voltage the standard DIN VDE 0274 is applicable. The stranded rope conductors (25 to 70 mm²) are made of drawn pure aluminium wires. These wires comply with DIN 48 200 parts prior to being manufactured into the conductor rope. For the construction and characteristics of the conductor rope DIN 48 201 Part 5 applies with the exception of the values for number of strands, diameter and electrical resistance of the rope which is laid down in DIN VDE 0274. The insulation of individual conductor ropes consists of black XLPE type 2XI1 to DIN VDE 0207 Part 22 which, to improve resistance to sunlight and weather, has an addition of at 2% carbon black. In addition to single-core overhead line cables, bundles of 4 cables are also standardized (4-core insulated overhead line cables). In this 4-core

Fig. 14.15
Insulated overhead line cables
NFA2X 4 × 70 RM 0.6/1 kV

cable the phase conductors are marked by 1, 2 or 3 longitudinal ribs along the length of the cable. The neutral conductor (N) being the fourth core has the same cross-sectional area as the phase conductors. The neutral does not bear an identifying mark.

Bundles of four cores each with a conductor cross section of 70 mm² are also available combined with either one or two additional cores of smaller cross section (35 mm²). These additional conductors carry identification markings of four and five ribs respectively.

Application and Installation

Insulated overhead line cables have advantages particularly over difficult terrain. They are frequently used in woodland (narrow avenues are possible) and are also used for the extension or refurbishment of existing urban networks. In these areas four-core overhead line cables are predominant for systems of three phase with PEN. Bundles of four-cores with one or two additional cores of reduced cross section are used where street lighting is feed from the same main line.

Single-core overhead line cables are frequently used for the supply of single-phase loads.

Suspension and support of the cables is possible by the use of wooden or concrete poles as well as from roof supports. For this system DIN VDE 0211 applies and for domestic feeds DIN VDE 0100 Part 732 applies. The specific characteristics of insulated overhead line cables provides full insulation without breaks both in open terrain as well as in buildings where mechanical damage to the insulation is most unlikely to occur. Recommendations for installation in different situations is given in Table 14.2.

For the current carrying capacity of overhead line cables (see Table 14.3) instead of the values in DIN VDE 0298, however, Table 4 of DIN VDE 0274 applies. The conductor temperature, in the event of short-circuit for mechanical reasons must not exceed 130 °C.

For the suspension of insulated overhead line cables special supporting equipments are required. In the suspension clamps each individual core of the bundle of four is securely wedged in a polymer chock thus transmitting a friction grip of the rope core indirectly via the insulation. Where the cable incorporates cores of smaller cross-section these are not held by the clamp. For the suspension from poles or roof sup-

Table 14.2   Recommendations for installations

Type of proximity or crossing	Recommendation for installation
On pole	No gap necessary
In woodland or near single trees	No distance specified, mechanical damage must be avoided
From roof areas	Touching when swinging and under maximum sag to be prevented
From chimney stacks	Mechanical damage when swinging must be prevented, the distance above opening of chimney stack 2.5 m
From antenna and sirens	Touching when swinging also with maximum sag must be prevented
From accessible parts of buildings e. g. flat roofs	not less than 0.6 m
From bridges or similar	No distance specified, mechanical damage must be prevented
From telecommunication overhead lines: Bare wires / Aerial cables / Fixing points of telecommunication equipment	0.5 m / 0 5 m / 1.5 m } Distance at crossing above or below

Table 14.3
Current-carrying capacity of insulated overhead line cables

Cross-sectional area of aluminium conductor mm²	Current-carrying capacity A
25	107
35	132
50	165
70	205

Operating frequency up to 60 Hz
Wind speed 0.6 m/s
Ambient temperature 35 °C
Direct sunlight
Maximum conductor temperature 80 °C

ports, insulated suspension clamps are used which also accommodate a change of route direction of up to 30°. For branch points, insulated branch point clamps are used which allow connection without removing insulation from the cores. The electrical connection is made through a toothed contact plate.

When selecting suspension clamps, the maximum allowable rope tension (normally between 30 and 40 N/mm² of rope diameter) with a maximum support load not exceeding 6000 N must be observed.

The suspension spans are selected depending on the terrain to be between 300 to 500 meters. This span also depends on the height of the mast which may be up to 150 m, however for pole heights of 8 to 10 m the span is approximately 40 to 60 m.

15 High- and Extra-High-Voltage Cables

All cables are subjected to changes in load and therefore to temperature cycling during operation. The changes due to thermal expansion and contraction of both the conductors and insulation materials under the metal sheath, in mass-impregnated cables, produces small cavities (voids) in the insulation which when of a certain size start to produce partial discharge due to the influence of the dielectric field when this field is of a certain strength. At this stage not only the dielectric losses are increased but also where high voltages are concerned the service life of the mass-impregnated cable may be reduced. For this reason this type of cable to DIN VDE 0255 is permitted only for rated voltages of up to $U_0/U = 18/30$ kV. For higher rated voltages thermally stable cables with paper insulation (Table 15.1) or cables with polymer insulation must be used.

Fig. 15.1
High voltage cable with XLPE insulation
Type 2XS(FL)2Y 1×240 RM/35 64/110 kV

15.1 Cable with Polymer Insulation

In the past for rated voltages from $U_0/U = 36/60$ kV almost without exception thermally stable cables with

paper insulation were used but these are now increasingly superseded by cables with insulation of XLPE and in some countries also of EPR. The special advantage of the cables is that they are maintenance-free. The construction of the cables complies, except for dimensions, with the standard construction for rated voltages up to $U_0/U = 18/30$ kV. Based on re-

Table 15.1 Summary of type of construction and area of application for thermally stable cables with paper insulation

Cable with metal sheath			*Cable in steel pipe*		
Basic construction	Normally used rated voltages U kV	Standards	Basic construction	Normally used rated voltages U kV	Standards
Oil filled cables					
Low-pressure oil-filled cables with lead or aluminium sheath	60 to 380 ($U_m = 420$)	DIN VDE 0256, IEC 141-1	High-pressure oil-filled cable	110 to 380 ($U_m = 420$)	IEC 141-4
			Gas pressure cables		
			Internal gas pressure cables	110 to 150 ($U_m = 170$)	DIN VDE 0258, IEC 141-2
			External gas pressure cables		DIN VDE 0257, IEC 141-3

cent developments it is recommended, for the increase of safety of operation and in service life to build these high-voltage cables with protection ingres of moisture in both longitudinal and radial direction (Fig. 15.1).

For rated voltages up to $U_0/U = 64/110$ kV cables with polymer insulation are already in use to a large extent. The development of this cable for higher voltages is continuing.

15.2 Low-Pressure Oil-Filled Cable with Lead or Aluminium Sheath

For the oil-filled cables (Figs. 15.2 and 15.3) the paper insulation is impregnated with a thin oil. When heated the expanding oil can flow through longitudinal channels to oil expansion vessels which receive the oil under increasing pressure, conversely when the load is reduced and cooling occurs, the oil is forced back into the cable.

In oil-impregnated cables cavities can never occur. These cables are therefore insensitive to temperature cycles and are therefore thermally stable. The voltage withstand of oil-filled cables during operation is markedly higher than that of mass-impregnated cables (Fig. 15.4).

The voltage withstand is:

		For low-pressure oil filled cables	For mass-impregnated cables
Short-time withstand	kV/mm	50	50
Limiting continuous withstand	kV/mm	40	12 to 15
Working stress	kV/mm	7 to 14	max. 5

The insulation of oil-filled cables therefore only requires to be half the thickness of that of mass-impregnated cables for a given rated voltage. With the higher thermal stability of the oil-filled cable also a higher operating temperature can be used. Since, because of the reduced insulation thickness on oil-filled cables and thus the thermal resistance is less, these cables have a higher current carrying capacity by approximately 50% higher for a given cross-sectional area (Fig. 15.5).

The higher the operating voltage of a cable the more important becomes the dielectric loss factor $\tan\delta$. With mass-impregnated cable the loss factor varies considerably with variations in temperature. At the

Fig. 15.2 Oil-filled cable
NÖKUDEY 1×300 RM/V 12 H 64/110 kV

Fig. 15.3
Oil-filled cable NÖKDEFOY 3×150 SM 36/60 kV

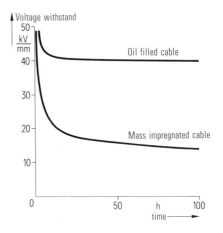

Fig. 15.4
Time-voltage withstand of oil-filled cable in comparison with mass-impregnated cable

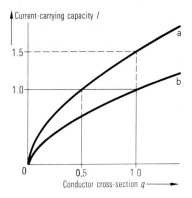

a oil-filled cable   b mass-impregnated cable

Fig. 15.5
Relationship of current-carrying capacity with respect to cross-sectional area

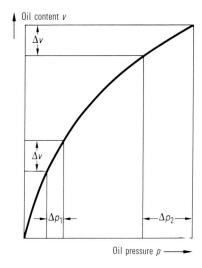

Fig. 15.7
Characteristic of an oil expansion vessel

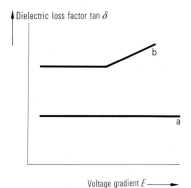

a oil-filled cable   b mass-impregnated cable

Fig. 15.6
Dielectric loss factor in respect to voltage gradient (typical relationship)

Fig. 15.8
Oil expansion vessel with cells

voltage at which the cavities start to ionise the loss factor often rises very considerably (ionisation knee). The dielectric loss factor $\tan \delta$ for oil-filled cables (Fig. 15.6) is considerably lower and is little influenced by temperature and voltage; formation of voids and ionisation do not occur. Oil-filled cables are, without having to increase the operating pressure, the only cable which can be used for the highest currently used a.c. operating voltage (up to $U_m = 420$ kV).

Oil-filled cables are manufactured as 3-core cables for cable rated voltage of $U = 60$ to 132 kV and as single-core cables up to the highest currently used operating voltages. The longitudinal channels for oil movement are achieved in 3-core cables by omitting the filler from the interstices between the laid-up cores and the metal sheath. The oil channels are therefore situated directly below the metal sheath which facilitates the connection of oil expansion vessels at any given straight joint, should this be necessary as on long cable runs (Figs. 15.7 and 15.8).

Single-core cables are provided with a longitudinal channel via a hollow conductor. Depending on the diameter of the oil channel, the viscosity of the oil used and the loading of the cable with double-sided feeding, cable runs of approximately 4000 m can be adequately supplied with oil. If only one end is connected to a vessel the relative length is halved.

For oil-filled cable runs where great differences in level occur and also for very long cable runs, sealed stop joints are fitted to divide the static pressures and also to allow the connection of expansion vessels along the cable run. With this system one achieves locked sections which with respect to oil content are completely independent of one another. For long cable runs, depending upon height differentials, the number of locked sections is reduced compared to those required for a level run.

The operating pressure of an oil-filled cable is normally between 1.5 and 6 bar. Since the strength of the lead sheath only permits low internal pressures, these cables have a pressure protection tape in the form of a helix wound directly over the lead. In 3-core oil-filled cables this tape is of steel but in single-core cables it is of non-magnetic material. For cables with aluminium sheath the tape is omitted.

Oil expansion vessels consist of cylindrical steel containers which contain oil-filled compressible cells. The cells are soldered air tight. All remaining space in the container is filled with oil. Depending on the expansion of the oil which results from load variations and seasonal temperatures of the ground, the number of vessels is calculated and thus the operating pressure is maintained within desired limits. Vessels of the same construction are built into the cable drum for the purpose of controlling the pressure inside the cable, within prescribed limits, during transport storage and installation thus catering for normal temperature fluctuations.

The operating pressure of an oil-filled cable, or any one of the several locking sections within an installation, is monitored by means of contact making manometers. If an excessive temperature rise occurs within the full length of the cable, the high-pressure contact is operated similarly and if damage or oil loss occures, the low pressure contact is operated. The functioning of the cable can thus be monitored from one central point and any fault can be signalled by either visual or audible alarm. Low-pressure oil-filled cables have the advantage over other high-voltage cable that this constant monitoring is possible at a relatively low operating pressure.

In case of small leakages, e.g. at pressure switches, metering links or valves any oil loss is replaced over a long period by the oil present in expansion vessels. Operation can therefore be continued until a suitable time occurs to make a repair.

For all voltage ranges outdoor sealing ends with porcelain insulators are available (Fig. 15.9).

Three-core oil-filled cables are sealed off with spreader joints. The individual cores are carried through corrugated flexible copper pipes to single-conductor sealing ends. The oil expansion vessels are connected to the splitter dividing box. With single-core oil-filled cables the pressure expansion vessels are connected to the individual sealing ends.

All low-pressure oil-filled cables can be connected via connecting sealing ends direct to transformers or switchgear (see page 394). Especially for the highest voltages the wide spaced through-bushings can be replaced by sealing ends showing particular advantage where space is limited, e.g. in caverns, enclosed switchgear etc.

Fig. 15.9
Outdoor sealing ends for single-core oil-filled cables

15.3 Thermally Stable Cable in Steel Pipe

Occasionally special requirements regarding mechanical strengths of cables are to be met. In areas subject to subsidence, e.g. movement of ground and also of the surface cables laid in the ground are subjected to pressure and tensile stresses. In long cable runs on bridges or scaffolding with long distances between support points it is necessary to have special mechanical protection or mechanical reinforcement. For these applications cables laid in steel pipe have advantages over other methods, furthermore it should be noted that steel pipe provides good screening where neighbouring control and telecommunication cables could otherwise be affected.

The pipe used to accommodate the cable can be installed independant of the cable installation. For this only parts of the underground cable run need to be accessible or opened at man holes otherwise the pipe can be sealed after being installed and tested. To cater for later extensions additional pipes can be laid in reserve.

The direction of cable runs in steel pipe must be planned in great detail such that sharp bands are avoided wherever possible. The installed length of cable cores is dependent upon the type of cable, the cross-sectional area of conductor and the type of terrain. Depending on circumstances every 300 to 800 meters jointing points are required and at these points a two to three meter length of larger diameter pipe is welded to the main pipe via steel adaptors. This construction makes it possible to reopen this joint at a later date and reclose it without endangering the cable or cutting the pipe. At the end of the cable pipe a spreader box is used from which the cable tails are lead to sealing ends; these cable tails being protected by non-magnetic flexible pipes. The steel pipe must have good corrosion protection because water tightness and mechanical withstand of the steel pipe are vital factors in this form of cable installation. The pipes are therefore protected by a layer of extruded PE covering.

It is possible to improve this corrosion protection by the use of "electric corrosion protection". To achieve this protection graphite electrodes are installed close to the pipe. Pipe and electrodes are connected to the output of a low-power rectifier set such that the graphite electrodes are anodes and the earthed steel pipe forms a cathode. Material transfer cannot occur.

The operating pressure and hence the air/oil tightness of the installation is monitored by contact manometer.

15.3.1 High-Pressure Oil-Filled Cable

The paper insulated cores which are screened with H foil are impregnated with low viscosity synthetic oil. A layer of copper tape in which the helix gap is closed with a plastic tape prevents the impregnation leaking and also prevents ingress of moisture during transportation and installation. Above this, a protection against damage during feeding into the pipes, a slow wound helix of non-magnetic gliding wire is added (Fig. 15.10).

On completion of final installation the pipe is then evacuated before filling with a low viscosity insulating oil and via a pressure regulating device, held under a working pressure of 16 bar. As the oil expands due to temperature rise caused by electrical load the excess flows into a storage container when a set operating pressure is reached. Conversely as the cable cools and oil pressure falls the required quantity is automatically returned, via a pump system, to maintain the set operating pressure. The vital component parts of the pressure regulating apparatus, such as pressure monitors, pumps and valves, are duplicated in the installation. Failure of any one item automatically initiates the switching in of the reserve item. The power supply for this system is normally taken from the network but a standby generator is also installed to cater for supply failure.

The operating pressure of 16 bar maintains the cable insulation void-free during any condition of opera-

Fig. 15.10
High pressure oil filled cable in steel pipe

tion. The a.c. voltage withstand is approximately 60 kV/mm and the peak voltage withstand 130 kV/mm.

These values are higher than those of low-pressure oil cables. Without sacrificing the safety in operation it is possible therefore to reduce the thickness of insulation. Because of this particularly economical type of construction cables can be produced for rated voltages of $U_0/U = 127/220$ kV and greater.

Fig. 15.11
Internal gas-pressure cable in steel pipe
NIvFSt 2Y 3×120 RM/V 64/110 kV

Fig. 15.12
Outdoor sealing ends with spreader box for internal gas-pressure cable in steel pipe

15.3.2 Internal Gas-Pressure Cable

The paper insulation over each core is impregnated with a non-migrating humidity proof mass. Because of thus no additional protective cover is necessary during transport and installation. This cable, in contrast to external gas-pressure cable does not have a sheath (Fig. 15.11). Above the outer layer of paper insulation copper tapes are applied overlapped with conducting paper which forms a screen for field limiting. Either individual non-laid-up cores each protected by a gliding wire or laid-up multi-core cables with flat steel-wire armour are fed into the steel pipe which, on completion, is then filled with Nitrogen. The operating pressure is set at 15 to 16 bar for rated voltages of $U/U_0 = 64/110$ kV. The gas can penetrate the insulation and fill all voids such that, even in the event of earth fault, ionisation is prevented. The voltage withstand of this cable core arrangement, the construction of which closely resembles that of mass-impregnated cable, is so improved by the gas pressure that the cable is suitable for higher operating voltage (see page 134).

15.3.3 External Gas-Pressure Cable (Pressure Cable)

The paper-insulated cores are each wrapped with H-foil and impregnated with a high viscosity synthetic oil. Above the foil is a lead sheath, which acts as a diaphragm, and this is strengthened by two layers of helically wound copper tape (Fig. 15.13). Above the laid-up cores is a flat steel-wire armour. After feeding cables into the pipe and the installation is completed, the pipe is filled with nitrogen at a pressure of 15 to 16 bar. The gas pressure allows the mass impregnation to expand under heat but, with the gas tight lead sheath acting as a membrane, forces it back to the original position when it cools. To ease this action of the sheath membrane the conductors are of oval cross-section instead of round.

Fig. 15.13
External gas-pressure cable in steel pipe
NPKDvFSt 2Y 3×240 OM/V 64/110 kV

Comprehensive solutions can be provided
quickly for complex planning tasks with the aid
of a data processing system

Planning of Cable Installations

16 Guide for Planning of Cable Installations

In planning an installation Table 16.1 may act as a guide.

The type of construction of the cable is to be selected to meet ambient conditions and to withstand the mechanical and thermal stresses. The installation requirements of both VDE and those of local regulatory authorities must be observed.

The short-circuit withstand of accessories must be checked accordingly. For the installation of sealing ends in either indoor or outdoor, the atmospheric conditions such as humidity, saline and dirt content as well as altitude above see level (if exceeding 1 000 m) are relevant. Special mechanical, chemical and moisture content of the soil are criteria to be considered when selecting cable joints.

Recommendations for transportation, installation and mounting methods can be found in Sections 29 and 30. Sections 17 to 25, of the part dealing with planning, contain instructions for the selection of rated voltage, conductor and screen cross-sectional areas and for the determination of key electrical data. Other considerations which may have an influence on planning are dealt with in Section 26 and distribution networks are dealt with in Section 27.

For help in the solution of special problems Siemens AG makes available their experience to assist in selecting the most suitable type of construction on technical and economic grounds together with the cross-sectional area of conductor. With the aid of special computer software, a solution can invariably be arrived at very quickly.

In the selection of a cable for a particular application the data listed in Table 16.2 (planning aid) are necessary. The more accurate and detailed this information the more accurate will be the result. The project engineer should have to rely as little as possible on pure assumptions or estimations.

Table 16.1 Guide for planning of cable installations

Action	Section
Selection of type of construction for cables and accessories	13
Consideration of conditions for transportation, installation and mounting	29, 30
Selection of cable rated voltage	17
Selection of conductor cross section to the following criteria when by the largest of the resulting values are to be used	
Current loading during normal operation	18
Fault current in case of short-circuit (mainly in networks with rated voltages greater than 1 kV)	19
Voltage drop (mainly in networks with rated voltages up to 1 kV)	24
Economics (Calculations appropriate for installations where large amounts of power are to be transmitted)	25
Electrical key data	20 to 23
Characteristics during operation	
Interference with communication cables	26
Industrial and urban networks	27

Table 16.2 Planning aid for cable installations.
For the selection of cable and determination of conductor and screen cross-sectional area the following data are necessary. To ease the handling of inquiries a check list is available on request.

1 Type of cable construction	1.1	Type designation
	1.2	Material for insulation (PVC, XLPE, mass-impregnated paper)
	1.3	Number of cores (single- or multi-core)
	1.4	Cross-sectional area of conductor q_n
	1.5	Conductor material (copper; aluminium)
2 Voltage	2.1	Nominal voltage of network U_n
	2.2	Maximum operating voltage $U_{b\,max}$
	2.3	System frequency f
	2.4	Type of current (3 ph, 1 ph, d.c.)
	2.5	Rated lightning impulse withstand voltage U_{rB}
3 Earthing conditions, treatment of star point (see Sections 17 and 19.1)	3.1	Insulated or with arc-suppression-coil earthed star point. If individual earth faults exceed 8 h and the total of all earth faults is greater than 125 h per year, the duration of the individual earth fault and duration of all earth faults per year must be stated.
	3.2	Direct earthing
	3.3	Earthing via additional impedance
4 Load capacity [1] in normal operation, operating conditions	4.1	Type of operation
	4.1.1	Load factor m, daily load fluctuation (in power supply systems approx. 0.7 to 0.8; in industrial networks 0.7 to 1.0). For intermittent operation a load diagram against time is required.
	4.1.2	Transmitted power (max. load to Fig. 18.1)
	4.1.3	Is a secure transmission essential (which means a minimum of two cables per connection)?
	4.2	Installation conditions
	4.2.1	Length of run in ground in pipe in ground in air (free air) in duct or tunnel
	4.2.2	Installation in ground depth of lay h cover of concrete tiles, plastic tiles, earthenware cover or laying in troughs with or without sand dimensions of troughs with drawings arrangement of single-core cables bundled or side by side dimensional drawings for massed group of cables

[1] Definition see page 150

Table 16.2   Continued

| 4 continued | 4.2.3 Installation in pipe in ground |
| | |

4 continued

4.2.3 Installation in pipe in ground
     depth of lay h
     pipe material PVC, PE, steel, concrete or earthen ware
     pipe diameter and thickness (of wall)

     Arrangement

     Diagram of groups of cables

4.2.4 Installation in air
     (e. g. indoors in large spaces such that the air temperature does not increase
     due to heat loss from the cables).
     Installation on the floor, wall, open duct or racking, dimensional drawing
     of groups (compare Tables 18.23 and 18.24).

4.2.5 Installation in covered channels, tunnels
     The air temperature in the channel is increased by heat loss from the
     cable
     Data of channel in line with Section 18.5:
     inside width b_T
     inside height h_T
     covering $\ddot{u}$
     dimensional drawing of overall arrangement and answer
     to questions in 4.3.1
     where forced ventilation is used the temperature of the outlet air or –
     for the calculation of the cooling required air quantity – the temperature
     of the ingoing air must be given (normally max. value of ambient tempera-
     ture).

4.3   Ambient conditions

4.3.1 Installation in ground
     ground temperature ϑ_E
     thermal resistivity of soil
        for moist area ϱ_E
        for dried-out area ϱ_x

4.3.2 Installation in air
     air temperature ϑ_U

4.4   External heat input

4.4.1 Heating by direct sunlight must be considered if sun protection is not pro-
     vided (see Section 18.4.2)

4.4.2 Heating by district heating pipes where laid in ground
     dimensional drawing to Fig. 16.1 and answers to questions in 4.3.1

4.4.3 Heating by other cables which run parallel or across
     type designation with data on cross-sectional area and rated voltage
     load current I_b
     distances and depths of lay with dimensional drawing

Table 16.2 Continued

5 Load-capacity in case of short-circuits (thermal and mechanical stress)	5.1	Calculation with the use of values from network calculation treatment of star point and indication of critical short-circuit currents (one, two or three pole) initial symmetrical short-circuit current I_k'' peak short-circuit current I_s continuous short-circuit current I_k'' short-circuit duration t_k
	5.2	Calculation with values from protective device (if values in 5.1 are not known) treatment of star point and indication of critical short-circuit current (one, two or three pole) breaking capacity S_a short-circuit duration t_k
6 Voltage drop		System frequency f Transmitted power S or loading current I_b Power factor $\cos \varphi$ Length of run l Type of current: 3 ph, 1 ph a.c. or d.c. Allowable voltage drop ΔU or Δu
7 Calculation of economy		Transmitted power S Length of run l Depreciation duration t Annual rate of interest p Amortization rate T Addition to amortization to cover maintainance and repair T_R Electricity price k_a Utilization time of power losses T_v Operation period T_B

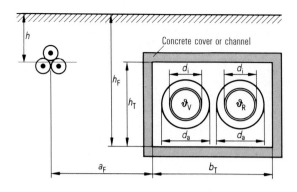

Cable

Type of construction
Depth of lay h … m

District heating duct

Depth of lay h_F … m
Distance a_F … m
Width b_T … m
Height h_T … m

	Feed pipe	Return pipe
Inner diameter of insulation	d_i … m	… m
Outer diameter of insulation	d_a … m	… m
Heat conductivity of insulation	λ … W/Km	… W/Km
Temperature of heating medium (as far as possible dependent on ambient temperature)	ϑ_V … °C	ϑ_R … °C

Fig. 16.1
Temperature rise of cable caused by district heating;
data for calculation

17 Cable Rated Voltages

17.1 Allocation of Cable Rated Voltages

The voltages for which a cable has been designed forms the basis of certain operating characteristics and test conditions and are termed the rated voltages.

As opposed to other electrical machinery or equipment cables have rated voltages stated as U_0/U where according to VDE:

U_0 the cable rated r.m.s. power-frequency voltage between each conductor and metallic cover or earth

U the cable rated r.m.s. power-frequency voltage between phase conductors in a three-phase network ($U = \sqrt{3}\, U_0$).

In IEC standards for cable an additional value for highest permissible voltage U_m is stated in brackets. The voltage designation is written as $U_0/U(U_m)$.

U_m is also the "highest r.m.s. power-frequency voltage for equipment" to DIN VDE 0101 and DIN VDE 0111a.

Cables with rated voltages U_0/U are according to DIN VDE 0298 and IEC 183 suitable for use on three-phase installation with a nominal voltage

$$U_n \leq U = \sqrt{3}\, U_0$$

in which the highest voltage of a system $U_{b\,max}$ does not exceed the values given in DIN VDE 0111 and IEC 71-1 (see Table 17.1).

Since the insulation of cables with polymer insulation having a rated voltage $U_0/U = 0.6/1$ kV and all radial field cables are designed for the voltage U_0, these are also suitable for installations

▷   in single-phase a.c. systems where both conductors are insulated from earth, with a system nominal voltage $U_n \leq 2\, U_0$,

▷   in single-phase a.c. systems where one conductor is earthed, with a system nominal voltage $U_n \leq U_0$.

The highest voltage of a system $U_{b\,max}$ for cables for single-phase alternating current are shown in Ta-

ble 17.1. These are derived from the values for three-phase installations by using the following formulae:

$$U_{b\,max} = 2\frac{U_m}{\sqrt{3}}, \text{ where neither conductor is earthed,}$$

$$U_{b\,max} = \frac{U_m}{\sqrt{3}}, \text{ where one conductor is earthed.}$$

To avoid confusion in installations having one conductor earthed e.g. in traction feed cables it must be observed that the highest voltage of system $U_{b\,max}$ for these cables must not be greater than the permissible voltage $\dfrac{U_m}{\sqrt{3}}$ to the metal cover.

In direct current systems having a maximum operating voltage of up to $U_{b\,max} = 1.8$ kV conductor/conductor and conductor/earth VDE permits the use of cable with $U_0 = 0.6$ kV.

In Germany the voltage rating of 3 kV has been made obsolete and is no longer included in VDE specifications. If in individual cases, e.g. in circuits for the starting of large slipring motors, operating voltages U_b are encountered which are higher than that allowed for cables with a rated voltage $U_0/U = 0.6/1$ kV it is acceptable according to DIN VDE 0271, to use PVC cables with concentric conductor or armour (e.g. NYCWY, NYFGY) having a cable rated voltage $U_0/U = 0.6/1$ kV. However with cable cross-sectional areas of 240 mm^2 and above, the insulation wall thickness is the same as is required to IEC 502 for cable with $U_0/U = 1.8/3$ kV. The permissible quantity of $U_m = 3.6$ kV must, however, not be exceed.

A comparison of cable rated voltages to VDE and IEC together with the permissible continuous "highest voltages for electrical equipment and machines" with relevant data to BS 77 shows that for these standards, apart from differing rated voltages the same highest permissible voltages apply (Table 17.2).

Table 17.1

Allocation of cable rated voltages $U_0/U(U_m)$ and highest voltages for equipment U_m to the nominal voltages U_n and the highest voltages of a system $U_{b\,max}$

Cable rated voltages $U_0/U\ (U_m)$	Systems for					
	Three-phase current		Single-phase current			
	Nominal voltage U_n	Highest voltage of a system $U_{b\,max}$	Non-earthed system		One conductor earthed	
			Nominal voltage $U_n \le 2\,U_0$	Highest voltage of a system $U_{b\,max}$	Nominal voltage $U_n \le U_0$	Highest voltage of a system $U_{b\,max}$
	kV	kV	kV	kV	kV	kV
0.6/1 (1.2)	1	1.2	1.2	1.4	0.6	0.7
1.8/3 (3.6)[1]	3	3.6				
3.6/6 (7.2)	6	7.2	7.2	8.3	3.6	4.2
6/10 (12)	10	12	12	14	6	7
8.7/15 (17,5)[1]	15	17.5				
12/20 (24)	20	24	24	28	12	14
18/30 (36)	30	36	36	42	18	21
26/45 (52)	45	52	Not applicable in these voltage ranges		Not applicable in these voltage ranges	
36/60 (72.5)	60	72.5				
64/110 (123)	110	123				
76/132 (145)	132	145				
87/150 (170)	150	170				
127/220 (245)	220	245				
220/380 (420)	380	420				
DIN VDE 0298 part 1 IEC 183 IEC 71-1	DIN VDE 0101 DIN VDE 0111		DIN VDE 0298 part 1			

[1] This voltage range to IEC 71-1 and IEC 183 is no longer contained in the VDE specifications for cable

17.2 Rated Lightning Impulse Withstand Voltage

The rated lightning impulse withstand voltages U_{rB} which must be considered for electrical equipment and machines in three-phase networks are listed in Table 17.2 as extracted from DIN VDE 0111 and IEC 71-1. Cables and accessories which comply with VDE or IEC standards are designed and tested to withstand these stresses. In the calculation of the insulation design for high-voltage cables the impulse withstand and switching overvoltages are very important and are simulated by a test impulse voltage wave form, e.g. 1.2/50 μs respectively 250/2 500 μs (1.2 and 250 being the wave front time with 50 and 2 500 being the decay time to half value).

17.3 Voltage Stresses in the Event of Earth Fault

In the event of an earth fault, the cable insulation is subjected to voltage stresses of shorter or longer duration dependant upon both the treatment of the star point and the design of the network protection (see also page 380). When applying cables different considerations must be given to the three types of stress A, B and C:

A Systems which in case of an earth fault are disconnected instantaneously i.e. within 1 s: these are mainly networks with a low-resistance earthed star point. For this stress type all cables are suitable.

Table 17.2 Allocation of voltages to VDE, IEC and British Standard (BS)

DIN VDE 0298, part 1 DIN VDE 0111, IEC 183		Three-phase networks to BS 77		DIN VDE 0111, list 2 DIN VDE 0298, part 1 IEC 71-1
Cable rated voltage U_0/U kV	Highest voltage for equipment U_m kV	Nominal voltage U kV	Highest voltage of a system $U_{b\,max}$ kV	Rated lightning impulse withstand voltage U_{rB} kV
Radial field cables				
0.6/1	1.2	–	–	–
1.8/3 [1]	3.6	3.3	3.6	40
3.6/6	7.2	6.6	7.2	60
6/10	12	11	12	75
8.7/15 [1]	17.5	–	–	95
12/20	24	22	24	125
18/30	36	33	36	170
26/45	52	–	–	At rated voltages from $U_0/U = 26/45$ kV the wall thickness of the insulation is selected and tested to meet the specified requirements
36/60	72.5	66	72.5	
–	–	88	100	
64/110	123	110	123	
76/132	145	132	145	
87/150	170	–	–	
127/220	245	220	245	
220/380	420	380	420	
Cables with non-radial field				
0.6/1	1.2	–	–	–
1.8/3 [1] 3/3 [2]	3.6 3.6	3.3	3.6	40
3.6/6 6/6 [2]	7.2 7.2	6.6	7.2	60
6/10 8.7/10 [2]	12 12	11	12	75

[1] In Germany no longer used and therefore not included in VDE standards
[2] Only for paper-insulated cables (e.g. to IEC 55). Not commonly used in Germany and hence not included in DIN VDE 0255

B Systems, which, under fault conditions, are operated for a short time only with one phase earthed: these are networks with an insulated star point or alternatively having earth fault compensation. According to IEC 183 this duration should not exceed one hour unless longer durations are specified in the relevant cable standards dependent upon the type of cable construction.

C Systems which in the event of a fault remain in operation for a longer period than described under B, with one phase earthed.

Table 17.3 Selection of medium-voltage cables according to stress types B and C under earth-fault conditions

Cable type	Cable rated voltage U_0/U kV	Operation with single earth fault to	
		Stress type B	Stress type C
		Single earth fault 8 h Sum of earth fault durations per year 125 h	Cable must be selected have rating voltage U_0/U kV
Non-radial field cables with PVC or EPR insulation	3.6/6	permissible	–
Belted cables with paper insulation	3.6/6	permissible	$6/6^{1)}$ or 6/10
	6/10	permissible	$8,7/10^{1)}$
	$8.7/15^{1)}$	permissible	–
Radial field cables with insulation of paper, PVC, PE, XLPE or EPR	3.6/6	permissible	6/10
	6/10	permissible	$8.7/15^{1)}$ or 12/20
	$8,7/15^{1)}$	permissible	12/20
	12/20	permissible	18/30
	18/30	permissible	Cables with correspondingly reinforced insulation are required (not covered by IEC- and VDE standards)

[1] No longer used in Germany and therefore no longer included in VDE standards

Selection of Cable

Stress type A: All cables are suitable
Stress type B and C: See Table 17.3

High-Voltage Cables with Paper Insulation
($U_0/U > 18/30$ kV),

which have been tested to VDE or IEC standards are suitable for stress type B providing any individual earth fault does not exceed a duration of approximately 8 h and the total sum of all earth-fault durations per year does not exceed 125 h approximately. These cables, however, are not designed for operation under stress type C. When it is required to install cables in a network or plant where longer earth-fault durations are to be expected, the cable insulation will require to be appropriately dimensioned and tested.

High-Voltage Cables with PE, XLPE
or EPR Insulation ($U_0/U > 18/30$ kV)

are normally dimensioned and tested for use in networks or plants with stress type A. If it is required that these cables will be operated for a limited time or longer with an earth fault on one phase, this must be taken into account when dimensioning and testing the cable.

Medium-Voltage Cables,

which comply with VDE or IEC standards are suitable for stress type B providing any individual earth-fault duration does not exceed approximately 8 h and the total sum of all earth-fault times in one year does not exceed approximately 125 h. If earth-fault durations are to exceed these values substantially, cables of the next highest voltage grade must be used (e.g. instead of $U_0/U = 6/10$ kV use $U_0/U = 12/20$ kV) or, in the case of belted cables, a cable with higher belt insulation must be used (e.g. instead of $U_0/U = 6/10$ kV use cable $U_0/U = 8.7/10$ kV) (see Table 17.3). This type of belted cable is not used in Germany and there for no provision is made for it in VDE standards. For cables having rated voltages greater than $U_0/U = 18/30$ kV the insulation wall thickness must be dimensioned appropriately.

For medium- and high-voltage cables it must be noted that their service life is affected if for frequent short periods and/or for longer periods the cables are operated with an earth fault on one phase.

Low-Voltage Cables,

which comply with the VDE and IEC standards are suitable for stress type C without limitation.

18 Current-Carrying Capacity in Normal Operation

18.1 Terms, Definitions and Regulations

Basically the terms definitions and regulations laid down in DIN VDE 0298 Part 2 and DIN VDE 0289 Part 8 apply.

Load Capacity

is the short term to express current-carrying capacity. With load capacity the permissible current I_z is being defined under certain operating conditions.

In addition to compliance with the above regulations the following is also relevant:

The values of current-carrying capacity for the reference operating conditions which are given in Tables 18.2 and 18.4 are rated values. These reference operating conditions (in DIN VDE 0298 Part 2 named as "normal" operating conditions) are in the same sense rated data to DIN IEC 50 (151).

The following equation applies

$$I_z = I_r \, \Pi f, \tag{18.1}$$

where Πf is the product of all factors which must be considered. For electricity utility operation or other cyclic types of operation the maximum load corresponds to load capacity which is defined as I_r or I_z.

Loading

is the short term for current loading. Loading relates to the currents which a cable be required to may carry under specific operational conditions.

In normal operation loading is the *operating current* I_b. In electricity utility operations or other cyclic types of operation the max. value of the loading is the operating current.

Permissible Operating Temperature

is the maximum permissible temperature at the conductor under normal operation. This value is used in the calculation of load capacity for normal operation. This is included in DIN VDE 0298 Part 2 in respect of load duration (load factor). For mass-impregnated cables, in addition, the temperature rise is limited to avoid the formation of voids in the insulation (Table 18.1).

Conductor Cross-Sectional Area

must be selected such that in normal operation the given loading I_b does not exceed the load capacity I_z

$$I_b \leq I_z. \tag{18.1a}$$

Decisive for this are the most unfavourable operating conditions at any point along the whole cable run during operation. This ensures that the conductor is not heated at any time and at any point above the permissible operating temperature.

Temperature Rise

of a cable is dependant upon construction, characteristics of materials used and operating conditions. An additional temperature rise must be considered where grouping with other cables or heat input from heating pipes, solar radiation etc. occurs.

Normal Operation

Normal operation includes all types of operation, such as, continuous operation, short-time operation, intermittent operation, cyclic operation, utility supply operation, providing the permissible operating temperature is not exceeded.

Overcurrents

include both overload currents and short-circuit currents (DIN VDE 0100 Part 430 and Part 200). These can cause, for a limited period, conductor temperatures which are higher than the permissible operating temperature. The cable in these cases must be protected against detremental temperature rise by overcurrent protection devices. If necessary the conductor cross-sectional area may have to be dimensioned to satisfy the conditions of short-circuit stresses as discussed in Section 19.3.

150

Table 18.1 Permissible operating temperatures and thermal resistivities

Type of construction	Standard	Permissible operating temperature	Permissible temperature rise installed in		Thermal resistivities of insulation
			Ground	Air	
		°C	K	K	Km/W
XLPE cable	DIN VDE 0272, DIN VDE 0273	90	–	–	3.5
PE cable	DIN VDE 0273	70	–	–	3.5 [1]
PVC cable	DIN VDE 0265, DIN VDE 0271	70	–	–	6.0 [2]
Mass-impregnated cable	DIN VDE 0255				
Belted cable					
0.6/1 kV		80	65	55	6.0
3.6/6 kV		80	65	55	6.0
6/10 kV		65	45	35	6.0
Single-core cable, S.L. and H cable					
0.6/1 kV		80	65	55	6.0
3.6/6 kV		80	65	55	6.0
6/10 kV		70	55	45	6.0
12/20 kV		65	45	35	6.0
18/30 kV		60	40	30	6,0

[1] Also applies for all outer sheaths of PE
[2] Also applies for all outer sheaths of PVC and protective covers of jute serving with bituminous compound

Overload currents can occur by operational overloading in what is otherwise a fault-free circuit. For these conditions permissible temperatures have not yet been defined. These will be dependent on both duration and frequency of the overload occurances; these again affect the heat deformation characteristics and accelerate ageing.

Short-circuit currents flow when a fault of neglegible impedance occurs between active conductors which in normal circumstances have different potentials. The permitted short-circuit temperatures are acceptable only for a duration of up to 5 seconds. In systems with an insulated neutral and in compensated networks, a line-to-earth short-circuit current is termed earth-fault current. Such earth-fault current cause voltage stresses in the fault-free conductors (see Section 17), to an extent that temperatures exceeding the permissible operating temperatures cannot be permitted.

Emergency Operation

is a type of operation quite common in USA and some other countries. Here currents are permitted which are higher than the load capacity in normal operation. The conductor "emergency operating temperature" which may on some occasions significantly exceed the permissible operating temperature are limited in duration for the individual faults both during any one year and during the service life of the cable. A definition and the question of what values of emergency operating temperature are acceptable for the different types of cable and also what reduction in service life is to be agreed is currently under discussion in the relevant IEC working groups.

151

Type of Operation

describes the temporal characteristics of the load capacity and the loading.

Continuous Operation

is an operation with constant current for a duration sufficient for the cable to reach a thermally stable condition but is otherwise not limited in time.

Utility Supply Operation

is described in Section 18.2.1.

Short Time and Intermittent Loading

is described in Section 18.6.

18.2 Operating Conditions and Design Tables

To assist in preparing a clear basis for design, regulatory and operating conditions are discussed under

▷ type of operation,

▷ conditions of installation,

▷ ambient conditions.

18.2.1 Operating Conditions for Installations in Ground

Type of Operation

The values included in the tables for installation in ground are based on the type of operation commonly experienced in electricity supply networks (supply utility loads). This load is defined by a 24 hour load diagram which illustrates maximum load and load factor (see Fig. 18.1).

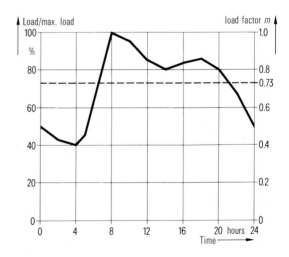

———— Relation of load to maximum load in %
– – – – Relation of average load to maximum load

Fig. 18.1
Daily load plot and determination of load factor *m*
(Example)

Maximum load and load factor of the given load are determined from the daily load plot or reference load plot. The daily load plot (24 hour load plot) is the shape of the load over 24 hours under normal operation. The reference load plot is the average load shape of selected, similar daily load plots.

The highest value of the maximum load read from the daily load plot is taken as operating current I_b. If the load fluctuates within time bands which are less than 15 minutes, then the mean value of the load peak over a 15 minutes period is taken as maximum load, i.e. a mean value must be determined over the range of time which contains the peak, this being then termed maximum load.

The load factor m is determined by plotting the load expressed as percent of maximum load on squared paper (see Fig. 18.1). The load factor m results in the total area below the curve which is equal to the area of the rectangular shape. By counting squares below the load curve the area can be determined reasonably accurately. This area should be entered on the diagram, thus enabling direct reading of the relationship between average load and maximum load and hence load factor m provided that, as in Fig. 18.1, the scale is selected such that 100% load is equal to unity on the load factor scale (see example 18.1, page 180).

The average load is the mean value of the daily load plot; the load factor being the quotient from the average load divided by the maximum load.

For this calculated load factor the given maximum load I_b must not exceed the load capacity I_z.

Installation Conditions

The *depth of lay* of a cable in ground is generally taken as 0.7 m which is the distance below the ground surface to the axis of the cable or the centre of a bunch of cables. If one calculates the load capacity of a cable laid in the ground it is found this reduces as depth increases, assuming the same temperature and soil-thermal-resistivity. With increasing depth of lay however, the ambient temperature is reduced and so, normally, is the soil-thermal-resistivity since the deeper regions of the ground are more moist and remain more consistant than the surface layers. For

the commonly used depth of lay for low-voltage and medium-voltage cables (0.7 to 1.2 m) it is therefore assumed that the necessary slight reduction in load capacity is compensated for by the slightly more favourable conditions.

For these reasons when the depth of lay varies within that range any variation in load capacity is ignored.

The quantities for cable load capacity are for the arrangements shown in Table 18.2 for one multi-core or one single-core cable in a d.c. system or for three single-core cables in a 3-phase system. With larger numbers of cables a reduction factor from Tables 18.15 to 18.21 must be applied. These reduction factors were derived for cables of equal size arranged side by side in one plane and loaded identically with the same maximum load and load factor. For cables of different constructions and/or operating with different load factor it is necessary to form appropriate reduction factors for each form of construction and/or load factor for the total number of cables in the trench and thus establish the factors most unfavourable for all cables.

Crossing of cable runs can cause difficulties especially when these are densely packed. At such points the cables must be laid with a sufficiently wide vertical and horizontal spacing. In addition to this the heat dissipation must be assisted by using the most favourable bedding material. A calculation of conductor heat output and temperature rise is advisable [18.1].

In situations of great grouping and where there is limited space, a sufficiently large bricked pit can elevate heat build-up. This pit can enable the cables to cross in air and the resultant temperature rise of the air in the pit and also the temperature rise of conductors can be calculated as indicated in Section 18.5.

The load capacity of multi-core PVC cables is calculated by multiplying the load capacity for 3-core cables in Table 18.5 by the rating factors for laying in the ground given in Table 18.25.

In the ground, cables are normally embedded in a layer of sand or a layer of sieved soil and are covered with either *bricks or tiles of concrete or plastic*. These bedding and covering arrangements (see Table 18.2) do not affect the load capacity. When inverted 'U'-shaped cover plates are installed, air may be trapped and therefore it is advisable to use a reduction factor of 0.9 in the case.

Table 18.2 Operating conditions, installation in ground

Reference operating conditions to evaluate the rated currents I_r	Site operating conditions[1] and calculation of current-carrying capacity $I_z = I_r \Pi f$
Type of operation	
Load factor of 0.7 and maximum load from tables for installation in ground	Rating factors f_1 to Table 18.15 or 18.16 f_2 to Table 18.17 to 18.21
Installation conditions	
Depth of lay 0.7 m	For depth of lay up to 1.2 m no conversion necessary

Arrangement:
1 multi-core cable
 	
1 single-core cable in d.c. system
 	
3 single-core cables in 3-phase system side by side with clearance of 7 cm
 	
3 single-core cables in 3-phase system bunched[2]
	Rating factors
	  for multi-core cables to Table 18.25
	  for grouping or bunched
	  f_1 to Table 18.15 or 18.16
	  f_2 to Table 18.17 to 18.21
	
	Calculation refer Section 18.4.4

Embedded in sand or soil backfill and if necessary with a cover of bricks, concrete plates or flat to slightly curved thin plastic plates	Rating factors for 'U'-shaped cover with trapped air $f = 0.9$ Installed in pipes $f = 0.85$ Calculation refer Section 18.4.6
Ambient conditions	
Ground temperature at installation depth 20 °C	Rating factors f_1 to Table 18.15 or 18.16 f_2 to Table 18.17 to 18.21 Calculation refer to Section 18.4.3
Soil-thermal resistivity of moist area 1 Km/W	
Soil-thermal resistivity of dry area 2.5 Km/W	
Protection from external heating e. g. from heating ducts	See Section 16, Table 16.2
Jointing and earthing of metal sheaths or screens at both ends (see Section 21)	

[1] Site operating conditions for installation in ground must always be calculated using the two rating factors f_1 and f_2 since both factors depend on the specific ground thermal resistivity and on the rating factor: $\Pi f = f_1 \cdot f_2$

[2] Cables touching in triangular formation are classed as "bunched"

When laying *cables in pipes* the heat insulation effect of the air layer between cable and pipe must be especially considered [18.2]. For installations in pipe systems a reduction of load capacity by a factor of 0.85 is recommended where an accurate calculation is not justifiable (see Section 18.4.5).

Ambient Conditions

The *ground temperature* ϑ_E is taken as the temperature at installation depth with the cable under no load conditions.

Figs. 18.2 and 18.3a indicate mean values of measured ground temperatures below a surface containing vegetation. The temperature at a depth of one meter below a concrete or asphalt surface which is subjected to solar radiation (Fig. 18.3b) may, during the summer months, achieve a level 5 °C higher than these measured values. Calculations with lower temperatures than 20 °C as given in the tables should not be made unless such a quantity is proved by measurements during the summer months. In desert areas the temperatures can be somewhat higher than those as shown in Fig. 18.4.

The soil-thermal-resistivity is largely dependant on density and water content of the relevant type of

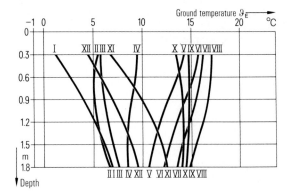

a) Below grass roots

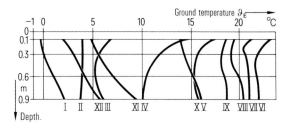

b) Below asphalt surface

Fig. 18.3
Ground temperature in Erlangen 1966 (months I to XII)

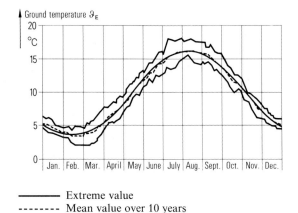

——— Extreme value

- - - - - Mean value over 10 years

Fig. 18.2
Ground temperature at a depth of 1 m, extreme values and mean value measured in Stuttgart-Hohenheim, 480 m above sea level, medium soil

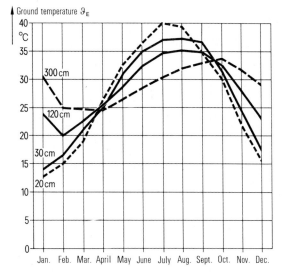

Fig. 18.4
Ground temperatures at various depths in Kuwait

soil. With differing types of soil and the effect of climatic conditions on water content (precipitation, ground temperature) the level of the water table as well as variations in cover of the surface and vegetation, both local and seasonal must be considered (see Fig. 18.5) [18.3].

Due to heat loss from the cable neighbouring cables and other heat dissipating items the soil may dry out. For the calculation of quantities in the tables and to simplify tabulation the region surrounding the cable has been distinguished between a moist area and a dry area.

The reference value of 1.0 Km/W was selected for the *soil-thermal resistivity* ϱ_E of the moist region. This quantity applies for normally sandy soil in a warm moderate climate (see DIN 50019) with a maximum ground temperature of 25 °C. Lower values are experienced in the colder seasons with sufficiently high precipitation and more favourable types of soil. Higher values must be selected for zones with higher ground temperatures, extensive dry periods or with almost zero precipitation. If detailed data are not available IEC 287 recommends quantities which should be used and these are reproduced in Table 18.3. Lower values of ambient can, where desired,

be used for the calculation of load carrying capacity for the winter period or during seasons of high rainfall.

Tables 18.15 to 18.21 provide rating factors for the individual soil-thermal resistivity of the moist region. In ground which has a content of rubble, slag, ash, organic material or waste etc. one must expect very much higher values of soil-thermal resistivity. In such instances it may be necessary to take measurements or to replace the soil in the vicinity of the cable.

For areas of built-up ground of normal types of soil which are not compacted and where increase of density is not to be expected for a considerable time the next higher value of soil-thermal resistivity from Tables 18.15 to 18.21 should be selected. The same applies where a cable run is situated in the rooting area of hedges or trees.

A *soil-thermal resistivity of 2.5 Km/W was selected for the dry region* taking into consideration that sand is frequently used as a bedding material. For certain types of soil or thermally stable bedding material with compacted dense soil lower quantities can be achieved. For individual cases quantities of resistivity and the resulting current-carrying capacity must be calculated separately (see Section 18.4.6).

Proximity to or the crossing of district heating lines often results in a dangerously high temperature rise in the cable, especially if the heating pipes are insufficiently insulated [18.4 to 18.5]. The continuous heat loss from the heating pipework can cause drying out of the soil. Because of this, sufficiently large clearances must be maintained between cables and pipes and also between cables.

District heating lines situated near to cables should be insulated on all sides. Minimum clearances which are given in [18.6] are estimated on the basis that the cable is loaded to approximately 60 to 70% of the load capacity and there is little grouping as is common practice in utility supply networks. At crossovers or at areas of parallel runs with district heating lines the current-carrying capacity will be reduced. Installing the cables in sufficiently large pits at these areas will increase their load capacity.

If insulation is arranged between the district heating lines and the cable, this is not fully effective and tends to reduce the heat dissipation of the cable.

To arrive at the measures required it is necessary to refer to the questions in Table 16.2, Section 4.4.2 as well as to Fig. 16.1.

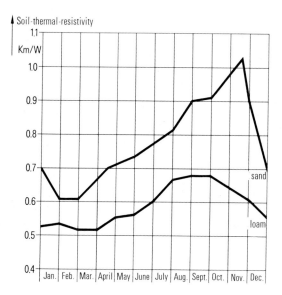

Fig. 18.5
Soil-thermal resistivity of virgin soil showing
seasonal variation
(measured at various locations) [18.5]

Table 18.3

Recommended calculation quantities to IEC 287 [18.2]

a) Ambient temperatures at sea level

Climate	Ambient temperature			
	of air		of ground at 1 m depth	
	Minimum °C	Maximum °C	Minimum °C	Maximum °C
tropical	25	55	25	40
subtropical	10	40	15	30
temperate	0	25	10	20

b) Soil-thermal resistivities

Weather conditions	Ground conditions	Soil-thermal resistivities Km/W
continuously moist	very moist	0.7
regular rainfall	moist	1.0
seldom rains	dry	2.0
Little or no rains	very dry	3.0

18.2.2 Operating Conditions, Installation in Air

Type of Operation

The quantities given in the table for installation in air apply for continuous operation. Because of the significantly shorter heating and cooling times compared with installations in ground in a public utility type of operation the highest load must not exceed the load capacity at continuous operation.

The load capacity in *intermittent operation* with shorter duty cycle times can be calculated by reference to Section 18.6.

Installation Conditions

The quantities one obtains for the rated load capacity I_r apply for the arrangements shown in Table 18.4 for multicore cables and for systems of three single-core cables installed in free air. The quantities are based on installation in free air with unhindered heat dissipation by radiation and convection and with the exclusion of external heat sources in an ambient air temperature which does not rise significantly. The requisite practical conditions for this are illustrated in Table 18.4. Rating factors for other installation conditions and for grouping of cables are given in Tables 18.23 and 18.24.

The load capacities of multicore PVC cables can be calculated by taking the quantities for three core cables from Table 18.6 and applying the rating factors from Table 18.25.

Where a cable is installed directly on a wall or on the floor, the load capacity must be reduced using a factor of 0.95. Factors for grouping are given in Tables 18.23 and 18.24. Where applicable in these tables the reduction factor of 0.95 for installation directly on a wall has already been taken into account.

The thermal resistance of the air in respect of a cable installed in free air can be calculated by reference to Section 18.4.2. By using the design data to Section 18.2 it is not necessary to know the air-thermal-resistance.

Ambient Conditions

The quantities given in the tables for installation in air are based on an air temperature of 30 °C. For other air temperatures the rating factors in Table 18.22 are to be used.

Table 18.4 Operating conditions, installed in air

Reference operating conditions to evaluate the rated current I_r	*Site operating conditions* and calculation of current-carrying capacity $I_z = I_r \Pi f$
Type of operation	
Continuous operation from tables for installation in air	Load capacity in intermittent operation to Section 18.6
Installation conditions	

Arrangement:
1 multi-core cable
 	
1 single-core d.c. cable
 	
3 single-core cables in 3-phase system side by side with clearance equal to cable dia.
 	
3 single-core cables in 3-phase system bunched [1]
	Rating factors for
	  multi-core cable to Table 18.25
	
	  grouping to Tables 18.23 and 18.24

Installation in free air i.e. unhindered heat dissipation by:
 	
  cable spaced away from wall, floor or ceiling ≥ 2 cm,
 	
  cables side by side with spacing minimum twice diameter,
 	
  cable runs above one another, vertical spacing twice cable diameter, minimum between layers of cables 30 cm	

Ambient conditions	

Air temperature 30 °C
 	
Sufficiently large and ventilated rooms in which the ambient temperature is not noticeably increased by losses from the cables
	Rating factors for
	  differing ambient temperatures to Table 18.21
	  grouping to Tables 18.23 and 18.24
	
	Load capacity when installed in channels or tunnels to Section 18.5

Protected from direct solar radiation etc.	Load capacity to Section 18.4.2

Jointing and earthing of metal sheath or screens at both ends (see Section 21)

[1] Cables touching in triangular formation are classed as "bunched"

Where it is necessary to assume an ambient air temperature for installing cables in air, provided that no higher values are known from experience or from measurement, the following are used:

Unheated cellar rooms	20 °C
Normal climate rooms (unheated in summer)	25 °C
Factory bays, work rooms etc.	30 °C

The above ambient temperatures are typical for mid European locations.

Temperatures exceeding the reference calculation quantity of 30 °C may well be experienced in rooms with inadequate protection from solar radiation, insufficient ventilation or rooms containing machines or plant having a high heat dissipation etc.

Under certain conditions the heat loss from cables may itself lead to an increase in ambient air temperature. This applies mainly to cable trenches, ducts, channels or tunnels (see Section 18.5).

If the air temperature in enclosed rooms is increased by the heat loss from the cables (e.g. in cable trenches, cable trays etc.) the rating factors in Table 18.22 for different air temperatures, together with the factors for grouping must be applied.

Other heat inputs, e.g. solar radiation, must be considered or prevented by the use of covers (see Section 18.4.2). If covers are used, however, the air circulation must not be hindered. A calculation of load capacity under conditions of solar radiation can be made by reference to Section 18.4.2.

18.2.3 Project Design Tables

The Tables 18.2 and 18.4 with the reference operating conditions and other differing conditions can be used as a guide for project design.

Tables 18.5 to 18.14 give quantities of load capacity of cables, i.e. rated currents I_r, based on specific operating conditions.

For conditions other than these specific operating conditions the rating factors for these are included in Tables 18.15 to 18.25.

Table 18.5 Load capacity, installed in *ground* $U_0/U = 0.6/1$ kV

	Mass-impregnated paper						PVC							XLPE			
Metal sheath	Lead			Aluminium			–					Lead		–			
Designation	N(A)KBA	N(A)KA		N(A)KLEY			N(A)YY N(A)YCWY			N(A)YY		NYKY		N(A)2XY			
Standard	DIN VDE 0255						DIN VDE 0271					DIN VDE 0265		DIN VDE 0272			
Permissible conductor temperature	80 °C						70 °C							90 °C			
Arrangement	1)			1)			2)		1)				1)	2)	1)		

Copper conductor nominal cross-sectional area mm² — Load capacity in A

mm²																	
1.5	–	–	–	–	–	–	40	32	26	–	–	31	27	48	30	32	39
2.5	–	–	–	–	–	–	54	42	34	–	–	41	35	63	40	43	51
4	–	–	–	–	–	–	70	54	44	–	–	54	46	82	52	55	66
6	–	–	–	–	–	–	90	68	56	–	–	68	58	103	64	68	82
10	–	–	–	–	–	–	122	90	75	–	–	92	78	137	86	90	109
16	–	–	–	–	–	–	160	116	98	107	127	121	101	177	111	115	139
25	133	147	172	135	146	169	206	–	128	137	163	153	131	229	143	149	179
35	161	175	205	162	174	200	249	–	157	165	195	187	162	275	173	178	213
50	191	207	241	192	206	234	296	–	185	195	230	222	192	327	205	211	251
70	235	254	294	237	251	282	365	–	228	239	282	272	236	402	252	259	307
95	281	303	350	284	299	331	438	–	275	287	336	328	283	482	303	310	366
120	320	345	395	324	339	367	499	–	313	326	382	375	323	550	346	352	416
150	361	387	441	364	379	402	561	–	353	366	428	419	362	618	390	396	465
185	410	437	494	411	426	443	637	–	399	414	483	475	409	701	441	449	526
240	474	507	567	475	488	488	743	–	464	481	561	550	474	819	511	521	610
300	533	571	631	533	544	529	843	–	524	542	632	–	533	931	580	587	689
400	602	654	711	603	610	571	986	–	600	624	730	–	603	1073	663	669	788
500	–	731	781	–	665	603	1125	–	–	698	823	–	–	1223	–	748	889

Aluminium conductor nominal cross-sectional area mm² — Load capacity in A

mm²																	
25	103	–	–	104	–	–	–	–	99	–	–	–	–	177	111	–	–
35	124	135	158	125	135	155	192	–	118	127	151	–	–	212	132	137	164
50	148	161	188	149	160	184	229	–	142	151	179	–	–	253	157	163	195
70	182	197	229	184	195	222	282	–	176	186	218	–	–	311	195	201	238
95	218	236	273	221	233	263	339	–	211	223	261	–	–	374	233	240	284
120	249	268	309	252	265	294	388	–	242	254	297	–	–	427	266	274	323
150	281	301	345	283	297	325	435	–	270	285	332	–	–	479	299	308	361
185	320	341	389	322	335	361	494	–	308	323	376	–	–	543	340	350	408
240	372	398	449	373	388	406	578	–	363	378	437	–	–	637	401	408	476
300	420	449	503	421	435	446	654	–	412	427	494	–	–	721	455	462	537
400	481	520	573	483	496	491	765	–	475	496	572	–	–	832	526	531	616
500	–	587	639	–	552	529	873	–	–	562	649	–	–	949	–	601	699

Table for rating factors	f_1	18.15						18.15							18.15			
	f_2	18.20	18.17 18.18	18.19	18.20	18.17 18.18	18.19	18.20	18.21	18.20	18.17 18.18	18.19	18.21	18.20	18.20	18.17 18.18	18.19

1) Cable in 3-phase operation
2) Load capacity in d.c. systems

Reference operating conditions and guide for site operating conditions see Table 18.2.

Table 18.6 Load capacity, installed in *air* $U_0/U = 0.6/1$ kV

Insulation material	Mass-impregnated paper						PVC [3]							XLPE			
Metal sheath	Lead			Aluminium			–					Lead		–			
Designation	N(A)KBA	N(A)KA		N(A)KLEY			N(A)YY N(A)YCWY			N(A)YY		NYKY		N(A)2XY			
Standard	DIN VDE 0255						DIN VDE 0271					DIN VDE 0265		DIN VDE 0272			
Permissible conductor temperature	80 °C						70 °C							90 °C			
Arrangement	1)			1)			2)		1)				1)	2)	1)		

Copper conductor nominal cross-sectional area mm² — Load capacity in A

mm²																	
1.5	–	–	–	–	–	–	26	20	18,5	20	25	20	18,5	32	24	25	32
2.5	–	–	–	–	–	–	35	27	25	27	34	27	25	43	32	34	42
4	–	–	–	–	–	–	46	37	34	37	45	37	34	57	42	44	56
6	–	–	–	–	–	–	58	48	43	48	57	48	43	72	53	57	71
10	–	–	–	–	–	–	79	66	60	66	78	66	60	99	73	77	96
16	–	–	–	–	–	–	105	89	80	89	103	89	80	131	96	102	128
25	114	138	167	114	136	163	140	118	106	118	137	118	106	177	130	139	173
35	140	168	203	139	166	199	174	145	131	145	169	145	131	218	160	170	212
50	169	203	246	168	200	239	212	176	159	176	206	176	159	266	195	208	258
70	212	255	310	213	251	299	269	224	202	224	261	224	202	338	247	265	328
95	259	312	378	262	306	361	331	271	244	271	321	271	244	416	305	326	404
120	299	364	439	304	354	412	386	314	282	314	374	314	282	487	355	381	471
150	343	415	500	350	403	463	442	461	324	361	428	361	324	559	407	438	541
185	397	479	575	402	462	522	511	412	371	412	494	412	371	648	469	507	626
240	467	570	678	474	545	594	612	484	436	484	590	484	436	779	551	606	749
300	533	654	772	542	619	657	707	–	481	549	678	–	492	902	638	697	864
400	611	783	912	628	726	734	859	–	560	657	817	–	563	1070	746	816	1018
500	–	893	1023	–	809	786	1000	–	749	940	–			1246	–	933	1173

Aluminium conductor nominal cross-sectional area mm² — Load capacity in A

mm²																	
25	89	–	–	88	–	–	128	91	83	–	–	–	–	137	100	–	–
35	108	130	157	107	128	154	145	113	102	113	131	–	–	168	122	131	163
50	131	157	191	130	155	186	176	138	124	138	160	–	–	206	147	161	200
70	165	198	240	166	195	234	224	174	158	174	202	–	–	262	189	205	254
95	201	243	294	203	238	284	271	210	190	210	249	–	–	323	232	253	313
120	233	283	343	237	277	328	314	244	221	244	291	–	–	377	270	296	366
150	267	323	390	272	316	370	361	281	252	281	333	–	–	433	308	341	420
185	310	374	450	314	363	421	412	320	289	320	384	–	–	502	357	395	486
240	366	447	535	372	432	489	484	378	339	378	460	–	–	605	435	475	585
300	420	515	613	428	494	548	548	–	377	433	530	–	–	699	501	548	675
400	488	623	733	503	589	627	666	–	444	523	642	–	–	830	592	647	798
500	–	718	833	–	669	687	776	–	–	603	744	–	–	966	–	749	926

Tables for rating factors	[4]	18.22						18.22							18.22			
	[5]	18.24	18.23	18.24		18.23		18.24			18.23		18.24		18.24		18.23	

[1] Cable in 3-phase operation
[2] Load capacity in d.c. systems
[3] Values up to 240 mm² harmonized in CENELEC
[4] for air temperature
[5] for grouping

Reference operating conditions and guide for site operating conditions see Table 18.4.

Table 18.7 Load capacity, installed in *ground* $\qquad$ $U_0/U = 3.6/6$ kV

Insulation material	Mass-impregnated paper							PVC		
Metal sheath	Lead				Aluminium			–		
Designation	N(A) KBA	N(A) EKBA	N(A)KA		N(A)KLEY			N(A)YFGY [1] N(A)YSY [2]		
Standard	DIN VDE 0255							DIN VDE 0271		
Permissible conductor temperature	80 °C							70 °C		
Arrangement	⊙	⊙	⊙⊙⊙ (trefoil)	⊙⊙⊙	⊙	⊙⊙⊙ (trefoil)	⊙⊙⊙	⊙	⊙⊙⊙ (trefoil)	⊙⊙⊙
Copper conductor nominal cross-sectional area (mm²)	Load capacity in A									
25	133	140	147	170	134	146	167	126	140	159
35	161	167	175	202	162	174	197	158	167	190
50	190	198	207	239	192	206	231	187	198	223
70	234	243	254	291	237	252	279	230	242	272
95	281	291	304	347	284	300	326	275	289	323
120	321	332	345	392	323	339	364	313	328	364
150	362	374	387	437	363	379	400	352	366	396
185	409	422	438	492	410	425	437	397	413	443
240	474	490	508	563	474	488	487	460	478	505
300	532	550	571	629	532	541	522	518	536	560
400	601	631	655	709	600	607	564	587	605	610
500	–	705	732	780	–	666	603	–	–	–
Aluminium conductor nominal cross-sectional area (mm²)	Load capacity in A									
25	103	108	–	–	–	–	–	–	–	–
35	124	129	135	156	125	135	154	122	129	147
50	147	154	161	185	149	160	182	145	154	174
70	182	189	197	226	184	196	220	178	188	213
95	218	226	236	270	221	234	260	214	225	254
120	250	256	268	307	251	265	292	243	256	287
150	281	291	301	343	283	297	323	274	286	316
185	320	329	341	386	321	335	337	310	324	355
240	372	384	398	447	373	388	405	361	377	409
300	419	432	449	501	420	434	441	408	425	457
400	481	503	520	572	481	495	487	468	488	509
500	–	570	588	638	–	552	529	–	–	–
Tables for rating factors f_1	18.15							18.15		
f_2	18.20		18.17 18.18	18.19	18.20	18.17 18.18	18.19	18.20	18.17 18.18	18.19

[1] three core $\qquad$ [2] single core

Reference operating conditions and guide for site operating conditions see Table 18.2.

Table 18.8 Load capacity, installed in *air* $U_0/U = 3.6/6$ kV

Insulation material	Mass-impregnated paper							PVC		
Metal sheath	Lead				Aluminium			–		
Designation	N(A)KBA	N(A)EKBA	N(A)KA		N(A)KLEY			N(A)YFGY [1] N(A)YSY [2]		
Standard	DIN VDE 0255							DIN VDE 0271		
Permissible conductor temperature	80 °C							70 °C		
Arrangement	⊙	⊙	⚛	⊙⊙⊙	⊙	⚛	⊙⊙⊙	⊙	⚛	⊙⊙⊙
Copper conductor nominal cross-sectional area (mm²)	Load capacity in A									
25	115	125	139	164	116	138	161	105	122	143
35	142	152	170	200	142	167	196	131	147	174
50	169	182	204	242	171	201	236	157	178	210
70	212	227	257	305	216	253	297	197	222	263
95	259	276	315	373	264	308	355	241	271	321
120	301	320	364	432	305	356	406	277	312	370
150	344	364	417	492	349	404	456	316	354	413
185	394	415	479	565	400	463	512	362	406	472
240	465	491	570	669	473	545	588	427	480	553
300	527	554	654	763	539	617	645	487	547	625
400	608	653	781	900	622	723	722	565	643	711
500	–	740	892	1016	–	808	783	–	–	–
Aluminium conductor nominal cross-sectional area (mm²)	Load capacity in A									
25	89	97	–	–	–	–	–	–	–	–
35	109	117	131	155	109	129	152	101	114	135
50	131	141	158	188	133	157	184	122	138	164
70	165	176	199	236	167	197	231	153	173	205
95	201	214	244	290	205	240	280	187	210	251
120	234	249	283	337	237	278	323	215	244	290
150	268	283	324	385	272	317	365	246	277	327
185	308	324	373	443	313	364	414	283	318	375
240	365	384	447	529	372	432	483	335	379	444
300	415	436	514	605	425	494	539	384	434	505
400	485	520	619	723	498	587	618	450	517	587
500	–	597	717	828	–	668	684	–	–	–
Tables for rating factors [3]	18.22							18.22		
Tables for rating factors [4]	18.24		18.23		18.24		18.23	18.24		18.23

[1] three core   [2] single core   [3] for air temperature   [4] for grouping

Reference operating conditions and guide for site operating conditions see Table 18.4.

Table 18.9 Load capacity, installed in *ground* $U_0/U = 6/10$ kV

	(1)	(2)	(3)	(4)	(5)	(6)	(7)	(8)	(9)	(10)	(11)	(12)	(13)	(14)	(15)	(16)	(17)
Insulation material	Mass-impregnated papers								PVC			PE			XLPE		
Metal sheath	Lead					Aluminium			–			–			–		
Designation	N(A)KBA	N(A)HKBA	N(A)EKBA	N(A)KA	N(A)KA	N(A)KLEY	N(A)KLEY	N(A)KLEY	N(A)YSEY [1] N(A)YHSY [2]			N(A)2YSY			N(A)2XSY N(A)2XS2Y		
Standard	DIN VDE 0255								DIN VDE 0271			DIN VDE 0273			DIN VDE 0273		
Permissible conductor temperature	65 °C	70 °C				65 °C	70 °C		70 °C			70 °C			90 °C		
Arrangement	⊙	⊙	⊙	trefoil	○○○	⊙	trefoil	○○○	⊙	trefoil	○○○	⊙	trefoil	○○○	⊙	trefoil	○○○
Copper conductor nominal cross-sectional area mm²	Load capacity in A																
25	117	132	133	142	162	121	141	159	133	138	155	–	146	166	–	157	179
35	143	158	159	169	194	149	168	189	160	164	185	166	174	197	178	187	212
50	171	188	189	200	229	178	198	221	189	193	217	195	205	231	210	220	249
70	212	231	233	245	279	220	242	266	230	236	264	238	251	281	256	269	303
95	257	278	281	293	332	266	288	312	275	281	313	286	299	333	307	321	358
120	293	315	321	333	376	304	326	347	312	318	353	325	339	375	349	364	404
150	332	354	360	373	419	341	364	379	350	354	384	364	377	408	392	405	441
185	377	399	407	422	470	358	409	416	394	399	429	412	425	455	443	457	493
240	437	460	471	489	539	444	469	464	455	460	490	477	490	519	513	528	563
300	493	516	530	549	599	498	519	497	512	515	543	–	549	575	–	593	626
400	561	582	608	630	674	561	580	536	584	579	590	–	614	618	–	665	676
500	–	–	678	703	744	–	632	568	–	–	–	–	682	678	–	739	743
Aluminium conductor nominal cross-sectional area mm²	Load capacity in A																
25	91	102	103	–	–	94	–	–	–	–	–	–	–	–	–	–	–
35	110	122	123	130	150	115	130	147	123	127	143	–	135	153	–	144	164
50	132	146	147	155	178	138	154	174	146	150	169	151	159	181	162	171	194
70	165	180	181	190	217	171	188	211	179	183	207	185	195	220	199	209	236
95	200	216	218	227	259	207	225	249	213	219	246	222	232	261	238	249	281
120	229	246	250	259	294	237	256	279	243	248	278	252	264	296	271	283	318
150	259	276	280	290	329	266	286	308	272	277	306	283	294	325	304	316	350
185	295	313	318	329	370	302	322	342	307	312	343	321	333	365	345	358	393
240	343	362	370	384	428	350	373	387	356	363	395	373	387	420	401	416	453
300	389	408	417	433	479	395	417	421	402	408	441	–	435	468	–	469	507
400	449	466	485	501	546	451	474	464	464	465	490	–	493	514	–	532	559
500	–	–	548	566	610	–	526	501	–	–	–	–	555	572	–	599	622
Tables for rating factors f_1	18.15								18.16			18.15			18.15		
f_2	18.21	18.21	18.20	18.17 18.18	18.19	18.21	18.17 18.18	18.19	18.21	18.17 18.18	18.19	18.20	18.17 18.18	18.19	18.20	18.17 18.18	18.19

[1] three core
[2] single core

Reference operating conditions and guide for site operating conditions see Table 18.2.

Table 18.10 Load capacity, installed in *air* $U_0/U = 6/10\,\text{kV}$

Insulation material	Mass-impregnated paper								PVC			PE			XLPE		
Metal sheath	Lead					Aluminium			–			–			–		
Designation	N(A)KBA	N(A)HKBA	N(A)EKBA			N(A)KLEY			N(A)YSEY[1] N(A)YHSY[2]			N(A)2YSY			N(A)2XSY N(A)2XS2Y		
Standard	DIN VDE 0255								DIN VDE 0271			DIN VDE 0273			DIN VDE 0273		
Permissible conductor temperature	65 °C	70 °C				65 °C	70 °C		70 °C			70 °C			90 °C		
Arrangement	⊙	⊙	⊙	❁	⊙⊙⊙	⊙	❁	⊙⊙⊙	⊙	❁	⊙⊙⊙	⊙	❁	⊙⊙⊙	⊙	❁	⊙⊙⊙

Copper conductor nominal cross-sectional area mm² — Load capacity in A

mm²																	
25	99	112	114	126	147	99	124	145	114	120	140	–	133	158	–	162	191
35	120	135	138	153	179	121	151	175	138	145	170	143	161	190	173	195	231
50	144	161	165	184	216	146	181	211	165	174	205	170	192	228	206	234	277
70	181	200	205	231	272	183	228	262	204	217	256	212	240	284	257	292	345
95	221	245	251	282	332	223	277	316	247	264	311	258	291	344	313	354	418
120	254	281	289	327	385	257	319	361	284	304	359	297	335	396	360	407	481
150	290	320	328	373	438	294	362	403	322	343	401	338	378	440	410	460	537
185	332	365	375	430	502	336	414	453	367	393	457	386	432	500	469	527	612
240	389	425	440	510	593	395	487	520	430	464	536	455	509	585	553	621	716
300	442	484	501	584	675	450	550	569	490	528	607	–	579	660	–	709	811
400	509	555	589	696	793	518	641	635	574	619	690	–	665	728	–	815	901
500	–	–	665	791	893	–	714	682	–	–	–	–	750	810	–	921	1006

Aluminium conductor nominal cross-sectional area mm² — Load capacity in A

mm²																	
25	76	87	89	–	–	77	–	–	–	–	–	–	–	–	–	–	–
35	93	104	106	118	138	94	117	136	106	112	132	–	124	148	–	151	178
50	112	125	128	143	168	113	141	164	128	135	159	132	149	178	160	181	215
70	140	156	160	179	211	142	177	206	158	168	200	165	186	222	199	226	269
95	172	190	195	219	258	173	216	250	192	205	243	200	226	269	242	275	327
120	198	219	225	255	300	201	249	287	221	237	281	231	261	310	280	317	377
150	226	249	256	291	342	229	284	324	250	268	316	262	295	348	318	359	424
185	260	286	293	335	394	263	326	367	286	307	363	301	338	398	365	412	485
240	305	334	345	400	469	311	386	428	336	365	429	356	401	469	431	489	573
300	349	382	394	460	536	356	441	476	385	418	488	–	459	534	–	559	652
400	407	444	470	553	639	416	522	545	456	496	568	–	533	603	–	651	741
500	–	–	537	637	729	–	592	599	–	–	–	–	609	680	–	744	838

| Tables for rating factors | [3] | 18.22 | | | | | | | | 18.22 | | | 18.22 | | | 18.22 | | |
| | [4] | 18.24 | | | 18.23 | | 18.24 | | 18.23 | 18.24 | | 18.23 | 18.24 | | 18.23 | 18.24 | | 18.23 |

[1] three core
[2] single core
[3] for air temperature
[4] for grouping

Reference operating conditions and guide for site operating conditions see Table 18.4.

Table 18.11 Load capacity, installed in *ground* $U_0/U = 12/20\ \text{kV}$

Insulation material	Mass-impregnated paper						PE		XLPE	
Metal sheath	Lead				Aluminium		–		–	
Designation	N(A)HKBA	N(A)EKBA	N(A)KA		N(A)KLEY		N(A)2YSY		N(A)2XSY N(A)2XS2Y	
Standard	DIN VDE 0255						DIN VDE 0273		DIN VDE 0273	
Permissible conductor temperature	65 °C						70 °C		90 °C	
Arrangement	⊙	⊙	⧉	⊙⊙⊙	⧉	⊙⊙⊙	⧉	⊙⊙⊙	⧉	⊙⊙⊙
Copper conductor nominal cross-sectional area (mm²)	Load capacity in A									
25	123	126	139	153	138	149	–	–	–	–
35	148	151	166	184	165	179	176	198	189	213
50	175	180	196	219	194	212	208	233	223	250
70	220	222	240	269	237	256	254	283	273	304
95	264	268	287	321	282	300	302	335	325	361
120	298	304	327	363	319	334	343	378	368	407
150	336	343	366	404	355	364	381	412	410	445
185	380	388	414	454	399	400	430	460	463	498
240	440	453	479	519	456	445	496	525	534	569
300	496	511	539	578	505	478	556	583	601	633
400	559	591	618	650	563	520	623	628	674	686
500	–	661	689	713	615	556	692	689	750	756
Aluminium conductor nominal cross-sectional area (mm²)	Load capacity in A									
25	95	97	–	–	–	–	–	–	–	–
35	114	117	128	142	127	139	–	–	–	–
50	136	140	152	170	151	166	161	181	173	195
70	171	173	186	210	185	203	197	221	211	237
95	205	208	223	250	221	240	235	263	252	282
120	233	237	254	284	250	270	267	297	287	320
150	262	267	285	317	280	297	298	327	320	353
185	298	304	323	358	315	329	337	369	362	396
240	346	355	377	414	365	373	391	423	421	457
300	391	403	425	463	407	406	440	473	474	511
400	448	471	491	529	462	450	499	521	538	566
500	–	534	555	588	513	489	562	579	606	630
Tables for rating factors f_1	18.15						18.15		18.15	
f_2	18.21		18.17 18.18	18.19	18.17 18.18	18.19	18.17 18.18	18.19	18.17 18.18	18.19

Reference operating conditions and guide for site operating conditions see Table 18.2.

Table 18.12 Load capacity, installed in *air* $U_0/U = 12/20$ kV

Insulation material	Mass-impregnated paper						PE		XLPE	
Metal sheath	Blei			Aluminium			–		–	
Designation	N(A) HKBA	N(A) EKBA	N(A)KA		N(A)KLEY		N(A)2YSY		N(A)2XSY N(A)2XS2Y	
Standard	DIN VDE 0255						DIN VDE 0273		DIN VDE 0273	
Permissible conductor temperature	65 °C						70 °C		90 °C	
Arrangement	⊙	⊙	⊛	⊙⊙⊙	⊛	⊙⊙⊙	⊛	⊙⊙⊙	⊛	⊙⊙⊙

Copper conductor nominal cross-sectional area (mm^2)	Load capacity in A									
25	106	109	119	136	119	134	–	–	–	–
35	128	132	145	165	143	162	164	193	199	233
50	153	158	175	199	172	194	197	230	238	279
70	192	196	218	249	215	240	244	287	296	347
95	232	238	266	304	261	289	295	347	358	420
120	264	272	308	351	300	329	340	398	412	483
150	299	309	350	398	339	366	383	444	466	540
185	340	352	401	456	387	411	438	504	532	614
240	397	414	476	536	453	470	515	589	627	718
300	449	471	543	608	510	515	586	665	715	813
400	513	552	645	714	592	577	671	734	819	904
500	–	623	733	799	661	627	757	817	927	1011

Aluminium conductor nominal cross-sectional area (mm^2)	Load capacity in A									
25	82	85	–	–	–	–	–	–	–	–
35	99	102	112	128	111	126	–	–	–	–
50	119	123	136	155	134	152	152	179	184	217
70	150	153	170	194	167	189	189	223	229	270
95	180	185	207	237	204	229	230	271	278	328
120	206	212	239	274	235	262	265	312	320	378
150	233	241	273	312	266	295	299	351	363	425
185	266	275	313	358	305	334	342	400	415	485
240	312	325	374	425	360	388	406	471	493	573
300	355	371	427	484	410	432	463	535	563	652
400	411	440	512	575	483	495	536	604	652	740
500	–	503	589	654	546	547	612	683	746	838

Tables for rating factors	1)	18.22						18.22		18.22	
	2)	18.24		18.23				18.23		18.23	

1) for air temperature   2) for grouping

Reference operating conditions and guide for site operating conditions see Table 18.4.

Table 18.13 Load capacity, installed in *ground* $U_0/U = 18/30\ \text{kV}$

Insulation material	Mass-impregnated paper						PE		XLPE	
Metal sheath	Lead			Aluminium			–		–	
Designation	N(A) HKBA	N(A) EKBA	N(A)KA		N(A)KLEY		N(A)2YSY		N(A)2XSY N(A)2XS2Y	
Standard	DIN VDE 0255						DIN VDE 0273		DIN VDE 0273	
Permissible conductor temperature	60 °C						70 °C		90 °C	
Arrangement	⊙	⊙	⊗	⊙⊙⊙	⊗	⊙⊙⊙	⊗	⊙⊙⊙	⊗	⊙⊙⊙

Copper conductor nominal cross-sectional area mm²	Load capacity in A									
35	138	142	156	168	155	164	–	–	–	–
50	164	169	187	202	185	196	210	234	226	251
70	207	209	232	250	229	240	257	284	276	306
95	247	252	280	301	274	284	306	337	329	363
120	281	287	319	343	312	319	347	381	373	410
150	316	324	358	385	347	353	386	416	415	449
185	356	367	404	435	388	386	435	465	468	503
240	411	428	468	501	443	430	503	532	541	576
300	462	483	526	557	490	463	564	590	608	641
400	521	558	603	627	546	505	632	638	684	697
500	–	623	672	686	594	541	703	702	762	768

Aluminium conductor nominal cross-sectional area mm²	Load capacity in A									
35	107	110	121	130	120	128	–	–	–	–
50	127	131	145	157	144	154	163	182	175	196
70	161	163	180	195	178	191	199	222	214	238
95	193	196	217	235	215	226	238	264	256	284
120	219	224	249	268	245	256	270	299	290	322
150	247	252	279	302	274	285	302	330	324	355
185	279	287	316	343	308	319	341	371	366	400
240	325	336	368	399	355	361	396	427	426	461
300	366	380	415	448	396	394	446	477	479	516
400	419	445	480	510	449	438	505	527	545	572
500	–	504	541	567	498	476	569	587	614	638

Tables for rating factors	f_1	18.15						18.15		18.15	
	f_2	18.22		18.17 18.18	18.19	18.17 18.18	18.19	18.17 18.18	18.19	18.17 18.18	18.19

Reference operating conditions and guide for site operating conditions see Table 18.2

Table 18.14 Load Capacity, installed in *air* $U_0/U = 18/30$ kV

Insulation material	Mass-impregnated paper						PE		XLPE	
Metal sheath	Lead				Aluminium		–		–	
Designation	N(A) HKBA	N(A) EKBA	N(A)KA		N(A)KLEY		N(A)2YSY		N(A)2XSY N(A)2XS2Y	
Standard	DIN VDE 0255						DIN VDE 0273		DIN VDE 0273	
Permissible conductor temperature	60 °C						70 °C		90 °C	
Arrangement	⊙	⊙	⊛	⊙⊙⊙	⊛	⊙⊙⊙	⊛	⊙⊙⊙	⊛	⊙⊙⊙

Copper conductor nominal cross-sectional area mm²	Load capacity in A									
35	119	124	135	150	134	148	–	–	–	–
50	142	147	162	182	161	177	199	232	241	279
70	179	183	203	228	200	220	248	288	299	348
95	216	221	246	277	242	263	300	348	362	421
120	246	254	284	318	278	299	344	400	416	483
150	278	288	323	361	314	334	388	446	469	540
185	315	328	370	413	357	373	442	507	536	615
240	366	385	437	484	417	425	520	590	630	718
300	414	437	499	548	469	467	590	666	717	812
400	470	512	591	640	543	526	675	737	823	904
500	–	576	670	716	603	572	763	821	929	1011

Aluminium conductor nominal cross-sectional area mm²	Load capacity in A									
35	92	95	105	116	104	115	–	–	–	–
50	110	114	126	141	125	139	155	180	187	217
70	140	142	157	177	156	173	192	224	232	270
95	168	172	191	216	189	209	233	272	281	328
120	192	198	221	249	218	239	268	313	323	378
150	217	224	252	283	247	269	302	351	365	425
185	247	257	289	324	282	303	346	401	418	485
240	289	302	343	384	332	351	408	471	494	572
300	328	344	393	437	377	392	465	535	564	649
400	378	408	469	517	443	450	538	605	654	737
500	–	466	538	588	501	499	615	683	747	835

Tables for rating factors	[1]	18.22						18.22		18.22	
	[2]	18.24		18.23					18.23		18.23

[1] for air temperature   [2] for grouping

Reference operating conditions and guide for site operating conditions see Table 18.4

Table 18.15 Rating factor f_1 for installation in ground (*not* applicable to PVC cables with $U_0/U = 6/10$ kV)

Permissible conductor temperature °C	Ground temperature °C	Soil-thermal resistivity															2.5 Km/W
		0.7 Km/W					1.0 Km/W					1.5 Km/W					
		Load factor					Load factor					Load factor					Load factor
°C	°C	0.50	0.60	0.70	0.85	1.00	0.50	0.60	0.70	0.85	1.00	0.50	0.60	0.70	0.85	1.00	0.5 to 1.00
90	5	1.24	1.21	1.18	1.13	1.07	1.11	1.09	1.07	1.03	1.00	0.99	0.98	0.97	0.96	0.94	0.89
	10	1.23	1.19	1.16	1.11	1.05	1.09	1.07	1.05	1.01	0.98	0.97	0.96	0.95	0.93	0.91	0.86
	15	1.21	1.17	1.14	1.08	1.03	1.07	1.05	1.02	0.99	0.95	0.95	0.93	0.92	0.91	0.89	0.84
	20	1.19	1.15	1.12	1.06	1.00	1.05	1.02	1.00	0.96	0.93	0.92	0.91	0.90	0.88	0.86	0.81
	25						1.02	1.00	0.98	0.94	0.90	0.90	0.88	0.87	0.85	0.84	0.78
	30								0.95	0.91	0.88	0.87	0.86	0.84	0.83	0.81	0.75
	35													0.82	0.80	0.78	0.72
	40																0.68
80	5	1.27	1.23	1.20	1.14	1.08	1.12	1.10	1.07	1.04	1.00	0.99	0.98	0.97	0.95	0.93	0.88
	10	1.25	1.21	1.17	1.12	1.06	1.10	1.07	1.05	1.01	0.97	0.97	0.95	0.94	0.92	0.91	0.85
	15	1.23	1.19	1.15	1.09	1.03	1.07	1.05	1.03	0.99	0.95	0.94	0.93	0.92	0.90	0.88	0.82
	20	1.20	1.17	1.13	1.07	1.01	1.05	1.03	1.00	0.96	0.92	0.91	0.90	0.89	0.87	0.85	0.78
	25						1.03	1.00	0.97	0.93	0.89	0.88	0.87	0.86	0.84	0.82	0.75
	30								0.95	0.91	0.86	0.85	0.84	0.83	0.81	0.78	0.72
	35													0.80	0.77	0.75	0.68
	40																0.64
70	5	1.29	1.26	1.22	1.15	1.09	1.13	1.11	1.08	1.04	1.00	0.99	0.98	0.97	0.95	0.93	0.86
	10	1.27	1.23	1.19	1.13	1.06	1.11	1.08	1.06	1.01	0.97	0.96	0.95	0.94	0.92	0.89	0.83
	15	1.25	1.21	1.17	1.10	1.03	1.08	1.06	1.03	0.99	0.94	0.93	0.92	0.91	0.88	0.86	0.79
	20	1.23	1.18	1.14	1.08	1.01	1.06	1.03	1.00	0.96	0.91	0.90	0.89	0.87	0.85	0.83	0.76
	25						1.03	1.00	0.97	0.93	0.88	0.87	0.85	0.84	0.82	0.79	0.72
	30								0.94	0.89	0.85	0.84	0.82	0.80	0.78	0.76	0.68
	35													0.77	0.74	0.72	0.63
	40																0.59
65	5	1.31	1.27	1.23	1.16	1.09	1.14	1.11	1.09	1.04	1.00	0.99	0.98	0.96	0.94	0.92	0.85
	10	1.29	1.24	1.20	1.14	1.06	1.11	1.09	1.06	1.02	0.97	0.96	0.95	0.93	0.91	0.89	0.82
	15	1.26	1.22	1.18	1.11	1.04	1.09	1.06	1.03	0.98	0.94	0.93	0.91	0.90	0.88	0.85	0.78
	20	1.24	1.20	1.15	1.08	1.01	1.06	1.03	1.00	0.95	0.90	0.90	0.88	0.86	0.84	0.82	0.74
	25						1.03	1.00	0.97	0.92	0.87	0.86	0.84	0.83	0.80	0.78	0.70
	30								0.94	0.89	0.83	0.82	0.81	0.79	0.77	0.74	0.65
	35													0.75	0.72	0.70	0.60
	40																0.55
60	5	1.33	1.28	1.24	1.17	1.10	1.15	1.12	1.09	1.05	1.00	0.99	0.98	0.96	0.94	0.92	0.84
	10	1.30	1.26	1.21	1.14	1.07	1.12	1.09	1.06	1.02	0.97	0.96	0.94	0.93	0.90	0.88	0.80
	15	1.28	1.23	1.19	1.12	1.04	1.09	1.06	1.03	0.98	0.93	0.92	0.91	0.89	0.87	0.84	0.76
	20	1.25	1.21	1.16	1.09	1.01	1.06	1.03	1.00	0.95	0.90	0.89	0.87	0.86	0.83	0.80	0.72
	25						1.03	1.00	0.97	0.92	0.86	0.85	0.83	0.82	0.79	0.76	0.67
	30								0.93	0.88	0.82	0.81	0.79	0.78	0.75	0.72	0.62
	35													0.73	0.70	0.67	0.57
	40																0.51

For mass-impregnated cables in line with Section 18.1 for temperatures below 20 °C, an increase of load capacity is only permitted under certain conditions in line with the quantities for the permissible temperature rise in Table 18.1.

The rating factor f_1 must only be used together with factor f_2 in Tables 18.17 to 18.21.

Table 18.16 Rating factor f_1 for installation in ground (*not* applicable to PVC cables with $U_0/U = 6/10$ kV)

Soil-thermal resistivity — Load factor. Resistivity groups: 0.7 Km/W, 1.0 Km/W, 1.5 Km/W (each with load factors 0.50, 0.60, 0.70, 0.85, 1.00) and 2.5 Km/W (load factor 0.5 to 1.0).

Arr. a (Number of Systems)	b (Cables)	c	Ground temp. °C	0.7: 0.50	0.60	0.70	0.85	1.00	1.0: 0.50	0.60	0.70	0.85	1.00	1.5: 0.50	0.60	0.70	0.85	1.00	2.5: 0.5 to 1.0
1	1	1	5	1.31	1.27	1.23	1.16	1.09	1.14	1.12	1.09	1.05	1.00	0.99	0.98	0.96	0.94	0.92	0.85
			10	1.29	1.25	1.21	1.14	1.07	1.12	1.09	1.06	1.02	0.97	0.96	0.95	0.93	0.91	0.89	0.81
			15	1.27	1.22	1.18	1.11	1.04	1.09	1.06	1.03	0.98	0.94	0.93	0.91	0.90	0.87	0.85	0.77
			20	1.24	1.20	1.15	1.08	1.01	1.06	1.03	1.00	0.95	0.90	0.89	0.88	0.86	0.84	0.81	0.73
			25						1.03	1.00	0.97	0.92	0.87	0.86	0.84	0.83	0.80	0.77	0.69
			30								0.94	0.89	0.83	0.82	0.80	0.79	0.76	0.73	0.64
			35													0.75	0.72	0.70	0.59
			40																0.54
4	3	3	5	1.29	1.24	1.20	1.13	1.06	1.11	1.08	1.05	1.01	0.96	0.95	0.94	0.93	0.90	0.88	0.81
			10	1.26	1.22	1.17	1.11	1.03	1.08	1.05	1.03	0.98	0.93	0.92	0.91	0.89	0.87	0.84	0.77
			15	1.24	1.19	1.15	1.08	1.00	1.05	1.03	0.99	0.95	0.90	0.89	0.87	0.86	0.83	0.81	0.73
			20	1.21	1.17	1.12	1.05	0.97	1.03	0.99	0.96	0.91	0.86	0.85	0.84	0.82	0.79	0.77	0.68
			25						0.99	0.96	0.93	0.88	0.83	0.82	0.80	0.78	0.76	0.73	0.64
			30								0.90	0.84	0.79	0.78	0.76	0.74	0.71	0.68	0.59
			35													0.70	0.67	0.64	0.53
			40																0.47
10	5	6	5	1.26	1.21	1.17	1.10	1.03	1.08	1.05	1.02	0.97	0.93	0.92	0.90	0.89	0.86	0.84	0.76
			10	1.23	1.19	1.14	1.07	1.00	1.05	1.02	0.99	0.94	0.89	0.88	0.87	0.85	0.83	0.80	0.72
			15	1.21	1.16	1.12	1.04	0.96	1.02	0.99	0.96	0.91	0.86	0.85	0.83	0.81	0.79	0.76	0.68
			20	1.18	1.14	1.09	1.01	0.93	0.99	0.96	0.93	0.87	0.82	0.81	0.79	0.77	0.75	0.72	0.63
			25						0.96	0.93	0.89	0.84	0.78	0.77	0.75	0.73	0.70	0.68	0.58
			30								0.86	0.80	0.74	0.73	0.71	0.69	0.66	0.63	0.52
			35													0.64	0.61	0.58	0.46
			40																0.38
–	8	10	5	1.23	1.19	1.14	1.07	0.99	1.05	1.02	0.99	0.94	0.89	0.88	0.86	0.85	0.82	0.80	0.72
			10	1.21	1.16	1.11	1.04	0.96	1.02	0.99	0.96	0.91	0.85	0.84	0.83	0.81	0.78	0.76	0.67
			15	1.18	1.13	1.09	1.01	0.93	0.99	0.96	0.92	0.87	0.82	0.81	0.79	0.77	0.74	0.72	0.63
			20	1.15	1.11	1.06	0.98	0.90	0.96	0.92	0.89	0.84	0.78	0.77	0.75	0.73	0.70	0.67	0.57
			25						0.92	0.89	0.85	0.80	0.74	0.73	0.71	0.69	0.66	0.63	0.52
			30								0.82	0.76	0.70	0.68	0.66	0.64	0.61	0.57	0.45
			35													0.60	0.56	0.52	0.38
			40																0.29
–	10	–	5	1.22	1.17	1.13	1.05	0.98	1.03	1.00	0.97	0.92	0.87	0.86	0.84	0.83	0.80	0.78	0.69
			10	1.19	1.15	1.10	1.02	0.94	1.00	0.97	0.94	0.89	0.83	0.82	0.81	0.79	0.76	0.73	0.65
			15	1.17	1.12	1.07	0.99	0.91	0.97	0.94	0.90	0.85	0.79	0.78	0.77	0.75	0.72	0.69	0.60
			20	1.14	1.09	1.04	0.96	0.88	0.94	0.90	0.87	0.81	0.76	0.74	0.73	0.71	0.68	0.65	0.54
			25						0.90	0.87	0.83	0.78	0.71	0.70	0.68	0.66	0.63	0.60	0.48
			30								0.79	0.73	0.67	0.66	0.63	0.61	0.58	0.54	0.41
			35													0.56	0.52	0.48	0.33
			40																0.22

Arrangement a ⊙ ⊙ ⊙ ⊙ ⊙ ⊙ or (trefoils) 7cm 25 cm

Arrangement b (trefoils) 7cm

Arrangement c (trefoils) 7cm

All clearances 7 cm

The rating factor f_1 must only be used together with rating factor f_2 in Tables 18.17 to 18.21.

Table 18.17 Rating factor f_2 for installation in ground. Single-core cables in three-phase system, bunched

7cm

Type of construction	Number of systems	Soil-thermal resistivity											
		0.7 Km/W			1.0 Km/W			1.5 Km/W			2.5 Km/W		
		Load factor			Load factor			Load factor			Load factor		
		0.5	0.6	0.7	0.5	0.6	0.7	0.5	0.6	0.7	0.5	0.6	0.7
XLPE cables 0.6/1 to 18/30 kV	1	1.09	1.04	0.99	1.11	1.05	1.00	1.13	1.07	1.01	1.17	1.09	1.03
	2	0.97	0.90	0.84	0.98	0.91	0.85	1.00	0.92	0.86	1.02	0.94	0.87
	3	0.88	0.80	0.74	0.89	0.82	0.75	0.90	0.82	0.76	0.92	0.83	0.76
	4	0.83	0.75	0.69	0.84	0.76	0.70	0.85	0.77	0.70	0.82	0.78	0.71
	5	0.79	0.71	0.65	0.80	0.72	0.66	0.80	0.73	0.66	0.81	0.73	0.67
	6	0.76	0.68	0.62	0.77	0.69	0.63	0.77	0.70	0.63	0.78	0.70	0.64
	8	0.72	0.64	0.58	0.72	0.65	0.59	0.73	0.65	0.59	0.74	0.66	0.59
	10	0.69	0.61	0.56	0.69	0.62	0.56	0.70	0.62	0.56	0.70	0.63	0.57
PE cables 6/10 to 18/30 kV	1	1.01	1.02	0.99	1.06	1.05	1.00	1.10	1.07	1.01	1.17	1.09	1.03
	2	0.95	0.90	0.84	0.98	0.91	0.85	1.00	0.92	0.86	1.02	0.94	0.87
	3	0.88	0.80	0.74	0.89	0.82	0.75	0.90	0.82	0.76	0.92	0.83	0.76
	4	0.83	0.75	0.69	0.84	0.76	0.70	0.85	0.77	0.70	0.86	0.78	0.71
	5	0.79	0.71	0.65	0.80	0.72	0.66	0.80	0.73	0.66	0.82	0.73	0.67
	6	0.76	0.68	0.62	0.77	0.69	0.63	0.77	0.70	0.63	0.78	0.70	0.64
	8	0.72	0.64	0.58	0.72	0.65	0.59	0.73	0.65	0.59	0.74	0.66	0.59
	10	0.69	0.61	0.56	0.69	0.62	0.56	0.70	0.62	0.56	0.70	0.63	0.57
PVC cables 0.6/1 to 6/10 kV	1	1.01	1.02	0.99	1.04	1.05	1.00	1.07	1.06	1.01	1.11	1.08	1.01
	2	0.94	0.89	0.84	0.97	0.91	0.85	0.99	0.92	0.86	1.01	0.93	0.87
	3	0.86	0.79	0.74	0.89	0.81	0.75	0.90	0.83	0.76	0.91	0.83	0.77
	4	0.82	0.75	0.69	0.84	0.76	0.70	0.85	0.77	0.71	0.86	0.78	0.71
	5	0.78	0.71	0.65	0.80	0.72	0.66	0.80	0.73	0.66	0.81	0.73	0.67
	6	0.75	0.68	0.62	0.77	0.69	0.63	0.77	0.70	0.64	0.78	0.70	0.64
	8	0.71	0.64	0.58	0.72	0.65	0.59	0.73	0.65	0.59	0.73	0.66	0.60
	10	0.68	0.61	0.55	0.69	0.62	0.56	0.69	0.62	0.56	0.70	0.63	0.57
Mass-impregnated cables 0.6/1 to 18/30 kV	1	0.94	0.95	0.97	0.99	0.99	1.00	1.06	1.04	1.01	1.15	1.08	1.02
	2	0.88	0.88	0.84	0.93	0.91	0.85	0.97	0.92	0.86	1.01	0.93	0.87
	3	0.84	0.79	0.74	0.87	0.81	0.75	0.90	0.82	0.76	0.91	0.83	0.76
	4	0.82	0.74	0.69	0.84	0.76	0.70	0.85	0.77	0.71	0.86	0.78	0.71
	5	0.78	0.70	0.65	0.79	0.72	0.65	0.80	0.73	0.66	0.81	0.73	0.67
	6	0.75	0.68	0.62	0.76	0.69	0.63	0.77	0.70	0.63	0.78	0.70	0.64
	8	0.71	0.64	0.58	0.72	0.64	0.58	0.72	0.65	0.59	0.73	0.66	0.59
	10	0.68	0.61	0.55	0.69	0.61	0.56	0.69	0.62	0.56	0.70	0.62	0.56
		Load factor			Load factor			Load factor			Load factor		
		0.85	1.0		0.85	1.0		0.85	1.0		0.85	1.0	
All types of construction	1	0.93	0.87		0.93	0.87		0.94	0.87		0.94	0.87	
	2	0.77	0.71		0.77	0.71		0.77	0.71		0.78	0.71	
	3	0.67	0.61		0.67	0.61		0.68	0.61		0.68	0.61	
	4	0.62	0.56		0.62	0.56		0.62	0.56		0.63	0.56	
	5	0.58	0.52		0.58	0.52		0.58	0.52		0.59	0.52	
	6	0.55	0.50		0.55	0.50		0.56	0.50		0.56	0.50	
	8	0.51	0.46		0.52	0.46		0.52	0.46		0.52	0.46	
	10	0.49	0.44		0.49	0.44		0.49	0.44		0.49	0.44	

Table 18.18 Rating factor f_2 for installation in ground.
Single-core cables in three-phase system, bunched

25 cm

Type of construction	Number of systems	Soil-thermal resistivity											
		0.7 Km/W			1.0 Km/W			1.5 Km/W			2.5 Km/W		
		Load factor			Load factor			Load factor			Load factor		
		0.5	0.6	0.7	0.5	0.6	0.7	0.5	0.6	0.7	0.5	0.6	0.7
XLPE cables 0.6/1 to 18/30 kV	1	1.09	1.04	0.99	1.11	1.05	1.00	1.13	1.07	1.01	1.17	1.09	1.03
	2	1.01	0.94	0.89	1.02	0.95	0.89	1.04	0.97	0.90	1.06	0.98	0.91
	3	0.94	0.87	0.81	0.95	0.88	0.82	0.97	0.89	0.82	0.99	0.90	0.83
	4	0.91	0.84	0.78	0.92	0.84	0.78	0.93	0.85	0.79	0.95	0.86	0.79
	5	0.88	0.80	0.74	0.89	0.81	0.75	0.90	0.82	0.75	0.91	0.83	0.76
	6	0.86	0.79	0.72	0.87	0.79	0.73	0.88	0.80	0.73	0.89	0.81	0.74
	8	0.83	0.76	0.70	0.84	0.76	0.70	0.85	0.77	0.70	0.86	0.78	0.71
	10	0.81	0.74	0.68	0.82	0.74	0.68	0.83	0.75	0.68	0.84	0.76	0.69
PE cables 6/10 to 18/30 kV	1	1.01	1.02	0.99	1.06	1.05	1.00	1.10	1.07	1.01	1.15	1.09	1.03
	2	0.97	0.94	0.89	1.00	0.95	0.89	1.04	0.97	0.90	1.06	0.98	0.91
	3	0.93	0.87	0.81	0.95	0.88	0.82	0.97	0.89	0.82	0.99	0.90	0.83
	4	0.91	0.84	0.78	0.92	0.84	0.78	0.93	0.85	0.79	0.95	0.86	0.79
	5	0.88	0.80	0.74	0.89	0.81	0.75	0.90	0.82	0.75	0.91	0.83	0.76
	6	0.86	0.79	0.72	0.87	0.79	0.73	0.88	0.80	0.73	0.89	0.81	0.74
	8	0.83	0.76	0.70	0.84	0.76	0.70	0.85	0.77	0.70	0.86	0.78	0.71
	10	0.81	0.74	0.68	0.82	0.74	0.68	0.83	0.75	0.68	0.84	0.76	0.69
PVC cables 0.6/1 to 6/10 kV	1	1.01	1.02	0.99	1.04	1.05	1.00	1.07	1.06	1.01	1.11	1.08	1.01
	2	0.97	0.95	0.89	1.00	0.96	0.90	1.03	0.97	0.91	1.06	0.98	0.92
	3	0.94	0.88	0.82	0.97	0.88	0.82	0.97	0.89	0.83	0.98	0.90	0.84
	4	0.91	0.84	0.78	0.92	0.85	0.79	0.93	0.86	0.79	0.95	0.87	0.80
	5	0.88	0.81	0.75	0.89	0.82	0.76	0.90	0.82	0.76	0.91	0.83	0.77
	6	0.86	0.79	0.73	0.87	0.80	0.74	0.88	0.81	0.74	0.89	0.81	0.75
	8	0.83	0.76	0.70	0.84	0.77	0.71	0.85	0.78	0.71	0.86	0.78	0.72
	10	0.82	0.75	0.69	0.82	0.75	0.69	0.83	0.76	0.69	0.84	0.76	0.70
Mass-impregnated cables 0.6/1 to 18/30 kV	1	0.94	0.95	0.97	0.99	0.99	1.00	1.06	1.04	1.01	1.15	1.08	1.02
	2	0.90	0.91	0.88	0.95	0.94	0.89	1.00	0.96	0.89	1.05	0.97	0.90
	3	0.87	0.86	0.80	0.91	0.87	0.81	0.95	0.88	0.81	0.97	0.89	0.82
	4	0.86	0.82	0.76	0.89	0.83	0.77	0.91	0.83	0.77	0.92	0.84	0.78
	5	0.84	0.79	0.73	0.86	0.79	0.73	0.87	0.80	0.73	0.89	0.81	0.74
	6	0.83	0.77	0.71	0.84	0.77	0.71	0.85	0.78	0.71	0.86	0.78	0.72
	8	0.80	0.73	0.67	0.81	0.74	0.68	0.82	0.74	0.68	0.83	0.75	0.68
	10	0.78	0.71	0.65	0.79	0.71	0.65	0.80	0.72	0.66	0.81	0.73	0.66

Type of construction	Number of systems	Load factor			Load factor			Load factor			Load factor		
		0.85	1.0		0.85	1.0		0.85	1.0		0.85	1.0	
Alle types of construction	1	0.93	0.87		0.93	0.87		0.94	0.87		0.94	0.87	
	2	0.82	0.75		0.82	0.75		0.82	0.75		0.83	0.75	
	3	0.74	0.67		0.74	0.67		0.74	0.67		0.74	0.67	
	4	0.70	0.64		0.70	0.64		0.70	0.64		0.71	0.64	
	5	0.67	0.60		0.67	0.60		0.67	0.60		0.67	0.60	
	6	0.65	0.59		0.65	0.59		0.65	0.59		0.65	0.59	
	8	0.62	0.56		0.62	0.56		0.62	0.56		0.62	0.56	
	10	0.60	0.54		0.60	0.54		0.61	0.54		0.61	0.54	

Table 18.19 Rating factor f_2 for installation in ground. Single-core cables in three-phase systems side by side

⊙ ⊙ ⊙ ⊙ ⊙ ⊙
|←7cm
All Clearances 7 cm

Type of construction	Number of systems	Soil-thermal resistivity											
		0.7 Km/W			1.0 Km/W			1.5 Km/W			2.5 Km/W		
		Load factor			Load factor			Load factor			Load factor		
		0.5	0.6	0.7	0.5	0.6	0.7	0.5	0.6	0.7	0.5	0.6	0.7
XLPE cables 0.6/1 to 18/30 kV	1	1.08	1.05	0.99	1.13	1.07	1.00	1.18	1.09	1.01	1.19	1.11	1.03
	2	1.01	0.93	0.86	1.03	0.94	0.87	1.05	0.95	0.88	1.06	0.96	0.88
	3	0.92	0.84	0.77	0.93	0.85	0.77	0.95	0.86	0.78	0.96	0.86	0.79
	4	0.88	0.80	0.73	0.89	0.80	0.73	0.90	0.81	0.74	0.91	0.82	0.74
	5	0.84	0.76	0.69	0.85	0.77	0.70	0.87	0.78	0.70	0.87	0.78	0.71
	6	0.82	0.74	0.67	0.83	0.75	0.68	0.84	0.75	0.68	0.85	0.76	0.69
	8	0.79	0.71	0.64	0.80	0.71	0.65	0.81	0.72	0.65	0.81	0.72	0.65
	10	0.77	0.69	0.62	0.78	0.69	0.63	0.78	0.70	0.63	0.79	0.70	0.63
PE cables 6/10 to 18/30 kV	1	0.98	0.98	0.99	1.04	1.03	1.00	1.11	1.07	1.01	1.19	1.11	1.03
	2	0.93	0.92	0.86	0.98	0.94	0.87	1.02	0.95	0.87	1.06	0.96	0.88
	3	0.89	0.84	0.77	0.93	0.85	0.77	0.95	0.86	0.78	0.96	0.86	0.79
	4	0.87	0.80	0.73	0.89	0.80	0.73	0.90	0.81	0.74	0.91	0.82	0.74
	5	0.84	0.76	0.69	0.85	0.77	0.70	0.86	0.77	0.70	0.87	0.78	0.71
	6	0.82	0.74	0.67	0.83	0.75	0.68	0.84	0.75	0.68	0.85	0.76	0.69
	8	0.79	0.71	0.64	0.80	0.71	0.65	0.81	0.72	0.65	0.81	0.72	0.65
	10	0.77	0.69	0.62	0.78	0.69	0.63	0.78	0.70	0.63	0.79	0.70	0.63
PVC cables 0.6/1 to 6/10 kV	1	0.96	0.97	0.98	1.01	1.01	1.00	1.07	1.05	1.01	1.16	1.10	1.02
	2	0.92	0.89	0.86	0.96	0.94	0.87	1.00	0.95	0.88	1.05	0.97	0.89
	3	0.88	0.84	0.77	0.91	0.85	0.78	0.95	0.86	0.79	0.96	0.87	0.79
	4	0.86	0.80	0.73	0.89	0.81	0.74	0.90	0.82	0.74	0.91	0.82	0.75
	5	0.84	0.76	0.70	0.85	0.77	0.70	0.87	0.78	0.71	0.87	0.79	0.71
	6	0.82	0.74	0.68	0.83	0.75	0.68	0.84	0.76	0.69	0.85	0.76	0.69
	8	0.79	0.71	0.65	0.80	0.72	0.65	0.81	0.72	0.65	0.81	0.73	0.66
	10	0.77	0.69	0.63	0.78	0.70	0.63	0.79	0.70	0.63	0.79	0.71	0.64
Mass-impregnated cables 0.6/1 to 18/30 kV	1	0.93	0.94	0.95	1.00	1.00	1.00	1.09	1.06	1.01	1.19	1.10	1.03
	2	0.89	0.89	0.86	0.95	0.93	0.87	1.01	0.95	0.88	1.05	0.97	0.89
	3	0.86	0.84	0.77	0.90	0.85	0.78	0.95	0.86	0.79	0.96	0.87	0.79
	4	0.84	0.80	0.73	0.88	0.81	0.74	0.91	0.82	0.74	0.91	0.82	0.75
	5	0.82	0.77	0.70	0.86	0.77	0.70	0.87	0.78	0.71	0.87	0.79	0.71
	6	0.81	0.74	0.68	0.83	0.75	0.68	0.85	0.76	0.69	0.85	0.76	0.69
	8	0.78	0.71	0.65	0.80	0.72	0.65	0.81	0.73	0.66	0.82	0.73	0.66
	10	0.77	0.69	0.63	0.78	0.70	0.63	0.79	0.70	0.64	0.79	0.71	0.64

Type of construction	Number of systems	Load factor			Load factor			Load factor			Load factor		
		0.85	1.0		0.85	1.0		0.85	1.0		0.85	1.0	
All types of construction	1	0.91	0.85		0.92	0.85		0.92	0.85		0.93	0.85	
	2	0.77	0.71		0.78	0.71		0.78	0.71		0.79	0.71	
	3	0.69	0.62		0.69	0.62		0.69	0.62		0.69	0.62	
	4	0.65	0.58		0.65	0.58		0.65	0.58		0.65	0.58	
	5	0.61	0.55		0.61	0.55		0.62	0.55		0.62	0.55	
	6	0.59	0.53		0.60	0.53		0.60	0.53		0.60	0.53	
	8	0.57	0.51		0.57	0.51		0.57	0.51		0.57	0.51	
	10	0.55	0.49		0.55	0.49		0.55	0.49		0.55	0.49	

Table 18.20 Rating factor f_2 for installation in ground.
Three-core [1] cables in three-phase systems

7cm

Type of construction	Number of cables	Soil-thermal resistivity											
		0.7 Km/W			1.0 Km/W			1.5 Km/W			2.5 Km/W		
		Load factor			Load factor			Load factor			Load factor		
		0.5	0.6	0.7	0.5	0.6	0.7	0.5	0.6	0.7	0.5	0.6	0.7
XLPE cables [2] 0.6/1 and 6/10 kV	1	1.02	1.03	0.99	1.06	1.05	1.00	1.09	1.06	1.01	1.11	1.07	1.02
	2	0.95	0.89	0.84	0.98	0.91	0.85	0.99	0.92	0.86	1.01	0.94	0.87
	3	0.86	0.80	0.74	0.89	0.81	0.75	0.90	0.83	0.77	0.92	0.84	0.77
	4	0.82	0.75	0.69	0.84	0.76	0.70	0.85	0.78	0.71	0.86	0.78	0.72
	5	0.78	0.71	0.65	0.80	0.72	0.66	0.81	0.73	0.67	0.82	0.74	0.67
	6	0.75	0.68	0.63	0.77	0.69	0.63	0.78	0.70	0.64	0.79	0.71	0.65
	8	0.71	0.64	0.59	0.72	0.65	0.59	0.73	0.66	0.60	0.74	0.66	0.60
	10	0.68	0.61	0.56	0.69	0.62	0.56	0.70	0.63	0.57	0.71	0.63	0.57
PE cables 6/10 kV	1	0.99	1.00	0.99	1.03	1.03	1.00	1.08	1.06	1.01	1.14	1.08	1.02
	2	0.91	0.89	0.84	0.96	0.91	0.85	0.99	0.92	0.86	1.01	0.94	0.87
	3	0.85	0.80	0.74	0.89	0.81	0.75	0.90	0.83	0.77	0.92	0.84	0.77
	4	0.82	0.75	0.69	0.84	0.76	0.70	0.85	0.78	0.71	0.86	0.78	0.72
	5	0.78	0.71	0.65	0.80	0.72	0.66	0.81	0.73	0.67	0.82	0.74	0.67
	6	0.75	0.68	0.63	0.77	0.69	0.63	0.78	0.70	0.64	0.79	0.71	0.65
	8	0.71	0.64	0.59	0.72	0.65	0.59	0.73	0.66	0.60	0.74	0.66	0.60
	10	0.68	0.61	0.56	0.69	0.62	0.56	0.70	0.63	0.57	0.71	0.63	0.57
PVC cables [2] 0.6/1 and 3.6/6 kV	1	0.91	0.92	0.94	0.97	0.97	1.00	1.04	1.03	1.01	1.13	1.07	1.02
	2	0.86	0.87	0.85	0.91	0.90	0.86	0.97	0.93	0.87	1.01	0.94	0.88
	3	0.82	0.80	0.75	0.86	0.82	0.76	0.91	0.84	0.77	0.92	0.84	0.78
	4	0.80	0.76	0.70	0.84	0.77	0.71	0.86	0.78	0.72	0.87	0.79	0.73
	5	0.78	0.72	0.66	0.81	0.73	0.67	0.81	0.74	0.68	0.82	0.75	0.68
	6	0.76	0.69	0.64	0.77	0.70	0.64	0.78	0.71	0.65	0.79	0.72	0.65
	8	0.72	0.65	0.59	0.73	0.66	0.60	0.74	0.67	0.61	0.75	0.67	0.61
	10	0.69	0.62	0.57	0.70	0.63	0.57	0.71	0.64	0.58	0.71	0.64	0.58
Mass-impregnated cables: Belted cables 0.6/1; 3.6/6 kV S.L. cables 3.6/6; 6/10 kV	1	0.94	0.95	0.97	1.00	1.00	1.00	1.06	1.05	1.01	1.13	1.07	1.02
	2	0.89	0.89	0.85	0.94	0.92	0.86	0.99	0.93	0.87	1.01	0.94	0.88
	3	0.84	0.81	0.76	0.89	0.83	0.77	0.91	0.84	0.78	0.92	0.85	0.79
	4	0.82	0.77	0.71	0.85	0.78	0.72	0.86	0.79	0.73	0.87	0.80	0.73
	5	0.80	0.73	0.67	0.81	0.74	0.68	0.82	0.75	0.69	0.83	0.76	0.69
	6	0.77	0.70	0.65	0.79	0.71	0.65	0.79	0.72	0.66	0.80	0.73	0.66
	8	0.73	0.66	0.61	0.74	0.67	0.61	0.75	0.68	0.62	0.75	0.68	0.62
	10	0.70	0.63	0.58	0.71	0.64	0.58	0.72	0.65	0.59	0.72	0.65	0.59

		Load factor			Load factor			Load factor			Load factor		
		0.85	1.0		0.85	1.0		0.85	1.0		0.85	1.0	
All types of construction [2]	1	0.94	0.89		0.94	0.89		0.94	0.89		0.95	0.89	
	2	0.77	0.72		0.78	0.72		0.78	0.72		0.79	0.72	
	3	0.68	0.62		0.68	0.62		0.69	0.62		0.69	0.62	
	4	0.63	0.57		0.63	0.57		0.63	0.57		0.64	0.57	
	5	0.59	0.53		0.59	0.53		0.59	0.53		0.60	0.53	
	6	0.56	0.51		0.56	0.51		0.57	0.51		0.57	0.51	
	8	0.52	0.47		0.52	0.47		0.52	0.47		0.53	0.47	
	10	0.49	0.44		0.50	0.44		0.50	0.44		0.50	0.44	

[1] In 3 phase systems these quantities apply also for cables 0.6/1 kV with 4 or 5 conductors
[2] In d.c. systems the quantities also apply for single-core cables for 0.6/1 kV

Table 18.21   Rating factor f_2 for installation in ground.
Three-core cables in three-phase systems

7cm

Type of construction	Number of cables	Soil-thermal resistivity											
		0.7 Km/W			1.0 Km/W			1.5 Km/W			2.5 Km/W		
		Load factor			Load factor			Load factor			Load factor		
		0.5	0.6	0.7	0.5	0.6	0.7	0.5	0.6	0.7	0.5	0.6	0.7
PVC cables 0.6/1 kV [1]	1	0.90	0.91	0.93	0.98	0.99	1.00	1.05	1.04	1.03	1.14	1.09	1.04
PVC cables 6/10 kV	2	0.85	0.85	0.85	0.93	0.92	0.89	0.98	0.95	0.90	1.03	0.96	0.90
Mass-impregnated-	3	0.80	0.79	0.78	0.87	0.86	0.80	0.93	0.86	0.80	0.95	0.87	0.81
belted cables 6/10 kV	4	0.77	0.77	0.74	0.85	0.81	0.75	0.89	0.82	0.75	0.90	0.82	0.76
H-cables 6/10 to	5	0.75	0.75	0.70	0.84	0.77	0.71	0.85	0.77	0.71	0.86	0.78	0.72
18/30 kV and	6	0.74	0.73	0.67	0.81	0.74	0.68	0.82	0.74	0.68	0.83	0.75	0.69
Mass-impregnated	8	0.73	0.69	0.63	0.77	0.70	0.64	0.77	0.70	0.64	0.78	0.71	0.64
S.L. cables 12/20 and 18/30 kV	10	0.71	0.66	0.60	0.74	0.67	0.61	0.74	0.67	0.61	0.75	0.67	0.61
		Load factor			Load factor			Load factor			Load factor		
		0.85	1.0		0.85	1.0		0.85	1.0		0.85	1.0	
All types of	1	0.96	0.91		0.96	0.91		0.97	0.91		0.97	0.91	
construction	2	0.81	0.76		0.82	0.76		0.82	0.76		0.82	0.76	
	3	0.72	0.66		0.72	0.66		0.73	0.66		0.73	0.66	
	4	0.67	0.61		0.67	0.61		0.68	0.61		0.68	0.61	
	5	0.63	0.57		0.63	0.57		0.63	0.57		0.64	0.57	
	6	0.60	0.55		0.60	0.55		0.61	0.55		0.61	0.55	
	8	0.56	0.51		0.56	0.51		0.57	0.51		0.57	0.51	
	10	0.53	0.48		0.54	0.48		0.54	0.48		0.54	0.48	

[1] Two- and three-core PVC cables for $U_0/U = 0.6/1$ kV in single-phase a.c. and in d.c. systems

Tabelle 18.22
Rating factors f_ϑ for differing air temperatures

Type of construction	Permissible conductor temperature °C	Permissible temperature rise K	Air temperature								
			10 °C	15 °C	20 °C	25 °C	30 °C	35 °C	40 °C	45 °C	50 °C
			Rating factor								
XLPE cables	90	–	1.15	1.12	1.08	1.04	1.0	0.96	0.91	0.87	0.82
PE- and PVC cables	70	–	1.22	1.17	1.12	1.06	1.0	0.94	0.87	0.79	0.71
Mass-impregnated cables											
Belted cables											
0.6/1 to 3.6/6 kV	80	55	1.05	1.05	1.05	1.05	1.0	0.95	0.89	0.84	0.77
6/10 kV	65	35	1.0	1.0	1.0	1.0	1.0	0.93	0.85	0.76	0.65
Single-core, S.L. and H-cables											
0.6/1 to 3.6/6 kV	80	55	1.05	1.05	1.05	1.05	1.0	0.95	0.89	0.84	0.77
6/10 kV	70	45	1.06	1.06	1.06	1.06	1.0	0.94	0.87	0.79	0.71
12/20 kV	65	35	1.0	1.0	1.0	1.0	1.0	0.93	0.85	0.76	0.65
18/30 kV	60	30	1.0	1.0	1.0	1.0	1.0	0.91	0.82	0.71	0.58

$$\frac{1}{1.04} = 0.96.$$

Table 18.23 Rating factors f_H for groups in air [1].
Single-core cables in three-phase systems

Arrangement of cables	Number of cable trays or cable racks	Installed in one plane Clearance = cable diameter d distance from wall ≥ 2 cm				Installation in bunches Clearance = $2d$ distance from wall ≥ 2 cm			
		Number of systems				Number of systems			
		1	2	3		1	2	3	
On the floor	–	0.92	0.89	0.88		0.95	0.90	0.88	
On cable trays	1	0.92	0.89	0.88		0.95	0.90	0.88	
	2	0.87	0.84	0.83		0.90	0.85	0.83	
	3	0.84	0.82	0.81		0.88	0.83	0.81	
	6	0.82	0.80	0.79		0.86	0.81	0.79	
On cable racks	1	1.00	0.97	0.96		1.00	0.98	0.96	
	2	0.97	0.94	0.93		1.00	0.95	0.93	
	3	0.96	0.93	0.92		1.00	0.94	0.92	
	6	0.94	0.91	0.90		1.00	0.93	0.90	
On supports or on the wall	–	0.94	0.91	0.89		0.89	0.86	0.84	
Arrangement for which a reduction is not required [1]		In installations in one plane with increased clearance the increased sheath or screen losses counteract the otherwise reduced temperature rise. Therefore indications as to reduction-free arrangements cannot be made here.							

[1] In confined spaces or where much grouping occurs the losses of the cables increase the air temperature and therefore additional rating factors for differing air temperatures from Table 18.22 must be applied

Table 18.24 Rating factors f_H for groups in air [1].
Multi-core cables and single-core cables in d.c. systems

Arrangement of cables cables	Number of cable trays or cable racks	Clearance = cable diameter d distance from wall ≥ 2 cm						Side by side without clearance and touching wall					
		Number of cables						Number of cables					
		1	2	3	6	9		1	2	3	6	9	
On the floor	–	0.95	0.90	0.88	0.85	0.84		0.90	0.84	0.80	0.75	0.73	
On cable trays	1	0.95	0.90	0.88	0.85	0.84		0.95	0.84	0.80	0.75	0.73	
	2	0.90	0.85	0.83	0.81	0.80		0.95	0.80	0.76	0.71	0.69	
	3	0.88	0.83	0.81	0.79	0.78		0.95	0.78	0.74	0.70	0.68	
	6	0.86	0.81	0.79	0.77	0.76		0.95	0.76	0.72	0.68	0.66	
On cable racks	1	1.00	0.98	0.96	0.93	0.92		0.95	0.84	0.80	0.75	0.73	
	2	1.00	0.95	0.93	0.90	0.89		0.95	0.80	0.76	0.71	0.69	
	3	1.00	0.94	0.92	0.89	0.88		0.95	0.78	0.74	0.70	0.68	
	6	1.00	0.93	0.90	0.87	0.86		0.95	0.76	0.72	0.68	0.66	
On supports or on the wall	–	1.00	0.93	0.90	0.87	0.86		0.95	0.78	0.73	0.68	0.66	
Arrangement for which a reduction is not required [1]		Number of cables arranged above each other is not restricted						Number of cables arranged side by side is not restricted					

[1] In confined spaces or where much grouping occurs the losses of the cables increase the air temperature and therefore additional rating factors for differing air temperatures from Table 18.22 must be applied

Table 18.25
Rating factors[1], multi-core cables with conductor cross-sectional area of 1.5 to 10 mm². Installation in ground or in air

Number of loaded cores	Installed in	
	Ground	Air
5	0.70	0.75
7	0.60	0.65
10	0.50	0.55
14	0.45	0.50
19	0.40	0.45
24	0.35	0.40
40	0.30	0.35
61	0.25	0.30

[5] These factors are to be applied to ratings in Table 18.5, multi-core cables in the ground and to ratings in Table 18.6, multi-core cables in air, both in 3-phase operation

18.2.4 Use of Tables

If the transmitted power is known the operating current I_b (loading) can be calculated using the equations from Table 18.26 where U_b is the operating voltage of the network and $\cos \varphi$ the power factor.

Table 18.26
Equations for the calculation of operating current I_b from the transmitted power

Type of Network	Apparent Power S VA	Active Power P W	Reactive Power Q var
Direct current	–	$\dfrac{P}{U_b}$	–
Single-phase a.c.	$\dfrac{S}{U_b}$	$\dfrac{P}{U_b \cos \varphi}$	$\dfrac{Q}{U_b \sin \varphi}$
Three phase	$\dfrac{S}{\sqrt{3}\, U_b}$	$\dfrac{P}{\sqrt{3}\, U_b \cos \varphi}$	$\dfrac{Q}{\sqrt{3}\, U_b \sin \varphi}$

From the 24 hour day load diagram and as referred to in Sections 18.1 and 18.2.3 the maximum load is also the operating current I_b. Where the installation is to be in ground the 24 hour load diagram is to be used to determine the load factor m. Where the installation is to be in air this is not required.

Example 18.1

● In a three-phase network with $U_b = 10$ kV an apparent power of 10 MVA is to be transmitted. The operating current I_b is determined from

$$I_b = \frac{S}{\sqrt{3}\, U_b} = \frac{10 \times 10^6 \text{ VA}}{\sqrt{3} \times 10 \times 10^3 \text{ V}} = 577 \text{ A.}$$

From the 24 hour load diagram (Fig. 18.6) with the maximum load equal to operating current $I_b = 577$ A, the average load is first calculated. This is done by taking the area below the load curve plotted from current and time values and calculating an average value over the 24 hour period:

$$4\,\text{h}\,\frac{300\,\text{A}+260\,\text{A}}{2}+4\,\text{h}\,\frac{260\,\text{A}+577\,\text{A}}{2}+6\,\text{h}\,\frac{577\,\text{A}+400\,\text{A}}{2}+4\,\text{h}\,\frac{400\,\text{A}+450\,\text{A}}{2}+6\,\text{h}\,\frac{450\,\text{A}+300\,\text{A}}{2} = 403\,\text{A}.$$
$$\overline{\hspace{6cm}24\,\text{h}\hspace{6cm}}$$

From this the load factor becomes $m = \dfrac{403}{577} = 0.7$.

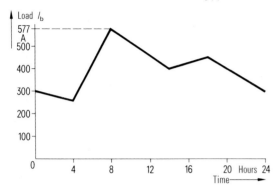

Fig. 18.6 Schematic daily load diagram

● The calculated operating current $I_b = 577$ A with the load factor $m = 0.7$ is to be transmitted using XLPE cables type

NA2XS2Y $3 \times \ldots/\ldots$ 6/10 kV

under the specified operating conditions in Table 18.2. From Part 2, Table 5.6.5 it is found that the largest cross-sectional area is not sufficient to carry 577 A; therefore 2 cables in parallel are required.

For $\varrho_E = 1.0$ Km/W and $m = 0.7$, the rating factor from Table 18.15: $f_1 = 1.0$,
For 2 cables, the group rating factor from Table 18.20: $f_2 = 0.85$.

In order to make a direct comparison with the tabulated currents I_r the calculation is made with a fictitious value of operating current I_{bf}. With $N = 2$ parallel connected cables

$$I_{bf} = \frac{I_b}{N\,\Pi f} = \frac{577}{2 \times 1.0 \times 0.85} = 339 \text{ A per cable}$$

(where Πf is the product of all relevant rating factors).

From Part 2, Table 5.6.5 two cables with Aluminium conductors and a cross-sectional area of 185 mm^2 will be adequate.

The load capacity for one cable is:

$$I_z = I_r\,\Pi f = 347 \times 1.0 \times 0.85 = 295 \text{ A}.$$

● The load capacity of two cables

NA2XS2Y 3×185 SE/25 6/10 kV

is required to be determined when installed in ducts under the following operating conditions:

Load factor m = 1.0
Soil-thermal resistivity $\varrho_E = 1.5$ Km/W.

Ground temperature $\vartheta_E = 30$ °C.

The rating factors for these conditions:

from Table 18.15	$f_1 = 0.81$,
and for two cables from Table 18.20	$f_2 = 0.72$,
for laying in pipe from Table 18.2	$f_R = 0.85$.

The load capacity per cable becomes

$$I_z = I_r\,\Pi f = 347 \times 0.81 \times 0.72 \times 0.85 = 172 \text{ A}.$$

18.3 Calculation of Load Capacity

A cable is heated by losses generated by current in the conductors and, when on a.c., by losses generated in the metal coverings as well as by dielectric losses in the insulation. The dielectric losses can be ignored, however, in PVC cables up to $U_0/U = 3.5/6$ kV, in mass-impregnated cables up to $U_0/U = 18/30$ kV and in cables with PE or XLPE insulation up to $U_0/U = 64/110$ kV. Under steady-state conditions the dissipated heat is equal to the sum of all losses in the cable. Heat losses are conducted to the surface of a cable and thence, when a cable is in air, transmitted to the ambient by convection and radiation (Section 18.4.2). Where a cable is installed in the ground, the heat loss is conducted from the cable surface through the surrounding soil to the atmosphere (Section 18.4.3). The difference between conductor temperature and ambient temperature is approximately proportional to the total losses. The law of heat flow is analogous to Ohm's law, where the heat flow Φ corresponds to electric current I, the temperature difference $\Delta\vartheta_L$ corresponds to voltage difference U and the total thermal resistance ΣT corresponds to electrical resistance R thus:

from $\qquad U = IR$

the analogy $\qquad \Delta\vartheta_L = \Phi \Sigma T.$ $\qquad$ (18.2)

The heat flow Φ (losses) is the sum of the heat losses P'_i attributed to load current and the losses P'_d related to the supply voltage. For heat to be transferred from its place of origin to the ambient it must overcome the thermal resistance T''_K of the cable and the thermal resistance T''_4 to the ambient. In considering heat transfer from a cable surface to the ambient T''_4 may be the thermal resistance of the air T'_L or the thermal resistance of the ground T'_E.

Using the analogy between the flow of heat and the flow of electric current (Equation 18.2) an aquivalent circuit diagram can be drawn (Fig. 18.7) for heat losses flowing from a cable and the resulting temperature rises produced. Heat transfer by radiation and convection from a cable installed in free air is represented by two resistors connected in parallel with each other but in series with the thermal resistances of the cable. When installed in the ground the two resistors are replaced by a single resistor being the soil-thermal resistance.

The heat losses P'_i which are related to load current arise in the conductor, in the metal parts and in the armour, whereas the dielectric heat losses P'_d are generated in the insulation. These losses are represented

in the analogy by currents fed in at appropriate points. Due to these losses the conductor temperature ϑ_L is increased by $\Delta\vartheta_L$ and the surface temperature of the cable ϑ_O is increased by $\Delta\vartheta_O$ relative to the ambient temperature ϑ_U.

For a cable with current flowing in n conductors the losses due to current are

$$P'_i = nI^2 R'_{wr} \qquad (18.3)$$

and the dielectric losses (see Section 22) are

$$P'_d = n\omega C'_b \left(\frac{U}{\sqrt{3}}\right)^2 \tan\delta. \qquad (18.4)$$

The effective resistance (a.c. resistance) R'_w (see Section 20) is practically constant at the permissible operating temperature and can be expressed by the equation

$$R'_{wr} = R'_\vartheta + \Delta R' = R'_\vartheta (1 + y_s + y_p)(1 + \lambda_1 + \lambda_2) \quad (18.5)$$

whilst the d.c. resistance at permissible operating temperature ϑ_{Lr} is

$$R'_\vartheta = R'_{20} [1 + \alpha_{20}(\vartheta_{Lr} - 20)]. \qquad (18.6)$$

and the additional resistance is

$$\Delta R' = R'_{wr} - R'_\vartheta \qquad (18.7)$$

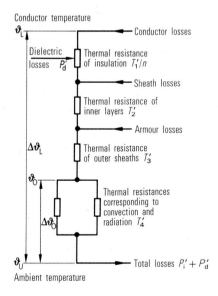

a) Cable in free air

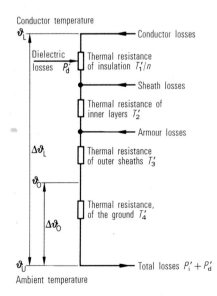

b) Cable in ground

Fig. 18.7
Equivalent circuit
for heat flow
in a cable

giving a measurable rise in conductor resistance caused by current dependant a.c. losses. These losses arise in each conductor due to skin effect and proximity effect (y_s and y_p) and by induction and eddy currents in the metal sheath (λ_1) as well as by eddy currents and magnetic reversal in the armour (λ_2). If these factors are incorporated in equation 18.2 for the temperature rise of each conductor the following equation applies:

$$\Delta \vartheta_L = [I^2 R'_\vartheta + (P'_d/2)]\, T'_1 +$$
$$+ [I^2 R'_\vartheta (1 + \lambda_1) + P'_d]\, n\, T'_2 + \qquad (18.8)$$
$$+ [I^2 R'_\vartheta (1 + \lambda_1 + \lambda_2) + P'_d]\, n(T'_3 + T'_4).$$

The actual thermal resistance of the cable (see also Section 18.4.1) is given by:

$$T'_K = (T'_1/n) + T'_2 + T'_3. \qquad (18.9)$$

The partial resistances of the insulation are represented by T'_1 and for the inner and outer protective covers as T'_2 and T'_3 respectively. (The thermal resistances of the metallic elements are small enough to be ignored).

To make the equations clearer and to simplify their application in design work, fictitious thermal resistances are introduced. The fictitious thermal resistance T'_{Ki} for heat losses due to the current, results from equation 18.2 and equation 18.8 with

$$T'_{Ki} = \frac{\dfrac{T'_1}{n} + (1 + \lambda_1)\, T'_2}{1 + \lambda_1 + \lambda_2} + T'_3 \qquad (18.10)$$

and the fictitious thermal resistance T'_{Kd} relating to the dielectric losses from equation 18.4, assumes that these originate at a mid point in the insulation, with

$$T'_{Kd} = \frac{T'_1}{2n} + T'_2 + T'_3. \qquad (18.11)$$

From these relationships the load capacity I_z can be found for a permissible operating temperature ϑ_{Lr} and an ambient temperature ϑ_U

$$I_z = \sqrt{\frac{\vartheta_{Lr} - \vartheta_U - \Delta \vartheta_d}{n R'_{wr}(T'_{Ki} + T'_4)}} \qquad (18.12)$$

with the temperature rise due to dielectric losses

$$\Delta \vartheta_d = P'_d(T'_{Kd} + T'_4). \qquad (18.13)$$

Where the individual thermal resistances, loss factors, or effective resistances are not given, they can be derived using the methods provided in the literature referred to later [18.2, 18.7 and 18.8].

In the following the effective resistances are calculated or derived for the permissible operating temperature ϑ_{Lr}.

If the operating voltage U_b is liable to deviate significantly from the rated voltage U of the cable then the dielectric losses must be calculated using U_b rather than U in equation 18.4.

The thermal resistance of the surroundings T'_4 is governed by operating conditions described in Section 18.2. For *installation in free air* the thermal resistance of the air T'_L is calculated as shown in Section 18.4.2 and has been used to determine the load capacity in air under specified conditions with an ambient temperature of 30 °C, as can be seen in the tables and text in Section 18.2:

$$I_r = \sqrt{\frac{\vartheta_{Lr} - 30 - \Delta \vartheta_d}{n R'_{wr}(T'_{Ki} + T'_{Lu})}}. \qquad (18.14)$$

The load capacity for installation arrangements other than in free air or for groups, is calculated using the rating factors (Table 18.23 and 18.24). Rating factors f for ambient temperatures ϑ_u other than 30 °C are calculated by using equations 18.2 and 18.14, assuming constant effective resistance and, thermal resistance (see also Table 18.22) with

$$f = \frac{I_z}{I_r} = \sqrt{\frac{\vartheta_{Lr} - \vartheta_U - \Delta \vartheta_d}{\vartheta_{Lr} - 30 - \Delta \vartheta_d}}. \qquad (18.15)$$

Normally the dieletric temperature rise $\Delta \vartheta_d$ in cables up to $U = 30$ kV is neglegible apart from PVC cables with rated voltages of $U \geq 10$ kV. For these cables however it is common practice when calculating rating factors in air to neglect the dielectric heat rise which with the exception of a few cases is little more than 2 K.

For *installations in the ground* T'_4 represents the thermal resistance of the soil. As indicated in Section 18.4.3 the equation 18.12 has to be extended because of drying out of the soil and cyclic loading. Values for load capacity can be taken from the tables in Section 18.2. The load capacity for non-specified operating conditions must be calculated according to Sections 18.4.3 to 18.4.5 or alternatively by the use of conversion factors in Tables 18.15 to 18.21.

18.4 Thermal Resistances

18.4.1 Thermal Resistance of the Cable

The thermal resistance of the cable T_K' takes into consideration the thermal insulating effect of electrical insulation and cable sheaths (Fig. 18.8) and must be calculated by using construction data and thermal resistivities [18.2, 18.7, 18.8].

For single-core cables with a metal sheath for example:

$$T_K' = T_1' + T_3'$$
$$= \frac{\varrho_1}{2\pi} \ln \frac{d_1}{d_L} + \frac{\varrho_3}{2\pi} \ln \frac{d}{d_M}. \qquad (18.16)$$

ϱ_1 thermal resistivity of insulation
ϱ_3 thermal resistivity of outer sheath material
d_L conductor diameter
d_1 diameter over insulation or under metal sheath or screen
d_M diameter over metal sheath or screen
d overall diameter

The thermal insulating effect of metal covers is very small and can be ignored. Values for the thermal resistivity of materials used in cables can be found in Table 18.1. These values are assumed to be constant over the temperature range up to the permissible conductor operating temperature and so is the resulting thermal resistance.

The fictitious thermal resistances T_{Ki}' to equation 18.10 and T_{Kd}' to equation 18.11 for commonly used cable types of constructions are shown in Fig. 18.9.

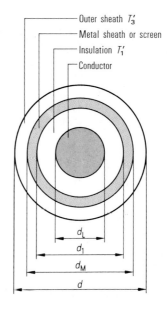

Outer sheath T_3'
Metal sheath or screen
Insulation T_1'
Conductor

d_L
d_1
d_M
d

Fig. 18.8
Thermal resistances T_1' and T_3' of a single-core cable

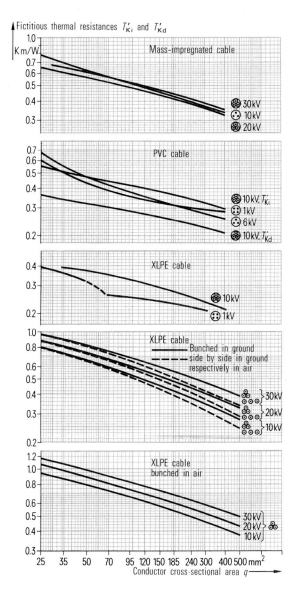

Fictitious thermal resistances T'_{Ki} and T'_{Kd}
K m/W

Mass-impregnated cable

- 30 kV
- 10 kV
- 20 kV

PVC cable

- 10 kV, T'_{Ki}
- 1 kV
- 6 kV
- 10 kV, T'_{Kd}

XLPE cable

- 10 kV
- 1 kV

XLPE cable
Bunched in ground
side by side in ground
respectively in air

- 30 kV
- 20 kV
- 10 kV

XLPE cable bunched in air

- 30 kV
- 20 kV
- 10 kV

25 35 50 70 95 120 150 185 240 300 400 500 mm²
Conductor cross-sectional area q ⟶

⊙ Cables with or without common screen
⊛ Cables with individual core screens
⊙ Single-core cables

Fig. 18.9
Fictitious thermal resistances of commonly used cable
constructions. T'_{Ki} from equation 18.10 and T'_{Kd} for
PVC-cable for $U_0/U = 6/10$ kV from equation 18.11.
Cables with XLPE insulation have been calculated
with PVC sheaths

Example 18.2

The cable data mentioned in the examples are taken
from Part 2 (English version is in preparation). These
values were calculated on the basis of the latest constructional design of the relevant cables and therefore
they may slightly deviate from the data indicated in
the Tables 18.5 to 18.14 in respect of the current-
carrying capacities.

The conductor resistances for the cable selected for
the example

NA2XS2Y 1×150 RM/25 12/20 kV

are taken from Part 2, Table 5.6.6 a and b:

Direct current resistance of conductor
at 20 °C $\qquad R'_{20} = 0.206 \ \Omega/\text{km}$
Effective resistance at 90 °C
– bunched installed
  in ground or air $\qquad R'_{wr} = 0.269 \ \Omega/\text{km}$
– side by side installed
  in ground $\qquad R'_{wr} = 0.285 \ \Omega/\text{km}$

The specific details of construction are:

Diameter of aluminium conductor $\qquad d_L = 14.5$ mm

Thickness of inner conducting layer $\qquad 0.7$ mm

Thickness of insulation of XLPE $\qquad 5.5$ mm

Thickness of outer conducting layer
including the protective cover under
the screen $\qquad 0.8$ mm

Diameter under the screen $\qquad d_1 = 28.5$ mm

Diameter of single screen wire $\qquad 0.5$ mm

Increase in length due to helically wound construction of screen wire $\qquad z = 0.05$ (5%)

Thickness of transverse helical tape $\qquad \delta = 0.2$ mm

Width of transverse helical tape $\qquad b = 5.0$ mm

Increase in length due to helically wound construction of transverse tape $\qquad z = 0.30$ (30%)

Geometric cross-sectional area of screen $q_M = 25$ mm²

Electrical conductivity of screen,
mean value $\qquad \varkappa = 56 \cdot 10^6 \ 1/\Omega \ \text{m}$

Diameter over screen $\qquad d_M = 29.9$ mm

Thickness of protective layers and
separating layer above the screen $\qquad 0.4$ mm

Thickness of outer PE sheath $\qquad 2.5$ mm

Outer overall diameter $\qquad d = 35.7$ mm

Using the thermal resistivities given in Table 18.1 we get:

$$T_1' = \frac{\varrho_1}{2\pi} \ln \frac{d_1}{d_L} = \frac{3.5}{2\pi} \ln \frac{28.5}{14.5} = 0.376 \frac{\text{Km}}{\text{W}}, \quad (18.16)$$

$$T_3' = \frac{\varrho_3}{2\pi} \ln \frac{d}{d_M} = \frac{3.5}{2\pi} \ln \frac{35.7}{29.9} = 0.099 \frac{\text{Km}}{\text{W}}, \quad (18.16)$$

$$T_K' = T_1' + T_3' = 0.376 + 0.099 = 0.475 \text{ Km/W}. \quad (18.16)$$

For the calculation of the fictitious thermal resistance T_{Ki}' of the cable, the sheath loss factor λ_1 must be used in the calculation according to [18.7]. λ_2 and T_2' are zero since armour and protective cover between screen and armour are missing. For a trefoil installation in the ground this gives:

$$T_{Ki}' = \frac{\dfrac{T_1'}{n} + (1 + \lambda_1) T_2'}{(1 + \lambda_1 + \lambda_2)} + T_3' \quad (18.10)$$

$$= \frac{0.376}{1 + 0.0160} + 0.099 = 0.469 \text{ Km/W}.$$

These values, together with values for other types of installation, are shown for comparison in Table 18.27. Values for the fictitious thermal resistance of the cables T_{Ki}' differ from one another due to their dependance on the magnitude of the sheath loss factor λ_1.

Table 18.27
Comparison of fictitious thermal resistances T_{Ki}' between calculated values from equation 18.10 and graphical results from Fig. 18.9

Arrangement	λ_1 to [18.7]	T_{Ki}' to equation 18.10	T_{Ki}' to Fig. 18.9 [1]
In ground, bunched	0.0160	0.469	≈ 0.52
In free air, bunched	0.0163	0.601 [2]	≈ 0.645
In ground, side by side	0.0776	0.448	≈ 0.5

[1] Values for cables with PVC sheath
[2] Calculated to Section 18.4.2

18.4.2 Thermal Resistance of Air

Horizontal Installation in Free Air

Heat from cables installed in air is dissipated by convection and radiation. In the equivalent circuit, Fig. 18.7, the thermal resistance T_{Lu}' of air is indicated by two thermal resistances in parallel representing convection and radiation. The thermal resistance of air can be expressed by [18.7; 18.9]:

$$T_{Lu}' = \frac{1}{\pi d (f_k \alpha_k + f_s \alpha_s)}. \quad (18.17)$$

Consider a cable which is not influenced by other sources of heat (solar radiation) and which does not increase the temperature of its surroundings. If such a cable is arranged horizontally in free air, so that it dissipates its losses into its surroundings by natural convection and unhindered radiation, the coefficient of heat transfer α_k, in dry air at an atmospheric pressure of 1013 hPa, is:

$$\alpha_k = k' \frac{0.0185}{kd} + k'' 1.08 \left(\frac{\Delta \vartheta_0}{kd}\right)^{\frac{1}{4}} \quad (18.18)$$

with

$$k' = 0.919 + \frac{\vartheta_m}{369}, \qquad k'' = 1.033 - \frac{\vartheta_m}{909}, \quad (18.19)$$

$$\vartheta_m = \frac{\vartheta_0 + \vartheta_U}{2}, \qquad \Delta \vartheta_0 = \vartheta_0 - \vartheta_U \quad (18.20)$$

and the thermal transfer coefficient α_s for radiation

$$\alpha_s = \frac{\varepsilon_0 \sigma [(273 + \vartheta_0)^4 - (273 + \vartheta_U)^4]}{\Delta \vartheta_0}, \quad (18.21)$$

where $\sigma = 5.67 \times 10^{-8}$ W/m^2 K^4 (Stephan-Bolzmann constant) and ε_0 the emissivity of the cable surface. With the factors k' and k'' for the mean temperature account is taken to the variable quantities of the air.

Fig. 18.10 [18.10; 18.11] facilitates the selection of auxiliary values for f_s, f_k and k for the arrangements selected as specified operating conditions (Table 18.4) [18.7; 18.10].

The cable shown in Fig. 18.10a radiates freely in all directions. The heat is transferred by radiation from the cable surface to the walls of the room in which the cable is situated. A decisive factor in the temperature rise of the cable surface at a constant rate of loss is the temperature of these walls which normally one would expect to be at ambient temperature.

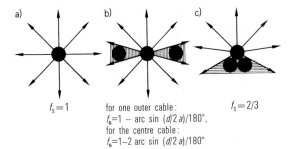

Heat dissipation by radiation

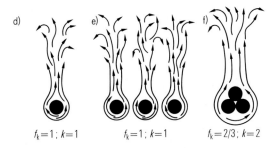

Heat dissipation by convection

Fig. 18.10 Heat dissipation, installed in free air

The emissivity of a cable surface can be taken as $\varepsilon_O = 0.95$.

The same considerations also apply for the arrangements 18.10b and 18.10c. However any obstruction to the thermal transfer must be considered. In Fig. 18.10b three single-core cables of a three-phase system are shown where only the thermal radiation from the centre cable is indicated. It is seen that the neighbouring cables obstruct heat transfer to the surroundings in the areas shown shaded. The reduction in heat dissipation is approximately directly proportional to the part of the cable surface embraced by the shaded angles.

In Fig. 18.10c three single-core cables are shown bunched in trefoil. The obstruction in this arrangement is greater than that of Fig. 18.10b since approximately one third of the cable surface considered does not radiate heat to the surroundings.

Fig. 18.10d illustrates free heat dissipation by convection. The heated air initially flows around the cable (laminar limiting layer) then rises upwards in laminar form mixing with cooler air from the surroundings in an area of turbulence. A decisive factor in the temperature rise of the cable surface is, in this instance, apart from the cable diameter and amount of losses, the temperature of the surrounding air. The selected clearances shown in Fig. 18.10e which are equal to the cable diameter do not obstruct the heat flow since the thickness of the flowing air stream is comparatively small.

In the bunched arrangement the cooling area of the cable is reduced to approximately two thirds. By reducing the cooling surface area the thermal flow within the cable is also hindered and because of this the thermal resistance of the cable is effectively increased [18.10]. This restriction in heat flow was taken into account when calculating the values shown in Fig. 18.9.

The temperature rise of the cable surface is:

$$\Delta \vartheta_O = \frac{(\vartheta_{Lr} - \vartheta_U - \Delta \vartheta_d)\, T'_{Lu}}{T'_{Ki} + T'_{Lu}} + P'_d\, T'_{Lu}. \tag{18.22}$$

and the temperature rise $\Delta \vartheta_d$ of the conductor caused by dielectric losses is:

$$\Delta \vartheta_d = P'_d (T'_{Kd} + T'_{Lu}). \tag{18.23}$$

The thermal resistance of the cable T'_{Lu} can be calculated e.g.:

▷ Calculation of temperature rise $\Delta \vartheta_O$ to equation 18.22 and 18.23 with $T'_{Lu} = 0.5$ Km/W;

▷ Calculation of thermal resistance T'_{Lu} to equation 18.17 to 18.21

▷ the calculations must be repeated n times until the difference between $(T'_{Lu})_n$ and $(T'_{Lu})_{n-1}$ is suffiently small.

For a multi-core cable without dielectric losses and with a 30 °C ambient temperature, the external thermal resistance can be reasonably accurately obtained from the curves shown in Fig. 18.11a. Where the dielectric loss can be ignored one obtains from equation 18.22

$$\frac{\Delta \vartheta_{Lr}}{\Delta \vartheta_O} = \frac{T'_{Ki} + T'_{Lu}}{T'_{Lu}}. \tag{18.22a}$$

By a graphical method, assuming a cable having a fictitious thermal resistance of $T'_{Ki} = 0.35$ Km/W with

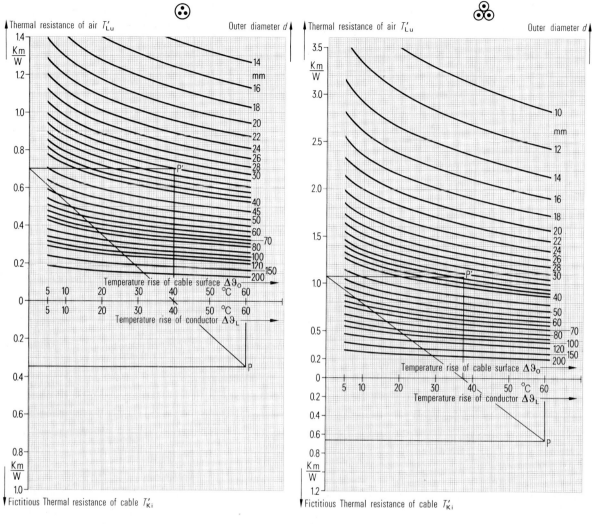

Fig. 18.11 a
Thermal resistance of air for a cable installed horizontally in free air
($\vartheta_U = 30\,°C$; $\varepsilon_O = 0.95$)

Fig. 18.11 b
Thermal resistance of air for three cables bunched in free air installed horizontally
($\vartheta_U = 30\,°C$; $\varepsilon_O = 0.95$)

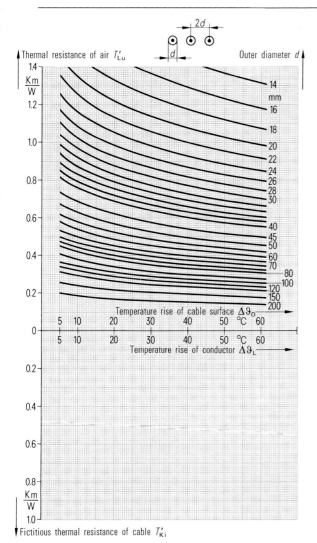

Thermal resistance of air T'_{Lu}

Outer diameter d

Fig. 18.11 c
Thermal resistance of air for three single-core cables installed side by side in free air

a permissible temperature rise of $\Delta\vartheta_{Lr} = 60$ K, entering these values as coordinates in Fig. 18.11a gives the point P. Through point P a straight line must be drawn such that point P', with the thermal resistance of air T'_{Lu} and the temperature difference value $\Delta\vartheta_0$ as coordinates, lies on the curve corresponding to the cable diameter $d = 32$ mm.

The following values are obtained from the graph:

$$T'_{Lu} = 0.7 \text{ Km/W},$$

$$\Delta\vartheta_0 = 40 \text{ K}$$

For *bunched single-core cables* [18.10] the thermal resistance of the insulation and the outer sheath is increased due to obstructed heat dissipation. For cables to Fig. 18.8 without a thermally conducting metal sheath one derives:

$$T'_1 = f_\varphi \frac{\varrho_1}{2\pi} \ln \frac{d_1}{d_L}, \qquad T'_3 = f_\varphi \frac{\varrho_3}{2\pi} \ln \frac{d}{d_M}, \qquad (18.24)$$

$$f_\varphi = \frac{\pi}{\pi - \varphi}, \qquad \varphi = \frac{\pi}{6} + \frac{\pi}{180°} \arcsin\left(\frac{d_L}{2d}\right). \qquad (18.25)$$

For cables with metallic covering and hence improved heat dissipation the following applies with additional reference to Table 18.29:

$$\frac{1}{T'_1} = \frac{1}{f_\varphi\, T_1^*} + \frac{1}{f_M\, T'_M}, \qquad (18.26)$$

$$T_1^* = \frac{\varrho_1}{2\pi} \ln \frac{d_1}{d_L}, \qquad (18.27)$$

$$T'_M = \sqrt{\frac{\varrho_M d_1}{4\delta_M^*} \pi T_1^*} \coth\left[\varphi \sqrt{\frac{\varrho_M}{4\delta_M^*} \frac{d_1}{\pi T_1^*}}\right] \qquad (18.28)$$

The thermal resistance of the outer protective covering T'_3 is calculated using equation 18.24 with equation 18.25.

The thermal resistivity ϱ_M of the metallic covering has to be taken into account and may be selected from Table 18.28.

Table 18.28 Thermal resistivity ϱ_M

Material	Thermal resistivity ϱ_M Km/W
Copper	$2.7 \cdot 10^{-3}$
Aluminium	$4.8 \cdot 10^{-3}$
Lead	$28.7 \cdot 10^{-3}$
Steel	$19.1 \cdot 10^{-3}$

Table 18.29   Values required for the calculation of the effective thermal thickness of a sheath or screen

	Sheath-, screen factor f_M	Thermal effective thickness of sheath or screen δ_M^*	Mean diameter of metal sheath or screen d_{Mm}
Metal sheath	1	δ_M	$d_M - \delta_M$
n Tapes with spacing	$\dfrac{\pi d_{Mm}}{nb(1+z)}$	$\delta\sqrt{1-\dfrac{1}{1+z^2}}$	$d_M - \delta$
Wire screen with a 5% increase in length due to the helix and with n transverse helical tapes	$\dfrac{\pi d_{Mm}}{nb(1+z)}$	$\delta\sqrt{1-\dfrac{1}{1+z^2}} \quad (\approx \delta)^{1)}$	$d_M - \delta$
Tape overlapped (as roof tiles) applied with $z=0.05$	1	$\approx \delta$	$d_M - 2\delta$
Two tapes applied without spacing.	1	$\delta\sqrt{1-\dfrac{1}{1+z^2}\dfrac{nb(1+z)}{\pi d_{Mm}}}$	$d_M - 2\delta$

$^{1)}$ At approximate consideration heat dissipation through the wires

Expression:
d_M Diameter over the metal sheath or screen (transverse helical tape) $\delta_M^* \approx \delta$
b Width of tape (transverse helical tape)
n Number of tapes (transverse helical tape)
z Increase in length, due to the helical wound construction of tape (transverse helical tape)
δ Thickness of each tape (transverse helical tape)
δ_M Thickness of metal sheath

Example 18.3

For three single-core cables NA2XS2Y 1×150 RM/25 12/20 kV bunched in free air, the following applies (Dimensions see example 18.2, page 185):

$$\varphi = \frac{\pi}{6} + \frac{\pi}{180°} \arcsin\left(\frac{d_L}{2d}\right) = \frac{\pi}{6} + \frac{\pi}{180°} \arcsin\left(\frac{14.5}{2 \times 35.7}\right) = 0.728, \tag{18.25}$$

$$f_\varphi = \frac{\pi}{\pi - \varphi} = \frac{\pi}{\pi - 0.728} = 1.302, \tag{18.25}$$

$$T_1^* = \frac{\varphi_1}{2\pi} \ln\frac{d_1}{d_L} = \frac{3.5}{2\pi} \ln\frac{28.5}{14.5} = 0.376 \frac{Km}{W}, \tag{18.16}$$

$$\delta_M^* \approx \delta = 0.2 \text{ (Table 18.29)}$$

$$f_M = \frac{\pi d_{Mm}}{nb(1+z)} = \frac{\pi(d_M - \delta)}{nb(1+z)} = \frac{\pi(29.9 - 0.2)}{1 \times 5(1+0.30)} = 14.35 \text{ (Table 18.29)},$$

$$T'_M = \sqrt{\frac{\varrho_M d_1}{4\delta^*_M} \pi T^*_1} \coth\left[\varphi \sqrt{\frac{\varrho_M d_1}{4\delta^*_M} \frac{1}{\pi T^*_1}}\right] \tag{18.28}$$

$$= \sqrt{\frac{2.7 \times 10^{-3} \times 28.5 \times 10^{-3}}{4 \times 0.2 \times 10^{-3}} \pi 0.376} \times \coth\left[0.728 \sqrt{\frac{2.7 \times 10^{-3} \times 28.5 \times 10^{-3}}{4 \times 0.2 \times 10^{-3}} \frac{1}{\pi 0.376}}\right]$$

$$= 1.646 \frac{Km}{W},$$

$$\frac{1}{T'_1} = \frac{1}{f_\varphi T^*_1} + \frac{1}{f_M T'_M} = \frac{1}{1.302 \times 0.376} + \frac{1}{14.35 \times 1.646} = \frac{1}{0.480}, \tag{18.26}$$

$$T'_1 = 0.480 \frac{Km}{W},$$

$$T'_3 = f_\varphi \frac{\varphi_3}{2\pi} \ln \frac{d}{d_M} = 1.302 \frac{3.5}{2\pi} \ln \frac{35.7}{29.9} = 0.129 \frac{Km}{W},$$

$$T'_{Ki} = \frac{T_1}{1 + \lambda_1} + T_3 = \frac{0.480}{1 + 0.0163} + 0.129 = 0.601 \frac{Km}{W}. \tag{18.10}$$

For the three single-core cables bunched in free air one arrives after several iterations

$$\Delta \vartheta_0 = \frac{(\vartheta_{Lr} - \vartheta_U) T'_{Lu}}{T'_{Ki} + T'_{Lu}} = \frac{(90 - 30) 1.06}{0.601 + 1.06} = 38.3 \text{ K}, \tag{18.22}$$

$$\vartheta_0 = \Delta \vartheta_0 + \vartheta_u = 38.3 + 30 = 68.3 \text{ K}, \tag{18.22a}$$

$$\vartheta_m = \frac{\vartheta_0 + \vartheta_U}{2} = \frac{\Delta \vartheta_0 + 2\vartheta_U}{2} = \frac{38.3 + 2 \times 30}{2} = 49.15 \,^\circ C, \tag{18.20}$$

$$k' = 0.919 + \frac{\vartheta_m}{369} = 0.919 + \frac{49.15}{369} = 1.05, \tag{18.19}$$

$$k'' = 1.033 - \frac{\vartheta_m}{909} = 1.033 - \frac{49.15}{909} = 0.98. \tag{18.19}$$

According to Fig. 18.10c and f: $k = 2$, $f_s = 2/3$, $f_k = 2/3$

$$\alpha_k = k' \frac{0.0185}{kd} + k'' 1.08 \left(\frac{\Delta \vartheta_0}{kd}\right)^{\frac{1}{4}} = 1.05 \frac{0.0185}{2 \times 35.7 \times 10^{-3}} + 0.98 \times 1.08 \left(\frac{38.3}{2 \times 35.7 \times 10^{-3}}\right)^{\frac{1}{4}} = 5.366 \frac{W}{Km^2}, \tag{18.18}$$

$$\alpha_s = \frac{\varepsilon_0 \sigma [(273 + \vartheta_0)^4 - (273 + \vartheta_U)^4]}{\Delta \vartheta_0} = \frac{0.95 \times 5.67 [(273 + 68.3)^4 - (273 + 30)^4]}{38.3 \times 10^8} = 7.229 \frac{W}{Km^2}, \tag{18.21}$$

$$T'_{Lu} = \frac{1}{\pi d (f_k \alpha_k + f_s \alpha_s)} = \frac{1}{\pi 35.7 \times 10^{-3} (\frac{2}{3} 5.366 + \frac{2}{3} 7.229)} = 1.06 \frac{Km}{W}, \tag{18.17}$$

$$I_r = \sqrt{\frac{\vartheta_{Lr} - \vartheta_U}{n R_{wr} (T'_{Ki} + T'_{Lu})}} = \sqrt{\frac{90 - 30}{1 \times 0.269 \times 10^{-3} (0.601 + 1.06)}} = 366 \text{ A}. \tag{18.14}$$

The dielectric losses are ignored, see Section 18.3.

Vertical Installation

All known methods of calculation and descriptions of values for load capacity relate to cables installed in the horizontal plane. For vertically installed cables neither theoretical nor experimental investigation is known.

Whether the mounting is vertical or horizontal has, in principle, no influence on the heat dissipated by radiation.

The heat dissipated by convection from a vertical cylinder of length l and diameter d is, with all other conditions equal, more favourable than that of the cylinder in horizontal position, provided that

$$\frac{l}{d} \geq 2.77$$

refer [18.12; 18.13]. Since this relationship is always satisfied in respect of cables installed in a vertical plane it follows that they can normally withstand heavier loads than when they are installed horizontally. Thus the same load capacity can be applied to both conditions.

Atmospheric Pressure

The heat dissipation by convection decreases with decrease in atmospheric pressure [18.9]. For high altitudes the thermal heat transfer constant for convection must be modified as follows:

$$\alpha_k = k' \frac{0,0185}{kd} + k'' 1,08 \left(\frac{\Delta\vartheta_0}{kd}\right)^{\frac{1}{4}} \left(\frac{p}{1013 \text{ hPa}}\right)^{\frac{1}{4}},$$

$$(18.18\,a)$$

with k' and k'' to equation 18.19 and k to Fig 18.10.

Altitude m	0	1000	2000	3000	4000
Atmospheric pressure p hPa	1013	899	795	701	616

Altitudes up to 2000 m create a negligible reduction in load capacity.

Example 18.4

At altitudes of 3000 m above sea level an atmospheric pressure of 701 hPa is used for the calculation.
For the cable

NA2XS2Y 1×150 RM/25 12/20 kV

this gives, after several iterations and, using dimensions from example 18.2 (page 185) and T'_{Ki} from example 18.3 (page 190)

$$\Delta \vartheta_O = \frac{(\vartheta_{Lr} - \vartheta_U)\, T'_{Lu}}{T'_{Ki} + T'_{Lu}} = \frac{(90 - 30)\, 1.134}{0.601 + 1.134} = 39.22 \text{ K}, \tag{18.22}$$

$$\vartheta_O = \Delta \vartheta_O + \vartheta_u = 39.22 + 30 = 69.22 \text{ K}, \tag{18.22 a}$$

with $P'_d = 0$

$$\vartheta_m = \frac{\Delta \vartheta_O + 2\, \vartheta_U}{2} = \frac{39.22 + 2 \times 30}{2} = 49.61 \,^\circ\text{C}, \tag{18.20}$$

$$k' = 0.919 + \frac{\vartheta_m}{369} = 0.919 + \frac{49.61}{369} = 1.0534, \tag{18.19}$$

$$k'' = 1.033 - \frac{\vartheta_m}{909} = 1.033 - \frac{49.61}{909} = 0.9784, \tag{18.19}$$

$$\alpha_k = k'\frac{0.0185}{kd} + k''\, 1.08 \left(\frac{\Delta \vartheta_O}{kd}\right)^{\frac{1}{4}} \left(\frac{p}{1013}\right)^{\frac{1}{2}} = 1.0534\, \frac{0.0185}{2 \times 35.7 \times 10^{-3}} + 0.9784 \times 108 \left(\frac{39.22}{2 \times 35.7 \times 10^{-3}}\right)^{\frac{1}{4}} \left(\frac{701}{1013}\right)^{\frac{1}{2}}$$

$$= 4.528\, \frac{\text{W}}{\text{Km}^2}, \tag{18.18 a}$$

$$\alpha_s = \frac{\varepsilon_O\, \sigma\, [(273 + \vartheta_O)^4 - (273 + \vartheta_U)^4]}{\Delta \vartheta_O} = \frac{0.95 \times 5.67\, [(273 + 69.22)^4 - (273 + 30)^4]}{39.22 \times 10^8} = 7.261\, \frac{\text{W}}{\text{Km}^2}. \tag{18.21}$$

$$T'_{Lu} = \frac{1}{\pi d (f_k\, \alpha_k + f_s\, \alpha_s)} = \frac{1}{\pi\, 35.7 \times 10^{-3}\, (\frac{2}{3}\, 4.528 + \frac{2}{3}\, 7.261)} = 1.134\, \frac{\text{Km}}{\text{W}}, \tag{18.17}$$

$$I_z = \sqrt{\frac{\vartheta_{Lr} - \vartheta_U}{n\, R'_{wr}\, (T'_{Ki} + T'_{Lu})}} = \sqrt{\frac{90 - 30}{1 \times 0.269 \times 10^{-3}\, (0.601 + 1.134)}} = 359 \text{ A}. \tag{18.12}$$

The load capacity at this altitude is therefore reduced by a factor

$$f = \frac{I_z}{I_r} = \frac{359}{366} = 0.981.$$

Ambient Temperature

The thermal resistance of air around a cable varies only to a small extent at constant conductor temperature and rising ambient temperature. Normally it is sufficient therefore to use equation 18.15 for the calculation of load capacity for other ambient temperatures. This formula was used to obtain the values in Table 18.22.

Example 18.5

For an ambient temperature deviating from $30\,°C$ e.g. $\vartheta_U = 45\,°C$ the conversion factor is

$$f = \frac{I_z}{I_r} = \sqrt{\frac{\vartheta_{Lr} - \vartheta_U}{\vartheta_{Lr} - 30\,°C}} = \sqrt{\frac{90 - 45}{90 - 30}} = 0.87 \quad (18.15)$$

and the load capacity

$$I_z = f I_r = 0.87 \times 366 = 318 \text{ A}.$$

Solar Radiation

Cables subjected to solar radiation are subject to an additional temperature rise

$$\Delta \vartheta_S = \alpha_O \, d \, E \, T_S' \quad (18.29)$$

and the cable surface temperature rise in relation to ambient temperature is

$$\Delta \vartheta_{OS} = \frac{(\vartheta_{Lr} - \vartheta_U - \Delta \vartheta_d + \alpha_O \, d \, E \, T_{Ki}') \, T_S'}{T_{Ki}' + T_S'} + P_d' \, T_S'. \quad (18.30)$$

The load capacity I_z is found from

$$I_z = \sqrt{\frac{\vartheta_{Lr} - \vartheta_U - \Delta \vartheta_d - \Delta \vartheta_S}{n \, R_{wr} \, (T_{Ki}' + T_S')}} \quad (18.31)$$

and the thermal resistance T_S' of air, taking into account the solar radiation, by iteration using the equations 18.17 to 18.21. For this the term T_{Lu}' in equation 18.17 must be replaced by T_S' and in equations 18.18 to 18.20 the calculation is made using $\Delta \vartheta_{OS}$ from equation 18.30 instead of $\Delta \vartheta_O$ and ϑ_{OS} instead of ϑ_O.

The absorption coefficient α_O of solar radiation for the cable surface can be found in Table 18.30 [18.2].

Table 18.30
Absorptivity of cable surfaces to solar radiation

Material of the outer protective cover	Absorptivity α_O
Asphalted jute	0.8
PVC	0.6
PE	0.4
Polychloroprene	0.8
Lead	0.6

The intensity of solar radiation E on a horizontal plane is 1.35 kW/m^2 maximum (solar constant). Normally the actual values are less and depend on the degree of lattitude, season, weather conditions, time of day etc. [18.14; 18.15]. Should local values not be available the value of $E = 1 \text{ kW/m}^2$ can be used in calculations [18.2].

Example 18.6

Three single-core cables bunched in free air

NA2XS2Y 1×150 RM/25 12/20 kV

are exposed to solar radiation with intensity $E = 1 \text{ kW/m}^2$.

The calculation of thermal resistance T'_S is made by iteration:

$$\Delta\vartheta_{os} = \frac{(\vartheta_{Lr} - \vartheta_U + \alpha_O \, d \, E \, T'_{Ki}) \, T'_S}{T'_{Ki} + T'_S} = \frac{(90 - 30 + 0.4 \times 35.7 \times 10^{-3} \times 1.0 \times 10^{+3} \times 0.601) \, 1.035}{0.601 + 1.035} = 43.39 \text{ K}, \quad (18.22)$$

$$\vartheta_o = \Delta\vartheta_{os} + \vartheta_U = 43.39 + 30 = 73.39 \text{ K},$$

$$\vartheta_m = \frac{\Delta\vartheta_{os} + 2\vartheta_U}{2} = \frac{43.39 + 2 \times 30}{2} = 51.7 \text{ °C}, \tag{18.20}$$

$$k' = 0.919 + \frac{\vartheta_m}{369} = 0.919 + \frac{51.7}{369} = 1.0591, \tag{18.19}$$

$$k'' = 1.033 - \frac{\vartheta_m}{909} = 1.033 - \frac{51.7}{909} = 0.9761, \tag{18.19}$$

$$\alpha_k = k' \frac{0.0185}{kd} + k'' \, 1.08 \left(\frac{\Delta\vartheta_{os}}{kd}\right)^{\frac{1}{4}} = 1.0591 \frac{0.0185}{2 \times 35.7 \times 10^{-3}} + 0.9761 \times 1.08 \left(\frac{43.39}{2 \times 35.7 \times 10^{-3}}\right)^{\frac{1}{4}} = 5.509 \frac{\text{W}}{\text{Km}^2}, \tag{18.18}$$

$$\alpha_s = \frac{\varepsilon_O \, \sigma \, [(273 + \vartheta_O)^4 - (273 + \vartheta_U)^4]}{\Delta\vartheta_{os}} = \frac{0.95 \times 5.67 \, [(273 + 73.39)^4 - (273 + 30)^4]}{43.39 \times 10^8} = 7.408 \frac{\text{W}}{\text{Km}^2}, \tag{18.21}$$

$$T'_S = \frac{1}{\pi d(f_k \, a_k + f_s \, \alpha_s)} = \frac{1}{\pi \times 35.7 \times 10^{-3} \, (\frac{2}{3} 5.509 + \frac{2}{3} 7.408)} = 1.035 \frac{\text{Km}}{\text{W}}, \tag{18.17}$$

$$\Delta\vartheta_S = \alpha_O \, d \, E \, T'_S = 0.4 \times 35.7 \times 10^{-3} \times 1.0 \times 10^3 \times 1.035 = 14.8 \text{ K}, \tag{18.29}$$

$$I_z = \sqrt{\frac{\vartheta_{Lr} - \vartheta_U - \Delta\vartheta_S}{n \, R_{wr} \, (T'_{Ki} + T'_S)}} = \sqrt{\frac{90 - 30 - 14.8}{1 \times 0.269 \times 10^{-3} (0.601 + 1.035)}} = 320 \text{ A}, \tag{18.12}$$

The load capacity with solar radiation intensity $E = 1.0 \text{ kW/m}^2$ is reduced by the factor

$$f = \frac{I_z}{I_r} = \frac{320}{366} = 0.87.$$

Arrangement of Cables

Heat dissipation of cables is affected when they are in contact with surfaces (walls, floors, ceilings). At the point of contact the flow of air is hindered and therefore the heat dissipation by convection is reduced. Heat dissipation by radiation is influenced by the emissivity and the temperature of the adjacent area in contact, should this differ from ambient. In direct contact heat may also be transmitted by conduction, so that the thermal conductivity of the adjacent area is important (Fig. 18.12).

Quantities of load capacity for cables in contact with surfaces established by experiment are normally less than for installations in free air. In the VDE specifications this type of installation is taken into consideration by use of a reduction factor of 0.95.

Grouping cables can also hinder heat dissipation as has been shown in the calculations of thermal resistance of air for three single-core cables (Fig. 18.10).

In Fig. 18.13 various arrangements of cables and the relevant rating factors to DIN VDE 0298 are shown. On the left hand side it can be seen that by arranging cables close together or by mounting on a solid surface the convection reduction is made worse. Similar comments apply to the vertical arrangement of cables. The reduction is however relatively greater than that for the horizontal arrangement since the upper cable lies in the path of warm air flow from lower cables. The chimney effect (improvement of heat dissipation by convection through moving air flows) is somewhat reduced. The greatest reduction occurs in densely filled troughs or racks as can be frequently found in cable trenches of large power installations (Table 18.23 and Section 18.5).

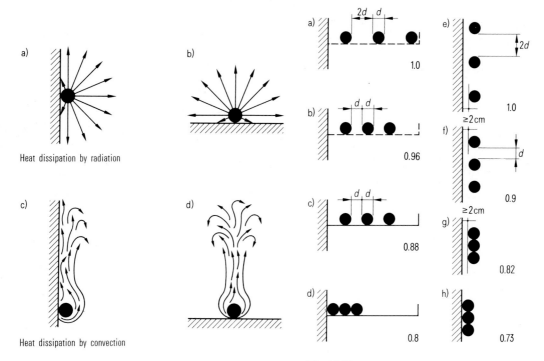

Heat dissipation by radiation

Heat dissipation by convection

Fig. 18.12
Obstruction of heat dissipation by adjacent surfaces

Fig. 18.13
Reduction factors for various arrangements of multi-core cables in air

18.4.3 Thermal Resistance of the Soil

Temperature Field of a Cable in the Ground

The heat loss P' generated in a cable flows through the surrounding soil to the surface of the ground where it is then dissipated into the atmosphere. To depict the temperature field of the cable in ground one normally assumes a constant ground temperature ϑ_E and a soil-thermal resistivity ϱ_E with a negligible thermal transfer resistance at the ground surface. Also it is assumed that the total heat loss generated in the cable (source) is directed to zero in an imaginary cable situated as a mirror image in relation to the ground surface. The temperature rise at the point P relative to the temperature of the ground ϑ_E, specifically the surface of the ground (Fig. 18.14), is obtained from [18.16]

$$\Delta\vartheta_P = P' \frac{\varrho_E}{2\pi} \ln \frac{c'_P}{c_P}. \qquad (18.32)$$

The isothermal lines are determined by the condition $\Delta\vartheta_P = $ constant and therefore the relation c'_P/c_P must also be constant.

$$k_P = \frac{c'_P}{c_P}. \qquad (18.33)$$

For a given heat loss P' and temperature rise $\Delta\vartheta_P$ one obtains from (18.32) and (18.33) the geometric constant for the isotherms

$$k_P = \exp\left(\frac{2\pi \Delta\vartheta_P}{\varrho_E P'}\right). \qquad (18.34)$$

The expressions for determining the isothermal line through point P for a cable with a diameter $d = 2r$ and a depth of lay h (Fig. 18.14) are as follows:

radius of the isotherm

$$r_P = \frac{2h_0 k_P}{k_P^2 - 1}, \qquad (18.35)$$

depth of the isotherm

$$h_P = h_0 (k_P^2 + 1)/(k_P^2 - 1) = h_0 + e_P, \qquad (18.36)$$

eccentricity of the isotherm

$$e_P = 2h_0/(k_P^2 - 1) = h_P - h_0. \qquad (18.37)$$

One can visualise a temperature field comprising a series of lines which at distance e (eccentricity of the

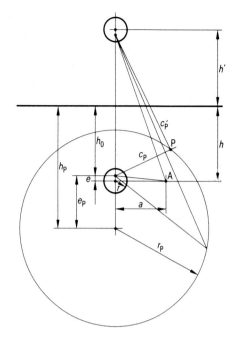

Fig. 18.14
Temperature field of a cable of diameter $d = 2r$ and depth of lay h.
Further explanations in the text

cable) run parallel to the cable axis. The depth of this line source is h_0 and the isothermal lines are eccentric to this by a distance e_P.

For the depth of the line source

$$h_0 = \sqrt{h^2 - r^2} \qquad (18.38)$$

and the eccentricity of the cable

$$e = h - h_0 = h - \sqrt{h^2 - r^2}. \qquad (18.39)$$

If the depth h_P and the radius r_P of the isotherms are known, their geometric constant can be found from

$$k_P = \frac{h_0 + h_P - r_P}{h_0 - h_P + r_P} \qquad (18.40)$$

with

$$h_0 = \sqrt{h_P^2 - r_P^2}. \qquad (18.41)$$

197

For a point P on the surface of the ground becomes $c'_P/c_P = 1$ and the temperature rise by definition is zero.

If a point A at a distance a from the cable axis and h from the surface of the ground is imagined the temperature rise at this point can be determined from equation 18.32 and equation 18.38:

$$\Delta \vartheta_a = P' \frac{\varrho_E}{2\pi} \ln k_a$$

with the geometric constant for mutual heating

$$k_a = \sqrt{\frac{(2h-e)^2 + a^2}{e^2 + a^2}}$$

$$= \sqrt{\frac{(h + \sqrt{h^2 - r^2})^2 + a^2}{(h - \sqrt{h^2 - r^2})^2 + a^2}}. \tag{18.42}$$

Definition of Soil-Thermal Resistance

The temperature rise of the cable surface is obtained by putting $a = r$ in equation 18.42 and, after some manipulation as well as putting $d = 2r$, to

$$\Delta \vartheta_E = P' \frac{\varrho_E}{2\pi} \ln k \tag{18.43}$$

with the geometric constant for the cable

$$k = \frac{2h}{d} + \sqrt{\left(\frac{2h}{d}\right)^2 - 1}. \tag{18.44}$$

This also defines the soil-thermal resistance of a cable, that is the thermal resistance between the cable surface and earth (Fig. 18.15a):

$$T'_E = \frac{\varrho_E}{2\pi} \ln k$$

$$= \frac{\varrho_E}{2\pi} \ln \left(\frac{2h}{d} + \sqrt{\left(\frac{2h}{d}\right)^2 - 1}\right). \tag{18.45}$$

Correspondingly the soil-thermal resistance between the ground surface and the isotherm through a point P is established (Fig. 18.15b):

$$T'_{EP} = \frac{\varrho_E}{2\pi} \ln k_P, \tag{18.46}$$

where the geometric constant of this isothermal line is determined using equation 18.32 and equation 18.40.

a) Cable, T'_E to equation 18.41

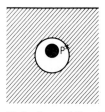

b) Isothermal line through point P, T'_{EP} to equation 18.46

Fig. 18.15
Soil-thermal resistance (the range considered in formula is shaded in each case)

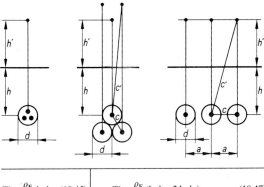

$T'_E = \frac{\varrho_E}{2\pi} \ln k$ (18.45)	$T'_E = \frac{\varrho_E}{2\pi} (\ln k + 2 \ln k_a)$ (18.47)
$k = \frac{2h}{d} + \sqrt{\left(\frac{2h}{d}\right)^2 - 1}$ (18.44)	or $k \approx \frac{4h}{d}$ (18.49)
—	$k_a \approx \frac{2h}{d}$ (18.48) $\quad$ $k_a \approx \sqrt{\left(\frac{2h}{a}\right)^2 + 1}$ (18.50)

a) Multi-core cable

b) Single-core cables, bunched

c) Single-core cables, side by side

Fig. 18.16
Soil-thermal resistance for continuous operation $m = 1.0$ without drying out, $\varrho_E =$ constant

For three single-core cables in a three-phase system – producing equal losses in all three cables –:

$$T'_E = \frac{\varrho_E}{2\pi} (\ln k + 2 \ln k_a), \tag{18.47}$$

with the geometric constant k of one cable as in Fig. 18.16. In bunched installations – arranged in trefoil – the geometric constant k_a for grouping can be found approximately from

$$k_a \approx 2h/d. \qquad (18.48)$$

Normally the depth of lay is very large in relation to the radius of the cable. The eccentricity of the cable then becomes neglegible and one obtains simplifications of the equation in Fig. 18.16 for continuous operation without drying out of the ground. This means the load is constant in time and also the soil-thermal resistivity is constant.

For a relationship where $h/d \geq 5$ the value found by calculation using equation 18.49 deviates by less than 1‰ from the value given by equation 18.44.

Daily Load Curve and Characteristic Diameter d_y

With cyclic operation the load capacity is greater than for continuous operation.

In continuous operation (Fig. 18.17) one obtains, after a warm-up period following the switch on, a constant temperature distribution in the ground which falls in a near logarithmic manner from the

cable surface to the ambient temperature. For a cyclicly changing load over a long period, after the switch on period one sees – between fixed temperature limits – a temperature curve varying against time. Near to the cable the temperature change is most extreme but this decreases with increase of distance from the cable.

If one considers the thermal field of a cable in the ground (Fig. 18.18) the areas within the isothermal lines can be depicted, for calculation purposes, by partial heat resistances and capacitors so that a chain of RC components is developed. A calculation of temperature rise and also of the load capacity is possible utilising this equivalent diagram. Accurate results, however, can only be achieved by very involved calculation.

For daily load curves including the pattern for urban utility supply networks, a method is used which provides a sufficiently accurate result with a reduced amount of calculation and is suitable for the load factors ranging $m = 0.5$ to 1.0. This type of operation is described in more detail in Section 18.2.3.

To simplify calculations the so called characteristic diameter d_y is introduced (Fig. 18.19). The tempera-

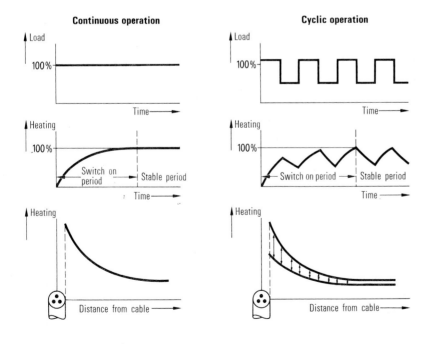

Fig. 18.17
Heating of the ground by continuous operation and cyclic operation

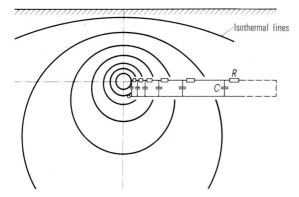

Fig. 18.18
Theoretical development of isothermal lines in the ground and substitution of layers between the individual lines by a chain of R/C components

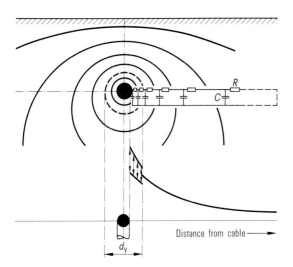

Fig. 18.19
Heating of the ground in cyclic operation

the daily load characteristic curve. Outside the characteristic diameter a constant temperature exists (i.e. the thermal capacitors are charged up during the warm-up phase and do not enter into the calculation for the steady state condition).

The loss factor μ for the determination of the mean current heat loss is

$$\mu = 0.3\, m + 0.7\, m^2. \tag{18.51}$$

The characteristic diameter d_y is dependant upon the thermal characteristics of the ground, the frequency of w equal fluctuations over 24 hour period and on the loss factor μ.

For the characteristic diameter d_y in m with a load factor to satisfy $0.5 \leq m \leq 1.0$ and for a sinusoidal load variation:

$$d_y = \frac{0.205}{\sqrt{w}\left(\dfrac{\varrho_E}{\mathrm{Km/W}}\right)^{0.4}}, \tag{18.52}$$

for rectalinear load variation

$$d_y = \frac{0.493\sqrt{\dfrac{\mu}{w}}}{\left(\dfrac{\varrho_E}{\mathrm{Km/W}}\right)^{0.4}}, \tag{18.53}$$

for an average shape of load variation which is neither sinusoidal nor rectalinear

$$d_y = \frac{0.103 + 0.246\sqrt{\mu}}{\sqrt{w}\left(\dfrac{\varrho_E}{\mathrm{Km/W}}\right)^{0.4}}. \tag{18.54}$$

Table 18.31
Loss factors and characteristic diameters for a soil-thermal resistivity of 1.0 Km/W and daily load curve with maximum load

Load factor m	Loss factor μ from equation 18.51	Characteristic diameter d_y in m		
		Sinusoidal load from equation 18.52	Rectalinear load from equation 18.53	Mixed load from equation 18.54
0.5	0.325	0.205	0.281	0.243
0.6	0.432	0.205	0.324	0.265
0.7	0.553	0.205	0.367	0.286
0.8	0.688	0.205	0.409	0.307
0.9	0.837	0.205	0.451	0.328

ture rise outside the characteristic diameter is determined by the average loss with dependance on the load factor however, the highest degree of temperature rise within the area embraced by the characteristic diameter is dependant on the maximum value of load. Within the characteristic diameter the temperature varies with time to a curve which closely follows

The geometric constant k_y of the circle with the characteristic diameter d_y is obtained from the analogy of equation 18.44:

$$k_y = \frac{2h}{d_y} + \sqrt{\left(\frac{2h}{d_y}\right)^2 - 1} \approx \frac{4h}{d_y}. \qquad (18.55)$$

Drying-Out of the Soil and Boundary Isotherm d_x

By reference to Fig. 18.20 it will be seen that at a certain load, which is limited only by the maximum permissible operating temperature, the surface of cables of different types of construction will assume different surface temperatures. While the surface temperature of a mass-impregnated cable may be approximately 45 °C, the surface of an XLPE cable can reach 75 °C (at 20 °C ground temperature, degree of loading 1.0 and assuming the soil does dry-out). The difference is significant. It is known that sandy soils tend to dry-out when the cable surface temperature is approximately 30 °C installed in a 20 °C ground temperature. The danger of drying-out is higher where XLPE cables are used than where mass-impregnated cables are installed. This danger also increases with increasing load factor.

This drying-out area (Fig. 18.21) is indicated by an isothermal line excentric to the cable – the boundary isotherm – having a diameter d_x.

Based on the stipulations given in DIN VDE 0298 Part 2 the limiting temperature rise $\Delta\vartheta_x$ can be derived from the equation

$$\Delta\vartheta_x = 15 + \frac{(1-m)\,100}{3} \qquad (18.56)$$

and this results in

$\Delta\vartheta_x = 15$ K for continuous operation with $m = 1.0$,

$\Delta\vartheta_x = 25$ K for utility load operation with $m = 0.7$,

$\Delta\vartheta_x = 32$ K for daily load curve with $m = 0.5$.

Within the boundary isothermal line the soil-thermal resistivity can be taken as $\varrho_x = 2.5$ Km/W representing the almost completely dried-out sandy soil or the sand used as bedding material. Outside of the boundary isotherm the value $\varrho_E = 1.0$ Km/W is used which represents almost all natural types of soil in European latitudes.

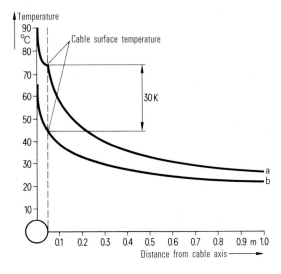

a Three XLPE cables
   NA2XS2Y $1 \times 150/25$ RM 12/20 kV

b One mass-impregnated cable
   NAEKEBY 3×150 RM 12/20 kV

Fig. 18.20
Maximum heating of the ground by different cables

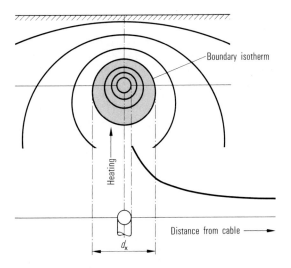

Fig. 18.21
Heating of the ground at differing thermal resistivities due to drying-out of the soil

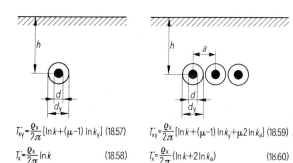

$$T'_{xy} = \frac{\varrho_x}{2\pi}[\ln k + (\mu-1)\ln k_y] \quad (18.57)$$

$$T'_x = \frac{\varrho_x}{2\pi}\ln k \quad (18.58)$$

a) Multi-core cable

$$T'_{xy} = \frac{\varrho_x}{2\pi}[\ln k + (\mu-1)\ln k_y + \mu 2\ln k_a] \quad (18.59)$$

$$T'_x = \frac{\varrho_x}{2\pi}(\ln k + 2\ln k_a) \quad (18.60)$$

b) Single-core cables side by side with $d_y \leq 2a$

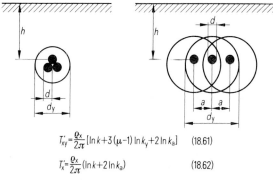

$$T'_{xy} = \frac{\varrho_x}{2\pi}[\ln k + 3(\mu-1)\ln k_y + 2\ln k_a] \quad (18.61)$$

$$T'_x = \frac{\varrho_x}{2\pi}(\ln k + 2\ln k_a) \quad (18.62)$$

c) Single-core cables bunched

d) Single-core cables side by side with $d_y \geq 2a$

k and k_a to Fig. 18.16, k_y to equation 18.55

Fig. 18.22
Formulae for calculation of the fictitious soil-thermal resistances T'_{xy} and T'_x with daily load cycle $m < 1.0$ and drying-out of the soil

Fictitious Soil-Thermal Resistances T'_x and T'_{xy}

The fictitious soil-thermal resistances T'_x and T'_{xy} take into account the cyclic performance of a daily load curve and drying-out of the soil. These can be calculated using the equations in Fig. 18.22 but can also be taken, for some arrangements, from Fig. 18.23. For the calculation of load capacity these resistances are to be incorporated into equation 18.63.

Load Capacity

Previously the values for load capacity in the ground were calculated using the rules for continuous operation but did not take account of drying-out of the soil, which was permitted only for defined public utility load type of operation. For continuous operation the recommendation was to use either the factor 0.75 or a factor which corresponded to a sufficiently high selected soil-thermal resistivity.

Load capacity is now – as explained in the previous section – (DIN VDE 0298 Part 2) calculated using a method which takes into account drying-out of the soil together with maximum load and load factor, which is derived from a daily load curve.

Load capacity can be found from

$$I_z = \sqrt{\frac{\vartheta_{Lr} - \vartheta_E - P'_d(T'_{Kd} + T'_x) + [(\varrho_x/\varrho_E) - 1]\Delta\vartheta_x}{nR'_{wr}(T'_{Ki} + T'_{xy})}},$$
$$(18.63)$$

with the individual terms or values determined as follows:
Load factor m to Fig. 18.1,
loss factor μ to equation 18.51,
characteristic diameter d_y to equation 18.54 or Table 18.2 for the thermal resistivity $\varrho_E = 1$ Km/W,
limiting temperature rise $\Delta\vartheta_x$ to equation 18.56,
geometric factors k, k_a and k_y to Fig. 18.16 and equation 18.55,
thermal resistances T'_x and T'_{xy} to Fig. 18.22 providing it is established, where necessary, that $d_y > 2a$ or $d_y \leq 2a$.

The diameter of the dry area d_x is not essential for the calculation but it must be verified whether the assumption that the soil is drying-out does apply, that means $d_x > d$ respectively $\vartheta_0 > \vartheta_x$:

$$\vartheta_0 = \vartheta_{Lr} - P'_i T'_{Ki} - P'_d T'_{Kd}, \quad (18.64)$$

$$\vartheta_x = \vartheta_E + \Delta\vartheta_x. \quad (18.65)$$

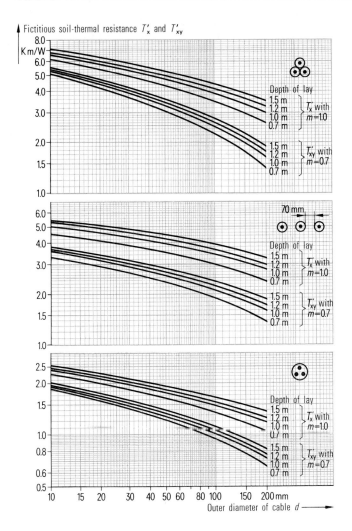

Fictitious soil-thermal resistance T'_x and T'_{xy}

Fig. 18.23
Fictitious soil-thermal resistance T'_x at $m=1.0$, T'_{xy} at $m=0.7$ relative to outer diameter d of cable and depth of lay h for a soil-thermal resistivity of $\varrho_x = 2.5$ Km/W and $\varrho_E = 1.0$ Km/W

The ohmic losses in equation 18.64 must be determined using the load capacity calculated for drying-out the soil. If the surface temperature is found to be less than the temperature of the boundary isotherm, the calculation for load capacity must be repeated but under the assumption that the soil does not dry-out. The calculation routine described above must be repeated with $\varrho_x = \varrho_E$. With the new value for capacity $\vartheta_O \le \vartheta_x$ must be satisfied.

To simplify this calculation the characteristic diameter is to be determined using the thermal resistivity of the moist area. A comparison of diameters is therefore avoided and the result is on the safe side since the lower thermal resistivity results in a maximum value for the characteristic diameter.

Example 18.7

Three single-core cables

NA2XS2Y 1 × 150 RM/25 12/20 kV

are installed in ground under different operating conditions.
Dimensions and thermal resistane of the cable can be taken from example 18.2 on page 185
or from Part 2, Table 5.6.6a.

Bunched installation for the specified operating conditions to Table 18.2.

Type of operation: Supply utility operation with $m = 0.7$ or any equivalent load variation
(Fig. 18.6) with a frequency of load cycles $w = 1$.

$$\mu = 0.3\,m + 0.7\,m^2 = 0.3 \times 0.7 + 0.7 \times 0.7^2 = 0.553, \tag{18.51}$$

$$d_y = \frac{0.103 + 0.246\,\sqrt{\mu}}{\sqrt{w}\left(\dfrac{\varrho_E}{\text{Km/W}}\right)^{0.4}} = \frac{0.103 + 0.246\,\sqrt{0.553}}{\sqrt{1}\,(1.0)^{0.4}} = 0.286 \text{ m}, \tag{18.54}$$

$$k = \frac{2h}{d} + \sqrt{\left(\frac{2h}{d}\right)^2 - 1} = \frac{2 \times 0.7}{35.7 \times 10^{-3}} + \sqrt{\left(\frac{2 \times 0.7}{35.7 \times 10^{-3}}\right)^2 - 1} = 78.42\,\frac{\text{Km}}{\text{W}}, \tag{18.44}$$

$$k_y = \frac{2h}{d_y} + \sqrt{\left(\frac{2h}{d_y}\right)^2 - 1} = \frac{2 \times 0.7}{0.286} + \sqrt{\left(\frac{2 \times 0.7}{0.286}\right)^2 - 1} = 9.69\,\frac{\text{Km}}{\text{W}}, \tag{18.55}$$

$$k_a \approx \frac{2h}{d} = \frac{2 \times 0.7}{35.7 \times 10^{-3}} = 39.23\,\frac{\text{Km}}{\text{W}}, \tag{18.48}$$

$$T'_{xy} = \frac{\varrho_x}{2\pi}\left[\ln k + 3(\mu - 1)\ln k_y + 2\ln k_a\right] \tag{18.61}$$

$$= \frac{2.5}{2\pi}\left[\ln 78.42 + 3\,(0.553 - 1)\ln 9.69 + 2\ln 39.23\right] = 3.445\,\frac{\text{Km}}{\text{W}},$$

$$T'_x = \frac{\varrho_x}{2\pi}\left[\ln k + 2\ln k_a\right] = \frac{2.5}{2\pi}\left[\ln 78.42 + 2\ln 39.23\right] = 4.656\,\frac{\text{Km}}{\text{W}}, \tag{18.62}$$

$$\Delta\vartheta_x = 15 + \frac{(1 - m)\,100}{3} = 15 + \frac{(1 - 0.7)\,100}{3} = 25 \text{ K}. \tag{18.56}$$

Since the calculation is made with the specified operating conditions to Table 18.2, this
gives in equation 18.63 the rated value of the load capacity I_r with
$P'_d = 0$, $T'_{Ki} = 0.469$ Km/W as in Section 18.4.1
and $R_{wr} = 0.269\ \Omega/\text{km}$ as in Part 2, Table 5.6.6a.

$$I_r = \sqrt{\frac{\vartheta_{Lr} - \vartheta_E + [(\varrho_x/\varrho_E) - 1]\,\Delta\vartheta_x}{n\,R'_{wr}(T'_{Ki} + T'_{xy})}} = \sqrt{\frac{90 - 20 + [(2.5/1.0) - 1]\,25}{1 \times 0.269 \times 10^{-3}\,(0.469 + 3.445)}} = 320 \text{ A}, \tag{18.63}$$

$$\vartheta_0 = \vartheta_{Lr} - P'_i\,T'_{Ki} = \vartheta_{Lr} - n\,I_r^2\,R'_{wr}\,T'_{Ki} = 90 - 1 \times 320^2 \times 0.269 \times 10^{-3} \times 0.469 = 77.1\ ^\circ\text{C}, \tag{18.64}$$

with $P'_d = 0$ and P'_i to equation 18.3

$$\vartheta_x = \vartheta_E + \Delta\vartheta_x = 20 + 25 = 45\ ^\circ\text{C}. \tag{18.65}$$

The assumption that $\vartheta_0 > \vartheta_x$ is therefore verified.

Bunched installation with m = 1

For $m = 1$ then $\mu = 1$ and $T'_{xy} = T'_x$

$$\Delta \vartheta_x = 15 + \frac{(1-1)\,100}{3} = 15 \text{ K}. \tag{18.56}$$

$$I_z = \sqrt{\frac{\vartheta_{Lr} - \vartheta_E + [(\varrho_x/\varrho_E) - 1]\,\Delta \vartheta_x}{n\,R'_{wr}\,(T'_{Ki} + T'_x)}} \tag{18.63}$$

$$= \sqrt{\frac{90 - 20 + [(2.5/1) - 1]\,15}{1 \times 0.269 \times 10^{-3}\,(0.469 + 4.656)}} = 259 \text{ A}.$$

The current-carrying capacity $I_r = 320$ A is identical to the quantity given in DIN VDE 0298 Part 2 as to be seen in Table 18.11.

From the quantities for I_r and I_z the rating factor is

$$f = \frac{I_z}{I_r} = \frac{259}{320} = 0.81.$$

The same value is obtained by using factors from Table 18.15 ($f_1 = 0.93$) and Table 18.17 ($f_2 = 0.87$) with

$$f = f_1 \times f_2 = 0.93 \times 0.87 = 0.81.$$

Installation in Ground Side by Side

The individual thermal resistance can also, in this case, be calculated using the equations in Fig. 18.16 and Fig. 18.22. It is however easier to take these from the graphs in Fig. 18.23 and 18.9 or Part 2, Table 5.6.6 b giving

$$T'_x = 3.794 \text{ Km/W},$$
$$T'_{xy} = 2.583 \text{ Km/W},$$
$$T'_{Ki} = 0.448 \text{ Km/W}.$$

For $R'_{wr} = 0.285$ Ω/km (according to Part 2, Table 5.6.6 b) and $m = 0.7$ is

$$I_r = \sqrt{\frac{90 - 20 + [(2.5/1) - 1]\,25}{1 \times 0.285 \times 10^{-3}\,(0.448 + 2.583)}} = 353 \text{ A} \tag{18.63}$$

and for $m = 1.0$

$$I_z = \sqrt{\frac{90 - 20 + [(2.5/1) - 1]\,15}{1 \times 0.285 \times 10^{-3}\,(0.448 + 3.794)}} = 277 \text{ A}. \tag{18.63}$$

The two quantities give a resulting rating factor of

$$f = \frac{I_z}{I_r} = \frac{277}{353} = 0.79.$$

The same value is obtained using the rating factors of Table 18.15 ($f_1 = 0.93$) and Table 18.19 ($f_2 = 0.85$)

$$f = f_1 \times f_2 = 0.93 \times 0.85 = 0.79.$$

For $m = 0.7$

$$\vartheta_0 = 90 - 1 \times 353^2 \times 0.285 \times 10^{-3} \times 0.448$$
$$= 74.1 \,^\circ \text{C}, \tag{18.64}$$
$$\vartheta_x = 20 + 25 = 45 \,^\circ \text{C},$$

and for $m = 1.0$

$$\vartheta_0 = 90 - 1 \times 277^2 \times 0.285 \times 10^{-3} \times 0.448$$
$$= 80.2 \,^\circ \text{C}, \tag{18.64}$$
$$\vartheta_x = 20 + 15 = 35 \,^\circ \text{C}.$$

In both cases therefore $\vartheta_0 > \vartheta_x$.

Calculation of diameter d_x and depth of lay h_x of the dry area for a bunched installation and for $m = 1.0$

Assuming: $d_x > d_y$

$$h_0 = \sqrt{h^2 - r^2} = \sqrt{0.7^2 - (35.7 \times 10^{-3}/2)^2} \approx 0.7 \text{ m}, \tag{18.41}$$

$$k_x = \exp\left[\frac{2\pi\,\Delta \vartheta_x}{3\,\varrho_E\,\mu\,P'_i}\right] \tag{Fig. 18.24}$$

$$= \exp\left[\frac{2\pi \cdot 15}{3 \times 1 \times 1.0 \times 259^2 \times 0.269 \times 10^{-3}}\right] = 5.70,$$

$$d_x = 4h_0\,\frac{k_x}{k_x^2 - 1} = 4 \times 0.7\,\frac{5.70}{5.70^2 - 1} = 0.51 \text{ m}, \tag{18.66}$$

$$h_x = h_0\,\frac{k_x^2 + 1}{k_x^2 - 1} = 0.7\,\frac{5.70^2 + 1}{5.70^2 - 1} = 0.74 \text{ m}. \tag{18.67}$$

With $d_y = 0.286$ m the assumption $d_x > d_y$ is proven.

Diameter of the Dry Area

The diameter of the dry area with respect to the characteristic diameter can be determined once the load capacity is known. For this calculation the necessary geometric factors k_x for a multi-core cable as well as for three single-core cables are shown in Fig. 18.24. The diameter for the dry area is obtained from

$$d_x = 4 h_0 \frac{k_x}{k_x^2 - 1} \qquad (18.66)$$

the depth of the boundary isotherm is given by

$$h_x = h_0 \frac{k_x^2 + 1}{k_x^2 - 1} \qquad (18.67)$$

and the depth of the line source h_0 is derived from equation 18.38.

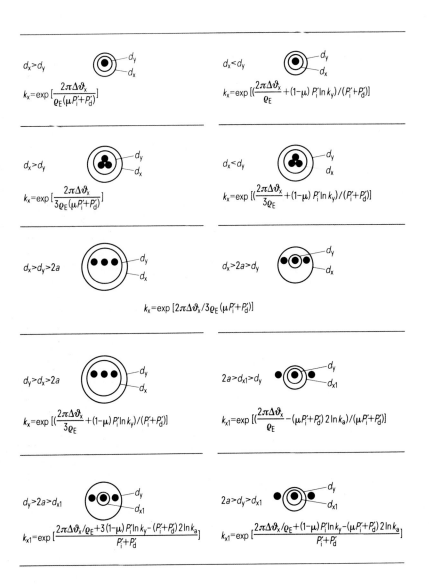

Fig. 18.24
Geometric constants of the dry area for one three-core cable and three single-core cables

18.4.4 Grouping in the Ground

Fictitious Additional Thermal Resistances $\Delta T'_x$ and $\Delta T'_{xy}$ due to Grouping

Cables grouped in a common cable trench or installed with insufficient spacing from one another result in mutual heating. Thus the load capacity is subsequently reduced. Reduction factors for the normally used spacings are shown in Tables 18.17 to 18.21. The load capacity for large spacings, for groups of cable etc. must be calculated for the individual situations.

For the calculation of load capacity the superposition of temperature fields is considered here also. Interference with ground heat conductivity, due to variations in homogeneity caused by the cables, is neglected. Because of the commonly used clearance of 7 cm between cables this can be done without introducing significant error.

For the fictitious additional thermal resistances $\Delta T'_x$ and $\Delta T'_{xy}$ of multi-core cables (calculated with ϱ_x) due to grouping, the following are applicable when considering daily load variations:

$$\Delta T'_{xy} = \frac{\varrho_x}{2\pi}\left[\sum_2^{N_y}\delta_i + \mu\sum_{N_y+1}^{N}\delta_i + (N_y - 1)(\mu - 1)\ln k_y\right].$$

(18.68)

Considering dielectric losses and continuous operation with $m=1$ in equation 18.68 μ must be made equal to 1 and this gives

$$\Delta T'_x = \frac{\varrho_x}{2\pi}\sum_2^{N}\delta_i.$$

(18.69)

Similarly for three single-core cables in a three-phase system:

$$\Delta T'_{xy} = \frac{\varrho_x}{2\pi}\left[\sum_4^{N_y}\delta_i + \mu\sum_{N_y+1}^{N}\delta_i + (N_y - 3)(\mu - 1)\ln k_y\right],$$

(18.70)

$$\Delta T'_x = \frac{\varrho_x}{2\pi}\sum_4^{N}\delta_i.$$

(18.71)

N_y is the number of cables within the circle of characteristic diameter d_y.

The grouping factor $\Sigma\delta_i$ is for a number of cables 1, 2, 3 ..., i, ... N (Figs. 18.14 and 18.25):

$$\sum_i \delta_i = \sum_i \ln \frac{c'_i}{c_i}.$$

(18.72)

The cable lying in the centre is heated most and is the reference cable designated 1. In most instances the eccentricity of the cable is neglegible. For two cables (Fig. 18.25a) the grouping factor is

$$\sum \delta = \ln \frac{\sqrt{(h_2 + h_1)^2 + a^2}}{\sqrt{(h_2 - h_1)^2 + a^2}},$$

(18.73)

independant of which cable is heating the other. If two cables are arranged at the same depth (Fig. 18.25b) then according to equation 18.50:

$$\sum \delta = \sqrt{\left(\frac{2h}{a}\right)^2 + 1} \approx \ln k_a.$$

(18.74)

For six cables as in Fig 18.25c then

$$\sum_2^6 \delta_i = 2\ln\sqrt{\left(\frac{2h}{a}\right)^2 + 1} + 2\ln\sqrt{\left(\frac{2h}{2a}\right)^2 + 1} +$$
$$+ \ln\sqrt{\left(\frac{2h}{3a}\right)^2 + 1}.$$

(18.75)

In groups of bunched single-core cables the distance between centres of bunches b can be used to simplify calculation. Since in equation 18.77 the number of loaded cores per cable is considered with $n=1$, to take account of all losses the figure 3 must be introduced into the group factor. If in Fig. 18.25c for example the six cables are replaced by six three-phase systems each comprising three bunched single-core cables this gives.

$$\sum_4^{18} \delta_i = 3\left[2\ln\sqrt{\left(\frac{2h}{b}\right)^2 + 1} + 2\ln\sqrt{\left(\frac{2h}{2b}\right)^2 + 1} +\right.$$
$$\left. + \ln\sqrt{\left(\frac{2h}{3b}\right)^2 + 1}\right].$$

(18.76)

Values for grouping factors for a cable laying at the end of the group may be taken also from Fig. 18.26. For a cable on the inside the group factors for the number of cables laying to the right and to the left must be summated.

With the aid of equation 18.64 and equation 18.65 it must be verified whether the soil actually dries out.

207

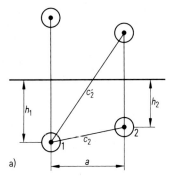

a)

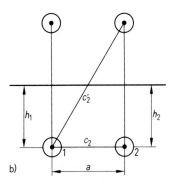

b)

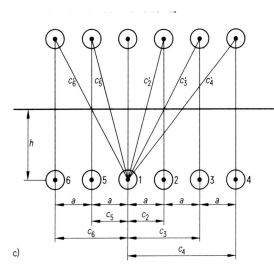

c)

Fig. 18.25
Groups of cables referred to in the text

Load Capacity

For N cables of the same type, having the same loading and the same losses installed in the same trench the load capacity is

$$I_z = \sqrt{\frac{\vartheta_{Lr} - \vartheta_E - P'_d(T'_{Kd} + T'_x + \Delta T'_x) + [(\varrho_x/\varrho_E) - 1]\,\Delta\vartheta_x}{n\,R'_{wr}(T'_{Ki} + T'_{xy} + \Delta T'_{xy})}}.$$

(18.77)

Extension of the Dry Area

As described in Section 18.43 the dry area may be represented in special cases by a circular or nearly circular area with a diameter, equal to the diameter d_x. More accurately the boundary of the dry area can be determed by calculating the temperature rise in all points $P(x, y)$ which accurately correspond with the temperature rise of the boundary isotherm $\Delta\vartheta_x$. This is effected by inserting to the relationships given in Fig. 18.27 in the formula

$$\Delta\vartheta_x = \sum_{i=1}^{N} (\mu_i\,P'_{ii} + P'_{di}) \times$$

$$\times \frac{\varrho_E}{2\pi}\ln\sqrt{\frac{(y + h_i - e_i)^2 + (x_i - x)^2}{(y - h_i + e_i)^2 + (x_i - x)^2}} \quad (18.78)$$

a fixed coordinate e.g. y and x is altered continuously until, with the given values for losses P'_{ii} and P'_{di} the calculated value of temperature rise at the point $P(x, y)$ exactly corresponds with the given value of $\Delta\vartheta_x$. In most instances the eccentricity e can be neglected (see Section 18.4.3).

In Fig. 18.27 the cables 1, 2, ... i ... N are shown with their mirror images to the ground surface. The x axis is located at the ground surface. It is assumed that the circles with the characteristic diameter lay within the dry area. Since the characteristic diameter based on the lower value of thermal resistivity of the moist region will be somewhat too large, the results will be on the safe side. The same applies if the extent of the dry area becomes smaller than the circle with the characteristic diameter.

With the aid of equation 18.78 the isotherms in the moist area can also be determined (Fig. 18.28).

For this $\Delta\vartheta_x$ must be replaced by the temperature rise of the selected isothermal line. The isotherms in the dry area can not be established using this relatively simple method.

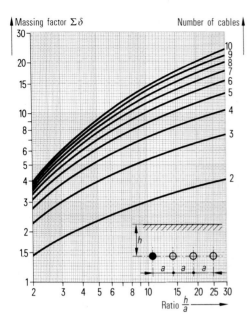

Fig. 18.26
Grouping factor $\Sigma\delta$ relative to depth of lay h and to spacing distance a, and the number of cables in the trench in relation to a cable on the end of the row

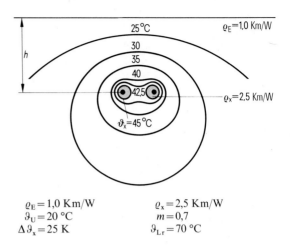

$$\varrho_E = 1,0 \text{ Km/W} \qquad \varrho_x = 2,5 \text{ Km/W}$$
$$\vartheta_U = 20 \text{ °C} \qquad m = 0,7$$
$$\Delta\vartheta_x = 25 \text{ K} \qquad \vartheta_{Lr} = 70 \text{ °C}$$

Fig. 18.28
Temperature field of two cables NYY 3×150 $0.6/1$ kV

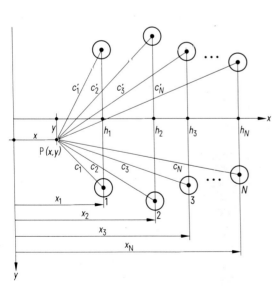

Fig. 18.27
Heating of a point $P(x, y)$ by cable $i = 1, 2, 3, ..., N$

Example 18.8

Four circuits of bunched single-core cables of the type

$$\text{NA2XS2Y} \quad 1 \times 150 \text{ RM/25} \quad 12/20 \text{ kV}$$

are arranged in the same trench. The clearance is 7 cm. The cables are to operate to the specified conditions in Table 18.3.

The centre spacing of two bunches is (Fig. 18.29)

$$b = 2\,d + 70 \text{ mm} = 2 \times 35.7 \text{ mm} + 70 \text{ mm} = 141.4 \text{ mm}.$$

The electrical and thermal data for a single bunch was calculated in example 18.7 (see page 204, Section 18.4.3). The calculation for a group of such cables is made in respect of a bunch laying in the centre (cables 1, 2 and 3):

$$\Delta T'_{xy} = \frac{\varrho_x}{2\pi}\left[\sum_{2}^{N_y}\delta_i + \mu\sum_{N_y+1}^{N}\delta_i + (N_y-1)(\mu-1)\ln k_y\right].$$

$$(18.68)$$

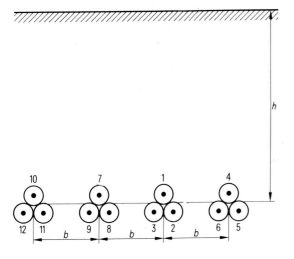

Fig. 18.29
Arrangement of installation for example 18.8

The characteristic diameter $d_y = 0.286$ m is greater than $2b \approx 0.280$ m. Therefore with $N_y = 9$ and with equation 18.72 as well as equation 18.76 one obtains

$$\Delta T'_{xy} = \frac{\varrho_x}{2\pi} \left[\sum_4^9 \delta_i + \mu \sum_{9+1}^{12} \delta_i + (9-3)(\mu-1)\ln k_y \right]$$

$$= \frac{2.5}{2\pi} \left[3 \times 2 \ln \sqrt{\left(\frac{2h}{b}\right)^2 + 1} + \right. \tag{18.68}$$

$$+ \mu\, 3 \times 1 \ln \sqrt{\left(\frac{2h}{2b}\right)^2 + 1} +$$

$$\left. + (9-3)(\mu-1)\ln \frac{4h}{d_y} \right]$$

$$= \frac{2.5}{2\pi} \left[6 \ln \sqrt{\left(\frac{2 \cdot 0.7}{0.14}\right)^2 + 1} + \right.$$

$$+ 0.553 \times 3 \ln \sqrt{\left(\frac{0.7}{0.14}\right)^2 + 1} +$$

$$\left. + 6(0.553-1)\ln \frac{4 \cdot 0.7}{0.286} \right]$$

$$= \frac{2.5}{2\pi} [13.85 + 0.553 \times 4.89 - 6.12]$$

$$= 4.15 \text{ Km/W}.$$

The grouping factor $\Sigma \delta_i$ can also be determined from Fig. 18.26. The factors in this figure are given for the outermost bunch. The factors for the cables 4 to 6 as well as the factors for cables 7 to 12, in each case relative to cables 1 to 3, must be determined and then summated. For bunches of three single-core cables the value established in this way must be multiplied by 3.

For $\dfrac{h}{b} = \dfrac{0.7 \text{ m}}{0.14 \text{ m}} = 5$

$$\sum_4^9 \delta_i \approx 3 \times 2 \times 23 = 13.8,$$

$$\sum_{10}^{12} \delta_i = \sum_4^{12} \delta_i - \sum_4^9 \delta_i \approx 3\,(3.95 - 2.3) = 4.95,$$

$$I_z = \sqrt{\frac{\vartheta_{Lr} - \vartheta_E + [(\varrho_x/\varrho_E)-1]\,\Delta\vartheta_x}{n R'_{wr}\,(T'_{Ki} + T'_{xy} + \Delta T'_{xy})}} \tag{18.77}$$

$$= \sqrt{\frac{90 - 20 + [(2.5/1.0)-1]\,25}{1 \times 0.269 \times 10^{-3}\,(0.469 + 3.445 + 4.15)}}$$

$$= 223 \text{ A}.$$

Nearly the same result can be obtained by reference to Section 18.2 and Part 2 using the following:

from Part 2 Table 5.6.6a: $I_r = 320$ A,
from Table 18.15: $f_1 = 1.0$,
from Table 18.17: $f_2 = 0.7$,
and $I_z = I_r \times f_1 \times f_2 = 320 \times 1.0 \times 0.7 = 224$ A.

Example 18.9

Two three-core cables type

NYSEY 3×185 RM/25 6/10 kV

are arranged in the same trench with a clearance between them of 7 cm and are to be operated to the specified conditions in Table 18.3.

The other cables in Fig. 18.30 are not loaded. The following electrical and thermal values are available

from Part 2, Table 5.1.18: $R'_{wr} = 0.121$ Ω/km,
$\quad I_r = 394$ A,
$\quad P'_d = 3.7$ W/m,
$\quad d = 68.8$ mm,
$\quad$ (reference diameter)

$\quad T'_{Ki} = 0.364$ Km/W,
$\quad T'_{Kd} = 0.253$ Km/W,

$\quad T'_{xy} = 1.071$ Km/W,
$\quad T'_x = 1.474$ Km/W.

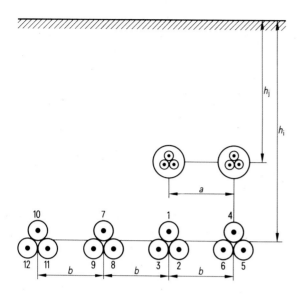

Fig. 18.30
Arrangement of installation for example 18.9

The load capacity of one cable is

$$I_r = \sqrt{\frac{\vartheta_{Lr} - \vartheta_E - P'_d(T'_{kd} + T'_x) + [(\varrho_x/\varrho_E) - 1]\Delta\vartheta_x}{n R'_{wr}(T'_{Ki} + T'_{xy})}} \quad (18.63)$$

$$= \sqrt{\frac{70 - 20 - 3.7(0.253 + 1.474) + [(2.5/1.0) - 1] 25}{3 \times 0.121 \times 10^{-3}(0.364 + 1.071)}}$$

$$= 394 \text{ A}$$

$a = 70 \text{ mm} + d \approx 70 \text{ mm} + 69 \text{ mm} = 139 \text{ mm}$

$$\frac{h_j}{a} = \frac{0.7 \text{ m}}{0.139 \text{ m}} = 5.04.$$

From Fig. 18.26 $\sum_1^2 \delta_i = 2.31$.

For $m = 0.7$ from Table 8.31 $\mu = 0.553$ and $d_y = 0.286$ m.

From example 18.7 $k_y = 9.69$ and hence

$$\Delta T'_{xy} = \frac{\varrho_x}{2\pi}\left[\sum_2^{N_y}\delta_i + \mu\sum_{N_y+1}^N\delta_i + (N_y - 1)(\mu - 1)\ln k_y\right] \quad (18.68)$$

$$= \frac{2.5}{2\pi}[2.31 + 0 + (2 - 1)(0.553 - 1)\ln 9.69]$$

$$= 0.515 \text{ Km/W}.$$

With $N_y = 2$ and $\sum_2^{N_y}\delta_i = \sum_2^2\delta_i = 2.31$

$$\Delta T'_x = \frac{\varrho_x}{2\pi}\sum_2^N\delta_i = \frac{2.5}{2\pi} \times 2.31 \quad (18.69)$$

$$= 0.919 \text{ Km/W};$$

$$I_z = \sqrt{\frac{70 - 20 - 3.7(0.253 + 1.474 + 0.919) + [(2.5/1.0) - 1] 25}{3 \times 0.121 \times 10^{-3}(0.364 + 1.071 + 0.515)}}$$

$$= 331 \text{ A}.$$

Current-Carrying Capacity of Dissimilar Cables

Apart from the N cables being similar to each other (indicated by the index i), as shown in Fig. 18.29, there are other M cables being similar to each other (indicated by the index j) accommodated in the same trench (Fig. 18.30). If the loading of these M cables is the same this results in mutual heating, for instance, the trefoil-arranged single-core cables indicated by 1, 2, 3 are heated by:

$$\Delta\vartheta_{ij} = P'_{dj}\frac{\varrho_x}{2\pi}\sum_1^M\delta_{ij} + \quad (18.79)$$

$$+ P'_{ij}\left\{\frac{\varrho_x}{2\pi}\left[\sum_1^{M_{yij}}\delta_{ij} + \mu\sum_{M_yij+1}^M\delta_{ij} + M_{yij}(\mu - 1)\ln k_y\right]\right\}.$$

The grouping factor for M multi-core cables is

$$\sum^M\delta_{ij} = \sum^M\ln\sqrt{\frac{(h_i + h_j)^2 + a_{ij}^2}{(h_i - h_j)^2 + a_{ij}^2}}, \quad (18.80)$$

for N bunched single-core cables operating under three-phase conditions the rating factor becomes

$$\sum^N\delta_{ij} = 3\sum^N\ln\sqrt{\frac{(h_i + h_j)^2 + b_{ij}^2}{(h_i - h_j)^2 + b_{ij}^2}}. \quad (18.81)$$

If all N cables are loaded at the same level the current-carrying capacity of the trefoil-arrangement (or cable) in question is:

$$I_{zi} = \sqrt{\frac{\vartheta_{Lri} - \vartheta_E - P'_{di}(T'_{kdi} + T'_{xi} + \Delta T'_{xi}) + [(\varrho_x/\varrho_E) - 1]\Delta\vartheta_x - \Delta\vartheta_{ij}}{n R'_{wri}(T'_{kji} + T'_{xyi} + \Delta T'_{xyi})}}. \qquad (18.82)$$

If the load capacity of the M cables is to be investigated in respect of the heating from the group of N cables the indices in the equation above need to be interchanged.

M_{yij} is the number of cables in group j, whose circle with characteristic diameter d_{yj} embraces the cable considered of the group i. Since it must be assumed that all cables N and M are situated in the same dry area the boundary isotherm must be determined by using the larger of the two load factors m_i or m_j (equation 18.56).

Using equations 18.64 and 18.65 the assumption that the soil is drying out must be verified. Should this not be the case then in all equations ϱ_x must be replaced by ϱ_E.

Example 18.10

The cables from example 18.8 (Fig. 18.29) are installed at a depth of $h_i = 1$ m (Fig. 18.30) but otherwise are operated under the same conditions.

From Fig. 18.23 it is found for $h_i = 1.0$ m, $m = 0.7$ and $d = 35.1$ mm

$$T'_{xy} = 3.7 \text{ Km/W}.$$

For the ratio

$$\frac{h_i}{b} = \frac{1.0 \text{ m}}{0.140 \text{ m}} = 7.14$$

from Fig. 18.26 the values 2.65 for two cables and 4.62 for three cables are found. For the load capacity of $N = 12$ single-core cables with a depth of lay increased to 1 m, but still ignoring the influence of the two three-core 10 kV cables M situated above, we have simplified (see example 18.8) in

$$\sum \delta_i = \sum_4^6 \delta_i + \sum_7^{12} \delta_i = 3(2.65 + 4.62) = 21.81,$$

$$\Delta T'_{xy} = \frac{2.5}{2\pi} 0.553 \times 21.81 = 4.80 \text{ Km/W and}$$

$$I_{zi} = \sqrt{\frac{90 - 20 + [(2.5/1.0) - 1] 25}{1 \times 0.269 \times 10^{-3}(0.469 + 3.7 + 4.80)}}$$
$$= 211 \text{ A}. \qquad (18.77)$$

The loading of the 10 kV cables is now to be 200 A each at a load factor of $m = 0.7$ and $d_y = 0.286$ m (Table 18.27). $d_y/2$ is therefore less than the smallest spacing $h_i - h_j = 0.3$ m at $a_{ij} = 0$ in Fig. 18.30 and hence $M_{yij} = 0$. Using equation 18.80 in relation to the axis of the bunched cables 1, 2 and 3 with $a_{ij} = 0$ compared with $a_{ij} \approx 0.14$ m (example 18.9) we get

$$\sum_1^M \delta_{ij} = \ln\sqrt{\frac{(1.0 + 0.7)^2 + 0^2}{(1.0 - 0.7)^2 + 0^2}} +$$

$$+ \ln\sqrt{\frac{(1.0 + 0.7)^2 + 0.14^2}{(1.0 - 0.7)^2 + 0.14^2}} \qquad (18.80)$$

$$= 3.37.$$

With the quantities from example 18.9 we get

$$P'_{ij} = n I_j^2 R'_{wrj} \qquad (18.3)$$
$$= 3 \times 200^2 \times 0.121 \times 10^{-3} = 14.52 \text{ W/m}.$$

The temperature rise caused by the two 10 kV cables is therefore

$$\Delta\vartheta_{ij} = 3.7\frac{2.5}{2\pi}3.37 + 14.52\frac{2.5}{2\pi}[0 + 0.553 \times 3.37 + 0]$$

$$= 15.7 \text{ K} \qquad (18.79)$$

and the load capacity

$$I_{zi} = \sqrt{\frac{90 - 20 + [(2.5/1) - 1] 25 - 15.7}{1 \times 0.269 \times 10^{-3}(0.469 + 3.7 + 4.8)}}$$
$$= 195 \text{ A}. \qquad (18.82)$$

Additionally, it must be ascertained that the 10 kV cables are not heated excessively by the 20 kV cables thus

$$P'_{ii} = n I_i^2 R'_{wri} \qquad (18.3)$$
$$= 1 \times 195^2 \times 0.269 \times 10^{-3}$$
$$= 10.23 \text{ W/m}.$$

Also in this example $d_y/2$ is less than the smallest distance $h_i - h_j = 0.3$ and therefore $N_{yji} = 0$.

$$\sum \delta_{ji} = 3 \left[\ln \sqrt{\frac{(1.0+0.7)^2+0}{(1.0-0.7)^2+0}} + 2 \ln \sqrt{\frac{(1.0+0.7)^2+0.14^2}{(1.0-0.7)^2+0.14^2}} + \ln \sqrt{\frac{(1.0+0.7)^2+(2\cdot0.14)^2}{(1.0-0.7)^2+(2\cdot0.14)^2}} \right] = 19.34, \quad (18.81)$$

$$\Delta\vartheta_{ji} = 0 + 10.23 \frac{2.5}{2\pi} [0 + 0.553 \times 19.34 + 0] = 43.53 \text{ K}, \quad (18.79)$$

$$I_{zi} = \sqrt{\frac{70 - 20 - 3.7(0.253 + 1.474 + 0.919) + [(2.5/1) - 1]\, 25 - 43.53}{3 \times 0.121 \times 10^{-3}(0.364 + 1.071 + 0.515)}} = 220 \text{ A}. \quad (18.82)$$

Load less
95 → 185 *Iron Loss*
Impedance

Since the loading with 200 A is smaller than the load capacity of 220 A, the 10 kV cables are not heated excessively. The interdependance of load capacity of a group of cables on loading of the other group can be seen by reference to Fig. 18.31. At the point of intersection of the curves the temperature of the conductors are at their maximum values of 90 °C and 70 °C respectively.

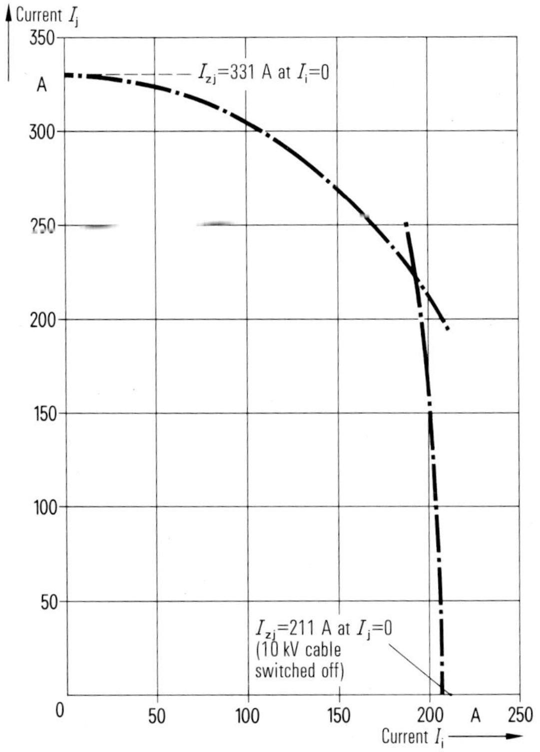

Fig. 18.31
Load capacity interdependance of two groups of cables from example 18.9

IEC 60502

18.4.5 Installation in Ducts and Pipes

As well as the thermal resistances described earlier additional thermal resistances are involved (Fig. 18.32)

▷ the thermal resistance T_P' of the space between the cable surface and the inner wall of the pipe and

▷ the thermal resistance T_R' of the pipe (with metal pipe T_R' is insignificant).

Thermal Resistance T_R' of the Pipe

The thermal resistance T_R' of the pipe is derived from the specific thermal resistance of ϱ_R of the pipe material, the outer diameter d_R and the thickness δ_R of the pipe wall with

$$T_R' = \frac{\varrho_R}{2\pi} \ln \frac{1}{1 - \frac{2\delta_R}{d_R}}. \quad (18.83)$$

Thermal Resistance T_P' of the Internal Space

The thermal resistance T_P' of the space whether filled with air or gas is determined by iteration [18.42, 18.43]:

$$T_P' = \frac{1}{d_E \left[a \left(\frac{\Delta\vartheta_P\, p^2}{100\, d_E} \right)^{\frac{1}{4}} + b + c\, \vartheta_m \right]}, \quad (18.84)$$

$$\Delta\vartheta_P = T_P'\, n_R\, (P_i' + P_d'). \quad (18.85)$$

The equivalent diameter d_E in m for cables with diameter d is

for one cable in a pipe $\quad d_E = d,$
for two cables in a pipe $\quad d_E = 1.65\, d,$
for three cables in a pipe $\quad d_E = 2.15\, d,$
for four cables in a pipe $\quad d_E = 2.50\, d.$

213

Duct = 0.85 derating on soil value.
Parallel circuits = depends on no. of circuits
+ spacing between them.

The mean temperature ϑ_m of the air space for n_R cables is calculated by approximation

$$\vartheta_m \approx \left[\vartheta_{Lr} - \vartheta_E - \Delta\vartheta_d + \left(\frac{\vartheta_x}{\vartheta_E} - 1\right)\Delta\vartheta_x\right] \times$$

$$\times \frac{n_R\left(\dfrac{T'_P}{2} + T'_R + T'_{xy}\right)}{T'_{Ki} + n_R\left(T'_P + T'_R + T'_{xy}\right)} +$$

$$+ P'_d\left[n_R\left(\frac{T'_P}{2} + T'_R + T'_{xy}\right)\right] +$$

$$+ \vartheta_E - \left(\frac{\varrho_x}{\varrho_E} - 1\right)\Delta\vartheta_x, \qquad (18.86)$$

P'_i is obtained from equation 18.3, P'_d is obtained from equation 18.4 and the dielectric temperature rise from

$$\Delta\vartheta_d = P'_d\left[T'_{Kd} + n_R\left(T'_P + T'_R + T'_x\right)\right]. \qquad (18.87)$$

The constants a, b, and c which depend on the type of pipe and arrangement can be taken from Table 18.32. The pressure p for cables in pipes is 1 bar.

For a temperature difference of $\Delta\vartheta_P = 20$ K between pipe inner wall and the cable surface for cables in an air filled pipe and $\Delta\vartheta_P = 10$ K for gas pressurised cables taking account of a limited range of diameters

$d_E = 25$ mm to 100 mm for cables in pipe

$d_E = 75$ mm to 125 mm for gas pressure cables

the simplified equation [18.42; 18.2]

$$T'_P = \frac{A}{1 + 100(B + C\vartheta_m)d_E} \qquad (18.88)$$

is used with constants A, B and C to Table 18.32. In addition iteration using equations 18.86 and 18.87 is required (d_E is to be applied in m). A rough calculation is possible with Fig. 18.33

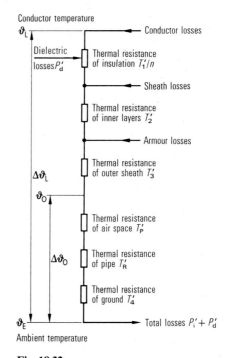

Conductor temperature

Conductor losses

Dielectric losses P'_d — Thermal resistance of insulation T'_1/n

Sheath losses

Thermal resistance of inner layers T'_2

Armour losses

Thermal resistance of outer sheath T'_3

Thermal resistance of air space T'_P

Thermal resistance of pipe T'_R

Thermal resistance of ground T'_4

Ambient temperature

Total losses $P'_i + P'_d$

Fig. 18.32
Equivalent circuit for the thermal flow from cables installed in a pipe in the ground

Table 18.32
Constants a, b, c, A, B, C for the calculation of thermal resistance T'_P for installation in ducts or pipes [1]

Type of pipe and arrangement	a	b	c	A	B	C
Cable in metal pipe	11.41	15.63	0.2196	5.2	1.4	0.0110
Cable in hard fibre pipe (fibre[2] duct)						
in air	11.41	4.65	0.1163	5.2	0.83	0.0063
in concrete	11.41	5.55	0.1808	5.2	0.91	0.0095
Cable in asbestos cement pipe						
in air	11.41	11.11	0.1033	5.2	1.2	0.0055
in concrete	11.41	10.20	0.2067	5.2	1.1	0.0110
Cable in earthenware pipe	–	–	–	1.87	0.46	0.0036
Gas-pressure cable in steel pipe (14 bar)	11.41	15.63	0.2196	0.95	0.00	0.0021
High-pressure oil-filled cable in steel pipe	–	–	–	0.26	0.28	0.0026

[1] For plastic pipes values not yet incorporated in IEC 287. It is recommended to use the values for hard fibre pipe as an approximate calculation. For installation of the pipes in ground the constants for pipes bedded in concrete may be used
[2] Bitumen impregnated wood fibre

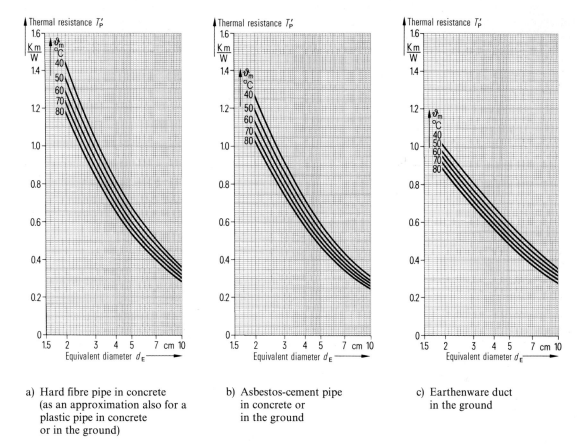

a) Hard fibre pipe in concrete
   (as an approximation also for a
   plastic pipe in concrete
   or in the ground)

b) Asbestos-cement pipe
   in concrete or
   in the ground

c) Earthenware duct
   in the ground

Fig. 18.33 Thermal resistance of the air space between a cable and a pipe

215

Load Capacity for an Installation of Pipes in Ground or in Air

The load capacity for cables laying in the ground can be calculated from

$$I_z = \sqrt{\frac{\vartheta_{Lr} - \vartheta_E - P'_d \left[T'_{Kd} + n_R(T'_P + T'_R + T'_x) + \Delta T'_x\right] + \left[(\varrho_x/\varrho_E) - 1\right]\Delta\vartheta_x}{n R'_{wr}\left[T'_{Ki} + n_R(T'_P + T'_R + T'_{xy}) + \Delta T'_{xy}\right]}}. \tag{18.89}$$

The number of cables in the pipe is n_R and n is the number of loaded conductors in each cable. The thermal resistances of the soil T'_x and T'_{xy} are calculated as in Section 18.4.3 using the diameter of the pipe d_R. The additional thermal resistances $\Delta T'_x$ and $\Delta T'_{xy}$ taking account of grouping are calculated as in Section 18.4.4.

If the load capacity in air is required the quantity of thermal resistance T'_{LU} for an installation in air as in Section 18.4.2 must be inserted while the thermal resistances T'_x, T'_{xy}, $\Delta T'_x$ and $\Delta T'_{xy}$ are omitted.

Load Capacity for an Installation in Ducts Banks

In some industrial installations the cables are installed in duct banks at 0.6 m depth or greater (Fig. 18.34). The ducts are firstly installed in layers with the aid of distance pieces and then bedding or filler material is compacted after each layer is positioned. The clearance between ducts must be selected wide enough to ensure proper filling. If normal sand is used for this the load capacity to equation 18.89 is appropriate. However a thermally stable bedding material (see Section 18.4.6), e.g. a suitable concrete mix may be selected provided that the thermal resis-

tivity does not exceed a specified value ϱ_B in the dried-out state. Normally $\varrho_B < \varrho_x$. For concrete, in cases where a specific quantity is not know it is normally satisfactory to use $\varrho_B = 1.2$ Km/W. The drying out of the soil outside of the block is the only factor which may reduce load capacity.

In duct banks the power cables are to be arranged only in the outer ducts as indicated in Fig. 18.35 [18.44]. The heat dissipation from the inner pipes of a duct bank into the ground is significantly less favourable by comparison to the outer pipes because of the obstruction caused by air in the outer pipes. If power and control cables are to be run together the power cables, because of the better heat dissipation, are preferably arranged in the upper layers.

It must be assumed that the soil dries out around the pipe block with dimensions x and y and the equivalent diameter d_B to equation 18.90. In the zone embraced by the equivalent diameter d_B and the diameter of the boundary isotherm d_x, therefore, one must calculate the corrective thermal resistances T'^+_B and T'^+_{By}, using the thermal resistivity ϱ_x.

Outside the diameter d_x calculations are made using the thermal resistivity ϑ_E for moist soil, which is introduced through the correction term in the top line.

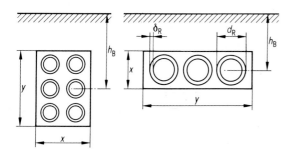

Fig. 18.34 Arrangement of duct banks

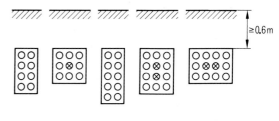

$\otimes$ unsuitable for power cables

Fig. 18.35
Examples of arrangement of pipes in duct banks

The equivalent diameter d_B of the duct bank with dimensions x and y is (Fig. 18.36) [18.2]

$$d_B = 2 \exp\left[\frac{1}{2}\frac{x}{y}\left(\frac{4}{\pi} - \frac{x}{y}\right)\ln\left(1 + \frac{y^2}{x^2}\right) + \ln\frac{x}{2}\right], \quad (18.90)$$

whereby one has to select $x < y$ and $\dfrac{y}{x} \leq 3$.

The geometric factor is

$$k_B = \frac{2h_B}{d_B} + \sqrt{\left(\frac{2h_B}{d_B}\right)^2 - 1}. \quad (18.91)$$

It is assumed that $d_x > d_B$ therefore drying-out of the soil occurs and for the corrective thermal resistances for multi-core cables and $d_x > d_y > d_B$

$$T_B'^+ = \frac{\varrho_x - \varrho_B}{2\pi}\left[N_B \ln k_B + \sum_{N_B+1}^{N} \delta_i\right], \quad (18.92)$$

$$T_{By}'^+ = \frac{\varrho_x - \varrho_B}{2\pi}\left[N_B \ln k_B + \mu \sum_{N_B+1}^{N} \delta_i + N_y(\mu - 1)\ln k_y\right]; \quad (18.93)$$

multi-core cables and $d_x > d_B > d_y$

$T_B'^+$ from equation 18.92,   (18.92a)
$T_{By}'^+ = \mu\, T_B'^+$;   (18.93a)

single-core cables and $d_x > d_y > d_B$

$T_B'^+$ from equation 18.92,   (18.92b)
$T_{By}'^+$ from equation 18.93;   (18.93b)

single-core cables and $d_x > d_B > d_y$

$T_B'^+$ from equation 18.92,   (18.92c)
$T_{By}'^+ = \mu\, T_B'^+$.   (18.93c)

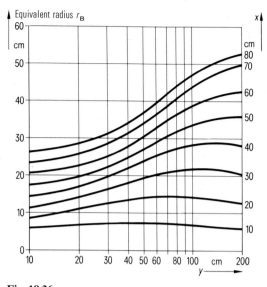

Fig. 18.36
Equivalent radius $r_B = d_B/2$ of a duct bank with dimensions x and y in Fig. 18.34, where $x < y$ provided that $y/x \leq 3$

Lastly it must also be investigated whether the assumption $d_x > d_B$ applies:

$$d_x \approx 4h_B \frac{k_x}{k_x^2 - 1}, \quad (18.66a)$$

with

$$k_x = \exp\frac{2\pi\,\Delta\vartheta_x}{\varrho_E N_B(\mu P_i' + P_d')} \quad (18.94)$$

and P_i' to equation 18.3 as well as P_d' to equation 18.4.

If $d_x < d_B$ the soil does not dry-out and in all equations ϱ_x must be replaced by ϱ_E.

For the load capacity an extension of equation 18.89 is used

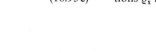

$$I_z = \sqrt{\frac{\vartheta_{Lr} - \vartheta_E - P_d'\left[T_{Kd}' + n_R(T_P' + T_R' + T_B') + \Delta T_B' + T_B'^+\right] + \Delta\vartheta_x\left[(\varrho_x/\varrho_E) - 1\right]}{n R_{wr}'\left[T_{Ki}' + n_R(T_P' + T_R' + T_{By}') + \Delta T_{By}' + T_{By}'^+\right]}}. \quad (18.95)$$

The thermal resistances T_B', T_{By}' and $\Delta T_{By}'$ are calculated in line with Section 18.4.3 and 18.4.4 with ϱ_x replaced by ϱ_B. This corresponds with the assumption that the thermal resistivity outside the pipes has the uniform quantity ϱ_B.

18.4.6 Soil-Thermal Resistivity

Cable in the Ground

An accurate knowledge of the thermal resistivity of the soil and the bedding materials not only allows optimum utilisation of the cable up to the permissible operating temperature but also prevents early aging or destruction due to excessive heating [18.18 to 18.20]. High soil-thermal resistivity – as a consequence of drying-out of the ground – are particularly dangerous for highly loaded cables in continuous operation in unfavourable ground conditions.

If the slightly wider surroundings of the cable are included in the consideration, three areas can be described (Fig. 18.37) which under certain conditions may have different thermal resistivities. The three areas can be distinguished as follows:

Virgin Soil

If this is undisturbed and is without significant inclusions of humus (moorland) the soil-thermal resistivity is normally, for European latitudes, no more than 1 Km/W. Care must be taken where the ground is made-up and is only partially consolidated with a mixture of slag, ashes and the like, included. In such cases it is advisable to measure the thermal and physical properties of the soil.

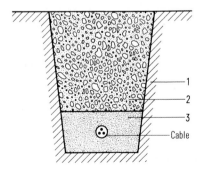

Area 1   Virgin soil
Area 2   Backfill
Area 3   Bedding material

Fig. 18.37
Thermal-resistance areas surrounding a cable laid in ground

Backfill

To avoid damage to a cable construction good ground, free of ingredients such as building rubble, clinker, etc., should be used for backfill and should be sufficiently compacted [18.6; 18.21]. Normally the excavated soil is suitable for this purpose. The physical and thermal characteristics can be approximately equal to those of the virgin soil (Area 1).

Bedding Material

Bedding materials, in line with the requirements discussed earlier [18.16; 18.21] should be free of stones and should comprise sand or other compactable type of soil with a maximum particle size of 10 mm. This should be laid in layers of 10 cm and compacted by

Table 18.33   Quantities of soil components

Basic element or material	Dry density t/m^3	Thermal resistivity Km/W
Granite	2.5 to 3.0	0.32 to 0.25
Basalt	2.9	0.6
Feldspar	2.5	0.43
Glimmer Mica		1.7
Gneiss	2.4 to 2.7	0.29
Limestone	2.5	0.78
Quartz	2.5 to 2.8	0.11
Sandstone	2.2	0.54
Slag	0.3 to 1.1	7 to 3.5
Organic materials, moist		4
Organic materials, dry		7
Water	1	1.68
Air		40

hand compactors up to a cover of 30 cm above the cable. Below the cables hard parts such as rocks or boulders should be replaced by filler material. Bog, peat, ash and building rubble as well as chemically contaminated earth should be replaced to a distance of 20 cm by filler material. Here also the previously excavated soil can be used providing it has suitable characteristics.

Physical and Thermal Characteristics of Soil

Soil comprises three basic components. It consists of *granular particles* of material which differ in their chemical and mineral constituents, size and form of particles, particle size distribution, density and moisture content. Between the more or less compacted particles there are cavities, or pores, which may be filled with either *water* or *air*. The air contained in the pores may itself contain water vapour depending on the temperature.

Heat is transferred in such amorphous materials by conduction.

A comparison of the thermal resistivities in Table 18.33 indicates the extent to which the total thermal resistance is related to the constituents of the soil.

The individual soil particles have molecular like powers of adhesion and attract a layer of condensed water. This hygroscopically bound water does not move and can be removed only by changing it into vapour, for example by heating to above 105 to 110 °C. Fine granular soils bind in this manner more water than coarse grained soils. The amount of bound water also depends on moisture content as well as pressure and the temperature of the air in the soil.

If sufficient water is present in the soil, the hygroscopically bound water is covered with an additional concentric skin of water (Fig. 18.38) which connects neighbouring particles as pore filling water. This improves heat conduction since, in comparison with air, water is a good conductor of heat and the pores become heat bridges. The amount of skin water is subject to great variations which are caused by storage of penetrating water and its evaporation. Especially in the temperature zone of cables a reduction of water content is to be expected even up to completely dried out. Even in this case it is important that the thermal resistivity remains sufficiently low. To meet this requirement it is necessary that the content of solid

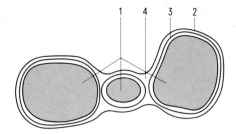

1 Soil particle
2 Skin of water
3 Hygroscopically bound water
4 Pore filling water

Fig. 18.38
Fine granular particles and water layer

material relative to the content of pores n is large. Such mixtures with reduced cavities have a high resulting dry density γ_d. The thermal resistivity reaches a minimum when all pores are filled with water i.e. at maximum water content w.

The above mentioned values can be determined by reference to DIN 4016 or from an information sheet prepared by Forschungsgesellschaft für das Straßenwesen [18.34], e.g. using the Dörr-Wäge-Method. The following relationships exist:

Water content $\qquad w = \dfrac{\gamma}{\gamma_d} - 1 \qquad$ (18.96)

Pore content $\qquad n = 1 - \dfrac{\gamma_d}{\gamma_s} \qquad$ (18.97)

where
γ_s particle density,
   i.e. the relationship of dry weight of solid material to the pore-free volume (in non- or weak binding soils $\gamma_s \approx 2.65 \ t/m^3$),

γ_d the dry density,
   i.e. the weight of the dry soil relative to the unit volume,

γ the density of moist soil,
   i.e. the weight of the moist soil relative to the unit volume.

To obtain the most densely compacted soil the pores between the larger particles should be filled with particles of a smaller group such that a less porous mixture is developed. Such an ideal grain distribution is shown on the distribution diagram as a parabolic curve (Fig. 18.39 curves 1 and 2) and can be treated analytically with equation

$$p = (d/d_{max})^x. \tag{18.98}$$

In this p represents the part of the weight of sieved material which passes through a mesh width of "equivalent diameter" d, d_{max} the diameter of the largest granule of the mixture, $x = 0.5$ according to Fuller, $x = 0.25$ to 0.4 according to Talbot and $x = 0.22$ to 0.514 according to Jahn [18.22].

The particle size distribution curve can be derived according to DIN 4016 or can be found in a paper of the Forschungsgesellschaft für das Straßenwesen [18.34], in which the sieved material is treated using a series of mesh widths. The point A in Fig. 18.39 on curve number 4 signifies that 73% of the total mass of sieved material has a granular diameter of ≤ 0.63 mm.

The steeper the particle size distribution curve the more uniform is the material i.e. it is made up of only a small variety of particle sizes (curve 4 in Fig. 18.39). Well graded soils, in which the smaller particles fill the pores between larger particles have a more flat or parabolic shape of curve. The relationship is expressed by

$$U = d_{60}/d_{10} \tag{18.99}$$

with d_{60} the particle diameter with 60% passing through the sieve and d_{10} the particle diameter with 10% passing through.

Soils having $U < 5$ (steep curve) are classed as uniform whilst soils having $U > 5$ (flat curve) are classed as non-uniform.

From the particle size distribution curve the ease of compaction can also be recognised. Easily compactable grades normally comprise well graded, weak or non-cohesive sands (also sand gravel mixtures) with $U > 7$. Soils are classified as non-cohesive where they have a low content of silt and clay (approximately $< 10\%$), do not tend to form clods and therefore remain loose and flowing. They permit cavity-free filling of the trench and especially in the vicinity of the cable. In mildly cohesive soils the individual particles adhere to one another and form a modular mass, they are therefore less suitable as a bedding material and

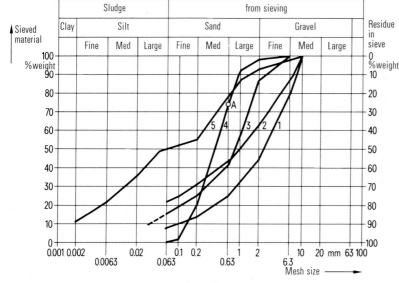

1 Particle size distribution curve to equation 18.98 with $x = 0.5$
2 Particle size distribution curve to equation 18.98 with $x = 0.3$
3 Crushed limestone (residue from splitting operation) Probe No 6 from Table 18.35
4 Building sand Probe No 1 from Table 18.35
5 Sand-loam-mixture

Fig. 18.39
Particle size distribution curve

require a more intensive compacting. Furthermore some kinds of soil, depending on water content, tend to swell and shrink which can lead to the formation of cracks and cavities in the vicinity of the cable where water content alters due to the temperature field of the cable.

The ease of densifying or compacting depends very much upon the water content during compacting. By use of the Proctor apparatus [18.34] the most favourable water content and the highest dry state density for compacting can be determined by sample investigation. For this test a probe is applied to several samples of soil, each having a different water content. Each sample is compressed in three layers in a cylindrical test vessel of say, 10 cm diameter and 12 cm height. The apparatus gives a consistant compacting effort, relative to the volume (60 Mpm/m^3 = 588.4 kJ/m^3) – known as the Proctor effort. This effort is derived from a weight of 2.5 kg falling through 30 cm with 25 blows for each of the three layers. The resulting dry density is depicted in curves shown in Fig. 18.40. This investigation shows the degree of compaction achievable depending on the type of soil – degree of non-uniformity – and materials. With a content of approximately 5 to 20% of silt this not only fills the pores between large granules, thus ensuring a higher dry density but also in conjunction with water acts as a lubricating agent when compacting.

The particle shape also influences the dry density. Round particles result in higher values in comparison to flat or crystaline shapes.

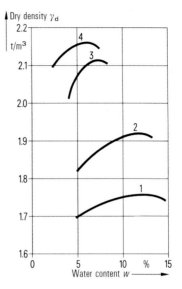

1 Sand
2 Sand and silt
3 Gravel, sand and powdered stone
4 Gravel, sand and silt

Fig. 18.40
Proctor curves $\gamma_d = f(w)$ of various types of soil

A marked influence of chemical-mineral composition on thermal resistivity is noticeable at high values of dry density and low water content. Sands and gravels containing quartz are, because of their reduced thermal resistivity (Table 18.33) preferred. Where soil contains, apart from large gravel and sand particles, sufficient silt, the water binding capacity and also the good adhesion of the larger particles is noticeable. Tests have shown an improved heat conductivity for this mixture. Because of its surface tension the silt forms a film over the larger granules and draws itself into the pores. Heat conducting bridges are formed from the solid constituents which remain present even when the soil is completely dried out. This phenomena can, however, only be readily observed when prior to drying-out a certain minimum water content is present.

A large number of tests have been conducted to establish an analytical relationship between the physical properties of soil and its thermal resistivity [18.23 to 18.27]. Direct measurement of thermal characteristics is however preferred to all other methods, since this provides the most accurate values. For the interpretation of heat conductivity processes in soils, it will be inevitable even in the future to occasionally make these thermal investigations.

The relationships between thermal resistivity, density and degree of humidity for two types of soil are shown in Fig. 18.41.

221

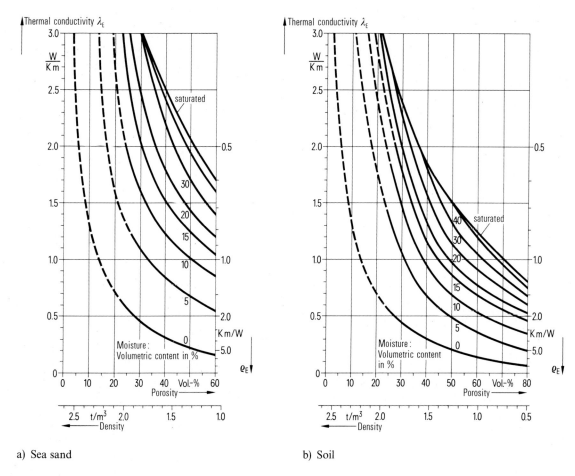

a) Sea sand

b) Soil

Fig. 18.41
Thermal conductivity λ_E and thermal resistivity of soil $\varrho_E = 1/\lambda_E$ relative to density and materials at 20 °C [18.40]

Influence of Moisture Content

The moisture content of soil is dependant on a number of natural factors [18.30]:

▷ Different types of soil have different capacities to absorb water and to retain it. The smaller the pores the better the water retention. Loam, i.e. clay containing soils dry-out much slower than sandy soils [18.28]. Crushed stones, gravel or made-up slag have no water holding capacity.

▷ The water table can re-wet the soil where the soil has sufficient suction [18.29]. The moisture con-

tent of the upper layers is dependant upon the water table level. In large-particle poor soils the suction can be zero but in coarse sands it can be 0.03 m to 1.0 m and in loam 1 m to 30 m.

▷ The surface contour can provide either a drainage (hill) or containment (valley) effect.

▷ Roots of shrubs and trees dry-out the soil during periods of low precipitation. Sandy soils are affected more so than loam.

▷ Road surfaces or other coverings prevent free

evaporation of water from the surface so that the moisture content below them may be present higher than in the non-covered surroundings.

▷ Precipitation provides a major proportion of moisture input to the ground.

▷ Solar radiation, both duration and intensity together with wind, surface characteristics and vegetation influence evaporation.

▷ The moisture content of the soil and hence its thermal resistivity follows an annual cycle which is controlled by the influences mentioned above and also depends on the type of soil and depth. High moisture content can also result from preceding weather conditions. Heavy rainfall or thaw can also influence deeper layers and can cause a rapid change of moisture content particularly in sandy soil. To predict moisture content is most difficult and can only be considered as a rough approximation. It requires observation and experience over many years.

In the vicinity of a cable (area 3, Fig. 18.37) the temperature field influences the water holding capacity whereas in the more distant surroundings (areas 2 and 1) the natural variations of water content described above are mainly related to the climate of the soil. The temperature field of the cable causes the vapour pressure in the vicinity of the cable to rise and the water vapour held in the air contained in the pores of the soil to move away from the cable. This action causes the capillary suction in the soil close to the cable to increase so that the water returns to the cable in liquid form. If the temperature of the cable surface exceeds a critical value of 30 °C for sandy soils or 50 °C for loam at ambients of approximately 15 °C to 20 °C, this circulation cycle is interrupted and the cable surroundings dry-out up to the critical isotherm. As tests have shown [18.18, 18.23 and 18.31] this cycle is mainly time dependant and can be suddenly interrupted by rainfall and may even be reversed. In some instances the dried-out zone extends only a few centimeter but can, in unfavourable conditions cover the total area of the bedding material and beyond. The selection as well as compaction of the bedding material is therefore of significant importance for the temperature rise and the load capacity of the cable.

Fig. 18.42 illustrates how weather conditions combined with a typical cable loss of 82 W/m can influence the drying-out process.

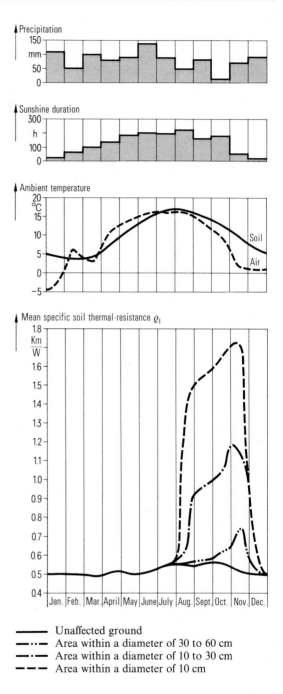

Fig. 18.42
Load test in open country (Erlangen 1968).
Influence of climate and cable heat loss on drying-out of the soil

The cable has a diameter of 40 mm and a depth of lay of 0.9 m. The made up soil consisted of a sand loam mixture with a uniformity index $U = 144$ and was used as back fill as well as bedding material (curve 5, Fig. 18.39). The drying-out process commenced in July as a (delayed) consequence of increasing duration of sunshine and soil temperature with reduced precipitation. The re-wetting commences approximately mid-November.

Measuring

Measurements [18.3, 18.30] are only truly meaningful if next to the thermal resistivity the moisture content, density and grain size distribution curve are also measured. Measurements in open country at depth of lay need to extend over several years to determine the annual differences of moisture content relative to weather conditions. These can be on a large scale and are therefore very costly and quantities gained from experience are normally used instead. A quantity specified in DIN VDE 0298 Part 2 of 1.0 Km/W is normally used except for areas such as

▷ suspected slag, waste or peat,
▷ continuously loaded high-voltage cabled or if
▷ basic investigations for general application are to be conducted.

The thermal resistivity can be measured e.g. by use of a needle probe. In Fig. 18.43 apparatus for laboratory measurements [18.33] is illustrated. The sample of moist soil is compacted, using one third of the Proctor force (200 kJ/m^3) (Fig. 18.43 a), together with the measuring probe. The reduced Proctor force is used in order to take account of the hindered compacting which is often the case in a cable trench. The probe is heated by means of a heating element wire (Fig. 18.43 d) while the increase in temperature is measured by means of a sensing resistance wire. Thermal resistivity is calculated from the temperature rise.

Measurements need to be conducted on a moist, partially dry as well as on a totally dried-out sample whilst the water content w as well as density γ need to be determined in each case. The graphical representation of the measured values (Figs. 18.44 and 18.46) characterises the type of soil investigated.

To dry or completely dry-out the soil sample, the vessel must be rearranged (Fig. 18.43 c) and placed open in a heating cabinet at 105 °C.

An equal distribution of moisture within a partly dried-out test sample can be achieved by heating for a sufficiently long period with the vessel closed (Fig. 18.43 b).

Apparatus which can be used on site in open country to measure thermal resistivity is commercially available with variable expenditure of measurement and time [18.32].

Basic Quantities for Calculation

To facilitate the calculation of soil-thermal resistances (Section 18.4.3) and to establish the load capacity tables (Section 18.2.3) the relevant quantities for ambient conditions had to be agreed as a basis.

The *thermal resistivity of the soil unaffected by heat from a cable* – the moist area – was fixed at 1.0 Km/W.

Measurements made previously in Germany rarely produced quantities in excess of 1.0 Km/W with the exception of very dry sandy soils, made up areas, or in areas which contained industrial waste e.g. building rubble [18.31, 18.35]. A large number of the quantities measured were below 1.0 Km/W due to the relatively high loam content within the soils having good water retention capacity, or due to a prevailing season of high precipitation at the time of measurement. If quantities of less than 1.0 Km/W are to be used these should be verified by sufficiently long periods of measurement and should embrace at least one dry period. It must be considered also that when the cable is installed the ground is disturbed. This means that the bedding material surrounding the cable and the back fill in the trench up to the ground surface is not so highly compacted as the original soil and the favourable characteristics of the undisturbed soil will not be fully achieved. Backfill and bedding material can, depending on the selection made, have characteristics inferior to those of the surrounding soil.

The *thermal resistivity of the dried-out soil* was agreed as 2.5 Km/W. Laboratory measurements made on test samples which had been dried-out at 105 °C indicated, depending on the type of soil and degree of compaction, quantities between 1.5 and 3.0 Km/W. Contaminated soils have much higher quantities [18.33]. The permissible operating temperature in modern cables ranges between 60 °C for 30 kV mass-impregnated cable and 90 °C for cable with insulation of XLPE. The surface temperature of such cables is however less than these quantities even after taking into consideration heat from neighbouring cables or

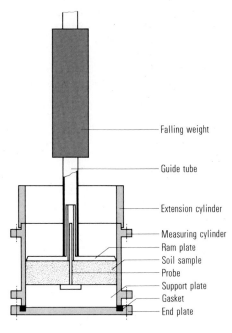

Falling weight

Guide tube

Extension cylinder

Measuring cylinder
Ram plate
Soil sample
Probe
Support plate
Gasket
End plate

a) Apparatus for compacting the soil sample

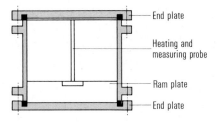

End plate

Heating and
measuring probe

Ram plate

End plate

b) Vessel closed at both ends to equalise
moisture distribution while in heating cabinet

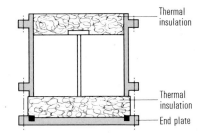

Thermal
insulation

Thermal
insulation

End plate

c) Sample prepared for measurements

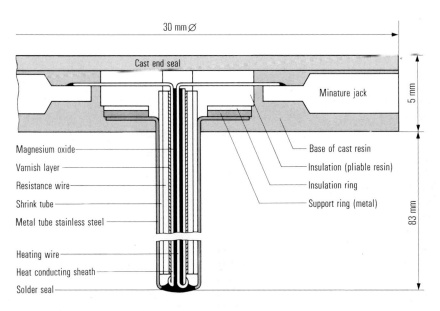

30 mm ∅

Cast end seal

Minature jack

5 mm

Magnesium oxide
Varnish layer
Resistance wire
Shrink tube
Metal tube stainless steel

Heating wire
Heat conducting sheath
Solder seal

Base of cast resin

Insulation (pliable resin)

Insulation ring

Support ring (metal)

83 mm

d) Cross section through heating and measuring probe

Fig. 18.43
Apparatus for determining thermal resistivity, water content and density

groups. Under practical operational conditions the surface temperature of the cable would always be less than 105 °C and drying out would be reduced such that it would appear permissible to use 2.5 Km/W as a standard quantity for the dry area.

The *boundary isotherm* which separates the moist from the dry area is affected by many influences such as type of soil, water retention capacity under local conditions, weather conditions, soil temperature as well as time related heating of the cable surface relative to soil temperature.

If all these effects are considered it appears possible to approximately double the temperature rise limit at $m = 0.5$ relative to the quantity for continuous operation and for intermediate quantities select a linear increase. The temperature rise limit $\Delta \vartheta_x$ can therefore be represented by the equation 18.56 in Section 18.4.3 such that with a quantity of 15 K for continuous operation at $m = 1.0$ this would relate to 25 K for a public utility load at $m = 0.7$ and 32 K for a daily load cycle with $m = 0.5$.

In Great Britain for low- and medium-voltage cables for both continuous operation and cyclic operation the quantities given in Table 18.34 are used. These are extracted from an E.R.A. report 69–30 Part 1 "Current rating standards for distribution cables", [18.36]. Where drying-out of the soil is expected and where a more accurate assessment of load capacity is necessary quantities are used as, e.g. in [18.37] for the moist region 1.2 Km/W, for the dry region 3.0 Km/W and for the boundary isotherm in loam 50 °C or sandy soil 35 °C where both these temperatures relate to a 15 °C soil temperature.

In [18.38] consideration is given to the different conditions prevailing in summer and winter and their effect on load capacity. For the winter months the quantities 0.9/3.0 Km/W at 10 °C ambient temperature and for the summer months 1.2/3.0 Km/W at 15 °C ambient temperature are recommended. The quantity of the boundary isotherm in both instances is 50 °C. These quantities are supported in [18.39] with the exception of the soil-thermal resistivity of the unloaded soil which is given as an increased quantity for the winter period of 1.05 Km/W.

Table 18.34 Soil-thermal resistivities from "Electric Cables Handbook" [18.36]
(Quantities in brackets where the ground surface is impervious to water)

Type of soil	Soil-thermal resistivity in Km/W at maximum loading		
	equally over whole year, therefore also during the dry period in the summer	in summer (Mar./April to mid.-Nov.), however outside the dry periods; also feeder cables which are only used in emergency	in winter (mid.-Nov. to Mar./April)
All soils with the exception of the following	1.5 (1.2)	1.2 (1.0)	1.0 (0.8)
Peat – bog	1.2 (1.2)	1.2 (1.0)	1.0 (0.8)
Clay bearing soil	1.5 (1.2)	1.2 (1.0)	0.9 (0.8)
Chalky soil with crushed sand as bedding material	1.2 (1.2)	1.2 (1.0)	1.2 (1.2)
Very stony soil or broken stone	1.5	1.3	1.2
Very dry sand	2.5	2.0	1.5
Made up soil	1.8	1.6	1.2

Bedding Material

The investigation and selection of bedding material is always recommended where the cables are to be operated under continuous load ($m=1.0$). A knowledge of the soil together with the physical and thermal characteristics of the bedding material makes it possible to establish a more appropriate load capacity. Generally the excavated soil is more favourable than the types of sand used by the building industry. Artificially produced mixes are particularly suited for cable runs which are operated at high thermal stress. The use of this for longer runs of continuously loaded high-voltage cable is related to a question of economy whereby it must be considered that in thermal bottle necks of short lengths – excessive grouping and crossing of cables or crossing of heating ducts – the cost of the material could play only a secondary role. Where building work is carried out at a later date the selected or specially mixed bedding material must neither be replaced by material having poorer properties nor must the volumetric weight be changed.

Of the types of soil which occur naturally the quartz containing sandy types have the most favourable granular distribution, e.g. a high uniformity index U and a reduced pore content n. The thermal conductivity as well as ease of working and compacting are improved by a content of fine granules $d<0.2$ mm and of silt $d<0.063$ mm. In Table 18.35 quantities are given of a number of measurements.

The highest thermal resistivity of 5.4 Km/W was found in household waste contained in sample number 16. Sands with a low uniformity index reached quantities of above 3.0 Km/W. The samples 7 to 10 are gravel sand mixtures containing different quantities of silt (powdered limestone). The thermal resistivities of these mixtures are shown in Fig. 18.44.

The particle size distribution curves of sand samples 11 to 15 are shown in Fig. 18.45 and the thermal resistivities relative to moisture content are shown

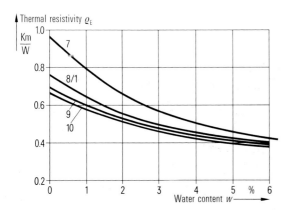

a) Relative to dry density
with various levels of silt content

    Sample number 7 (0% SchA) $\gamma_d = 1.975$ t/m^3
                     8/1 (5% SchA) $\gamma_d = 2.015$ t/m^3
                    9 (10% SchA) $\gamma_d = 2.03$ t/m^3
                10 (15% SchA) $\gamma_d = 2.04$ t/m^3

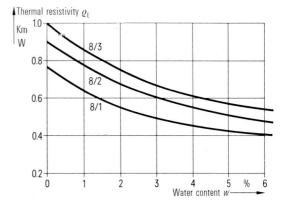

b) Relative to dry density
with constant level of silt content

    Sample number 8/1 (5% SchA) $\gamma_d = 2.015$ t/m^3
                      8/2 (5% SchA) $\gamma_d = 1.94$ t/m^3
                      8/3 (5% SchA) $\gamma_d = 1.84$ t/m^3

Fig. 18.44
Relationship of thermal resistivity ϱ_E to water content w of a gravel sand mixture with a silt granular content (SchA) of powdered limestone

Table 18.35
Soil physical key data and thermal resistivities of tested samples

Sample number		$p_{0.063}$ %	d_{10} mm	d_{60} mm	$U = d_{60}/d_{10}$	n	γ_d t/m³	w %	ϱ_E Km/W
1	Sand	0	0.135	0.46	3.4	0.34	1.75	0.05	1.80
1a	Sample No. 1 with 4% clay	4.0	0.115	0.42	3.7	0.335	1.76	0.0	1.35
2	Basalt wheathered ($\gamma_s = 2.75$ t/m³)	5	0.1	0.75	7.5	0.388	1.685	0.0	2.52
3		5	0.1	0.75	7.5	0.32	1.87	0.0	1.68
4		6	0.085	1.5	17.5	0.247	2.09	0.0	1.22
5		8	0.07	1.6	23	0.218	2.15	0.0	1.15
6	Crushed limestone ($\gamma_s = 2.75$ t/m³)	15	0.035	1.1	31.5	0.28	1.98	0.0	1.25
7	Gravel, sand in proportion 1:1	1	0.22	0.93	4.22	0.255	1.975	0.0 / 6.0	0.96 / 0.41
8	Gravel, sand in proportion 1:1 + 5% (SchA)	3	0.21	4.0	19	0.24	2.015	0.0 / 6.0	0.76 / 0.40
9	Gravel, sand in proportion 1:1 + 10% (SchA)	7	0.1	2.3	23	0.234	2.03	0.0 / 6.0	0.69 / 0.40
10	Gravel, sand in proportion 1:1 + 15% (SchA)	11	0.06	1.0	26.6	0.23	2.04	0.0 / 6.0	0.67 / 0.39
11	Sand	0	0.22	0.55	2.5	0.40	1.6	0.0	1.88
12	Sand	5	0.18	0.7	3.9	0.37	1.66	0.0	1.21
13	Sand	10	0.22	0.47	2.14	0.42	1.54	0.0	2.42
14	Sand	15	0.1	0.16	1.6	0.5	1.33	0.0	3.78
15	Sandy loam	not measured				0.38	1.63	0.0	1.77
16	Waste material	–	–	–	–	–	1.2	0.0	5.4

$p_{0.063}$	Content in % of granular size $d < 0.063$ mm (SchA) silt content of powdered limestone
d_{10}	Particle diameter at 10% sieve-let-through
d_{60}	Particle diameter at 60% sieve-let-through
U	Degree of uniformity
n	Pore content
γ_d	Dry density
γ	Density of moist sample
γ_s	Density of solid material
w	Water content
ϱ_E	Measured thermal resistivity

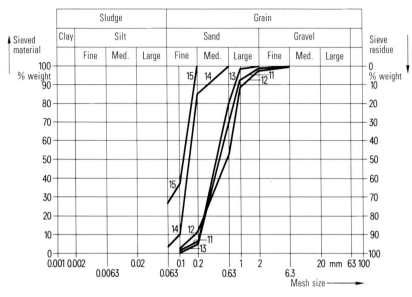

Fig. 18.45
Particle size distribution curves for soil samples 11 to 15 in Table 18.35

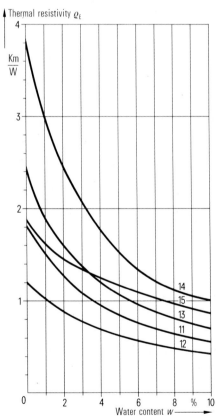

Fig. 18.46
Thermal resistivity ϱ_E for soil samples numbered 11 to 15 in Table 18.35

in Fig. 18.46. Sample number 15 contained almost 30% of particles $d < 0.063$ and was found difficult to compact.

In Great Britain the sand-gravel and sand-cement mixtures have become known under the heading of "thermally stable bedding materials".

Sand-Gravel Mixtures

The mixture ratio is intended to be 50:50 but deviations of up to 45:55 are acceptable. The grain size of the sand should not exceed 2.4 mm but a 5% content of up to 5 mm is acceptable. The dry density should not be less than 1.6 t/m³. No organic or clay content is permitted. The grain size of gravel should be between 2.4 and 10 mm. Sharp edged particles should not be present. The mixture should be delivered with a water content of between 7 and 10% and compacted to a dry density of 1.8 t/m³ to obtain a thermal resistivity of 1.2 Km/W in the dried out state.

The conditions required can also be fulfilled by a powdered stone-gravel mixture. Crushed gravel may only be used up to 50% of the total gravel content.

229

Sand-Cement Mixtures

In the set state the bedding material must be crumbly so that it will not damage the cable in the event of subsidence and also should it become necessary to subsequently remove it. It is recommended to use sands with a pore content ≤0.55 which approximate the particle size distribution curve (18.47, curve D)

Mesh width	mm	5	3.15	2.5	1.25	0.63	0.315	0.125
Sieve-let-through	%	95	89	84	75	54	18	1.6

The sand-cement proportions should be 14:1 by volume or 18 to 20:1 by weight. To achieve compaction to 1.6 t/m³ and a relevant thermal resistivity of 1.2 Km/W, a water-cement ratio of approximately 2:1 by weight is required.

For the development of suitable mixtures the rules for the manufacture of concrete in DIN 1045 can be used, since a pore reduced mix and also an ability for compaction are also required for concrete. To DIN 1045 the range between particle size distribution curves A and B result in a particularly good mixture whereas particles between B and C produce a mixture

which is still acceptable. The requirement for the mixture described above is approximately fulfilled by the curve A [18.41].

Calculation of Load Capacity

The calculation of load capacity, where thermally stable bedding material is used, is made to Section 18.4.6 using equations 18.92 to 18.95. Using the dimensions of the bedding material, designated x, y and characteristic diameter d_B, the geometric factor k_b as well as thermal resistances $T_B'^+$ and $T_{By}'^+$ with ϱ_B the thermal resistivity of the bedding material in the dried-out state can be calculated. The thermal resistances T_P' and T_R' arc not used where the cable is buried directly in the bedding material and docs not lay in a pipe.

18.5 Installation in Channels and Tunnels

18.5.1 Unventilated Channels and Tunnels

In unventilated and covered channels and tunnels, the heat generated in the cables is transmitted in the main only through the walls, base and top of the duct. Natural ventilation is mostly prevented by the compartmentalization which are unavoidable. These form heat barriers and cause the air surrounding the cables

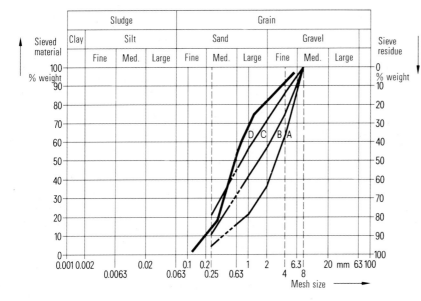

Fig. 18.47
Particle size distribution curves A, B, C for a concrete mixture to DIN 1045 and curve D for a sand-cement mixture

in the channel to increase in temperature such that the load capacity is reduced compared with that of free air.

The temperature rise of the air in the channel depends upon the dimensions of the channel and the magnitude of the losses of all the cables in it. The number of cables generating losses and the locations within the channel have no influence on the temperature rise of the air contained in the channel [18.45].

The equivalent circuit for the thermal path of heat flow from a cable enclosed in a channel is shown in Fig. 18.48. Heat is transmitted from the cable surface by radiation to the inner surfaces of the channel. Since these areas are large compared to the cable surface area, the heat transmission factor for radiation can be calculated with the emission factor $\varepsilon_0 \approx 0.95$, as for a cable installed in free air. As opposed to installation in free air the following additional thermal resistances must be taken into consideration:

▷ the thermal transfer resistance T'_{TK} for convection on the inner wall of the channel,

▷ the thermal resistance T'_{TE} of the channel walls and the surrounding soil,

▷ the thermal transfer resistance T'_{TO} at the ground surface, respectively channel surface.

The thermal transfer resistance T'_{TK} for convection at the channel inner wall is

$$T'_{TK} = \frac{1}{\alpha_i \, 2(b_T + h_T)} \qquad (18.100)$$

with the dimensions h_T and b_T to Fig. 18.49. The thermal transfer factor α_i is selected to DIN 4701 with 7.7 W/Km² (arithmetic mean from the quantities for walls, base and top cover) [18.46].

T'_{TK} is in series with T'_{KK} and both are in parallel with T'_{KS}. The thermal resistance of air T'_{LuT} of a cable in the channel is calculated to equation 18.17 and the equations

$$T'_{KS} = \frac{1}{\pi \, d f_s \alpha_s}, \qquad (18.101)$$

$$T'_{KK} = \frac{1}{\pi \, d f_k \alpha_k}, \qquad (18.102)$$

$$T'_{LuT} = \frac{1}{\dfrac{1}{\dfrac{1}{T'_{KK} + T'_{TK}} + \dfrac{1}{T'_{KS}}}}. \qquad (18.103)$$

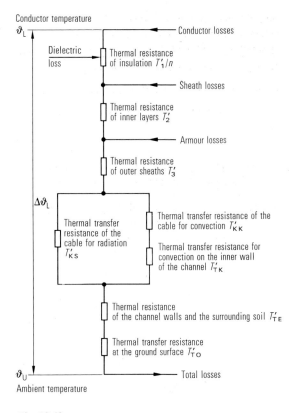

Fig. 18.48
Equivalent diagram for heat flow from a cable in a channel

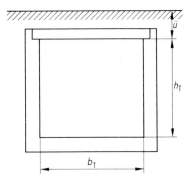

h_T Height of channel
b_T Width of channel
$ü$ Thickness of covering

Fig. 18.49   Covered channel in ground

231

The thermal transfer resistance T'_{TK} is very small in relation to the other thermal resistances and can be ignored in the following cases:

▷ for three-core cable with $d < 90$ mm,

▷ for single-core cable with $d < 45$ mm,

▷ in channels with a circumference $2(b_T + b_T) > 1$ m.

The thermal resistance T'_{TE} takes into consideration heat conduction from the inner wall of the channel through to the ground surface [18.47]. This is affected by the channel dimensions (Fig. 18.49) and the soil-thermal resistivity ϱ_E (to reduce complication the thermal resistivity of the channel material is taken also as ϱ_E):

$$T'_{TE} = \cfrac{\varrho_E}{\dfrac{4}{\pi}\left\{1 + \ln\left[\dfrac{\dfrac{h_T}{\ddot{u}}+1}{\sqrt{2}} + \sqrt{\left(\dfrac{\dfrac{h_T}{\ddot{u}}+1}{\sqrt{2}}\right)^2 - \dfrac{1}{2}}\right]\right\} + \dfrac{b_T}{\ddot{u}}}$$

(18.104)

The thermal transfer resistance T'_{TO} at the ground surface, or where applicable the channel surface, is approximated using the thermal resistance of an imaginary layer with a thickness δ and a soil-thermal resistivity ϱ_E. With a thermal transfer factor $\alpha_a = 20$ W/Km^2 this becomes [18.48]

$$\delta = \frac{1}{\alpha_a \varrho_E}.$$

(18.105)

The thickness of covering $\ddot{u}$ must be increased by the value δ

$$\ddot{u}^+ = \ddot{u} + \delta = \ddot{u} + \frac{1}{\alpha_a \varrho_E}$$

(18.106)

and inserted in equation 18.104.

This results in

$$T'_{TE} + T'_{TO} = \cfrac{\varrho_E}{\dfrac{4}{\pi}\left\{1 + \ln\left[\dfrac{\dfrac{h_T}{\ddot{u}^+}+1}{\sqrt{2}} + \sqrt{\left(\dfrac{\dfrac{h_T}{\ddot{u}^+}+1}{\sqrt{2}}\right)^2 - \dfrac{1}{2}}\right]\right\} + \dfrac{b_T}{\ddot{u}^+}}$$

(18.104a)

With the aid of Fig. 18.50 a quick result can be obtained for the two thermal resistances assuming a soil-thermal resistivity of 1 Km/W. For other quantities of soil-thermal resistivity the result from the graph must be multiplied by

$$\frac{\varrho_E}{1 \text{ Km/W}}.$$

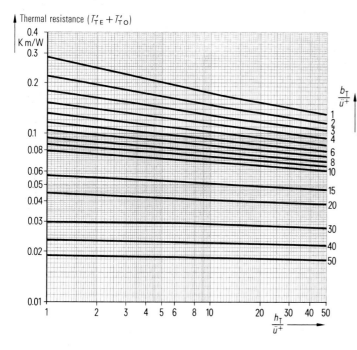

Fig. 18.50
Thermal resistance $(T'_{TE} + T'_{TO})$ of a channel relative to dimensions h_T, b_T and $\ddot{u}^+$ at $\varrho_E = 1$ Km/W

The temperature of the soil is dependant on the depth of lay, i.e. the measuring depth (see Sections 18.3 and 18.4). The temperature at a depth of approximately 10 m is constant and is equal to the mean annual temperature of the air (in Germany approximately 9 °C). In smaller depths the temperature follows the variations of air temperature with a certain time delay. Various depths are affected by seasonal variations whereas close to and at the ground surface the temperature can vary depending on time of day. The mean value of these temperatures during the summer months is higher than the temperature of the deeper layers. With cables on no load the air in a channel assumes a mean temperature resulting from the temperatures of the inner surfaces and the parts of the inner areas of the channel boundary faces. The base and walls of the channel assume approximately soil temperature at the depth of channel centre line. The inner surface of the channel lid reaches, because of the influence of air temperature and sunshine during summer months, a higher temperature by a value of $\Delta\vartheta_i$ (Fig. 18.51). Thus the mean temperature of the air in the channel becomes

$$\vartheta_{TE} = \vartheta_E + \frac{\Delta\vartheta_i}{2\left(\dfrac{h_T}{b_T}+1\right)} \tag{18.107}$$

and with loaded cables and the summation of losses of all cables in the channel as $\Sigma(\Gamma'_i + \Gamma'_d)$

$$\vartheta_T = \vartheta_{TE} + \Sigma(P'_i + P'_d)(T'_{TE} + T'_{TO}). \tag{18.108}$$

18.5.2 Arrangement of Cables in Tunnels

The cables are either mounted direct to the walls with the aid of cable clips or laid on racks or trays. The ventilation clearance between trays depends on their width; this should wherever possible be not less than 300 mm to provide for the installation of heavy cables. On trays and racks as well as where cables are fixed direct to the walls a clearance between highly loaded cables equal to the diameter of the cable should be maintained to keep heat transmission from cable to cable as low as possible.

The height of tunnels should not be less than 2.2 m. The width should be chosen such that a free passageway clearance of 60 to 80 cm is maintained. With trays installed at a vertical pitch of 30 cm their width should be limited to 50 cm to allow access for cable installation.

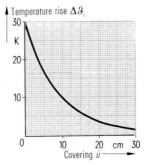

Fig. 18.51
Temperature rise $\Delta\vartheta_i$ of the inner face of channel top relative to thickness $\ddot{u}$ of the covering

In the design of an installation the following procedures can apply: initially a first approximation is made of the cross-sectional area for each individual cable at some 30 to 50% greater than the size required for installation in free air. For high currents it may be necessary to use several cables per current path. Secondly a sketch plan is made of the tunnel showing the required height, width, number of trays and arrangement of cables following the rules mentioned above.

From the proposed arrangement of cables shown in the sketch plan the rating factor for groups installed in air f_H to Tables 18.23 or 18.24 can be selected. The total losses in the tunnel are next calculated and the resultant increase in temperature of air in the tunnel is found from equation 18.108. The temperature of the tunnel air with cables under no load must be increased by this amount and a revised rating factor selected relative to this increased ambient temperature f_ϑ from Table 18.22 or from equation 18.15. When the load capacity I_r is multiplied by these factors the product must not be less than the load to be transmitted.

$$I_b \le I_r f_H f_\vartheta. \tag{18.109}$$

If this condition is not satisfied either the number of cables, the cross-sectional area or the tunnel dimensions must be increased. If these proposals are not possible or not practical then forced ventilation must be employed.

The time constant of a tunnel is great compared to the time constant of a cable. The temperature rise of the air in the tunnel can be determined therefore

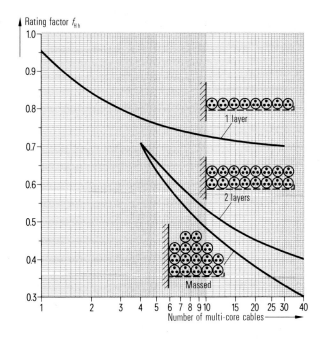

a) Relative to arrangement and number of cables on a cable tray f_{Hh}

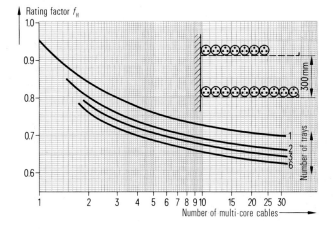

b) Relative to number of equally loaded trays above one another with single layer on each f_H

Fig. 18.52
Rating factors for grouping of multi-core cables – or bunched single-core cables of one circuit – on cable trays

by using the root mean square value I_q of the currents producing the losses over 24 hours:

$$I_q = \sqrt{\frac{I_{b1}^2 t_1 + I_{b2}^2 t_2 + \ldots + I_{bi}^2 t_i}{t_1 + t_2 + \ldots + t_i}} \qquad (18.110)$$

with $t_1 + t_2 + \ldots t_i = 24$ h. Where I_{b1}, $I_{b2} \ldots$ are the currents which flow during the times $t_1, t_2 \ldots$

For groups of larger numbers of cables than is allowed for in the tables rating factors to Fig. 18.52 can be used. These values are also valid for single-core cables if for each circuit instead of a multi-core cable the requisite number of single-core cables are bunched. In these cases the rating factors for load capacity I_r apply as for bunched cables.

If more than six trays are installed above one another the rating factor for six trays may be used in the calculation.

An approximation of the rating factor for bunching in air f_H, for cables touching one another, can be formed from the rating factor f_{Hh} (horizontal component) for groups on a cable tray to Fig. 18.52a and the rating factor f_{Hv} (vertical component) for groups of approximately equally loaded cable trays above one another.

It is

$$f_H = f_{Hh} \cdot f_{Hv}$$

with

$f_{Hv} = 0.95$ for two cable trays above one another
$f_{Hv} = 0.93$ for three cable trays above one another
$f_{Hv} = 0.9$ for six and more cable trays above one another.

Where the number of cables and the loading are not known the cross-sectional areas must be determined using an assumed total reduction. A final review will then enable a decision to be made as to whether forced ventilation is required and whether or not the rating factor applied initially was adequate.

18.5.3 Channels with Forced Ventilation

If natural ventilation proves to be inadequate, i.e. the air in the channel is overheated and the conductor temperature exceeds the permissible quantity forced ventilation is necessary where other means are not possible e.g. enlarging the size of channel.

Mostly the calculation is based on the total heat loss generated within the channel. Heat dissipated through the channel walls is not taken into consideration. In this way fans are not sized too small and thus some reserve capacity is available for future extensions.

The air rate required Q is dependent on the total heat loss generated by the cable $\Sigma(P_i' + P_d')$, the channel length l and the temperature rise of the cooling air $\Delta \vartheta_{Kü}$ between entry and exit. This is expressed by

$$Q = \frac{\Sigma(P_i' + P_d')\, l}{c_p \, \Delta \vartheta_{Kü}}. \qquad (18.111)$$

c_p being the specific heat of air at constant pressure but is dependent on temperature as well as humidity; an approximate calculation can be made with $c_p = 1.3$ kJ/Km^3.

The air velocity v is determined by taking the cross-sectional area of the channel calculated from height and width (see also Fig. 18.53)

$$v = \frac{Q}{A}. \qquad (18.112)$$

If noise nuisance is to be avoided the air velocity must not exceed 5 m/s.

The temperature rise of the cooling air must be chosen giving consideration to the temperature at the point of entry and the temperature which is permissible at the exit. In most instances the temperature of the input cooling air will be identical with the design ambient temperature ϑ_U. The hottest cable is considered in respect of permissible operating temperature ϑ_{Lr} in deciding the temperature rise of the cooling air using the formula

$$\Delta \vartheta_{Kü} \leq \vartheta_{Lr} - \vartheta_U - \Delta \vartheta \qquad (18.113)$$

with

$$\Delta \vartheta = (\vartheta_{Lr} - 30\,^{\circ}\mathrm{C})\left(\frac{I_b}{I_r}\right)^2. \qquad (18.114)$$

Since the moving air significantly improves heat dissipation from the cable the rating factor required for groups f_H need not be applied.

Example 18.11

In a tunnel with dimensions 2.2 m × 1.5 m the cables shown in Table 18.36 are to be installed and be loaded with the currents given in the table. The duration of operation is first of all planned for 8 hours full load per day. It is required to operate also at full load for 16 hours per day when under this condition forced ventilation may be provided. The ambient temperature ϑ_U of the air is 35 °C and the soil temperature ϑ_E at the depth equivalent to the tunnel centre, with cables unloaded, is 25 °C. The soil-thermal resistivity is 1.2 Km/W. The planned arrangement of cables is shown in Fig. 18.53.

For the 8 hour operation in respect of 24 hours the root mean square value of current in the cable NYFGY is:

$$I_q = \sqrt{\frac{I_b^2 t_1}{t_1 + t_2}} = I_b \sqrt{\frac{8}{24}}$$

for the losses

$$P_i' = P_{ir}' \left(\frac{I_q}{I_r}\right)^2 = P_{ir}' \frac{8}{24}\left(\frac{I_b}{I_r}\right)^2$$

$$= 44.9 \frac{8}{24}\left(\frac{205}{315}\right)^2 = 6.34 \text{ W/m}$$

and

$$\Sigma P_i' = 13 \times 6.34 = 82.4 \text{ W/m}.$$

The sum of all losses of all cables to Table 18.36 gives

$$82.4 + 55.0 + 27.5 + 37.2 = 202.1 \text{ W/m}.$$

From the curve, Fig. 18.50 with

$$\ddot{u}^+ = \ddot{u} + \delta = 0.15 + \frac{1}{20 \cdot 1.2} = 0.192 \text{ m},$$

$$\frac{h_T}{\ddot{u}^+} = \frac{2.2}{0.192} = 11.5 \quad \text{and}$$

$$\frac{b_T}{\ddot{u}^+} = \frac{1.5}{0.192} = 7.8$$

the thermal resistance $T_{TE}' + T_{TO}'$ results for ρ_E = 1 Km/W to 0.078 Km/W.

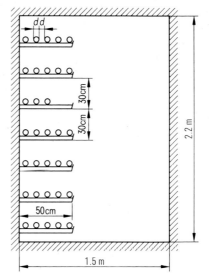

Fig. 18.53 Arrangement of cables for example 18.11

Table 18.36
Cable types and loading for Example 18.11 for 8 hour operation

Cable type		NYFGY 3×150SM	NYCY 4×240SM	NEKBY 3×70RM	NEKBY 3×120RM
U_0/U	kV	3.6/6	0.6/1	12/20	12/20
Number of cables		13	7	6	7
Loading I_b	A	205	285	120	170
Load capacity I_r	A	315	429	195	271
Ohmic losses P_{ir}'	W/m	44.9	53.4	36.3	40.5
Permissible operating temperature ϑ_{Lr}	°C	70	70	65	65
$\dfrac{I_b}{I_r}$		0.65	0.66	0.62	0.63
Eight-hour-operation P_i'	W/m	6.34	7.86	4.58	5.31
$\Sigma P_i'$	W/m	82.4	55.0	27.5	37.2
f_ϑ		0.79	0.79	0.76	0.76
$f_H f_\vartheta$		0.69	0.69	0.66	0.66
$\Delta\vartheta$	K	16.9	17.7	13.3	13.8
$\Delta\vartheta_{Kü}$	K	18.1	17.3	16.7	16.2

For grouping up to 5 cables on 7 trays a rating factor $f_H \approx 0.87$ is to be applied (see Table 18.24).

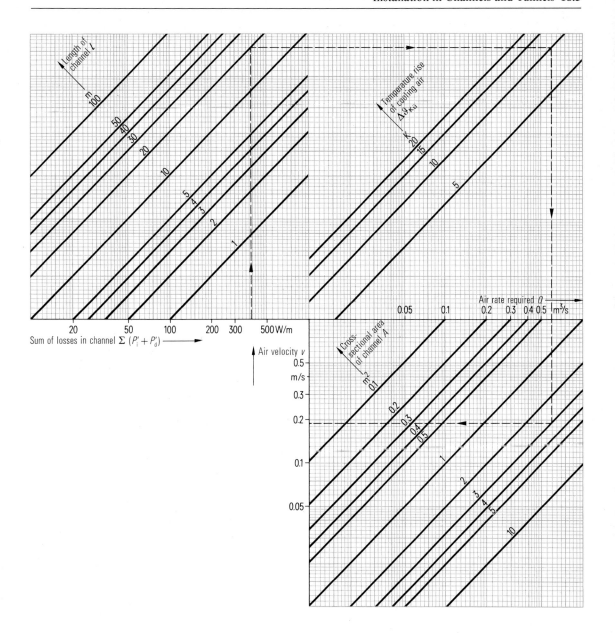

Q	Air rate required in m^3/s
v	Air velocity in m/s
$\Sigma(P_i' + P_d')$	Heat loss from all cables in W/m
$\Delta\vartheta_{K\ddot{u}}$	Temperature rise of cooling air in K
l	Length of channel in m
A	Cross-sectional area of channel in m^2

The dotted line applies to quantities used in example 18.11

Fig. 18.54
Calculation diagram: Cable channel with forced ventilation

With a soil-thermal resistivity of $\varrho_E = 1.2$ Km/W the corrected thermal resistance of the soil including the tunnel becomes

$$T'_{TE} + T'_{TO} = \frac{1.2}{1.0} 0.078 = 0.094 \text{ Km/W}.$$

The temperature in the tunnel with no load on the cables is

$$\vartheta_{TE} = \vartheta_E + \frac{\Delta\vartheta_i}{2\left(\frac{a}{b} + 1\right)} = 25 + \frac{5.1}{2\left(\frac{2.2}{1.5} + 1\right)} = 26 \,°C$$

(18.107)

with $\Delta\vartheta_i = 5.1$ K from Fig. 18.51 and the temperature with the loaded cables is

$$\vartheta_T = \vartheta_{TE} + \Sigma(P'_i + P'_d)(T'_{TE} + T'_{TO})$$

$$= 26 + 202.1 \times 0.094 = 45 \,°C.$$

(18.108)

The rating factor for the deviating ambient temperature is

$$f_\vartheta = \sqrt{\frac{\vartheta_{Lr} - \vartheta_T}{\vartheta_{Lr} - 30\,°C}} = \sqrt{\frac{70 - 45}{70 - 30}} = 0.79 \qquad (18.15)$$

and the overall rating factor is

$$f = f_H f_\vartheta = 0.87 \times 0.79 = 0.69.$$

The factors I_b/I_r are in each case smaller than the required reduction factors f_H and f_ϑ, i.e. the cross-sectional areas are dimensioned correctly.

When changing to 16 hour operation one gets

$$I_q = I_b \sqrt{\frac{16}{24}}$$

and

$$P'_i = P'_{ir} \frac{16}{24} \left(\frac{I_b}{I_r}\right)^2.$$

The total losses in the tunnel are therefore doubled

$$\Sigma P'_i = 2 \times 202.1 = 404.2 \text{ W/m}$$

and the temperature of the tunnel air is raised to

$$\vartheta_T = 26 + 404.2 \times 0.094 = 64 \,°C. \qquad (18.108)$$

Therefore the tunnel must be ventilated.

For the cable NYFGY

$$\Delta\vartheta = (70 - 30)\left(\frac{205}{315}\right)^2 = 16.9 \text{ K}, \qquad (18.114)$$

$$\Delta\vartheta_{Kü} \leq 70 - 35 - 16.9$$

$$\leq 18.1 \text{ K}. \qquad (18.113)$$

The quantities for all remaining cables can be taken from Table 18.36. With an appropriate quantity of $\Delta\vartheta_{Kü} = 10$ K, length of tunnel 20 m and profile of the tunnel of $1.5 \times 2.2 = 3.3$ m^2 the air rate required then becomes:

$$Q = \frac{\Sigma(P'_i + P'_d)l}{c_p \Delta\vartheta_{Kü}} = \frac{404.2 \times 20}{1.3 \times 10^3 \times 10} = 0.622 \text{ m}^3/\text{s}$$

(18.111)

and for the air velocity

$$v = \frac{Q}{A} = \frac{0.622}{3.3} = 0.19 \text{ m/s}. \qquad (18.112)$$

The same results can be obtained from Fig. 18.54.

18.6 Load Capacity of a Cable for Short-Time and Intermittent Operation

18.6.1 General

Under a constant current load a cable heats up until after a certain time a stable temperature is reached. This temperature rise occurs at a rate which approximates an exponential function or several combined exponential functions. The exact mathematical treatment of such heating and cooling processes requires a substantial effort of calculation [18.44; 18.49; 18.50]. In the following simplified calculation methods the results are less accurate but are based always on the safe side. These methods involve the introduction of a minimum time value, the calculation using the squared mean value or using adiabatic heat rise. These methods are particularly applicable for a cable installed in air. They can however be used for cables installed in ground because the heating process takes longer in ground than in air and the temperature rise will be less.

For a cable installed in the ground a method of calculation for operation of cables under utility load conditions is described in Section 18.4.3 and in those following.

18.6.2 Calculation with Minimum Time Value

A typical temperature rise curve is shown in Fig. 18.55. The plotted theoretical curve corresponds to the exponential function

$$\Delta\vartheta = (\vartheta_e - \vartheta_U)(1 - e^{-\frac{t}{\tau}}). \tag{18.113}$$

The time value τ (time constant) is one fifth of the time taken from the curve to almost reach the permissible final temperature. It is defined by the thermal resistance T and the heat capacity C:

$$\tau = TC. \tag{18.114}$$

Since at the begining of the temperature rise the temperature increases more rapidly than that given by the theoretical progression it is recommended, especially for short-time operation and intermittent operation, to use a minimum time value for τ. In this only the heat capacity of the conductor and the total thermal resistance of the cable in air is considered. The load factor calculated using the minimum time value provides a certain margin of safety, especially for longer load durations.

The minimum time value of a cable is calculated from equation 18.114 with

$$C = ncq, \tag{18.115}$$

$$T = \frac{\Delta\vartheta_r}{nI_r^2 R'_{wr}} \approx \frac{\Delta\vartheta_r\, q\varkappa_{20}}{nI_r^2\,[1+\alpha_{20}(\vartheta_{Lr}-20\,°C)]}, \tag{18.116}$$

$$\tau = B\left(\frac{q}{I_r}\right)^2, \tag{18.117}$$

$$B = \frac{\Delta\vartheta_r\,\varkappa_{20}\,c}{1+\alpha_{20}(\vartheta_{Lr}-20\,°C)}. \tag{18.118}$$

Table 18.38 and Fig. 18.56 are used for the determination of the minimum time value.

Table 18.37
Material quantities for calculation of minimum time value τ

Conductor material	α_{20} 1/K	$\varkappa_{20}$ 1/Ω m	c J/K m^3
Copper	0.00393	56×10^6	3.45×10^6
Aluminium	0.00403	34×10^6	2.50×10^6

Table 18.38
Quantity B for the calculation of the minimum time value τ for installation in air

Conductor material	Quantity B in A^{2}s/m^4 for permissible operating temperatures of				
	90 °C	80 °C	70 °C	65 °C	60 °C
Copper	9.09×10^{15}	7.82×10^{15}	6.46×10^{15}	5.75×10^{15}	5.01×10^{15}
Aluminium	3.98×10^{15}	3.42×10^{15}	2.83×10^{15}	2.52×10^{15}	2.20×10^{15}

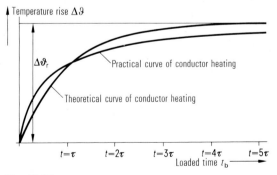

Fig. 18.55
Temperature rise at the conductor of a cable carrying constant current

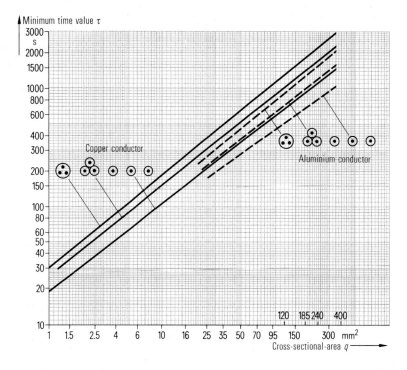

Fig. 18.56
Minimum time value τ relative to cross-sectional area q for PVC cables 0.6/1 kV

Table 18.39
Short-time current densities J_{th} for the calculation of current for short-time operation during some seconds

Type of construction	Permissible operating temperature ϑ_{Lr}	Conductor temperature ϑ_a prior to loading								
		90°C	80°C	70°C	65°C	60°C	50°C	40°C	30°C	20°C
	°C	Short-time current density in A/mm² for 1 s								
Copper conductor										
XLPE-cable	90	—	39.9	56.9	63.9	70.3	81.8	92.3	102	111
PVC-cable	70	—	—	—	29.0	41.2	58.8	72.7	84.7	95.6
Aluminium conductor										
XLPE-cable	90	—	26.5	37.7	42.4	46.6	54.3	61.3	67.7	73.8
PVC-cable	70	—	—	—	19.3	27.4	39.0	48.2	56.2	63.5

18.6.3 Adiabatic Heat Rise

The load capacity I_{KB} for operating times t_B in the second range (e.g. run-up currents of electric motors, short-time operation) can be calculated according to the rules for short-circuit capacity (see Section 19, equation 19.26)

$$I_{KB} = q J_{th} \sqrt{\frac{1 \text{ s}}{t_b}} \qquad (18.119)$$

with the conductor cross-sectional area q in mm^2.

Quantities of short-time current densities J_{th} are given in Table 18.39 in respect of starting temperatures ϑ_a (conductor temperature prior to short-time loading).

If the cable is not loaded prior to commencement of the short-time loading then

$$I_0 = 0, \quad \vartheta_a = \vartheta_U. \qquad (18.120)$$

If the cable is loaded with I_0 then equation 18.15 applies

$$I_0 \neq 0, \quad \vartheta_a = \vartheta_U + (\vartheta_{Lr} - 30 \text{ °C}) \left(\frac{I_0}{I_r}\right)^2. \qquad (18.121)$$

18.6.4 Root-Mean-Square Value of Current

DIN VDE 0100 Part 523 gives the following rules for the application of the root-mean-square value:

"If the rated currents of motors in multi-motor drives is exceeded for some periods by longer run-up times, frequency of starts or load peaks, the rated current of the drive (drive rated current) must be determined using the root-mean-square value for the calculation of conductor cross-sectional area."

The root-mean-square value can however only be used for calculation of conductor cross-sectional areas "when the on-time duration of peak current remains below the quantities given in the following table":

Cross-sectional area mm^2	Permissible on-time s
up to 6	4
10 to 25	8
35 to 50	15
70 to 150	30
185 and more	60

With this calculation the "average" heating effect of the current is used. The varying conductor temperature has a mean value which ultimately corresponds to the permissible temperature, but rises above this mean at the end of a load period and falls below at the end of a no-load period. The temperature variation of the conductor is greater the longer the load time t_b and the duty cycle t_s relative to the time value τ and the shorter the relative on-time duration since then the current lies above the root-mean-square value during the load period.

The root-mean-square value of a series of load periods of quantity I_{bi} which each flow for a period t_i is

$$I_q = \sqrt{\frac{I_{b1}^2 t_1 + I_{b2}^2 t_2 + \ldots + I_{bi}^2 t_i}{t_1 + t_2 + \ldots + t_i}} \qquad (18.110)$$

with $t_1 + t_2 + \ldots + t_i = 24$ h.

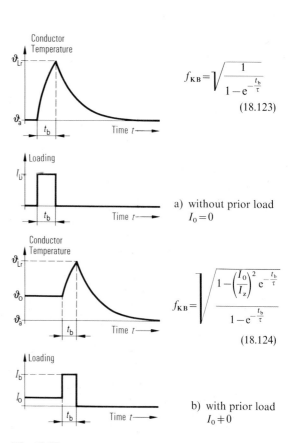

$$f_{KB} = \sqrt{\frac{1}{1 - e^{-\frac{t_b}{\tau}}}} \qquad (18.123)$$

a) without prior load
$I_0 = 0$

$$f_{KB} = \sqrt{\frac{1 - \left(\frac{I_0}{I_z}\right)^2 e^{-\frac{t_b}{\tau}}}{1 - e^{-\frac{t_b}{\tau}}}} \qquad (18.124)$$

b) with prior load
$I_0 \neq 0$

Fig. 18.57
Temperature rise in a conductor for short-time operation, rating factors f_{KB} ($t_b > 0.5\,\tau$)

18.6.5 Short-Time Operation

Short-time operation is a fixed load of short duration with $I_b > I_z$ where the duration is not sufficient for the cable to reach a thermally steady state and where the pause following is sufficiently long for the temperature to fall to the value prior to loading.

The sequence of current and temperature in short-time operation as well as the equations for the calculation of the loading factor f_{KB} with $t_b > 0.5\,\tau$ are shown in Fig. 18.57. For short-time operation without prior loading Fig. 18.58 can be used.

By taking the load capacity I_z for uninterrupted continuous operation the current permissible for short-time operation over the time t_b, without the conductor permissible operating temperature being exceeded, can be found from the following:

$$I_{KB} = I_z f_{KB} \qquad (18.122)$$

Example 18.12

For a cable

$$\text{NYY} \quad 4 \times 150\,\text{SM} \quad 0.6/1\,\text{kV}$$

the load capacity for short-time operation with a load period $t_b = 300$ s is to be determined. Installation conditions and ambient conditions to Table 18.4 apply. The load capacity in uninterrupted operation is according to Part 2, Table 5.15

$$I_z = I_r = 315 \text{ A},$$

$$B = \frac{\Delta \vartheta_r \varkappa_{20} c}{1 + \alpha_{20}(\vartheta_{Lr} - 20\,°\text{C})} \qquad (18.118)$$

$$= \frac{(70 - 30)\,56 \times 10^6 \times 3.45 \times 10^6}{1 + 0.00393\,(70 - 20)}$$

$$= 6.459 \times 10^{15} \frac{\text{A}^2\,\text{s}}{\text{m}^4},$$

$$\tau = B \left(\frac{q}{I_r}\right)^2$$

$$= 6.459 \times 10^{15} \left(\frac{150 \times 10^{-6}}{315}\right)^2 = 1465 \text{ s} \quad (18.117)$$

(to Fig. 18.56 is $\tau \approx 1500$ s).

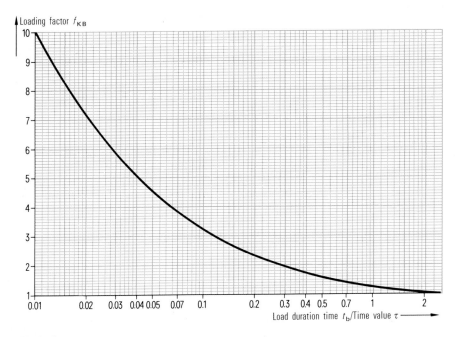

Fig 18.58
Rating factor f_{KB} for short-time operation without prior loading I_0 relative to the function t_b/τ

For a load period $t_b = 300$ s without prior loading

$$f_{KB} = \sqrt{\dfrac{1}{1 - e^{-\frac{t_b}{\tau}}}}$$

$$= \sqrt{\dfrac{1}{1 - e^{-\frac{300}{1465}}}} = 2.32 \qquad (18.119)$$

and

$$I_{KB} = I_z f_{KB} = 315 \times 2.32 = 731 \text{ A}. \qquad (18.122)$$

with a prior load of $I_0 = 150$ A and the same load period of $t_b = 300$ s we have

$$f_{KB} = \sqrt{\dfrac{1 - \left(\dfrac{I_0}{I_z}\right)^2 e^{-\frac{t_b}{\tau}}}{1 - e^{-\frac{t_b}{\tau}}}}$$

$$= \sqrt{\dfrac{1 - \left(\dfrac{150}{315}\right)^2 e^{-\frac{300}{1465}}}{1 - e^{-\frac{300}{1465}}}} = 2.10 \qquad (18.126)$$

and

$$I_{KB} = I_z \times f_{KB} = 315 \times 2.10 = 662 \text{ A}. \qquad (18.122)$$

18.6.6 Intermittent Operation

Intermittent operation is a periodically switched constant load where $I_b > I_z$ with a load period t_b followed by a pause. The current and temperature curves are shown in Fig. 18.59. The load capacity for intermittent operation is calculated from

$$I_{AB} = I_z f_{AB} \qquad (18.125)$$

$$f_{AB} = \sqrt{\dfrac{1 - e^{-\frac{t_s}{\tau}}}{1 - e^{-\frac{t_b}{\tau}}}} = \sqrt{\dfrac{1 - e^{-\frac{t_s}{\tau}}}{1 - e^{-\frac{t_s}{\tau}\frac{ED}{100}}}} \qquad (18.126)$$

or f_{AB} can be taken from Fig. 18.60.

With a load period t_b and a duty cycle t_s the relative on-time duration ED becomes

$$ED = \dfrac{t_b}{t_s} 100 \%. \qquad (18.127)$$

DIN VDE 530 Part 1 gives nominal values for electrical machines of ED = 15%, 25%, 40% and 60% and a duty cycle of 10 minutes if non other is specified.

For very short duty cycles calculations can be made using the root-mean-square method, described in Section 18.6.4, from

$$f_{AB} = \dfrac{I_{AB}}{I_z} = \sqrt{\dfrac{100}{ED}}. \qquad (18.128)$$

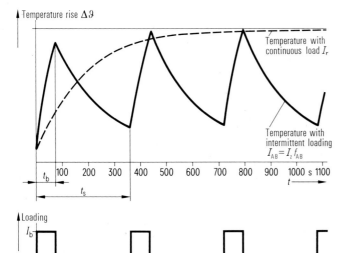

Fig. 18.59
Temperature variation at a conductor in a cable during intermittent operation

243

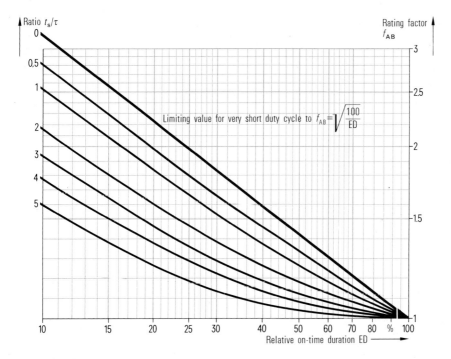

Fig. 18.60
Rating factor f_{AB} for intermittent operation in relation to relative on-time duration ED
and the ratio t_s/τ

Example 18.13

A cable

  NYY 4×150 SM $0.6/1$ kV

is operated intermittently with a load period $t_b = 72$ s,
a duty cycle $t_s = 360$ s at $35\,°C$ ambient temperature
while installed in free air (Fig. 18.60).

From Table 18.22 $f_\vartheta = 0.94$ and from example 18.12
$\tau = 1384$ s,

$$f_{AB} = \sqrt{\frac{1 - e^{-\frac{t_s}{\tau}}}{1 - e^{-\frac{t_b}{\tau}}}} \qquad (18.126)$$

$$= \sqrt{\frac{1 - e^{-\frac{300}{1465}}}{1 - e^{-\frac{72}{1465}}}} = 2.13$$

(see also Fig. 18.60 with $ED = \dfrac{72}{360}\,100 = 20\%$ and

$\dfrac{t_s}{\tau} = \dfrac{360}{1\,465} = 0.25$) and

$$I_{AB} = I_r f_\vartheta f_{AB} = 315 \times 0.94 \times 2.13 = 631 \text{ A}.$$

18.7 Symbols Used in Formulae in Section 18

All equations are written as quantitative equations. The symbols used in the formulae have a numerical value and also relate to a unit of measurement. The quantities are independant of the system of units, whichever compatable system of units is selected, e.g. Système International d'Unités (SI units).

Longitudinally related dimensions "per unit value" such as electrical key data, thermal resistances and losses have symbols marked with an apostrophe stroke (DIN 1304). In the text only the electrical key data are marked as unit value. For thermal resistances and losses, to simplify the expression, this is not used.

a	Axial spacing between two cables	m
b	Width of construction element	m
b	Distance between centres of bunches for bunched installations of single-core cables in 3-phase system	m
b_T	Inside width of channel or tunnel	m
c_i	Distance from axis, e.g. the line source representing the cable to a given point in the temperature field	m
c_i'	Distance from the mirror image of the line source to a point in the temperature field	m
c	Specific heat of conductor material	J/Km^3
c_p	Specific heat of air at constant pressure	J/Km^3
d	Outer diameter of a cable	m
d_1	Diameter over insulation – inside of metal sheath or screen	m
d_x	Diameter of the boundary isotherm – the dry area	m
d_y	Characteristic diameter	m
d_B	Equivalent diameter of a duct bank	m
d_E	Equivalent diameter of cables bunched in a pipe	m
d_L	Diameter of a conductor	m
d_M	Outer diameter of a metal sheath (screen)	m
d_{Mm}	Mean diameter of a metal sheath (screen)	m
d_R	Outside diameter of a pipe	m
e	Eccentricity of a cable	m
f	Frequency	Hz
f	Total rating factor for load capacity in ground	–
f_1	Part rating factor for load capacity in ground, dependant upon ground temperature, soil-thermal resistivity and load factor	–
f_2	Part rating factor for load capacity in ground, dependant upon the number of cables, the soil-thermal resistivity and load factor	–
f_k	Auxiliary variable value for correction of α_k	–
f_s	Auxiliary variable value for correction of α_s	–
f_H	Rating factor for load capacity in air, for groups	–
f_{Hh}	Rating factor for load capacity in air, for cables installed side by side (grouped)	–
f_{Hv}	Rating factor for load capacity in air for cable trays arranged above one another	–
f_{AB}	Rating factor for load capacity in intermittent operation	–
f_{KB}	Rating factor for load capacity in short time operation	–
f_ϑ	Rating factor for load capacity in air at different ambient temperatures	–
h	Depth of lay, for one cable: Distance from cable axis to surface of ground, for bunched single core cables: Distance from axis of bunch to surface of ground	m
h_T	Inside height of a cable tunnel or channel	m
h_x	Depth of the boundary isotherm – dry area	m
h_B	Depth of lay of a duct bank – distance of centre axis to surface of ground	m
k	Geometric factor of a cable	–
k_a	Geometric factor for the determination of temperature rise by a neighbouring cable	–

k_x	Geometric factor for the boundary isotherm – dry area	–
k_y	Geometric factor for the characteristic diameter	–
k_B	Geometric factor for the duct bank	–
l	Length	m
m	Load factor	–
n	Number of loaded cores in a cable	–
n_R	Number of cables installed in one pipe or duct	–
p	Atmospheric pressure	hPa
p	Pressure of the gas filling in gas pressure cables or of the air filling for cables in pipe	bar
q_n	Nominal cross-sectional area of a conductor	m^2
q_M	Geometric cross-sectional area of a sheath or screen	m^2
r	Radius of a cable $(r = d/2)$	m
r_B	Equivalent radius of a duct bank $(r_B = d_B/2)$	m
t	Time	s
t_b	Time of load duration in cyclic operation	s
t_s	Duty cycle time	s
$\tan \delta$	Dielectric loss factor	–
$\ddot{u}$	Thickness of covering of a cable channel or tunnel	m
$\ddot{u}^+$	Corrected value for $\ddot{u}$	m
v	Air velocity in forced ventilation	m/s
w	Number of equal load fluctuations during 24 hour utility supply operation	–
w	Water content of ground in %	–
x	Short-side dimension of a rectangular shaped duct bank	m
y	Long-side dimension of rectangular shaped duct bank	m
y_p	Proximity effect factor	–
y_s	Skin effect factor	–
z	Increase in length due to the helically wound construction of a cable element	–
A	Cross-sectional area of a cable channel or tunnel with forced ventilation	m^2
C	Thermal capacity of a conductor	J/Km
C_b'	Electrical capacity per unit length	F/m
E	Intensity of solar radiation	W/m^2
I_0	Prior loading before the commencement of short-time operation	A
I_b	Operating current, loading	A
I_{bf}	Fictitious value of operating current, value for calculation	A
I_q	Root mean square value of a sequence of current loads during a load cycle	A
I_r	Current-carrying capacity (rated value) for specified operating conditions	A
$I_{\ddot{u}}$	Overload current, loading where a permissible operating temperature is exceeded	A
I_z	Current-carrying capacity (load capacity) when deviating from specified operating conditions	A
I_{AB}	Current-carrying capacity in intermittent operation	A
I_{KB}	Current-carrying capacity in short-time operation	A
I_{th}	Short-time current density under short circuit (1 s)	A/mm^2
M	Number of equal and equally loaded cables in one trench	–
N	Number of equal and equally loaded cables in one trench	–
N_y	Number of equal and equally loaded cables embraced by a circle of diameter d_y	–
N_B	Number of equal and equally loaded cables in one duct bank embraced by a circle of diameter d_B	–
P	Effective power	W
P'	Total losses of all cables	W/m
P_d'	Dielectric loss of a cable	W/m
P_i'	Ohmic losses of a cable	W/m

Q	Reactive power	var
Q	Air rate required for ventilation of a cable channel or tunnel	m^3/s
R'_{20}	d.c. resistance of a conductor per unit length at 20 °C	Ω/m
$R'_{w\vartheta}$	Effective resistance per unit length of a conductor at operating temperature ϑ_L	Ω/m
R'_{wr}	Effective resistance per unit length of a conductor at permissible operating temperature ϑ_{Lr}	Ω/m
R'_{ϑ}	d.c. resistance per unit length of a conductor at operating temperature ϑ	Ω/m
$R'_{\vartheta r}$	d.c. resistance per unit length of a conductor at permissible operating temperature ϑ_{Lr}	Ω/m
$\Delta R'$	Additional resistance per unit length of a conductor due to alternating current loss	Ω/m
S	Apparent power	VA
T'	Thermal resistance	Km/W
T'_1	Thermal resistance of insulation including conductive layer	Km/W
T'_2	Thermal resistance of a protective cover between screen or metal sheath and armour	Km/W
T'_3	Thermal resistance of an outer protective cover	Km/W
T'_4	Thermal resistance of surrounding medium	Km/W
T'_x	Fictitious soil-thermal resistance (determined with ϱ_x) for continuous operation of a multi-core cable or the reference cable for three single-core cables in a three-phase system	Km/W
T'_{xy}	Fictitious soil-thermal resistance (determined with ϱ_x) of a multi-core cable or the reference cable of three single-core cables in a three-phase system with daily load cycle	Km/W
T'_B	Fictitious soil-thermal resistance (determined with ϱ_B) for continuous operation of a multi-core cable or the reference cable of three single-core cables in a three-phase system installed in duct bank	
T'^+_B	Corrective thermal resistance for continuous operation when installed in duct banks	Km/W
T'_{By}	Fictitious soil-thermal resistance (determined with ϱ_B) when installed in duct banks and subject to daily load cycle for multi-core cable or for the reference cable of three single-core cables in a three-phase system	Km/W
T'^+_{By}	Corrective thermal resistance when installed in duct banks with daily load cycle	Km/W
T'_E	Soil-thermal resistance of a multi-core cable or the reference cable of three single-core cables in a three-phase system for continuous operation without drying out (determined with ϱ_E)	Km/W
T'_K	Thermal resistance of a cable	Km/W
T'_{Kd}	Fictitious thermal resistance of a cable when considering the dielectric losses	Km/W
T'_{Ki}	Fictitious thermal resistance of a cable when considering the ohmic losses	Km/W
T'_{KK}	Thermal transfer resistance of a cable for convection	Km/W
T'_{KS}	Thermal transfer resistance of a cable for radiation	Km/W
T'_{Lu}	Thermal resistance of the air arround a cable	Km/W
T'_{LuT}	Thermal resistance of the air in a channel or tunnel	Km/W
T'_P	Thermal resistance of gas filling between a cable and a pipe inner wall	Km/W
T'_R	Thermal resistance of a pipe	Km/W
T'_S	Thermal resistance of the air around a cable subject to solar radiation	Km/W
T'_{TK}	Thermal transfer resistance for convection on the inner wall area of the channel or tunnel	Km/W
T'_{TE}	Thermal resistance of the soil from the inside top cover of a channel or duct to the ground surface	Km/W
T'_{TO}	Thermal transfer resistance at the ground surface	Km/W
$\Delta T'_x$	Additional fictitious thermal resistance (determined with ϱ_x), in continuous operation, due to grouping	Km/W
$\Delta T'_{xy}$	Additional fictitious thermal resistance (determined with ϱ_x), with daily load cycle, due to grouping	Km/W

$\Delta T'_{B}$	Additional fictitious thermal resistance (determined with ϱ_x), for installation in a pipe with continuous operation, due to grouping	Km/W
$\Delta T'_{By}$	Additional fictitious thermal resistance (determined with ϱ_x), for installation in a pipe with daily load cycle, due to grouping	Km/W
U	Cable rated voltage between phase conductors in a three-phase system $U = \sqrt{3}\,U_0$	V
U_0	Cable rated voltage between a phase conductor and the metal covering or earth	V
U_b	Operating voltage of a network	V
$U_{b\,max}$	Highest voltage of a (three-phase) system ($U_{b\,max} \leq U_m$)	V
U_m	Highest voltage for a cable (IEC) or highest voltage for equipment (VDE)	V
U_n	Nominal voltage of a three-phase network	V
U_{rb}	Rated lightning impulse withstand voltage	V
ΔU	Voltage drop	V
α_0	Absorption coefficient of solar radiation for the cable surface	–
α_{20}	Temperature coefficient for electrical resistance at 20 °C	1/K
α_a	Thermal transfer coefficient of the ground surface	W/Km^2
α_k	Thermal transfer coefficient for convection	W/Km^2
α_s	Thermal transfer coefficient for radiation	W/Km^2
γ	Density of the moist soil	kg/m^3
γ_s	Particle density	kg/m^3
γ_d	Dry density	kg/m^3
δ	Thickness, wall thickness of a construction element, general	m
δ	Thickness of an imaginary layer for the consideration of thermal resistance T'_{T0}	m
δ_M	Wall thickness of a metal sheath	m
δ_R	Wall thickness of a pipe	m
ε_0	Emissivity coefficient of a cable surface	–
ϑ_m	Mean temperature of air filling in a pipe with cables loaded	°C
ϑ_x	Temperature of the boundary isotherme	°C
ϑ_E	Temperature of the soil with cables not loaded, ambient temperature	°C
ϑ_L	Conductor temperature	°C
ϑ_{Lr}	Permissible operating temperature, rated value	°C
ϑ_O	Temperature of a cable surface	°C
ϑ_U	Temperature of ambient air	°C
ϑ_{TE}	Mean temperature of air in a channel or tunnel with cables not loaded	°C
ϑ_T	Temperature of air in a channel or tunnel with cables loaded	°C
$\varkappa$	Electrical conductivity	1/Ωm
$\varkappa_{20}$	Electrical conductivity at 20 °C	1/Ωm
λ_1	Sheath loss factor	–
λ_2	Armour loss factor	–
λ_E	Thermal conductivity of moist soil	W/Km
μ	Loss factor	–
ϱ_1	Thermal resistivity of insulation	Km/W
ϱ_2	Thermal resistivity of the inner protective layers between screen and sheath or armour	Km/W
ϱ_3	Thermal resistivity of an outer protective sheath	Km/W
ϱ_x	Thermal resistivity of dried-out soil, i.e. the dry area	Km/W
ϱ_B	Thermal resistivity of concrete	Km/W
ϱ_E	Thermal resistivity of moist soil, i.e. the moist area	Km/W
ϱ_M	Thermal resistivity of a metal sheath	Km/W
ϱ_R	Thermal resistivity of a pipe	Km/W
σ	Stefan-Boltzmann constant	5.67×10^{-8} W/K^{4}m^2
τ	Minimum time value (time constant) for the calculation of load capacity for intermittent operation	s

$\Sigma\delta$	Grouping factor for an installation in the ground	—
$\Delta\vartheta_a$	Temperature rise of the cable under consideration due to heat from a neighbouring cable	K
$\Delta\vartheta_d$	Temperature rise of a conductor above ambient due to dielectric losses	K
$\Delta\vartheta_i$	Temperature rise of the inside surface of a channel or tunnel top cover due to solar radiation	K
$\Delta\vartheta_x$	Limiting temperature rise of the boundary isotherm above ground temperature	K
$\Delta\vartheta_{K\ddot{u}}$	Temperature rise of cooling air	K
$\Delta\vartheta_L$	Temperature rise of a conductor above ambient	K
$\Delta\vartheta_{Lr}$	Temperature rise of a conductor at permissible operating temperature above ambient temperature	K
$\Delta\vartheta_O$	Temperature rise of a cable surface above ambient	K
$\Delta\vartheta_{OS}$	Temperature rise of a cable surface due to solar radiation and load above ambient temperature	K
$\Delta\vartheta_P$	Temperature rise of a cable surface relative to the temperature of the inner surface of duct or pipe, when a cable is installed in duct or pipe	K
$\Delta\vartheta_S$	Temperature rise of a cable above ambient temperature, due to solar radiation	K

18.8 Literature Referred to in Section 18

[18.1] Brakelmann, H.: Belastbarkeiten der Energiekabel. VDE-Verlag Berlin, Offenbach 1985

[18.2] IEC-Publication 287, 1982, Calculation of continuous current rating of cables (100% load factor), 2nd edition

[18.3] Stubbe, R.: Erdboden – ein Kriterium für die Strombelastbarkeit von Energiekabeln. Siemens-Elektrodienst 17 (1975) H. 2, S. 16 und 17

[18.4] Pöhler, W.: Beeinflussung von Leistungskabeln durch Ferndampfleitungen in Hameln. Elektriz.-Wirtschaft 69 (1970) H. 3, S. 236 bis 239

[18.5] Jürgens, W.: Schutzmaßnahmen bei Kreuzungen von Fernwärmeleitungen mit Höchstspannungskabeln. VDE-Fachberichte 22 (1962) S. I/65–I/71

[18.6] VDEW Kabelhandbuch, 3. Auflage, VDEW Verlag, Frankfurt 1977

[18.7] Hütte Energietechnik, Band 3. Elektrische Energietechnik, Netze. Springer Verlag Berlin, 1988

[18.8] Goldenberg, H.: The calculation of continuous current ratings and rating factors for transmission and distribution cables. E.R.A. report Ref. F/T 187, 1958

[18.9] Winkler, F.: Strombelastbarkeit eines waagerecht frei in Luft verlegten Starkstromkabels. Siemens-Z. 37 (1963) S. 394 bis 400

[18.10] Winkler, F.: Strombelastbarkeit von drei gebündelten waagerecht frei in Luft angeordneten einadrigen Starkstromkabeln. etz-a Bd. 97 (1976), S. 344 bis 349

[18.11] Stubbe, R.; Winkler, F.: Bemessung des Leiterquerschnittes. Technische Mitteilungen, Organ des Hauses der Technik e.V. Essen, Bd. 72, H. 7/8, 1979

[18.12] Gröber, Erk, Grigull: Die Grundgesetze der Wärmeübertragung. Springer-Verlag, Berlin, Göttingen, Heidelberg 1957

[18.13] Leyen, van, D.: Wärmeübertragung, Grundlagen und Berechnungsbeispiele aus der Nachrichtentechnik. Siemens AG, Berlin, München 1971

[18.14] Recknagel-Sprenger: Taschenbuch für Heizung und Klimatechnik. R. Oldenburg, München, Wien 1972

[18.15] Davis, M.W.: Nomographic Computation of the Ampacity Rating of Aerial Conductors. IEEE Trans (Power Apparatus and Systems) Vol. PAS-89, No. 3 (1970) P. 387 to 399

[18.16] Winkler, F.: Strombelastbarkeit von Starkstromkabeln in Erde bei Berücksichtigung der Bodenaustrocknung und eines Tageslastspieles. ETZ-Report 13, VDE-Verlag GmbH, Berlin 1978

[18.17] Heinhold, L.; Stubbe, R.: Kabel und Leitungen für Starkstrom, Teil 2, 4. Auflage. Siemens AG, Berlin, München 1989

[18.18] Ludwig, Ch.: Die Erwärmung erdverlegter Kabel bei langdauernder Belastung. Elektriz.-Wirtsch. Bd 61 (1962) S. 568 bis 574

[18.19] Soil-Thermal Characteristics in Relation to Underground Power Cables. Trans. Amer. Inst. Electr. Eng. Part III, Dec. 1960 P. 792–856

[18.20] Orchard, R.S.; Barnes, C.C.; Holingsworth, P.M.; Mochlinski, K.: Soil-thermal resistivity; A practical approach to its assessment and its influence on the current rating of buried cables. CIGRE 1960, Vol. 214

[18.21] Forschungsgesellschaft für das Straßenwesen e.V. Köln, Arbeitsgruppe Untergrund: Merkblatt über das Zufüllen von Leitungsgräben. Ausgabe 1970

[18.22] Jahn, A.: Versuche mit Tragschichten aus mechanisch und bituminös verfertigten Mineralgemischen. Bitumen, 17 (1955) H. 3, S. 53 bis 60

[18.23] Milne, A.G.; Mochlinski, K.: Characteristics of soil affecting cable ratings. Proc. Instn. electr. Eng. Vol. 111 (1964) P. 1017 to 1039

[18.24] Krischer, O.: Der Einfluß von Feuchtigkeit, Körnung und Temperatur auf die Wärmeleitfähigkeit körniger Stoffe. Beihefte zum Gesundheits-Ingenieur, Reihe 1, H. 33 (1934) S. 4 bis 5

[18.25] Makowski, M.W.; Mochlinski, K.: An evaluation of two rapid methods of assessing the thermal resistivity of soil. Proc. Inst. Electr. Eng. Part A Vol. 103 (1956) P. 453 to 470

[18.26] Donazzi, F.; Occhini, E.; Seppi, A.: Soil-thermal and hydrological characteristics in designing underground cables. Proc. IEE (1979) P. 506 to 516

[18.27] Arrighi, R.; Ridon, R.; Benard, P.; Causse, L.: Contribution to the study of the thermal environment of buried cables. CIGRE-report 1970, No. 21–06

[18.28] Heinemann, H.J.: Kabelerwärmung und Bodenaustrocknung. F & G-Rdsch. H. 54 (1967) S. 26 bis 37

[18.29] Rode, A.: Das Wasser im Boden. Akademie Verlag, Berlin 1959

[18.30] Geiling, L.: Die Veränderlichkeit des spezifischen Wärmewiderstandes des Erdbodens und ihre Berücksichtigung bei Planung und Betrieb von Kabelanlagen. VDE-Fachber. Bd. 22 (1962) S. I/73–I/82

[18.31] Sorms, R.: Thermische Belastbarkeit eines in Sand verlegten Kabels. ETZ-B Bd. 20 (1968) S. 315 bis 318

[18.32] Lücking, H.W.; Weltgen, J.: Der äußere Wärmewiderstand im Erdboden verlegter Kabel. F & G-Rdsch. H. 51 (1963) S. 165 bis 176

[18.33] Winkler, F.: Der Einfluß des Bettungsmaterials auf die Belastbarkeit von Energiekabeln. etz-a, 92 (1971) S. 131 bis 137

[18.34] Forschungsgesellschaft für das Straßenwesen e.V. Köln, Arbeitsgruppe Untergrund: Merkblatt für bodenphysikalische Prüfverfahren im Straßenbau. 3. Ausgabe 1963

[18.35] van Hove, C.: Kabelbelastbarkeit. Elektrizitätswirtschaft, Jg. 73 (1974), S. 81 bis 82

[18.36] McAllister, D.: Electric Cables Handbook. Granada Publishing, London, Toronto, Sydney, New York, 1982

[18.37] Armann, A.N.; Cherry, D.M.; Gosland, L.; Hollingsworth, D.M.: Influence of soil-moisture migration on power rating of cables in h.v. transmission systems. Proc. IEE. Vol. 111 (1964), P. 1000 to 1016

[18.38] Cox, H.N.; Coates, R.: Thermal analysis of power cables in soils of temperature-responsive thermal resistivity. Proc. IEE, Vol. 112, 1965, P. 2275 to 2283

[18.39] Endacott, J.D.; Flack, H.W.; Morgan, A.M. et al.: Thermal design parameters used for high capacity e.h.v. cable circuits in Great Britain. CIGRE-report 1970, No. 21–03

[18.40] VDI-Wärmeatlas. VDI-Verlag GmbH, Düsseldorf, 1984

[18.41] Kirgis, L.; Kirgis, H.: Tiefbau Taschenbuch. Franckh'sche Verlagshandlung, Stuttgart, 1967

[18.42] CIGRE-Report 233 (1964). Report on the work of study committee No. 2

[18.43] Neher, J.H. und McGrath, M.H.: The calculation of the temperature rise and load capability of cables systems. Trans. Amer. Inst. Electr. Eng. Vol. 76 (III) (1957), P. 752 to 772

[18.44] Underground system reference book, Edison Institut, New York, 1957

[18.45] Gasser, O.: Belastbarkeit von Kabeln bei Häufung in nichtbelüfteten Kanälen. VDE Fachberichte 17 (1953) S 39/II bis 43/II

[18,46] DIN 4701, Regeln für die Berechnung des Wärmebedarfs von Gebäuden. Beuth-Vertrieb GmbH, Berlin

[18.47] Slaninka, P.: Wärmewiderstand eines Kabelkanals. Bulletin VUKI, 18 (1965) S. 212 bis 221

[18.48] Krischer, O.: Das Temperaturfeld in der Umgebung von Rohrleitungen, die in die Erde verlegt sind. Gesundheits-Ingenieur 59 (1936) S. 537 bis 539

[18.49] Morello, A.: Transient Temperature Variations in Power Cables. L'Elettrotecnica, 45 (1958) P. 213 to 222

[18.50] Brookes, A.S.: Current ratings of cables for cyclic and emergency loads. Electra (1972) No. 24, P. 63 to 97

High-current test arrangement for development of
short-circuit proof supports for single-core cables as
well as for type testing to the relevant design rules
and specifications

19 Short-Circuit Conditions

19.1 General

Effects of Short-Circuit Current

A short circuit causes the following effects which are proportional to the square of the current;

▷ A temperature rise in the conducting components subjected to current flow such as conductor, screen, metal sheath, armour. Indirectly the temperature of adjoining insulation and protective covers (Sections 19.2 and 19.3) also increases,

▷ electro-magnetic forces between the current-carrying components (Section 19.4).

The temperature rise is important for its effect on ageing, heat pressure characteristics etc., and should be limited to a permissible short-circuit temperature (Sections 19.3.1 and 19.3.2). The thermo mechanical effects of the current must also be recognized (Section 19.3.3).

For a given short-circuit duty therefore the short-circuit capacity of a cable installation must be investigated with respect to all these parameters. The Table 19.1 can be used as a guide.

Table 19.1
Guide to the rules for the determination of short-circuit withstand of a cable installation

Effect of short-circuit current	Cable	Accessories	Installation
	Guidelines in section		
Thermal effect Permissible temperature rise	19.2.2 19.3.1 19.3.2	19.3.4	
Thermal expansion	19.3.3	19.3.4	19.3.3
Dynamic forces	19.4.2 19.4.3	19.4.4	19.4.3

For multi-core cables in most instances the thermal effect – related to the magnitude of fault current and clearance time – is the critical parameter, since the cable will normally have sufficient mechanical strength. With single-core cables however, in addition, the mechanical effect – related to the magnitude of the peak short-circuit current – is of such significance that, next to the thermal, the mechanical withstand of both cable and its supports must be investigated.

In addition, accessories must be rated with respect to thermal and mechanical short-circuit stresses.

The short-circuit withstand of a cable system is not quantitatively defined with regard to permissible number of repeated short circuits, degree of deformation or destruction or impairment quality. It is expected, however, that a cable installation will remain safe in operation and that any deformation remains within tolerable limits even after several short circuits (Section 19.3.2).

Magnitude of Short-Circuit Current

The magnitude of a short-circuit current can be determined by reference to [19.1] or to DIN VDE 0102 Parts 1 and 2. Often the breaking capacity S_a of the associated switchgear and network components is known. A survey made in Germany of short-circuit data and typical limiting values for medium- and high-voltage networks is given in Table 19.2 [19.2].

Guideline values for domestic installations which are based on measurements taken on appartments which were all electric or nearly all electric are [19.3]:

▷ at the input terminals in 75% of all cases less than 5 kA,

▷ at the meter in 99.5% of all cases less than 5 kA,

▷ at a socket outlet or fixed appliance in 91% of all cases less than 1 kA and none exceeded 2 kA.

From the breaking power S_a with the nominal voltage U_n one can calculate the symmetrical breaking current I_a

$$I_a = \frac{S_a}{\sqrt{3}\, U_n}. \tag{19.1}$$

If the initial symmetrical short-circuit capacity S_k'' is given in VA, the initial symmetrical short-circuit current is

$$I_k'' = \frac{S_k''}{\sqrt{3}\, U_n} \tag{19.2}$$

and the peak short-circuit current

$$I_s = \varkappa \sqrt{2}\, I_k''. \tag{19.3}$$

Breaking current I_a and the initial short-circuit current I_k'' are related as follows

$$I_a = \mu I_k''. \tag{19.4}$$

Guide values for the factors $\varkappa$ and μ are shown in Table 19.2.

Example 19.1

A network operating at 20 kV is designed for a breaking power of 250 MVA.

From Table 19.2 we get $\varkappa = 1.8$ and $\mu = 1.0$.

$$I_a = \frac{S_a}{\sqrt{3}\, U_n} = \frac{250 \times 10^6\,\text{VA}}{\sqrt{3} \times 20 \times 10^3\,\text{V}} = 7.2\ \text{kA} \tag{19.1}$$

$$I_k'' = I_a/\mu = 7.2/1 = 7.2\ \text{kA} \tag{19.4}$$

$$I_s = \varkappa \sqrt{2}\, I_k'' = 1.8 \sqrt{2} \times 7.2 = 18.3\ \text{kA} \tag{19.3}$$

The quantities I_k'' and I_s could also be obtained by reference to Fig. 19.1.

Critical Short-Circuit Currents

The critical short-circuit currents for conductors and screens can be taken from Table 19.3.

The short-circuit current quantities derived by calculation from the breaking or short-circuit current are applicable for a three-phase short circuit and form the basis for sizing conductors. The maximum stress on conductors normally occurs in a three-pole symmetrical short circuit.

In networks having low resistance star-point earthing in which the preconditions for star-point earthing to DIN VDE 0111 are fulfilled, the fault current during a line-to-earth or line-to-line short circuit may have a greater magnitude than a three-phase short circuit.

In a line-to-line short circuit without earth connection and in a three-phase short circuit only the conductors are stressed. The quantity of current for a line-to-line short circuit without earth fault is $(\sqrt{3}/2)$ times the quantity for a three-phase short circuit.

With a line-to-earth fault, or with a line-to-line short circuit with earth connection or via double earth connection the screens or metal sheaths of the cables are also stressed.

The magnitude of stress on the screens and metal sheath is dependant upon the treatment of the network starpoint and the resulting asymmetrical short-circuit current. A review of quantities is given in Table 19.4.

Table 19.2
Typical limiting quantities of short-circuit currents in various installations and networks within Germany

Net-work	Public utility supply	Industry	Limiting values
	Initial symmetrical short-circuit current I_k'' in kA		
6 kV	–	25 to 50	63
10 kV	16 to 20	20 to 40	63
20 kV	10 to 16	20 to 25	31.5
30 kV	10 to 16	20 to 25	31.5
110 kV	25 to 40	–	40
220 kV	40 to 50	–	63
380 kV	50 to 63	–	80

Factor μ for determination
of initial symmetrical short-circuit current

	≈ 1.0	≈ 0.8	–

Disconnecting time:
Instantaneous   0.1 to 0.2 s
Delayed   0.5 to 1.0 s

Factor for determination
of peak short-circuit current $\varkappa = 1.8$

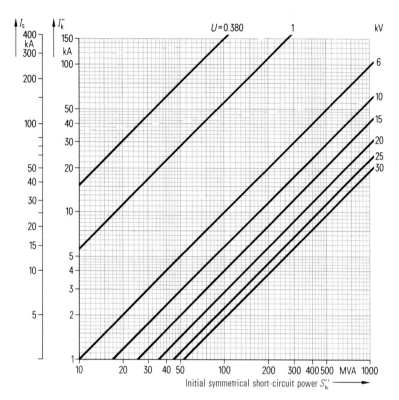

Fig. 19.1
Initial symmetrical short-circuit currents I_k'' (r.m.s.) and peak short-circuit currents I_s in relation to initial symmetrical short-circuit power S_k'' with $\varkappa = 1.8$

Initial symmetrical short-circuit power S_k'' ⟶

Table 19.3
Critical short-circuit and earth-fault currents for rating of conductors and screens

Treatment of star point		Maximum short-circuit current in phase conductor	Maximum short-circuit current in the screens and sheath
Networks with insulated star point		I_{k3}''	Double earth-fault current $I_{kEE}'' \leq \dfrac{\sqrt{3}}{2} I_{k3}''$
Networks with earth-fault compensation		I_{k3}''	Double earth-fault current $I_{kEE}'' \leq \dfrac{\sqrt{3}}{2} I_{k3}''$
Network with low-resistance star-point earthing	with direct earthing of all transformers	$I_{k2}'' < 1.5\, I_{k3}''$	$I_{k1}'' < 1.5\, I_{k3}''$
	with direct earthing of one or more transformers	I_{k3}''	$I_{k1}'' < I_{k3}''$
	earthing through additional impedances	I_{k3}''	$I_{k1}'' = 1$ to 2 kA

I_{k1}'' Initial short-circuit current with line-to-earth short circuit (earth short-circuit current)

I_{k2}'' Initial short-circuit current with line-to-line short circuit

I_{k3}'' Initial symmetrical short-circuit current with balanced three-phase short circuit (is also referred to as I_k'')

I_{kEE}'' Initial line-to-line short-circuit current via double earth connection (double earth-fault current)

255

Table 19.4
Star-point treatment in three-phase networks. Magnitude of earth-fault and line-to-earth short-circuit current

Voltage	Country	Cable network	Overhead network
Medium voltage (3 to 30 kV)	D	Insulated star point (industry) $I_F = 0.5$ to 100 A $I''_{kEE} = 1$ to 15 kA	–
		Earth-fault compensation $I_F = 10$ to 100 A $I''_{kEE} = 1$ to 15 kA	
		Low-resistance earthing of star point $I''_{k1} = 0.5$ to 2 kA	–
	A	Low-resistance earthing of star point $I''_{k1} = 0.5$ to 5 kA $\quad I''_{k1} < 1.5\,I''_{k3}$;	
High voltage (50 to 150 kV)	D	–	Earth-fault compensation $I_F = 10$ to 100 A $I''_{kEE} = 5$ to 35 kA
		Low-resistance earthing of star point $I''_{k1} < 10$ kA $\quad I''_{k1} < I''_{k3}$	
	A	Low-resistance earthing of star point $I''_{k1} = 10$ to 40 kA	
Extra-high voltage (220 to 800 kV)	D	Low-resistance earthing of star point $I''_{k1} = 5$ to 25 kA	
	A	Low-resistance earthing of star point $I''_{k1} = 20$ to 100 kA	

D   Within Germany
A   Outside Germany
I_F Earth-fault current
I''_{kEE} Initial line-to-line short-circuit current via double earth connection (double earth-fault current)
I''_{k1} Initial short-circuit current with line-to-line short circuit
I''_{k3} Initial symmetrical short-circuit current with balanced three-phase short-circuit

Earth-Fault Current

The earth-fault current I_F is the current which flows from the system to earth or to earthed parts at the point of fault on a system where only one earth fault point is present.

In a network with an insulated star point the earth-fault current I_F under earth-fault conditions is the capacitive current I_C. The magnitude of this current is determined by the sum of the capacitances to earth of the total network C_{EN}:

$$I_F = I_C = U_e\,3\,\omega\,C_{EN}. \qquad (19.5)$$

In cable networks I_C is, in most instances, significantly less than 100 A. Only in exceptional cases it is higher.

Generally, small local networks of up to 10 kV are operated with an *insulated star point*. If the earth-fault current does not exceed 10 A the earth-fault arc is self extinguishing and hence complications are avoided. However, where a network is operated so that the earth-fault current could reach 100 A and more an arc can not be expected to self extinguish. Intermittent arcing can generate high-transient voltages which again can lead to multiple faults (double-earth fault or line-to-line short circuit with earth connection) which would then be cleared by the short-circuit protective device. Additional temperature rises attributable to sustained earth faults can therefore, in networks with insulated star point, only be expected to occur in exceptional circumstances [19.4, 19.5].

In a network with an insulated star point as well as those with earth-fault compensation, however, double-earth faults can develop. The magnitude of such a double earth-fault current depends upon the characteristics of the network, the distance between the two points of fault and the cross-sectional area of the screen, but will never exceed the value $\frac{\sqrt{3}}{2}\,I''_k$.

In resonant earthed systems (with arc suppression coil) in the event of an earth fault a residual current will flow from the conductor to screen and to earth, the quantity of which is less than the capacitive earth-fault current. An *earth-fault compensation* is used in networks covering a large area. For this the star points of one or more transformers are earthed through arc suppression coils (*Peterson coil*).

By tuning the inductance L of the arc suppression coil to the capacitance C_{EN} of the network

$$\omega L = \frac{1}{3\omega C_{EN}} \qquad (19.6)$$

the earth-fault current I_F at the point of fault becomes an unbalanced residual current of low quantity and is made up of two components. One component is the in-phase current and relates to losses in the coil and connections while the other component is reactive resulting from the difference between inductive and capacitive quantities. If the in-phase current is ignored the following applies

$$I_F = I_{Rest} \approx U_e \left(3\omega C_{EN} - \frac{1}{\omega L} \right). \qquad (19.7)$$

In overhead networks or in mixed overhead and cable networks it is probable that earth-fault arcs are self extinguishing and have a momentary contact characteristic.

Only a minor proportion of all faults lead to faults of significant duration. The magnitude of earth-fault current can, in compensated networks exceed several hundred amperes [19.4, 19.6].

In networks with low-resistance star-point earthing the earth-fault current becomes the initial line-to-earth short-circuit current I_{k1}''. The quantity of current is determined mainly by the zero-phase-sequence impedance of the source equipment, where the star point is earthed. Double earth faults do not normally occur in those networks. There are three basic modes of operation for networks:

▷ Networks with direct star-point earthing of all transformers. The line-to-earth short-circuit current can, in this mode, exceed the quantity of that of a three-phase short circuit: $I_{k1\,max}'' < 1.5\, I_{k3}''$.

▷ Networks with direct star-point earthing of one or more transformers (e.g. one star-point earthing per station). The line-to-earth short-circuit current has a lower quantity: $I_{k1\,max}'' < I_{k3}''$

▷ Networks with star-point earthing through additional impedances for current limiting. This arrangement is used on medium-voltage networks. The line-to-earth short-circuit current is by this means limited to a maximum value of 1 to 2 kA.

19.2 Temperature Rise of Conductor under Line-To-Earth Short Circuit

19.2.1 Conductor and Sheath Currents under Line-To-Earth Short Circuit

If the supply is maintained under conditions of an earth fault the sheath or screen currents – induced by the operating current – are added to the operating currents in the conductor together with capacitive fault currents arising from the earth fault [19.7].

Earth-fault currents which derive from the cable in question, are small enough to be ignored in comparison to the earth-fault current component derived from the network.

Mostly for project design only the earth-fault current I_C is known or considered. The conductors are, however, subjected to charging currents and their part in affecting the heat rise of the cable must be taken into account. The ratio of capacities is $C_L/C_E \approx 1/4$ for overhead lines [19.8] and for cables of construction type A (Fig. 19.2 and Section 22) such that for mixed networks containing both overhead lines and cables the approximation $I_{CL} = I_C/4$ may be used. If the operating current, with its in-phase part I_{bw} and its reactive part I_{bb} is added to this increased earth-fault current the equations given in Fig. 19.2 result. For cables of construction types B and C the part of the charging current derived from the network must be considered in the same way as for type A construction. The equations for the conductor currents of all types of construction are the same ($\cos\varphi = 0.96$; $I_C/I_b = 0.77$; $I_C/I_{CL} = 4$). In Figure 19.3 the vector diagram of the currents are to be seen.

For the sheath and screen currents of construction types A and B the formulae are likewise identical. To the sheath and screen currents of construction type C, the current induced by the conductor current must be added. In considering a worst case it is assumed that the total earth-fault current flows only via the metalic parts of the cable and none flows through the ground. Also it is assumed that in mass-impregnated cables with lead-sheath and steel-tape armour the earth-fault current flows only through the lead sheath.

The highest possible stress will be in a feeder cable for the total network, should an earth fault occur near to the input end of the cable. This possibility is assumed in the following example. The calculated result therefore lies on the safe side as regards the magnitude of the permissible earth-fault current and

Types of construction A and B

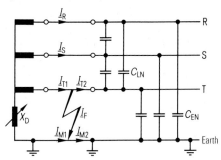

Type of construction C

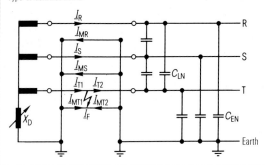

With insulated star point:

$$\underline{I}_R = I_{bw} + I_C/2\sqrt{3} - j(I_{bb} - 3I_C/4);$$
$$\underline{I}_S = I_{bw} - I_C/2\sqrt{3} - j(I_{bb} - 3I_C/4);$$
$$\underline{I}_{T1} = \underline{a}[I_{bw} - j(I_{bb} - 5I_C/4)];$$
$$\underline{I}_{T2} = \underline{a}[I_{bw} - j(I_{bb} - I_C/4)];$$

$$I_R = \sqrt{(I_{bw} + I_C/2\sqrt{3})^2 + (I_{bb} - 3I_C/4)^2};$$
$$I_S = \sqrt{(I_{bw} - I_C/2\sqrt{3})^2 + (I_{bb} - 3I_C/4)^2};$$
$$I_{T1} = \sqrt{I_{bw}^2 + (I_{bb} - 5I_C/4)^2};$$
$$I_{T2} = \sqrt{I_{bw}^2 + (I_{bb} - I_C/4)^2};$$

$$\underline{I}_{M1} = 0$$
$$\underline{I}_{M2} = j\underline{a}\,I_C; \quad I_{M2} = I_C;$$

$$\underline{I}_{MR} = -j\,I_g\Lambda + j\underline{a}\,I_C/3;$$
$$\underline{I}_{MS} = -j\underline{a}^2\,I_g\Lambda + j\underline{a}\,I_C/3;$$
$$\underline{I}_{MT1} = +j\underline{a}\,I_g\Lambda + j\underline{a}\,2I_C/3;$$
$$\underline{I}_{MT2} = -j\underline{a}\,I_g\Lambda + j\underline{a}\,I_C/3;$$
$$I_{MR} = \sqrt{(\Lambda I_g - I_C/6)^2 + (I_C/2\sqrt{3})^2};$$
$$I_{MS} = \sqrt{[(\sqrt{3}\Lambda I_g/2) + (I_C/2\sqrt{3})]^2 + [(\Lambda I_g/2) - (I_C/6)]^2};$$
$$I_{MT1} = I_g\Lambda + 2I_C/3;$$
$$I_{MT2} = I_g\Lambda - 2I_C/3;$$

With ideal compensation ($X_D = 1/3\,\omega C_E$) (all other formulae given above do not change):

$$\underline{I}_{T1} = \underline{I}_{T2} = \underline{a}[I_{bw} - j(I_{bb} - I_C/4)];$$

$$I_{T1} = I_{T2} = \sqrt{I_{bw}^2 + (I_{bb} - I_C/4)^2};$$

$$\underline{I}_{M1} = \underline{I}_{M2} = j\underline{a}\,I_C;$$
$$I_{M1} = I_{M2} = I_C;$$

$$\underline{I}_{MT1} = \underline{I}_{MT2} = -j\underline{a}\,I_g\Lambda + j\underline{a}\,I_C/3;$$
$$I_{MT1} = I_{MT2} = I_g\Lambda - I_C/3;$$

$$\underline{a} = -\frac{1}{2} + j\frac{\sqrt{3}}{2}; \quad \underline{a}^2 = -\frac{1}{2} - j\frac{\sqrt{3}}{2}; \quad \Lambda = \sqrt{\frac{1}{1 + (R'_M/X'_M)^2}}$$

Type A construction: "Belted cable" (e.g. NKBA), cables with polymer insulation and common screen, sheath or armour (e.g. NYFGY) as well as overhead lines.

Type B construction: Three-core radial field cables with common screen, sheath or armour (e.g. N2XS2Y).

Type C construction: Single-core cables with screen (e.g. N2XS2Y) or sheath and three-core cables with single-core screens (e.g. N2XSE2Y) as well as S.L. cables (e.g. NEKBA).

Phase conductors are marked R.S.T., sheath, screen and armour are marked M.

Fig. 19.2
Capacitive currents and operating currents in a cable under conditions of earth fault. C_{LN} and C_{EN} are network capacitances, line-line and line-earth; the capacitances of the cable under earth fault are ignored; it is assumed that the earth-fault current flows only via the sheaths or screens

the temperature rise. Further, if current flows through the ground the cable stresses will be reduced. Where the star point is insulated it is not readily possible to calculate the division of current between sheath or screen and earth. With ideal compensation under earth-fault condition no current will flow via the sheath or screen hence in the formulae for sheath currents in Fig. 19.2 $I_c = 0$. With non-ideal compensation, currents will also flow via sheath and screen but these are difficult to calculate with accuracy.

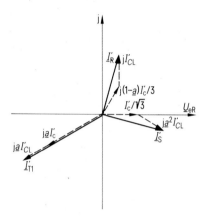

a) Charge and earth-fault currents to Section 22

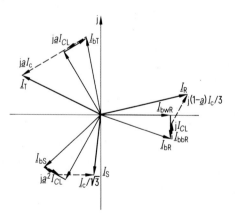

b) Capacitive current and operating currents to Fig. 19.2

Fig. 19.3   Vector diagram

19.2.2   Load Capacity under Line-To-Earth Short Circuit

Under earth-fault conditions the conductor currents as well as the sheath currents are all unbalanced. It is therefore suggested that calculations in general are made with a geometric mean value of currents I_g and I_{gM} using the known equations for the load capacity of cables.

The conductors are heated unequally due to the unbalanced currents whereby the conductors which carry currents in excess of the mean quantity become hotter and the temperature can exceed the calculated quantity.

It follows therefore, since in addition an increased voltage stress occurs at this time (see Section 17), that calculations should be made to ensure that the higher conductor temperature does not exceed the quantity for normal operation. By reference to Section 18 for the load capacity, the permissible temperature rise $\vartheta_{LR} - \vartheta_E$ which must not be exceeded is:

$$\vartheta_{Lr} - \vartheta_E = n I_g^2 R'_{we} (T'_{Kie} + T'_{xy}) -$$
$$- [(\varrho_x/\varrho_E) - 1]\, \Delta\vartheta_x, \tag{19.8}$$

with

$$I_g = \sqrt[3]{I_R I_S I_T}, \tag{19.9}$$

$$R'_{we} = R'_y (1 + \lambda_{1e} + \lambda_{2e}), \tag{19.10}$$

$$T'_{Kie} = \frac{(T'_1/n) + (1 + \lambda_{1e})\, T'_2}{1 + \lambda_{1e} + \lambda_{2e}} + T'_3. \tag{19.11}$$

For cables of type A construction with metal sheath or with a steel-tape armour, or for cables with screens, additional losses must be considered. Due the flow of earth-fault current in the sheath (screen), in addition to the sheath loss factor λ_1 for the eddy-current losses ohmic losses $I_M^2 R'_M$ must be included. This is embraced by a sheath loss factor under earth fault:

$$\lambda_{1e} = \lambda_1 + (I_M/I_g)^2 (R'_M/n\, R'_y). \tag{19.12}$$

For cables of type A construction having only steel wire armouring $\lambda_1 = 0$. The ohmic losses in the armour (using the resistance per unit length R'_A) caused by the earth-fault current are to be added to the eddy-current and hysteresis losses which are related to the armour loss factor λ_2:

$$\lambda_{2e} = \lambda_2 + (I_M/I_g)^2 (R'_A/n\, R'_y). \tag{19.13}$$

For cables of type B construction the additional losses caused by the earth-fault current must be calculated using one of the two equations 19.12 or 19.13 depending on whether the cables have either a screen and/or a flat steel wire armour.

For cables of type C construction voltages are induced in the sheath or screens which, in bonded sheath circuits result in circulating currents. Under earth-fault conditions the conductor currents are unequal. The current induced in a sheath I_{Mi} can be calculated in a simplified manner using the geometric mean value of the conductor currents I_g from:

$$I_{Mi} = I_g \Lambda \tag{19.14a},$$

$$\Lambda = 1/[1 + (R'_M/X'_M)^2]^{\frac{1}{2}}. \tag{19.14b}$$

The induced sheath currents of three-core cables having individually screened or sheathed cores, as well as three single-core cables bunched, are always equal. Where cables are installed side by side the sheath currents are unbalanced. To simplify calculation of the varying impedance per unit length X'_M between conductor and sheath the geometric mean value of the distances between centres can be used which then results in balanced quantities of sheath currents.

The above mentioned induced currents must be added to the sheath current due to the earth fault (Fig. 19.2). Since also these summation currents are of differing magnitude the total sheath losses are calculated using the geometric mean value of sheath current:

$$P'_{VM} = I^2_{gM} R'_M, \tag{19.15}$$

$$I_{gM} = \sqrt[3]{I_{MR} I_{MS} I_{MT}}. \tag{19.16}$$

(To achieve greater accuracy reference can be made to the original paper [19.7] where the complex parts of the earth-fault currents in the sheaths are considered.)

The sheath loss factor under earth-fault conditions for use in the general equation for temperature rise is

$$\lambda_{1e} = (I_{gM}/I_g)^2 (R'_M/n\,R'_y). \tag{19.17}$$

For the calculation of the permissible earth-fault current in networks having an insulated star point the currents in zone 1 (currents with index 1), which is between the transformer and the earth-fault point, must be used as it is in this zone where most heat will be developed. The currents in zone 2 (currents with index 2), which is beyond the earth-fault point, are less critical. In networks with accurate compensation the current divides equally between both zones.

With PVC cables for 10 kV on the left hand side of the temperature rise equation 19.8 the dielectric temperature rise must be deducted from the permissible temperature rise. In the core, of the cable being investigated, in which the fault occurs no dielectric losses are generated but in the other two cores they become three times greater than the losses in normal operation. Thus the dielectric temperature rise in a three-core cable (e.g. NYSEY) [18.2 and 18.16] is:

$$\Delta \vartheta_d = 2 \times 3\,U^2_e\,\omega\,C'_E \tan\delta\,(T'_{Kd} + T'_x), \tag{19.18}$$

for single-core cable

$$\Delta \vartheta_d = 2 \times 3\,U^2_e\,\omega\,C'_E \tan\delta\,(\Delta T'_x + T'_{Kd} + T'_x) \tag{19.19}$$

$$T'_{Kd} = (T'_1/2n) + T'_2 + T'_3, \tag{19.20}$$

$$T'_x = (\varrho_x/2\pi) \ln(4h/d), \tag{19.21}$$

$$\Delta T'_x = (\varrho_x/2\pi) \ln \sqrt{(2h/a)^2 + 1}. \tag{19.22}$$

U_e Voltage to earth
C'_E Capacitance per unit length of cable
h Depth of lay of cable
d Outer diameter of cable
a Axial distance of single-core cables
$\tan\delta$ Dielectric loss factor

Furthermore the quantity $\Delta \vartheta_g$ for the temperature rise caused by other cables in the same trench must be deducted from the permissible temperature rise. The temperature rise caused by other cables being in close proximity can be calculated by reference to Section 18.4.4. If however, the reduction factor for cable groups laying in the same trench was used, e.g. to DIN VDE 0298 part 2, then the permissible temperature rise must be multiplied by the square of this factor.

In Fig. 19.4, as an example, the calculated permissible capacitive line-to-earth fault current I_{Cz} of an aluminium conductor single-core XLPE insulated cable for 20 kV is shown in relation to the power factor $\cos\varphi$. As a parameter the ratio of operating current I_b to load capacity under normal operating conditions I_r was selected. According to DIN VDE 0298 part 2 $I_r = 320$ A for a bunched installation, the load factor $m = 0.7$ (public utility load) and other conditions are as for normal operation.

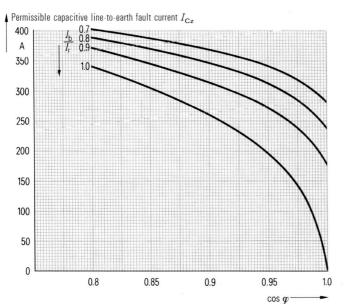

a) Network with earth-fault compensation

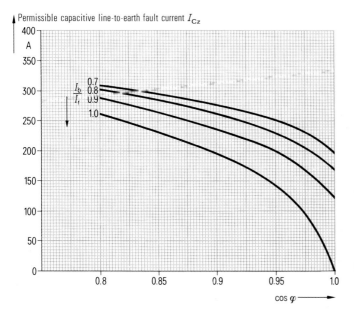

b) Network with insulated neutral

Fig. 19.4
Permissible capacitive line-to-earth fault currents I_{Cz} of a NA2XS2Y 1×150 RM/25 12/20 kV cable
in a bunched installation in the ground (refer to text)

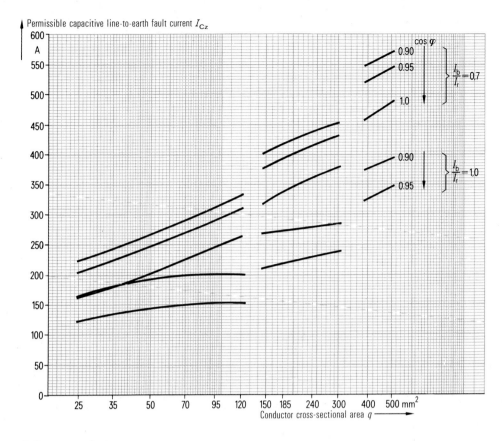

a) Copper conductor

Fig. 19.5
Permissible capacitive line-to-earth fault currents I_{Cz} for single-core XLPE insulated 12/20 kV cables in a bunched installation in the ground (see text) in a network with earth-fault compensation

Fig. 19.5 shows the permissible capacitive line-to-earth fault currents I_{Cz} for XLPE insulated 12/20 kV cables in a bunched installation in ground and under specified operating conditions (Table 18.2) according to DIN VDE 0298 part 2. The preloading is 70% ($I_b/I_r = 0.7$) and 100% ($I_b/I_r = 1.0$). A preloading of 100% and cos $\varphi = 1.0$ results in $I_{Cz} = 0$. For 6/10 kV cable the load capacity is reduced by approximately 2%. For 18/30 kV cable the load capacity is increased by approximately 2%.

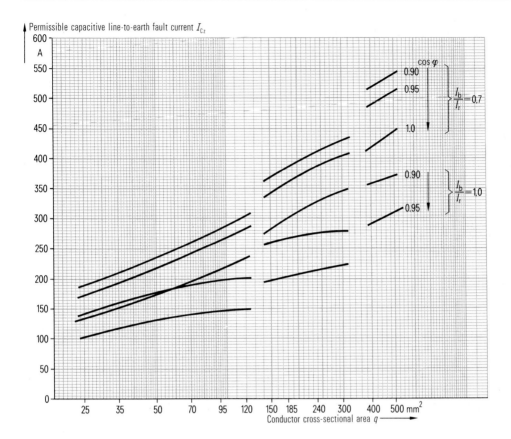

Permissible capacitive line-to-earth fault current I_{Cz}

Conductor cross-sectional area q ⟶

b) Aluminium conductor

Example 19.2

For three single-core NA2XS2Y 1×150 RM/25 12/20 kV cables, installed bunched in the ground, operated under the specified conditions in Table 18.2 with a utility load ($m = 0.7$), the permissible capacitive line-to-earth fault current I_{Cz} is to be calculated for a compensated network. The power factor $\cos \varphi$ is 0.8.

$$I_b = I_r = 320 \, \text{A},$$

$$I_{bw} = I_b \cos \varphi = 320 \times 0.8 = 256 \, \text{A},$$

$$\sin \varphi = \sqrt{1 - \cos^2 \varphi} = \sqrt{1 - 0.8^2} = 0.6,$$

$$I_{bb} = I_b \sin \varphi = 320 \times 0.6 = 192 \, \text{A}.$$

(Constructional and electrical data from Section 18.4.1.) After several iterations this results in $I_{Cz} = 344 \, \text{A}$ for the condition $\vartheta_{Lr} - 20 \, \text{K} \approx 70 \, \text{K}$. The calculation is:

$$I_R = \sqrt{(I_{bw} + I_C/2\sqrt{3})^2 + (I_{bb} - 3I_C/4)^2} \quad \text{(Fig. 19.2)}$$

$$= \sqrt{(256 + 344/2\sqrt{3})^2 + (192 - 3 \times 344/4)^2}$$

$$= 361.4 \, \text{A},$$

$$I_S = \sqrt{(I_{bw} - I_C/2\sqrt{3})^2 + (I_{bb} - 3 \times 344/4)^2} \quad \text{(Fig. 19.2)}$$

$$= \sqrt{(256 - 344/2\sqrt{3})^2 + (192 - 3 \times 344/4)^2}$$

$$= 170 \, \text{A},$$

$$I_T = \sqrt{I_{bw}^2 + (I_C/4 - I_{bb})^2} \quad \text{(Fig. 19.2)}$$

$$= \sqrt{256^2 + (344/4 - 192)^2}$$

$$= 277.1 \, \text{A},$$

$$I_g = (I_R \, I_S \, I_T)^{1/3} \quad (19.9)$$

$$= (361.4 \times 170 \times 277.1)^{1/3}$$

$$= 257.3 \, \text{A},$$

$$M' = \frac{\mu_0}{2\pi} \ln(2 \times 1.11 \, d/d_{Mm}) \quad \text{(s. Section 21)}$$

$$= \frac{4\pi \times 10^{-7}}{2\pi} \ln(2 \times 1.11 \times 35.7/29.2)$$

$$= 1.997 \times 10^{-7} \, \text{H/m},$$

$$X_M' = 2\pi f M'$$

$$= 2\pi \, 50 \times 1.997 \times 10^{-7}$$

$$= 0.0627 \, \Omega/\text{km},$$

$$R_{M20}' = \frac{(1 + z)}{\varkappa \cdot q_M}$$

$$= \frac{1 + 0.05}{56 \times 10^6 \times 25 \times 10^{-6}}$$

$$= 0.75 \, \Omega/\text{km},$$

$$R_{M\vartheta}' = R_{M20}' [1 + (\vartheta_M - 20) \alpha_{20}]$$

$$= 0.75 [1 + (79.8 - 20) \, 0.00393]$$

$$= 0.9263 \, \Omega/\text{km},$$

$$\Lambda = [1 + (R_{M\vartheta}'/X_M')^2]^{-\frac{1}{2}} \quad (19.14)$$

$$= [1 + (0.9263/0.0617)^2]^{-\frac{1}{2}}$$

$$= 0.0675,$$

$$\Lambda I_g = 0.0675 \times 257.3 \quad (19.14)$$

$$= 17.4 \, \text{A},$$

$$I_{MR} = \sqrt{(\Lambda I_g - I_C/6)^2 + (I_C/2\sqrt{3})^2} \quad \text{(Fig. 19.2)}$$

$$= \sqrt{(17.4 - 344/6)^2 + (344/2\sqrt{3})^2}$$

$$= 107.0 \, \text{A},$$

$$I_{MS} = \sqrt{(\sqrt{3}\Lambda I_g/2 + I_C/2\sqrt{3})^2 + (\Lambda I_g/2 - I_C/6)^2}$$

$$= \sqrt{(\sqrt{3} \times 17.4/2 + 344/2\sqrt{3})^2 + (17.4/2 - 344/6)^2}$$

$$= 124.3 \, \text{A}, \quad \text{(Fig. 19.2)}$$

$$I_{MT} = \sqrt{(\Lambda I_g - I_C/3)^2} \quad \text{(Fig. 19.2)}$$

$$= \sqrt{(17.4 - 344/3)^2}$$

$$= 97.3 \, \text{A},$$

$$I_{gM} = (I_{MR} \, I_{MS} \, I_{MT})^{1/3} \quad (19.16)$$

$$= (107.0 \times 124.3 \times 97.3)$$

$$= 109.0 \, \text{A},$$

$$\lambda_{1e} = (I_{gM}/I_g)^2 (R_M'/n R_y') \quad (19.12)$$

$$= (109.0/257.3)^2 (0.9263/0.2647)$$

$$= 0.6280,$$

$$R_{we}' = R_y'(1 + \lambda_{1e}) \quad (19.10)$$

$$= 0.2647 (1 + 0.6280)$$

$$= 0.4309 \, \Omega/\text{km},$$

$$T'_{\text{Kie}} = (T'_1/n)/(1+\lambda_{1\text{e}}) + T'_3 \qquad (19.11)$$

$$= [(0.376/1)/(1+0.6280) + 0.099]$$

$$= 0.330 \text{ Km/W},$$

$$\vartheta_L - \vartheta_E = n\, I_g^2\, R'_{\text{we}}(T'_{\text{Kie}} + T'_{xy}) - [(\rho_x/\rho_E) - 1]\,\Delta\vartheta_x$$

$$= 1 \times 257.3^2 \times 0.4309 \times 10^{-3}(0.330 + 3.445) -$$

$$- [(2.5/1) - 1]\,25$$

$$= 70.19 \text{ K}, \qquad (19.8)$$

$$\vartheta_{Lr} - \vartheta_E = (90 - 20)$$

$$= 70 \text{ K}.$$

19.3 Short-Circuit Thermal Rating

19.3.1 Guide for Project Design

Performance under Short-Circuit Conditions
It is required to determine the cross-sectional areas of conductor and screen in respect of short-circuit thermal stress using data in Fig. 19.6. A cross-sectional area is adequate when the thermally equivalent short-circuit current I_{th}, having a duration t_k, fulfills the following condition,

$$I_{\text{th}} \le I_{\text{thz}}. \qquad (19.23)$$

where I_{thz} is the thermal short-circuit capacity determined by thermal considerations.

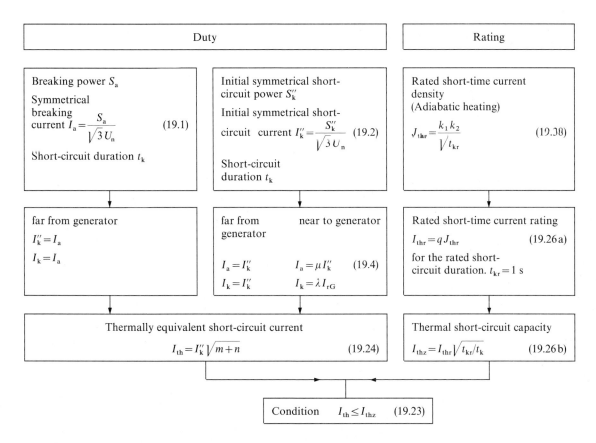

Fig. 19.6
Thermal duty I_{th} and rating I_{thz} under short-circuit conditions

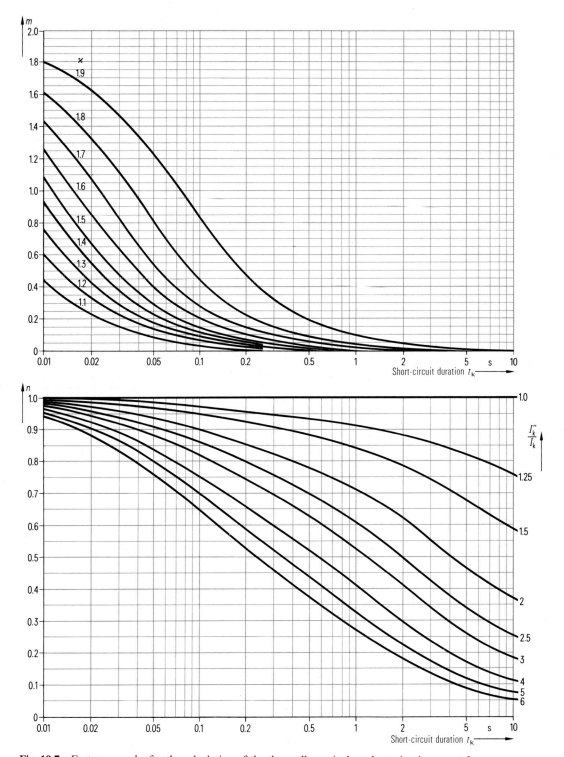

Fig. 19.7 Factors m and n for the calculation of the thermally equivalent short-circuit current I_{th}

Short-Circuit Duty

It is necessary for calculation to know the initial symmetrical short-circuit power S_k'' or alternatively the breaking power S_a of the supply switchgear (19.1 and 19.2 in Fig. 19.6).

To determine the thermal effect of the short-circuit current the r.m.s. value I_{th} is the decisive quantity which flows for the duration of the short circuit under the most unfavourable conditions in accordance with Table 19.3. The current I_{th} flowing for the same duration of time as the short-circuit duration would generate the same quantity of heat as the decaying actual short-circuit current. The actual short-circuit current contains a decreasing direct current component (factor m) on which is superimposed the alternating current component (factor n) (Fig. 19.7). From these factors the thermally equivalent short-circuit current is found

$$I_{th} = I_k'' \sqrt{m+n}. \qquad (19.24)$$

For several short circuits following in quick succession the resulting thermally effective short-circuit current is

$$I_{th} = \sqrt{\frac{1}{t_k} \sum_1^i I_{thi}^2 \, t_{ki}}, \qquad (19.25)$$

with

$$t_k = \sum_1^i t_{ki}.$$

The individual currents I_{thi} are found from equation 19.24 (see also DIN VDE 103).

The d.c. component (factor m) is a function of the peak factor $\varkappa$ and the short-circuit duration t_k, while the a.c. component (factor n) is a function of the short-circuit duration t_k and the ratio of I_k''/I_k (Fig. 19.7).

The progression of short-circuit current for a short-circuit far from the generator is shown in Fig. 19.8. In this instance $I_k = I_k''$ and the a.c. component $n = 1$; giving

$$I_{th} = I_k'' \sqrt{m+1}. \qquad (19.24\,a)$$

A short circuit is said to be far from the generator when each of the generators (or power plants) which feed the three-phase short circuit are participating in the initial short-circuit current at a level of no more than twice their rated current. This condition normally applies to networks which are not fed directly by generators. For short-circuit durations of $t_k > 0.1$ s the a.c. breaking current I_a is equal to the

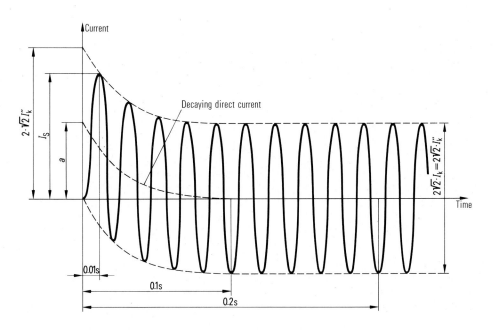

Fig. 19.8   Progression of short-circuit current, far-from-generator

267

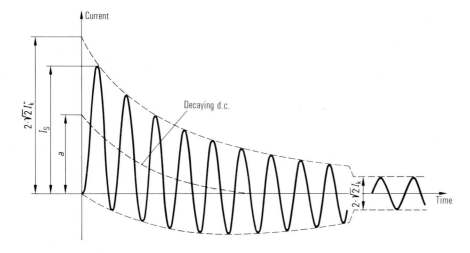

Fig. 19.9 Progression of short-circuit current, near-to-generator

initial symmetrical short-circuit current I_k''. By using Equation (19.2) these can be calculated from the breaking power of the supply circuit breaker or from the maximum short-circuit power of the network. Where the peak short-circuit current I_s is not known the peak factor can be taken as 1.8 (Table 19.2).

Whereas in supply utility networks most short circuits experienced are far-from-generator, short circuits near-to-generator can be experienced in private networks containing a power plant (industrial plant) or even within the power station itself. The progression of a short circuit near-to-generator is shown in Fig. 19.9.

The steady-state short-circuit current I_k is less than the initial symmetrical short-circuit current I_k''. Factor $n\langle 1$. Where the steady-state short-circuit current is not known then, to be on the safe side, calculations are made with $n=1$ and with equation (19.24a).

If only the initial symmetrical short-circuit current I_k'' is known then the peak factor can be taken as $\varkappa=1.8$. Where the symmetrical breaking current I_a is used as the base value (determined from breaking capacity) the initial symmetrical short-circuit current I_k'' can be found from equation (19.4) with $\mu\approx0.8$.

Example 19.3

Using quantities from Example 19.1 for a short-circuit far-from-generator in a 20 kV network having a breaking power of

$$S_a = 500 \text{ MVA}$$

thus with

$$I_k \approx I_a = 14.4 \text{ kA}.$$

for a peak factor

$$\varkappa = 1.8$$

and a short-circuit duration $t_k=0.5$ s and a steady-state short-circuit current $I_k=I_k''$ from Fig. 19.7 we get

$$m = 0.1\,(\varkappa=1.8; t_k=0.5\,\text{s}),$$

$$n = 1\,(I_k''/I_k=1; t_k=0.5\,\text{s}) \text{ and}$$

$$I_{th} = I_k'' \sqrt{m+1} = 14.4\sqrt{0.1+1} = 15.1 \text{ kA}.$$

$$(19.24\,\text{a})$$

Table 19.5 Permissible short-circuit temperature and rated short-time current density J_{thr}.
Cable with copper conductors

Construction	Permissible operating temperature ϑ_{Lr} °C	Permissible short-circuit temperature ϑ_e °C	Conductor temperature ϑ_a at the commencement of a short circuit								
			90 °C	80 °C	70 °C	65 °C	60 °C	50 °C	40 °C	30 °C	20 °C
			Rated short-time current density J_{thr} in A/mm² for $t_{kr}=1$ s								
Soft soldered joints	–	160	100	107	115	119	122	129	136	143	150
XLPE-cables	90	250	143	148	154	157	159	165	170	176	181
PE cables	70	150	–	–	109	113	117	124	131	138	145
PVC cables ≤ 300 mm²	70	160	–	–	115	119	122	129	136	143	150
> 300 mm²	70	140	–	–	103	107	111	118	126	133	140
Mass-impregnated Belted cables 0.6/1 to 3.6/6 kV	80	180	–	119	126	129	132	139	145	151	158
6/10 kV	65	165	–	–	–	121	125	132	138	145	152
Single-core, S.L. and H cables 0.6/1 to 3.6/6 kV	80	180	–	119	126	129	132	139	145	151	158
6/10 kV	70	170	–	–	120	124	127	134	141	147	154
12/20 kV	65	155	–	–	–	116	119	127	134	141	147
18/30 kV	60	140	–	–	–	–	111	118	126	133	140

Short-Circuit Capacity of a Conductor

In the event of a short circuit the heat generated by the short-circuit current is mainly stored in the conductor. Under this condition the conductor must not be heated beyond the permissible short-circuit temperature. Thus, the conductor temperature at commencement of the short circuit as well as the short-circuit duration must be considered.

The *conductor temperature at the commencement of a short circuit* ϑ_a is dependant upon the load prior to the short circuit. If the quantity of ϑ_a at the commencement is not known the permissible operating temperature from Table 19.5 or Table 19.6 should be used. For screens and metal sheaths the values in Table 19.8 apply.

The *permissible short-circuit temperatures* ϑ_e given in Tables 19.5 and 19.6 take into account the insulation and conducting layers which are in contact with the conductor. For soft soldered joints it is recommended

these do not exceed 160 °C, and for tinned conductors 200 °C. These quantities apply for short-circuit durations not exceeding 5 s.

The rated short-time current density J_{thr} for a conductor is defined for a rated short-circuit duration t_{kr} of 1 second (rated quantity of short-circuit capacity). This quantity is calculated by multiplying the rated short-time current density J_{thr} from Tables 19.5 and 19.6 by the nominal cross-sectional area q_n of the conductor.

The thermal short-circuit capacity I_{thz} for a short-circuit duration t_k is

$$I_{thz} = q_n J_{thr} \sqrt{\frac{t_{kr}}{t_k}}. \tag{19.26}$$

269

Table 19.6 Permissible short-circuit temperature and rated short-time current density J_{thr}.
Cable with Aluminium conductor

Construction	Permissible operating temperature ϑ_{Lr} °C	Permissible short-circuit temperature ϑ_e °C	Conductor temperature ϑ_a at the commencement of a short-circuit.								
			90 °C	80 °C	70 °C	65 °C	60 °C	50 °C	40 °C	30 °C	20 °C
			Rated short-time current density J_{thr} in A/mm^2 for $t_{kr}=1$ s								
XLPE cables	90	250	94	98	102	104	105	109	113	116	120
PE cables	70	150	–	–	72	75	77	82	87	91	96
PVC cables											
≤ 300 mm^2	70	160	–	–	76	78	81	85	90	95	99
> 300 mm^2	70	140	–	–	68	71	73	78	83	88	93
Mass-impregnated Belted cables											
0.6/1 to 3.5/6 kV	80	180	–	78	83	85	87	92	96	100	104
6/10 kV	65	165	–	–	–	80	83	87	92	96	100
Single-core, S.L. and H-cables											
0.6/1 to 3.5/6 kV	80	180	–	78	83	85	87	92	96	100	104
6/10 kV	70	170	–	–	80	82	84	89	93	97	102
12/20 kV	65	155	–	–	–	77	79	84	88	93	98
18/30 kV	60	140	–	–	–	–	73	78	83	88	93

Example 19.4

For the thermally equivalent short-circuit current $I_{th}=15.1$ kA and the short-circuit duration time $t_k=0.5$ s as used in Example 19.3 it is required to determine the conductor cross-sectional area for an XLPE insulated cable for $U_0/U=12/20$ kV. By rearranging the equation 19.26 and with the condition $I_{th} \leq I_{thz}$ the cross-sectional area is calculated for $t_{kr}=1$ s

$$q_n \geq \frac{I_{th}}{J_{thr}} \sqrt{\frac{t_k}{t_{kr}}}. \qquad (19.26\,a)$$

For an XLPE cable with an aluminium conductor, from Table 19.6 with an initial temperature $\vartheta_a=90$ °C and a final temperature of $\vartheta_e=250$ °C we find a rated short-time current density of $J_{thr}=94$ A/mm^2, which gives

$$q_n \geq \frac{15.1 \cdot 10^3}{94} \sqrt{\frac{0.5}{1}} \geq 114 \text{ mm}^2.$$

Therefore we select a cross-sectional area of 120 mm^2 with

$$I_{thz} = 120 \cdot 94 \sqrt{\frac{1}{0.5}} = 16.0 \text{ kA}. \qquad (19.26)$$

Guidelines for Use of Graphs Given in Figs. 19.10 to 19.14

These graphs can be used to determine the thermal short-circuit capacity I_{thz} relative to a conductor cross-sectional area and a short-circuit duration time t_k. The permissible initial symmetrical short-circuit current I_{kz}'' is derived from equations 19.23 and 19.24 giving

$$I_{kz}'' = \frac{I_{thz}}{\sqrt{m+n}}. \qquad (19.24\,b)$$

To assist in project design this permissible initial symmetrical short-circuit current is also shown on the graphs.

The permissible breaking time t_{kz} for a thermally equivalent short-circuit current of I_{th} (DIN VDE 0100 Part 430) is calculated with equation 19.26 with

$$t_{kz} = \left(k \frac{q_n}{I_{th}} \right)^2, \qquad (19.27)$$

$$k = J_{thr} \sqrt{t_{kr}}, \qquad (19.28)$$

$$t_{kr} = 1 \text{ s}.$$

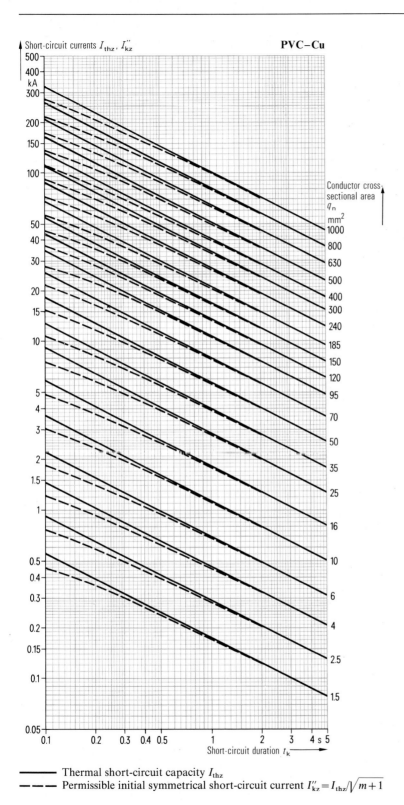

Short-circuit currents I_{thz}, I''_{kz}

PVC–Cu

Conductor cross-sectional area q_n mm²

Short-circuit duration t_k

——— Thermal short-circuit capacity I_{thz}

– – – Permissible initial symmetrical short-circuit current $I''_{kz} = I_{thz}/\sqrt{m+1}$

Fig. 19.10 a
Short-circuit currents of *PVC cables with copper conductors* for $\varkappa = 1.8$,
$\vartheta_a = 70\,°\text{C}$ and
$\vartheta_e = 160\,(140)\,°\text{C}$

271

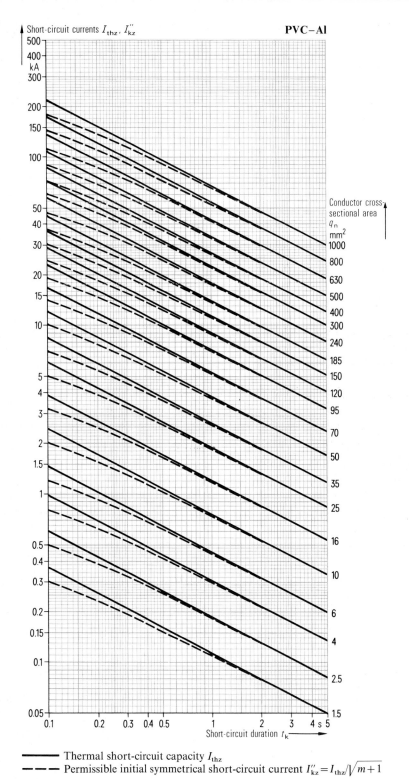

Short-circuit currents I_{thz}, I_{kz}'' **PVC–Al**

Fig. 19.10 b

Short-circuit currents of *PVC cables with aluminium conductors* for $\varkappa = 1.8$,
$\vartheta_a = 70\,°C$ and
$\vartheta_e = 160\ (140)\ °C$

Conductor cross-sectional area
q_n
mm^2

Short-circuit duration t_k

——— Thermal short-circuit capacity I_{thz}
– – – Permissible initial symmetrical short-circuit current $I_{kz}'' = I_{thz}/\sqrt{m+1}$

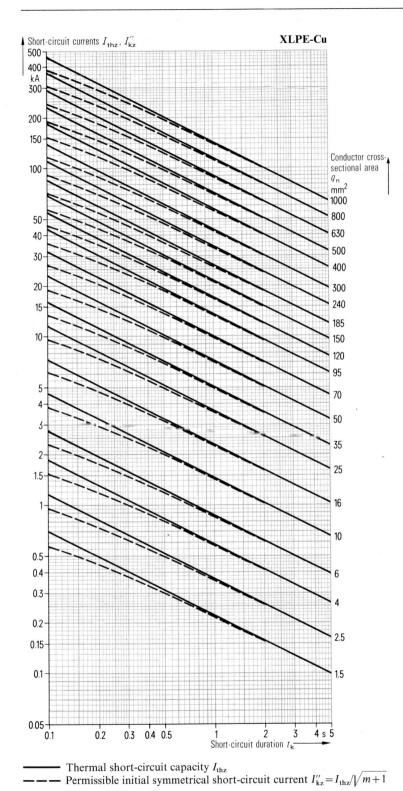

Short-circuit currents I_{thz}, I''_{kz} **XLPE-Cu**

Conductor cross-sectional area q_n mm²

Short-circuit duration t_k

——— Thermal short-circuit capacity I_{thz}
– – – Permissible initial symmetrical short-circuit current $I''_{kz} = I_{thz}/\sqrt{m+1}$

Fig. 19.11 a
Short-circuit currents of *XLPE cables with copper conductors* for $\varkappa = 1.8$,
$\vartheta_a = 90\,°C$ and
$\vartheta_e = 250\,°C$

273

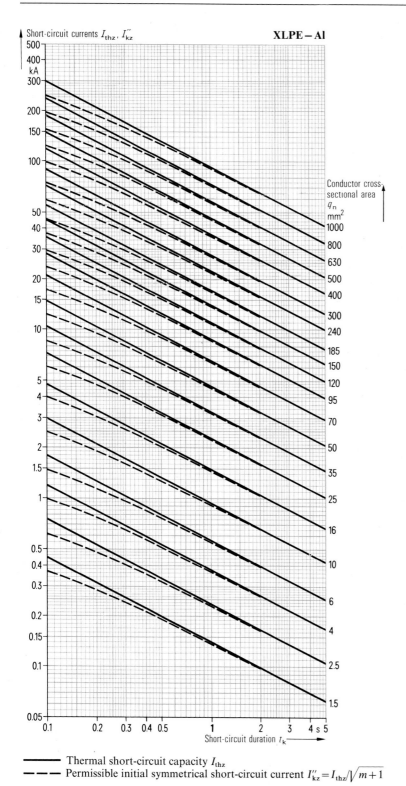

Short-circuit currents I_{thz}, I''_{kz}

XLPE – Al

Conductor cross-sectional area q_n mm²

Short-circuit duration t_k

——— Thermal short-circuit capacity I_{thz}

– – – Permissible initial symmetrical short-circuit current $I''_{kz} = I_{thz}/\sqrt{m+1}$

Fig. 19.11 b
Short-circuit currents of *XLPE cables with aluminium conductors* for $\varkappa = 1.8$,
$\vartheta_a = 90\,°C$ and
$\vartheta_e = 250\,°C$

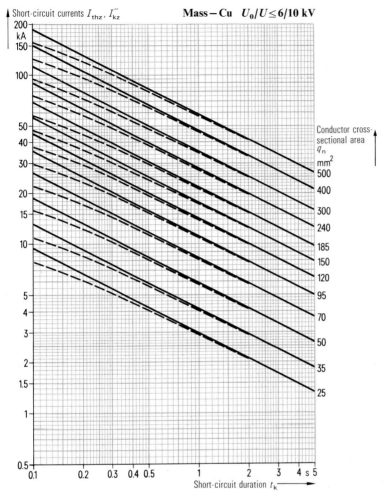

Short-circuit currents I_{thz}, I''_{kz} **Mass — Cu $U_0/U \leq 6/10$ kV**

Conductor cross-sectional area q_n mm^2

500
400
300
240
185
150
120
95
70
50
35
25

Short-circuit duration t_k ⟶

—— Thermal short-circuit capacity I_{thz}
‐ ‐ ‐ Permissible initial symmetrical short-circuit current $I''_{kz} = I_{thz}/\sqrt{m+1}$

Fig. 19.12 a
Short-circuit currents of *mass-impregnated cables for 0.6/1 kV to 6/10 kV with copper conductors* for $\varkappa = 1.8$ and $\vartheta_e - \vartheta_a = 100$ K

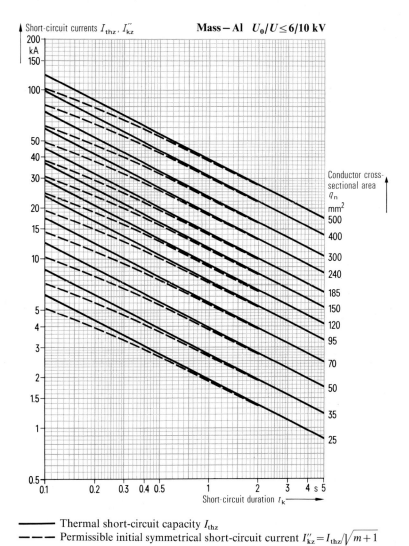

Short-circuit currents I_{thz}, I''_{kz} **Mass — Al** $U_0/U \leq 6/10$ kV

Conductor cross-sectional area q_n mm²

500
400
300
240
185
150
120
95
70
50
35
25

Short-circuit duration t_k ⟶

Fig. 19.12b
Short-circuit currents of *mass-impregnated cables for 0.6/1 kV to 6/10 kV with aluminium conductors*
for $\varkappa = 1.8$ and
$\vartheta_e - \vartheta_a = 100$ K

——— Thermal short-circuit capacity I_{thz}
– – – Permissible initial symmetrical short-circuit current $I''_{kz} = I_{thz}/\sqrt{m+1}$

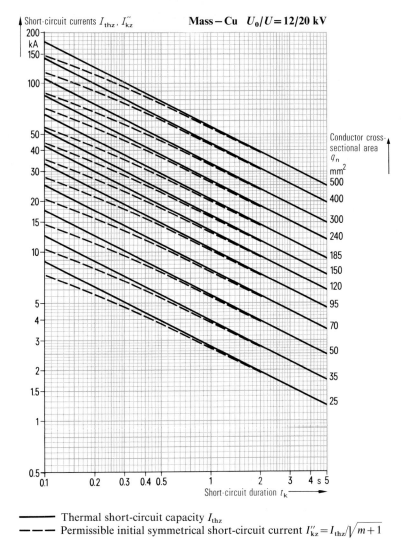

Short-circuit currents I_{thz}, I''_{kz} **Mass — Cu $U_0/U = 12/20$ kV**

Conductor cross-
sectional area
q_n
mm^2

Short-circuit duration t_k

─────── Thermal short-circuit capacity I_{thz}

– – – – Permissible initial symmetrical short-circuit current $I''_{kz} = I_{thz}/\sqrt{m+1}$

Fig. 19.13a
Short-circuit currents of *mass-im-pregnated cables for 12/20 kV with copper conductors*
for $\varkappa = 1.8$,
$\vartheta_a = 65\,°C$ and
$\vartheta_e = 155\,°C$

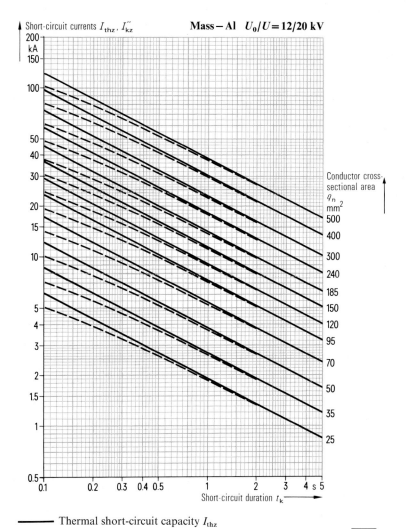

Short-circuit currents I_{thz}, I''_{kz}

Mass – Al $U_0/U = 12/20$ kV

Conductor cross-sectional area
q_n
mm^2

500
400
300
240
185
150
120
95
70
50
35
25

Short-circuit duration t_k

——— Thermal short-circuit capacity I_{thz}

– – – Permissible initial symmetrical short-circuit current $I''_{kz} = I_{thz}/\sqrt{m+1}$

Fig. 19.13b
Short-circuit currents of *mass-impregnated cables for 12/20 kV with aluminium conductors* for $\varkappa = 1.8$,
$\vartheta_a = 65\,°C$ and
$\vartheta_e = 155\,°C$

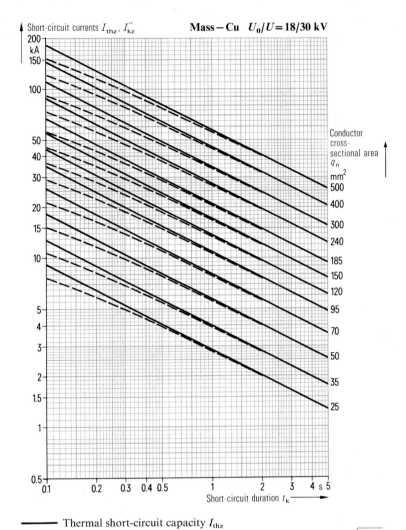

Short-circuit currents I_{thz}, I''_{kz}

Mass — Cu $U_0/U = 18/30$ kV

Conductor cross-sectional area q_{n} mm^2

Short-circuit duration t_{k}

Fig. 19.14a
Short-circuit currents of *mass-impregnated cables for 18/30 kV with copper conductors for* $\varkappa = 1.8$,
$\vartheta_{\text{a}} = 60\,°C$ and
$\vartheta_{\text{e}} = 140\,°C$

——— Thermal short-circuit capacity I_{thz}
– – – Permissible initial symmetrical short-circuit current $I''_{\text{kz}} = I_{\text{thz}}/\sqrt{m+1}$

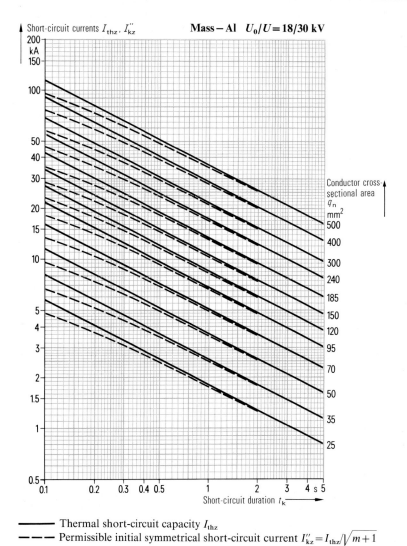

Short-circuit currents I_{thz}, I''_{kz}

Mass — Al $U_0/U = 18/30$ kV

Conductor cross-sectional area
q_n
mm²
500
400
300
240
185
150
120
95
70
50
35
25

Short-circuit duration t_k

——— Thermal short-circuit capacity I_{thz}

– – – Permissible initial symmetrical short-circuit current $I''_{\text{kz}} = I_{\text{thz}}/\sqrt{m+1}$

Fig. 19.14b
Short-circuit currents of *mass-impregnated cables for 18/30 kV with aluminium conductors for* $\varkappa = 1.8$,
$\vartheta_a = 60\,°C$ and
$\vartheta_e = 140\,°C$

**Short-Circuit Capacity of Screens,
Metal Sheaths and Armour**

In a network with an insulated star point or resonant earthing double earth faults are to be considered, while in a network with low resistance star-point earthing consideration are to be given to line-to-earth short-circuit faults (Table 19.4). The screens, the metal sheaths or the armour carry these (asymmetrical) short-circuit currents and become heated. Cables which have been selected on the basis of short-circuit capacity of the conductor for a balanced three-phase short circuit need to be investigated in respect of these currents.

The fault current carried by the screen, metal sheath or armour and also via the ground are divided in inverse proportion to the impedance of the individual current paths.

If the base points of an earth fault or double earth fault are situated outside the cable under consideration a proportion of the short-circuit current, the value of which is difficult to calculate, also flows via the ground. If the base point lies in the cable under consideration (Figs.19.15 and 19.16) it must be expected that almost the whole short-circuit current will be carried by the screen or sheath of the cable. The same applies for multi-core cables with polymer insulation and individual screening without good electrical contact between the screens and also to S.L. cables with a corrosion protection over each sheath. When checking the short-circuit capacity this matter must be considered.

If a cable has an insulating outer sheath e.g. of thermoplastic material, appreciable current will flow to earth only at the point of short circuit and only when the cable is installed in the ground. If a metal sheath is not insulated from earth, e.g. NKBA construction, then the greater the distance from the point of a fault the greater the portion of short-circuit current which flows to earth.

For a cable installed in air the short-circuit current at the point of fault does not divide. The current flows only through the screen or sheath; flashing over, e.g. to the cable rack, is not expected. Racks are insignificant as earth return conductors because their impedance is normally relatively high.

For the determination of short-circuit capacity, in principle the same rules apply as for conductors. For the duration of the short circuit however, a significant portion of the heat is transferred to the adjacent layers. The determination of a rated short-time cur-

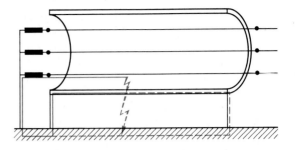

Fig. 19.15
Line-to-earth short circuit in a three-core cable with common screen in a low resistance earthed network. Base point in cable

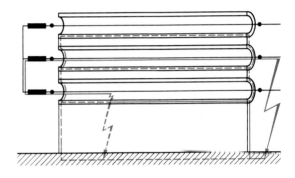

Fig. 19.16
Double earth fault in three screened single-core cables in insulated network. One base point in cable

rent density is therefore somewhat difficult and it is common practice to determine the short-circuit capacity I_{thz} relative to the short-circuit duration t_k by graphical means (see Figs. 19.17 and 19.18). A possible method of calculation is shown in Section 19.3.2.

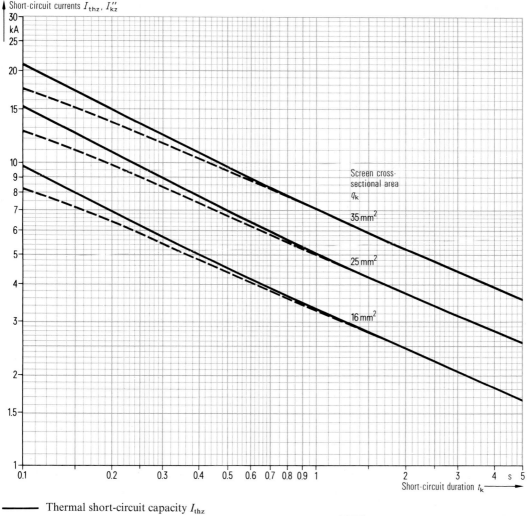

Short-circuit currents I_{thz}, I''_{kz}

Short-circuit duration t_k

——— Thermal short-circuit capacity I_{thz}

– – – Permissible initial symmetrical short-circuit current $I''_{kz} = I_{thz}/\sqrt{m+1}$

Fig. 19.17a
Short circuit capacity of *copper wire screens* of cables with polymer insulation for $\varkappa = 1.8$, $\vartheta_a = 60\,°C$, $\vartheta_e = 350\,°C$, $b = 0.12 \times 10^{-3}\,m/\sqrt{s}$ and screen wire diameter $d_D = 0.7$; 0.8; $0.9\,mm$ with nominal cross-sectional area of screens in sequence;
nominal cross-sectional area of screen q_k as parameter.
The graphs apply approximately also for copper tape screens

Guidelines for the Application of Curves Shown in Figs. 19.17 and 19.18

These graphs can be used to determine the thermal short-circuit capacity I_{thz} for a given cross-sectional area and short-circuit duration t_k. The initial symmetrical short-circuit current I''_k associated with this rating is obtained using equations (19.23) and (19.24)

with

$$I''_k = \frac{I_{thz}}{\sqrt{m+n}}. \qquad (19.24\,b)$$

To facilitate project work the value of symmetrical short-circuit current is also included in the graphs.

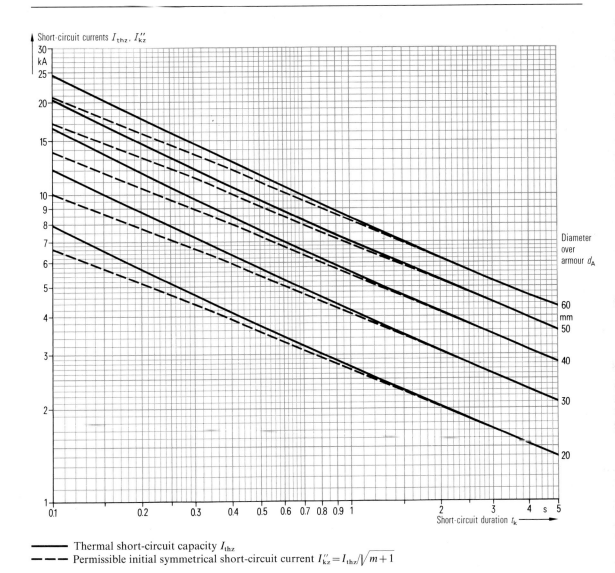

Short-circuit currents I_{thz}, I''_{kz}

Diameter over armour d_A

60 mm
50
40
30
20

Short-circuit duration t_k

———— Thermal short-circuit capacity I_{thz}
– – – – Permissible initial symmetrical short-circuit current $I''_{\text{kz}} = I_{\text{thz}}/\sqrt{m+1}$

Fig. 19.17 b
Short-circuit capacity of *flat steel-wire armour* of cables with polymer insulation for $\varkappa = 1.8$, $\vartheta_a = 60\,°C$, $\vartheta_e = 200\,°C$, $b = 0.17 \times 10^{-3}$ m/$\sqrt{s}$ and thickness of flat steel wires $\delta_A = 0.8$ mm.
Diameter over flat steel-wire armour d_A as parameter

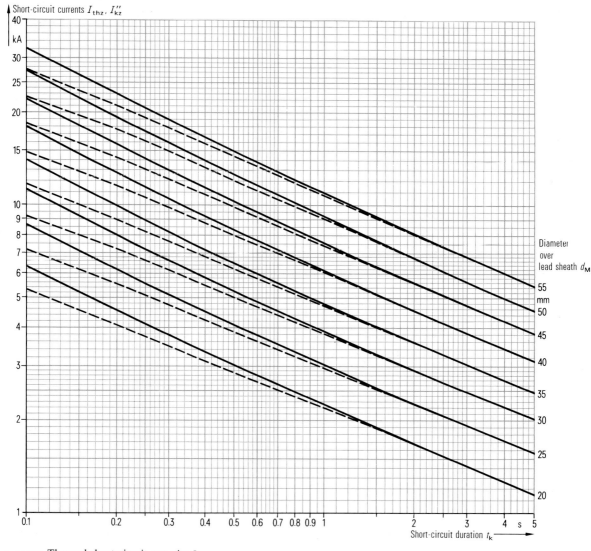

Short-circuit currents I_{thz}, I''_{kz}

kA

Diameter over lead sheath d_M

55 mm
50
45
40
35
30
25
20

Short-circuit duration t_k ⟶

――――― Thermal short-circuit capacity I_{thz}

― ― ― Permissible initial symmetrical short-circuit current $I''_{kz} = I_{thz}/\sqrt{m+1}$

Fig. 19.18
Short-circuit capacity of *lead sheaths* of mass-impregnated cable for $\varkappa = 1.8$, $\vartheta_a = 60\,°C$, $\vartheta_e = 200\,°C$, $b = 0.38 \times 10^{-3}\ m/\sqrt{s}$; diameter over lead sheath d_M as parameter.
Thickness of lead sheath to DIN VDE 0255

Example 19.5

An XLPE cable
NA2XS2Y 1 × 120 RM/16 12/20 kV
has a rated cross-sectional area of screen of 16 mm² to DIN VDE 0273. This can also be interpreted from the type reference. The effective cross-sectional area q_k which is in the path of the short circuit must be at least as large as the nominal cross-sectional area. For a short-circuit duration of $t_k = 0.5$ s the current capacity of the screen to Fig. 19.17a is

$$I_{thz} = 4.5 \text{ kA}$$

and the greatest permissible short-circuit current with a short circuit far-from-generator and with $\varkappa = 1.8$ becomes

$$I''_{kz} = I_{az} = I_{kz} = 4.3 \text{ kA}$$

19.3.2 Calculation of Short-Circuit Capacity

Adiabatic Temperature Rise Method

For short temperature rise periods such as during a short circuit it can be assumed that the total heat generated in an electrical conductor remains in that conductor [19.10 to 19.19]

$$I^2 R_\vartheta \, dt = K \, d\vartheta. \tag{19.29}$$

Where the resistance of the conductor is

$$R_\vartheta = R_{20} \frac{\Theta + \vartheta}{\Theta + 20\,°C}$$

$$= \frac{\varrho_{20}\, l(1+z)\, f_q}{q_{geo}} \cdot \frac{\Theta + \vartheta}{\Theta + 20\,°C}$$

$$= \frac{\varrho_{20}\, l}{q_{el}} \cdot \frac{\Theta + \vartheta}{\Theta + 20\,°C}, \tag{19.30}$$

and the heat capacity of the conductor is

$$K = c\,l(1+z)\, q_{geo} = c\,l\, q_g, \tag{19.31}$$

the electrical cross-sectional area

$$q_{el} = \frac{q_{geo}}{(1+z)\,f_q}, \tag{19.32}$$

and the cross-sectional area by weight

$$q_g = q_{geo}(1+z). \tag{19.33}$$

The solution of equation 19.29 for a given short-circuit duration t_k gives the short-circuit capacity:

$$I_{thz} = q_k\, k_1\, k_2\, \frac{1}{\sqrt{t_k}}, \tag{19.34}$$

with the effective cross-sectional area during the short circuit

$$q_k = \sqrt{q_{el}\, q_g} = \frac{q_{geo}}{\sqrt{f_q}} \approx q_{geo}, \tag{19.35}$$

the material coefficient

$$k_1 = \sqrt{\frac{c(\Theta + 20\,°C)}{\varrho_{20}}} \tag{19.36}$$

and the temperature factor

$$k_2 = \sqrt{\ln \frac{\Theta + \vartheta_e}{\Theta + \vartheta_a}}. \tag{19.37}$$

The rated short-time current density is defined as

$$J_{thr} = \frac{k_1\, k_2}{\sqrt{t_{kr}}} \tag{19.38}$$

with a rated short-circuit duration $t_{kr} = 1$ s.

The relevant quantities for conductor sheath and screen materials are shown in Table 19.7.

Table 19.7 Physical constants for conductor, sheath and screen materials

Material	Electrical resistivity ϱ_{20}	Temperature coefficient of electrical resistance α_{20}	$\Theta = \dfrac{1}{\alpha_{20}} - 20\,°C$	Thermal resistivity ϱ	Specific heat c	Material coefficient k_1
	$\Omega\,m$	$1/K$	K	Km/W	$J/K\,m^3$	$A\sqrt{s}/m^2$
Copper	1.7241×10^{-8}	3.93×10^{-3}	234.5	2.7×10^{-3}	3.45×10^6	226×10^6
Aluminium	2.8264×10^{-8}	4.03×10^{-3}	228	4.8×10^{-3}	2.5×10^6	148×10^6
Lead	21.4×10^{-8}	4.0×10^{-3}	230	28.7×10^{-3}	1.45×10^6	41.2×10^6
Steel	13.8×10^{-8}	4.5×10^{-3}	202	19.1×10^{-3}	3.8×10^6	78.2×10^6

Table 19.8 Quantities for short-circuit capacity of screens, sheaths and armour

Metallic cover	Copper screen		Steel armour (Galvanised)		Sheath	
	Round wire	Tape with gap	Round wire	Flat wire	of lead-alloy	of Aluminium
Effective wall thickness δ_K	$0.8\,d_D$	δ_B	$0.8\,d_D$	δ_A	δ_M	δ_M
Contacting layer *under* the metallic cover	Conducting varnish or adhesive, graphited or impregnated crepe paper or textile tape		Graphited or impregnated crepe paper or textile tape or inner sheath of natural rubber compound		Mass-impregnated paper	
Heat penetration value $\beta_i = \sqrt{c_i/\varrho_i}$	$300\ \mathrm{W}\sqrt{s}/(\mathrm{K\ m^2})$		$230\ \mathrm{W}\sqrt{s}/(\mathrm{K\ m^2})$		$590\ \mathrm{W}\sqrt{s}/(\mathrm{K\ m^2})$	
Contacting sheath *over* the metal cover	Textile tape or crepe paper		PVC-outer sheath		Corrosion protection	Corrosion protection
Heat penetration quantity $\beta_i = \sqrt{c_i/\varrho_i}$	$300\ \mathrm{W}\sqrt{s}/(\mathrm{K\ m^2})$		$705\ \mathrm{W}\sqrt{s}/(\mathrm{K\ m^2})$		$210\ \mathrm{W}\sqrt{s}/(\mathrm{K\ m^2})$	$640\ \mathrm{W}\sqrt{s}/(\mathrm{K\ m^2})$
Screen constant b of the contacting layers	$0.12 \times 10^{-3}\ \mathrm{m}/\sqrt{s}$		$0.17 \times 10^{-3}\ \mathrm{m}/\sqrt{s}$		$0.38 \times 10^{-3}\ \mathrm{m}/\sqrt{s}$	$0.34 \times 10^{-3}\ \mathrm{m}/\sqrt{s}$
Insulation	PVC	XLPE	PVC		Impregnated paper	
Initial temperature ϑ_a	60 °C	80 °C	60 °C		70 °C	
Permissible Temperature ϑ_e	350 °C	350 °C	200 °C		200 °C	
Material coefficient k_1	$226 \times 10^6\ \mathrm{A}\sqrt{s}/\mathrm{m^2}$		$78.2 \times 10^6\ \mathrm{A}\sqrt{s}/\mathrm{m^2}$		$41.2 \times 10^6\ \mathrm{A}\sqrt{s}/\mathrm{m^2}$	$148 \times 10^6\ \mathrm{A}\sqrt{s}/\mathrm{m^2}$
Temperature factor k_2	0.828	0.787	0.654		0.600	0.602

The *initial temperature* ϑ_a depends upon the operating conditions immediately prior to the short circuit. Normally the permissible operating temperature for conductors will be chosen as given in Tables 19.5 and 19.6. For screens, sheaths and armour the guidelines given in Table 19.8 are applicable.

Because of the short duration of stress during a short circuit and the rapid cooling thereafter, for a short-circuit duration of up to 5 s higher temperatures are permitted than those for normal operation.

The *permissible short-circuit temperatures* ϑ_e for conductors are given in DIN VDE 0298 Part 2 and are reproduced in Tables 19.5 and 19.6. The quantities for screens, sheaths and armour given in Table 19.8 apply and take into consideration the contacting coverings which have a thermal protective function [19.21, 19.29].

If the guidelines of IEC 724 [19.29] are followed the permissible short-circuit temperatures are selected such that a single short circuit will not cause damage to the cable, however repeated short circuits may lead to failure. The permissible conductor temperatures are selected in respect of the permissible temperatures of the insulation and conductor joints whereas the permissible screen, sheath or armour temperatures are limited by the insulation and outer sheath (Table 19.9). The lower of all these quantities is the determining factor. For protective covers of "suitable material" and of "sufficient thickness" between insulation and screen the permissible short-circuit temperature of the outer sheath can be selected.

With only a few exceptions the permissible short-circuit temperatures to VDE are identical to those to IEC. For mass-impregnated cables it had previously been the practice to use quantities below 250 °C. Installations already in existance were designed to these levels. Since hardly any new installations in the 0.6/1 kV level incorporate mass-impregnated cables it is not justified or required to increase the short-circuit capacity of existing installations and the original standard remains. For the screens of cables with polymer insulation, having suitable adjacent heat resisting layers, permissible short-circuit temperatures of 350 °C were established by testing (Table 19.8).

The short-circuit effective cross-sectional area q_k is the geometric mean value of the electrically effective cross-sectional area q_{el} and the cross-sectional area by weight q_g which in the equations 19.32 and 19.33 takes into consideration the longitudinal addition $(1+z)$ for the laying up of the core and other

Table 19.9

Permissible short-circuit temperatures of constructional elements of 0.6/1-kV-cable to IEC 724 [19.29]

Material of constructional element	Permissible short-circuit temperature ϑ_e for		
	Insulation	Outer-sheath	Metall-sheath and conductor jointing
	°C	°C	°C
Impregnated paper	250	–	–
PVC to 300 mm²	160	200	–
over 300 mm²	140	–	–
Butyl rubber	220	–	–
Natural rubber	200	–	–
XLPE	250	–	–
EPR	250	–	–
Silicone rubber	350	–	–
PE	–	150	–
CSP	–	220	–
SEI[1]	–	200	–
Soft soldered connection	–	–	160
Welded connection	–	–	250
Crimped connection	–	–	250
Sheath of lead	–	–	170
Sheath of lead alloy	–	–	200

[1] Elastomer compound according to IEC 502

individual construction elements (19.10). The factor f_q represents the electrical contact between the individual constructional elements of a screen, such as transverse conductivity. The transverse conductivity is variable and in new cables more favourable than for cables which have been in operation for some time. Since transverse conductivity is difficult to calculate, contact between constructional elements is neglected and $f_q = 1$ is used. As an approximation in equation 19.35 the calculation is made using the geometric cross-sectional area.

Conductors are designated with a nominal cross-sectional area q_n. In DIN VDE 0295 a quantity of electrical resistance R_{20} at 20 °C is allocated to the nominal cross-sectional areas. It is therefore generally recommended to base calculations on the quantity of nominal cross-sectional area q_n.

Table 19.10 Effective short-circuit cross-sectional area q_k and effective screen wall thickness δ_k

	Factor for the longitudinal addition (helix) $1+z=\dfrac{S(1+z)}{S}=\dfrac{\pi d_M}{B}$	Geometric cross-sectional area q_{geo}	Cross-sectional area by weight $q_g=q_{geo}(1+z)$ (19.33)	Electrical cross-sectional area $q_{el}=\dfrac{q_{geo}}{(1+z)f_q}$ (19.32)
n_B Tape helix with gap l (unwound)	$B=n_B(b+l)$	$n_B b\delta$ δ Tape thickness	$n_B b\delta(1+z)$	$\dfrac{n_B b\delta}{(1+z)f_q}$, $f_q=1.0$
n_B Tape helix overlapped (unwound)	$B=n_B(b-l)$	$n_B b\delta$ δ Tape thickness	$n_B b\delta(1+z)$	$\dfrac{n_B b\delta}{(1+z)f_q}$, $\dfrac{1}{1+z}<f_q<1.0$
Sheath	$z=0$	$\pi\delta(d_M-\delta)$	$\pi\delta(d_M-\delta)$	$\pi\delta(d_M-\delta)$, $z=0;\ f_q=1.0$
Wire screen with transverse helical tape	$1+z$ for the transverse helical tape $1+z_D$ for the wires (as specified by the designer)	$b\delta+n_D\dfrac{\pi d_D^2}{4}$	$b\delta(1+z)+$ $n_D\dfrac{\pi d_D^2}{4}(1+z_D)$	$\dfrac{b\delta}{1+z}+$ $n_D\dfrac{\pi d_D^2}{4(1+z_D)f_q}$, $\dfrac{1}{1+z_D}<f_q<1.0$
Circular conductor	$z=0$	$\approx q_n$	$\approx q_n$	$\approx q_n$

[1] Specified value for calculation

288

Cooling circumference A	Effective under short circuit: Cross-sectional area	Screen wall thickness
	$q_k = \sqrt{q_g \cdot q_{el}}$ (19.35)	$\delta_k = 2\dfrac{q_g}{A}$
$2 n_B b (1+z)$	$n_B b \delta$	$\delta_K = \delta$
$2\pi d_M$	$n_B b \delta$	$\dfrac{n_B b \delta (1+z)}{\pi d_M} =$ $\dfrac{b\delta}{b-l}$
$2\pi(d-\delta)$	$\pi \delta (d_M - \delta)$	$\delta_K = \delta$
only for the wires considered $F n_D \pi d_D (1+z_D)$, $F \approx 0.6$	$b\delta + n_D \dfrac{\pi d_D^2}{4}$	only for the wires considered $\dfrac{d_D}{2F} \approx 0.8\, d_D$
πd_L	q_n [1]	$\dfrac{d_L}{2}$

For screens the geometric cross-sectional area q_{geo} is stipulated in the VDE specifications. This must not be less than the nominal cross-sectional area included in the type reference. If in multi-core cables which are provided with a screen on each individual core contact is not ensured between screens then each screen must have the full geometric cross-sectional area.

Screens are always of copper having a conductivity of not less than 56×10^6 S/m (see DIN VDE 0271 or DIN VDE 0273). The nominal cross-sectional areas have been allocated to the conductor cross-sectional areas in such a way that the short-circuit capacity of the screen is always approximately that of a lead sheath of a S.L. cable for 12/20 kV (see Fig. 19.9). This comparison is made based on a non-adiabatic temperature rise.

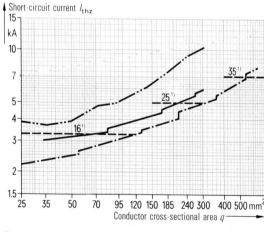

[1] Cross-sectional area of screen

- – – – PVC and XLPE cable for 6/10 kV to 18/30 kV
- ——— S.L. cable for 18/30 kV
- —·—· S.L. cable for 12/20 kV
- —··— Belted cable for 6/10 kV

Fig. 19.19
Comparison of short-circuit rating I_{thz} of copper-wire screens of cables with polymer insulation and of lead sheaths of mass-impregnated cables for a short-circuit duration $t_k = 1$ s and non adiabatic temperature rise

Example 19.6

The rated short-time current density J_{thr} of the conductors of the NA2XS2Y 1×120 RM/16 12/20 kV cable in Examples 19.4 and 19.5 is to be calculated for an initial temperature $\vartheta_a = 90\,°C$, a final temperature $\vartheta_e = 250\,°C$ using the values from Table 19.7 for adiabatic temperature rise, therefore:

$$k_1 = \sqrt{\frac{2.5 \times 10^6 (228 + 20)}{2.8264 \times 10^{-8}}} = 148 \times 10^6 \;\; A\sqrt{s}/m^2, \tag{19.36}$$

$$k_2 = \sqrt{\ln \frac{228 + 250}{228 + 90}} = 0.638, \tag{19.37}$$

$$J_{thr} = \frac{148 \times 10^6 \times 0.638}{\sqrt{1}} = 94.4 \times 10^6 \; A/m^2$$

$$\approx 94 \; A/mm^2 \tag{19.38}$$

The same quantity was used in Example 19.4 taken from Table 19.6.

For the copper wire screen using quantities from Table 19.7 with an

initial temperature $\vartheta_a = 80\,°C$ and a
final temperature $\vartheta_e = 350\,°C$:

$$k_1 = \sqrt{\frac{3.45 \times 10^6 (234.5 + 20)}{1.7241 \times 10^{-8}}} = 226$$

$$\times 10^6 \;\; A\sqrt{s}/m^2, \tag{19.36}$$

$$k_2 = \sqrt{\ln \frac{234.5 + 350}{234.5 + 80}} = 0.787. \tag{19.37}$$

The short-circuit rating of the screen for adiabatic temperature rise and a short-circuit duration of $t_k = 0.5$ s is therefore

$$I_{thz} = 16 \times 10^{-6} \times 226 \times 10^6 \times 0.787 \frac{1}{\sqrt{0.5}} = 4025 \; A \tag{19.34}$$

The load capacity for non-adiabatic temperature rise is calculated in Example 19.7.

Non-Adiabatic Temperature Rise Method

Electrical conductors having a relatively large cooling surface in relation to their cross-sectional area dissipate a significant proportion of heat to the ambient during the short circuit. For a given temperature rise therefore the actual short-circuit capacity is higher then the value given by an adiabatic calculation

[19.20, 19.22 to 19.27]. This is taken into consideration by the short-time rating factor k_3:

$$I_{thz} = q_k \, k_1 \, k_2 \, k_3 \, \frac{1}{\sqrt{t_k}}. \tag{19.39}$$

The short-time rating factor can be calculated by a simplified empirical method [19.22] with

$$k_3 = \sqrt{1 + \frac{A}{q_g \, c} \beta_i \sqrt{\frac{3}{2\pi} t_k^z}} \tag{19.40a}$$

$$= \sqrt{1 + \frac{1.382}{\delta_k \, c} \beta_i \, t_k^z} \tag{19.40b}$$

$$= \sqrt{1 + \frac{b}{\delta_k} t_k^z}. \tag{19.40c}$$

Equations for the cooling surface area in terms of the corresponding circumference A, the effective wall thickness δ_k during the short circuit as well as the cross-sectional area by weight q_g can be found in Table 19.10. The specific heat of the conductor c can be taken from Table 19.7 and the thermal penetration quantity β_i of the surrounding media together with screen constant b from Table 19.8. (The determination of heat penetration quantities is very difficult and the quantities in particular for coverings of only a few layers, with indeterminate inclusions of air, cannot be calculated reliably.) The quantities selected herein are on the safe side. If the characteristics of the media above and below the screen differ then the calculation should be made using the arithmetic mean of the heat penetration quantities.

The exponents z were established from tests conducted

$$z = 0.8 \quad \text{for} \quad 0.1 \; s \leq t_k \leq 1 \; s \quad \text{and}$$

$$z = 0.6 \quad \text{for} \quad 1.0 \; s < t_k \leq 5 \; s.$$

The quantities in Table 19.8 and the calculation methods mentioned above are used to prepare the graphic diagrams for screens (Fig. 19.17a), flat steel-wire armour (Fig. 19.17b) and lead sheath (Fig. 19.8) as well as for the short-time rating factor k_3 (Fig. 19.20).

The conductor diameter of circular conductors is calculated as an approximation from the nominal values of cross-sectional area. During the short circuit the effective screen wall thickness is therefore (Table 19.10).

$$\delta_k = \sqrt{\frac{q_n}{\pi}}. \tag{19.41}$$

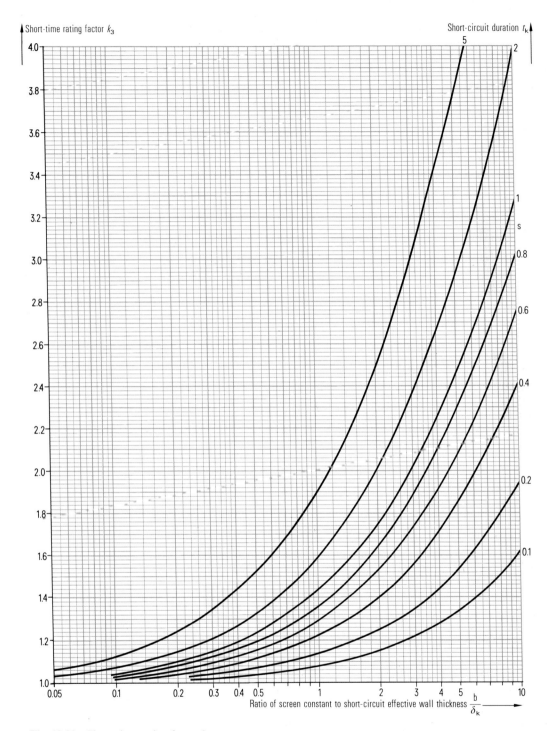

Fig. 19.20   Short-time rating factor k_3

Sector-shaped conductors, for a given nominal cross-section, have a perimeter approximately 11% greater than that of a circular conductor; that is

$$\delta_k = \frac{1}{1.11}\sqrt{\frac{q_n}{\pi}}.$$

Measurements of short-time temperature rise in PVC cable for 0.6/1 kV are published in [19.19].

A noticeable increase of short-circuit capacity in comparison with the adiabatic quantity is apparent for small cross-sectional areas $q_n < 10$ mm^2 or for longer short-circuit duration. Where the ratio of short-circuit duration to rated conductor cross-sectional area is 0.1 s/mm^2 or less the increase in load capacity is neglegible, so that it is appropriate to use the simple adiabatic calculation method.

Example 19.7

Following Example 19.6, the short-circuit rating of the screen is to be calculated assuming non-adiabatic temperature rise. The diameter of screen wire $d_D = 0.7$ mm, the effective wall thickness during short circuit from Table 19.8:

$$\delta_k = 0.8\,d_D = 0.8 \times 0.7 = 0.56 \text{ mm}$$
$$= 0.56 \times 10^{-3} \text{ m}.$$

If the commonly used coverings above and below the screen, which are poor conductors of heat and therefore have a low heat penetration quantity, are assumed then the calculation must be made with a screen constant $b = 0.12 \times 10^{-3}$ m/$\sqrt{s}$. Giving

$$k_3 = \sqrt{1 + \frac{b}{\delta_k}t_k^z} = \sqrt{1 + \frac{0.12 \times 10^{-3}}{0.56 \times 10^{-3}}0.5^{0.8}} = 1.06.$$

$$\text{(19.40c)}$$

This quantity can also be obtained from Fig. 19.20 from the ratio $b/\delta_k = 0.21$.

The short-circuit capacity for non-adiabatic temperature rise, when using the result from Example 19.7 is

$$I_{thz} = q_k k_1 k_1 k_3 \frac{1}{\sqrt{t_k}}$$
$$= 4025 \times 1.06 = 4267 \text{ A}$$
$$\approx 4.3 \text{ kA}.$$

This value was calculated for an initial temperature of $\vartheta_a = 80$ °C. The value of 4.5 kA in Example 19.5 from Fig. 19.16 is valid for an initial temperature of 60 °C and is acceptable for practical applications because of the small deviation.

Temperature Rise During a Short Circuit

The final temperature for a heating caused by a short-circuit current I_{th} is

$$\vartheta_e = \left[\exp\left(\frac{I_{th}\sqrt{t_k}}{q_k k_1 k_3}\right)^2\right](\Theta + \vartheta_a) - \Theta. \qquad (19.43)$$

For an adiabatic calculation $k_3 = 1$.

The committee IEC TC 20 are currently investigating methods to establish a more exact calculation method to give more accurate short-circuit temperatures [19.20].

On a broader basis theoretical and practical investigations will provide a more accurate knowledge of calculation methods and of the required construction of screens in cables with polymer insulation [19.27].

Example 19.8

If the screen of the cable from Example 19.7 carries only 3.5 kA instead of the 4.3 kA but all other conditions remain unchanged the screen temperature will now reach a final quantity of

$$\vartheta_e = \left[\exp\left(\frac{3500\sqrt{0.5}}{16 \times 10^{-6} \times 226 \times 10^6 \cdot 1.06}\right)^2\right] \times$$
$$\times (234.5 + 80) - 234.5 = 243 \text{ °C}. \qquad (19.43)$$

19.3.3 Thermo-Mechanical Forces and Expansion

General

Due to the high temperature rise during a short circuit a significant expansion of the conductor occurs. If the conductor of length l is free to expand then an increase in length Δl caused by a temperature rise $\Delta \vartheta$ will occur (Fig. 19.21).

$$\Delta l = l\,\alpha_{th}\,\Delta \vartheta. \qquad (19.44)$$

If the conductor is fixed at both ends the centre will move to one side or the other. Assuming the cable axis will move to simulate a sine wave shape – which

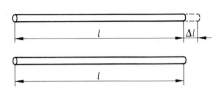

Δl longitudinal increase
h deflection

Length of bar	l	$=2.0$ m
Cross-sectional area		
of bar	q	$=300$ mm^2
Diameter of bar	d	$=19.54$ mm
Temperature increase	$\Delta \vartheta$	$=90\,°C-30\,°C=60$ K

Fig. 19.21 Thermal expansion of an aluminium bar

Longitudinal increase ($h=0$; $F=0$):

$$\Delta l = \alpha_{th}\, l\, \Delta \vartheta = 23.8 \times 10^{-6} \times 2 \times 60 = 2.86 \text{ mm}$$
$$(19.44)$$

Thermo-mechanical force ($\Delta l=0$; $h=0$):

$$F_{th} = q\, E\, \alpha_{th} \Delta \vartheta$$
$$= 300 \times 10^{-6} \times 65 \times 10^9 \times 23.8 \times 10^{-6} \times 60$$
$$= 27.8 \text{ kN} \qquad (19.46)$$

Deflection ($\Delta l=0$; $F=0$):

$$h = \frac{2l}{\pi}\sqrt{\alpha_{th}\Delta\vartheta}$$

$$= \frac{2 \times 2}{\pi}\sqrt{23.8 \times 10^{-6} \times 60}$$

$$= 48.1 \text{ mm}. \qquad (19.45)$$

Material	Copper	Aluminium
Coefficient of thermal expansion α_{th} in $1/K$	16.2×10^{-6}	23.8×10^{-6}
Modulus of elasticity E in N/m^2	115×10^9	65×10^9

can be assumed for a cable which is secured at equal spacings – the deflection h when movement longitudinally is prevented ($\Delta l=0$) is

$$h = \frac{2l}{\pi}\sqrt{\alpha_{th}\Delta\vartheta}. \qquad (19.45)$$

In a conductor restrained on all sides which can neither displace ($h=0$) nor expand ($\Delta l=0$) a force F_{th} is produced

$$F_{th} = q\, E\, \alpha_{th}\, \Delta\vartheta. \qquad (19.46)$$

Example 19.9

Fig. 19.21 is supplimented by a sample calculation to illustrate the effect of a temperature rise on a bar or conductor of aluminium.

A bar of length 2.0 m and diameter 19.54 mm with a temperature rise of 60 K expands 2.86 mm. If longitudinal movement and radial deflection is prevented a force of 28 kN is developed. If only longitudinal movement is prevented then the side ways deflection would be 2.5 × diameter.

Effect of Thermal Expansion in Cables

In Equation 19.46 the conductor is assumed to be an elastic body and therefore under conditions of temperature rise produces a linear dependant force. In reality the progression is dependant upon material, shape and construction of the conductor and with low temperature rises it is linear until with increasing temperature rise it reaches a limiting value. Forces of up to 50 N/mm^2 must be expected. Under the influence of such a force deformation takes place, i.e. an effective shortening, especially with aluminium conductors such that upon cooling a mechanical tension may develop in the conductor. In solid aluminium conductors of tempered material tensile forces of up to 40 N/mm^2 can be experienced, if the cable in normal operation carries little load.

Whether the temperature rise of a conductor produces a longitudinal movement or a compressive force largely depends on the type of conductor, the adhesion of the insulation to the conductor, the type of cable and also the method of fixing.

In multi-core cables the expansion of the conductors is hindered and therefore longitudinal forces are developed in the conductors and the increase in length of the conductors, and consequently of the cable, does not become fully effective. In general longitudinal forces develop when expansion of the conductors and deflection of the cable is hampered due to the type of fixing or when installed in ground.

The cables must be installed and secured in such a way that longitudinal expansion is equally divided over the full length of the cable and does not occur only at a few points. This is of particular importance for cables having large cross-sectional areas which in normal operation are heavily loaded and are operated in cyclic loading as well as possibly being subjected to short-circuit currents.

Multi-core cables installed over long straight runs in air must be arranged in a wavy line and at any point of transition fixed to leave a free loop. For installation in ground no special measures are necessary.

Single-core cables in ground (see Section 29) and in air must be installed in long straight runs in a wavy line. Cables installed in air must be fixed to supports at sufficiently large distances to permit deflection. Providing the cables are fixed taking into consideration electromagnetic forces under short circuit (see Section 19.4.3, page 305) and the thermo-mechanical forces in normal operation (see next section) then it can be assumed that the thermo-mechanical forces in the event of a short circuit can be accommodated.

During the installation of the cables the minimum bending radii (see Section 29) must be strictly observed so as to avoid the development of excessive radial stresses in the bends and hence avoid danger of damage to the insulation and outer sheath. Cables should therefore not be secured at the bends but at the point of transition to the straight run.

For cables installed in the ground or where cables in air are fixed at short distances, suitable measures must be taken to ensure that forces are not transmitted on to the cable joints (see Section 19.3.4 page 296).

Mounting of Single-Core Cables

Single-core cables must, to allow for the effect of heat expansion "be installed in such a way that damage, e.g. by pressure points caused by thermal expansion are avoided" (DIN VDE 0298 Part 1). This can be achieved by installing the cables in an approxi-

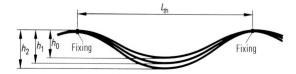

Fig. 19.22
Deflection of a cable by thermal expansion when sine-wave laying is assumed

mate sine-wave form and fixing at points as shown in Fig. 19.22.

With unloaded cables and at the lowest possible ambient temperature ϑ_0 a deflection of h_0 should be allowed to avoid tensile stress in the cable on cooling and to control the part and direction of deflection when heated so that no great longitudinal pressure can develop.

The length of bow l_{B0} (actual length of cable compared with the distance between fixing points l_{th}) at the minimum ambient temperature ϑ_0 and a deflection of h_0 is

$$l_{B0} = l_{th}\left[1 + \left(\frac{\pi h_0}{2\,l_{th}}\right)^2\right]. \tag{19.47}$$

If during installation the ambient temperature is ϑ_1 the length of bow is

$$l_{B1} = l_{B0}[1 + \alpha_{th}(\vartheta_1 - \vartheta_0)]$$
$$= l_{th}\left[1 + \left(\frac{\pi h_0}{2\,l_{th}}\right)^2\right]\cdot[1 + \alpha_{th}(\vartheta_1 - \vartheta_0)] \tag{19.48}$$

and the deflection

$$h_1 = \frac{2l_{th}}{\pi}\sqrt{\frac{l_{B1}}{l_{th}} - 1}. \tag{19.49}$$

If the cable is loaded such that the conductor reaches a permissible operating temperature $\vartheta_2 = \vartheta_{Ln}$ the length of bow is

$$l_{B2} = l_{B0}[1 + \alpha_{th}(\vartheta_2 - \vartheta_0)] \tag{19.50}$$

and the deflection

$$h_2 = \frac{2l_{th}}{\pi}\sqrt{\frac{l_{B2}}{l_{th}} - 1}. \tag{19.51}$$

Sufficient space must be provided on the cable racks to accommodate the maximum deflection h_2 of the cable under normal operation.

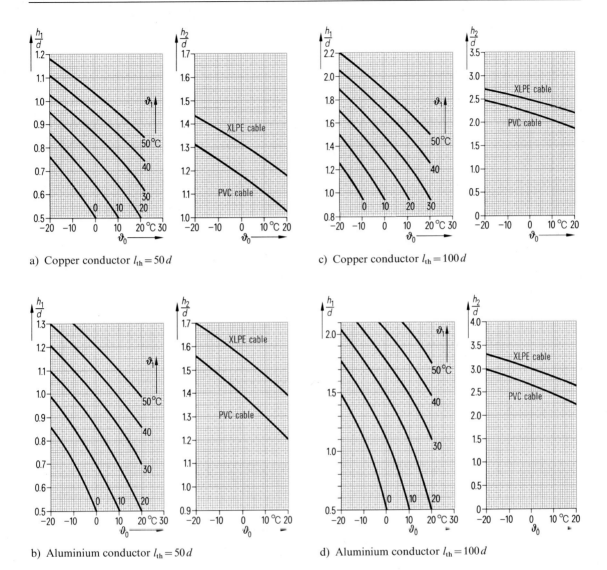

a) Copper conductor $l_{th}=50\,d$

c) Copper conductor $l_{th}=100\,d$

b) Aluminium conductor $l_{th}=50\,d$

d) Aluminium conductor $l_{th}=100\,d$

when installing at ambient temperature ϑ_1: $h_1/d=f(\vartheta_0)$
with full current loading up to a conductor temperature of $\vartheta_2=\vartheta_{Lr}$: $h_2/d=f(\vartheta_0)$

$\vartheta_{Lr}=90\,°C$ for XLPE cable
$\vartheta_{Lr}=70\,°C$ for PVC cable

Fig. 19.23   Determination of deflection of a cable due to thermal expansion

The fixing distance l_{th} can be selected to lie between $50\,d$ and $100\,d$. Guiding for the selection of deflection may be taken from Fig. 19.23. In this diagram the distance between fixings l_{th} is selected in relation to the cable diameter with $l_{th}/d=50$ and $l_{th}/d=100$. Using these quantities, on the one hand the bending radii of the cables will not be too severe and on the other hand the width of cable trays (and therefore the cost) will not be too excessive. The deflection at the lowest ambient temperature ϑ_0 is selected with $h_0=0.5\,d$.

Example 19.9

An XLPE cable

NA2XS2Y 1×120 RM/16   12/20 kV

having an outer diameter of $d = 34$ mm is to be installed at an ambient temperature of 20 °C. The lowest ambient temperature is to be -10 °C.

The calculation is first made to determine the free actual length of cable between clamps with $l_{th} = 50 d$. The deflection of the cable on no-load and at the lowest ambient temperature is to be $0.5 d$:

$$h_0 = 0.5 \, d = 0.5 \times 34 \times 10^{-3} = 17 \times 10^{-3} \text{ m},$$

$$l_{th} = 50 \, d = 50 \times 34 \times 10^{-3} = 1.7 \text{ m},$$

$$l_{B0} = l_{th} \left[1 + \left(\frac{\pi \, 17 \times 10^{-3}}{2 \times 1.7} \right)^2 \right] = l_{th} \, 1.000247, \quad (19.47)$$

$$l_{B1} = l_{th} \, 1.000247 [1 + 23.8 \times 10^{-6} (20 - (-10))]$$

$$= l_{th} \, 1.000961, \quad (19.48)$$

$$h_1 = \frac{2 \times 1.7}{\pi} \sqrt{\frac{1.7 \times 1.000961}{1.7} - 1} = 0.0335 \text{ m}$$

$$= 0.99 \, d, \quad (19.49)$$

$$l_{B2} = l_{th} \, 1.000247 [1 + 23.8 \times 10^{-6} (90 - (-10))]$$

$$= l_{th} \, 1.00263, \quad (19.50)$$

$$h_2 = \frac{2 \times 1.7}{\pi} \sqrt{\frac{1.7 \times 1.00263}{1.7} - 1} = 0.0555 \text{ m}$$

$$= 1.63 \, d. \quad (19.51)$$

From diagram b) in Fig. 19.23 the same quantities can be found:

for $\vartheta_0 = -10$ °C and $\vartheta_1 = 20$ °C is $\dfrac{h_1}{d} = 0.99 \, d,$

for $\vartheta_0 = -10$ °C and XLPE cable is $\dfrac{h_2}{d} = 1.63 \, d.$

The distance between rungs in ladder type cable trays is normally 0.3 m and accordingly the spacing between clamp points becomes 1.8 m. For a small difference in clamping distance a linear conversion is possible without noticeable error. The deflection which must be allowed for in the installation is

$$h_1 = 0.99 \, d \, \frac{1.8}{1.7} = 1.05 \, d = 36 \text{ mm}$$

and for dimensioning the cable tray width a deflection under full current load, up to the point where the permissible operating temperature is reached, must be allowed for, which is

$$h_2 = 1.63 \, d \, \frac{1.8}{1.7} = 1.73 \, d = 59 \text{ mm}.$$

If a clamping distance is selected as

$$l_B = 100 \, d = 3.4 \text{ m},$$

Then from the diagrams in Fig. 19.23 d

for $\vartheta_0 = -10$ °C and $\vartheta_1 = 20$ °C the quantity $h_1 = 1.68 \, d$ and

for $\vartheta_0 = -10$ °C and XLPE cable the quantity $h_2 = 3.15 \, d.$

The rung spacing of 0.3 m and a clamping distance of 3.6 m results in:

$$h_1 = 1.68 \, d \, \frac{3.6}{3.4} = 1.78 \, d = 60 \text{ mm and}$$

$$h_2 = 3.15 \, d \, \frac{3.6}{3.4} = 3.34 \, d = 113 \text{ mm}.$$

This illustrates that by selecting a greater clamping distance the number of clamps is reduced but wider cable trays are necessary to accommodate h_2.

19.3.4 Accessories

Accessories must be selected and installed having due regard for the stresses developed under a short circuit.

To avoid the effect of thermo-mechanical forces within the accessory or joint box the cores within the joint must be secured by binding and cable ends should be supported by clamps before the cable enters the sealing end especially where longitudinal forces cannot be controlled in a bow before the sealing end. These measures will also be somewhat effective in controlling the electromagnetic short-circuit forces (see Section 19.4.4 page 315).

In mass-impregnated cables the expansion of the impregnant can create a significant fluid pressure. The housing of the accessory must be strong enough to withstand this pressure. A limitation of short-circuit temperature may be necessary under certain circumstances (see Section 19.3.2 page 285).

The electrical resistance of joints in the conductor must not be increased by short-circuit stresses. Soft-solder joints are suitable only for short-circuit temperatures of up to 160 °C. Sheaths, screens or armour may be stressed to a higher short-circuit temperature (Table 19.8). When making connections to earth or when jointing tails by soft-soldering sufficiently large heat sinks must be built in so that the temperature at the joint does not exceed 160 °C under short-circuit conditions.

Screens are normally sized for a non-adiabatic temperature rise since their heat dissipation is more favourable than that of a compact conductor of equal cross-section. If at the end of a cable the screen wires are bunched up for connecting the screen to earth it must be noted that these bunched up screen wires will become significantly hotter than the individual spaced screen wires within the cable.

19.4 Mechanical Short-Circuit Capacity

19.4.1 Electromagnetic Forces

Introduction

With the increase of short-circuit and breaking capacities the dimensioning of the cables and fixing elements in respect of short-circuit currents is of increasing importance. To support this trend DIN VDE 0298 Part 2 contains a section "capacity in the event of short circuit" which for the time being considers the thermal aspects but as yet does not consider mechanical aspects in depth.

Little is known of the mechanical effects of short-circuit currents. With the aid of a number of test findings, coupled with newly developed methods of calculation a guide to project design is offered, which in many ways could be improved or extended. It is, however, better to use a method which may not be perfect but which provides results which are on the safe side than to use vague assumptions.

Since the thermal and mechanical effects cannot be separated in a practical situation, for all situations the permissible operating temperature in normal operation must be selected as the initial temperature.

It is also important to consider not only the withstand of cable and accessories but also of the fixings and cable supports as well as the arrangement of cables and method of fixing.

Effect of Electromagnetic Forces

Currents in conductors laying side by side produce electromagnetic forces between the conductors. If the currents in the two conductors flow in opposite directions they will produce a repulsion force. If the currents flow in the same direct the force is one of attraction (Fig. 19.24) [19.30].

The highest impact force is proportional to the square of the peak short-circuit current. This is followed by a pulsating oscillatory stress at a frequency of twice the operating frequency. Following general practice it is assumed that the stress attributed to the highest force which occurs is used in calculating the short-circuit withstand.

The forces caused by the short circuit must be considered when dimensioning the accessories, the cables and fixings. These can cause severe problems under certain circumstances.

Between two long conductors side by side the current force F_s occurs. For the arrangement shown in Fig. 19.24 this can be expressed as current force per unit length as

$$F'_s = \frac{\mu_0}{2\pi} \frac{I_s^2}{a}, \tag{19.52}$$

which increases with the square of the peak short-circuit current I_s and is decreased with increase in spacing between conductors. For circular conductors it is assumed that the current is concentrated at the axis of the conductor (line source); the axial distance a for spacing represents the distance between centres of the conductors.

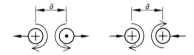

← — ⟶ effective force
+ • direction of current flow
⟩ ⟨ direction of magnetic field

Fig. 19.24 Effective force from currents

System	Short circuit	Factor α for radial, i.e. conductor stresses		Factor β [3] for tangential, i.e. tape stresses	
Single-phase a.c. Three phase	Line-to-earth	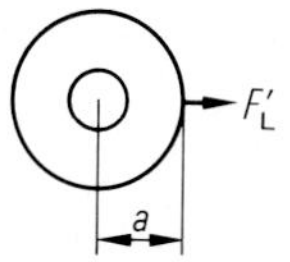 $\;F_L'$	$\dfrac{1}{2}$	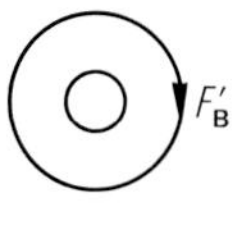 $\;F_B'$	$\dfrac{1}{4\pi}$
Single-phase a.c. Three phase	Line-to-line Line-to-line	$\;F_L'$ [1]	1	F_B'	$\dfrac{1}{2}$
	Balanced three phase	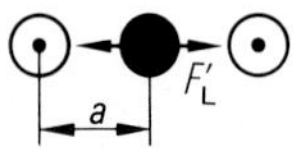 F_L'	$\dfrac{\sqrt{3}}{2}$	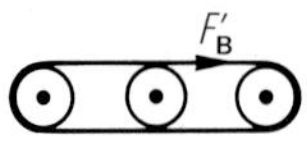 F_B'	0.404 [3]
		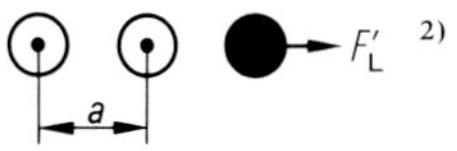 $\;F_L'$ [2]	0.808	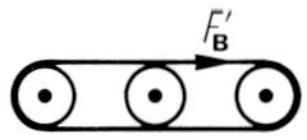 F_B'	
		F_L'	$\dfrac{\sqrt{3}}{2}$	F_B'	$\dfrac{1}{2}$
		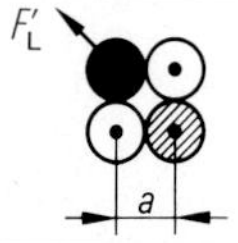 F_L'	$\dfrac{1}{\sqrt{2}}$	 F_B'	$\dfrac{1}{2}$

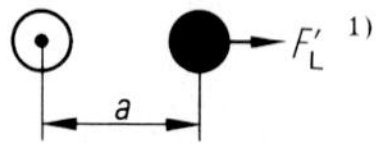

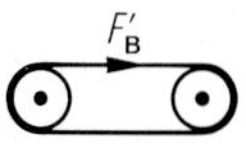

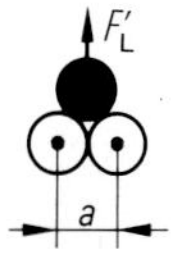

→   Direction of major effective force

●   Cable or conductor under consideration

◍   Protective conductor

[1] The factors $\alpha = 1$ and $\beta = \frac{1}{2}$ are valid for a line-to-line short circuit also for all other arrangements mentioned in a 3 ph system
[2] Included for theoretical interest since only the arrangement with the highest effective force is used for design purposes
[3] The factor β apply also for the fixing of cables to racks with tape; for the centre cable where three are installed side by side $\beta = \sqrt{3}/4$

Fig. 19.25 Factors for the calculation of stresses under short circuit.

The equation 19.52 is applicable for a line-to-line short circuit on single-phase or three-phase a.c. and for an arrangement of conductors as shown in Fig. 19.24. For the *effective force on the conductor* for other arrangements

$$F_L' = \alpha F_s'. \tag{19.53}$$

For the effective force in a common cover over all conductors within items such as clamps, binder tapes and the like

$$F_B' = \beta F_s'. \tag{19.54}$$

The factors α and β are to be taken from Fig. 19.25 in order to calculate the highest quantity of effective force.

In Fig. 19.25 the direction of the major effective force is indicated in respect of a single-core cable or a conductor in a multi-core cable.

From the effective force F'_L can be found:

▷ the pressure per unit area of the conductor on the insulation

▷ the pressure per unit area of the core on a surrounding cover

▷ the force on individual fixings for single-core cables on supports

▷ the force acting on a fixed clamp which surrounds all single-core cables, e.g. the tensile stress in the screwed rods of a two-part split clamp.

The tensile force in a binder, clamp or the like which surrounds all the single-core cables of a system is calculated, represented by F'_B.

Equation 19.54 is also used for the calculation of tensile forces in covers which embrace all conductors of a multi-core cable (see Section 19.4.2).

Example 19.10

Given

Breaking power	$S_a = 750$ MVA,
Short-circuit duration	$t_k = 0.5$ s,
Peak factor	$\varkappa = 1.8$,
Operating voltage	$U_b = 10$ kV.

In the following it is required to calculate the short-circuit currents and forces for a cable.

$$I''_k \approx I_a = \frac{S_a}{\sqrt{3}U_b} = \frac{750 \times 10^6}{\sqrt{3} \times 10 \times 10^3} = 43.3 \text{ kA.} \quad (19.1)$$

From Fig. 19.7 for $\varkappa = 1.8$ and $t_k = 0.5$ s; $m = 0.1$ also for $I''_k/I_k \approx 1.0$; $n = 1.0$.

From this the current which determines the thermal duty is

$$I_{th} = I''_k\sqrt{m+n} \quad (19.24a)$$
$$= 43.3\sqrt{0.1+1.0} = 45.4 \text{ kA.}$$

From Fig. 19.11a for an XLPE cable with a copper conductor and a cross-sectional area of 240 mm² is selected. For a breaking time $t_k = 0.5$ s the short-circuit currents are

$$I_{thz} = 48.5 \text{ kA}$$
and
$$I''_{kz} = 46.2 \text{ kA.}$$

The type of cable selected is

N2XS2Y 1 × 240 RM/25 6/10 kV,

with an outside diameter $d = 36$ mm and with a mean screen diameter of $d_{Mm} = 29.1$ mm.

Line-To-Earth Short Circuit

The effective force between conductor and screen (it is assumed that the total line-to-earth short-circuit current returns through the screen) with $a = 0.5\,d_{Mm}$

is

$$I_s = \sqrt{2} \times 1.8 \times 43.3 = 110.2 \text{ kA,} \quad (19.3)$$

$$F'_L = \alpha \frac{\mu_0}{2\pi} \frac{I_s^2}{0.5 d_{Mm}} \quad (19.53)$$
$$= 0.5 \times 0.2 \times 10^{-6} \frac{(110.2 \times 10^3)^2}{0.5 \times 29.1 \times 10^{-3}}$$
$$= 83.5 \text{ kN/m,}$$

with

$$\mu_0 = 4\pi \times 10^{-7} \text{ H/m}$$

and

$$\frac{\mu_0}{2\pi} = 0.2 \times 10^{-6} \text{ H/m.}$$

This force attempts to push the screen away from the conductor or widen the space between them.

The tensile stress generated is

$$F'_B = \beta \frac{\mu_0}{2\pi} \frac{I_s^2}{0.5 d_{Mm}} \quad (19.54)$$
$$= \frac{1}{4\pi} 0.2 \times 10^{-6} \frac{(110.2 \times 10^3)^2}{0.5 \times 29.1 \times 10^{-3}}$$
$$= 13.3 \text{ kN/m}$$

and this is spread over the screen wires to the binder tape above and possibly through the separating layer to the outer sheath.

Line-To-Line Short Circuit

With two identical single-core cables installed side by side with a spacing of $a = 2d$ operating in a single-phase system the electromagnetic force per unit length is

$$F'_s = \frac{\mu_0}{2\pi} \frac{I_s^2}{2d} = 0.2 \times 10^{-6} \frac{(110.2 \times 10^3)^2}{2 \times 36 \times 10^{-3}} \quad (19.52)$$
$$= 33.7 \text{ kN/m,}$$

and the effective force on the conductors with $\alpha = 1$ is

$$F'_L = \alpha \, F'_s = 1 \times 33.7 = 33.7 \text{ kN/m} \qquad (19.53)$$

The effective force in a binder surrounding the cables is

$$F'_B = \beta \, F'_s = 0.5 \times 33.7 = 16.9 \text{ kN/m}. \qquad (19.54)$$

In a three-phase system the currents I''_k and I_s with a line-to-line short circuit it can be assumed to have a quantity equal to $\sqrt{3}/2$ times the quantities of a balanced three-phase short circuit (see Table 19.3).

Balanced Three-Phase Short Circuit

In an installation of single-core cables side by side with an axial spacing of $a = 2d$ the electromagnetic force per unit length (see page 299)

$$F'_s = 33.7 \text{ kN/m},$$

and the effective force on the outer conductors is

$$F'_L = \alpha \, F'_s = 0.808 \times 33.7 = 27.2 \text{ kN/m}. \qquad (19.53)$$

The effective force on the inner conductor

$$F'_L = \alpha \, F'_s = \frac{\sqrt{3}}{2} \, 33.7 = 29.2 \text{ kN/m} \qquad (19.53)$$

and the effective force in a binder embracing all cables is

$$F'_B = \beta \, F'_s = 0.404 \times 33.7 = 13.6 \text{ kN/m}. \qquad (19.54)$$

In a bunched installation of single-core cables (in standard trefoil arrangement) with $a = d$

$$F'_s = \frac{\mu_0}{2\pi} \frac{I_s^2}{d} = 0.2 \times 10^{-6} \frac{(110.2 \times 10^3)^2}{36 \times 10^{-3}}$$
$$= 67.5 \text{ kN/m},$$

and the effective force on one conductor is

$$F'_L = \alpha \, F'_s = \frac{\sqrt{3}}{2} \, 67.5 = 58.5 \text{ kN/m}. \qquad (19.53)$$

The effective force in a binder or a tape clamp embracing all cables is

$$F'_B = \beta \, F_s = 1/2 \times 67.5 = 33.8 \text{ kN/m}. \qquad (19.54)$$

19.4.2 Multi-Core Cable

The short-circuit repulsion forces F'_L acting on the conductors cause a pressure on the inner conducting layer and also the insulation. This pressure spreads in such a way that the cores are pressed against the screen, the armour and/or the outer sheath. As a consequence surface pressure forces F'_F result on the conducting layers and the insulation which can lead to deformation. In the coverings surrounding these cores the short-circuit forces produce a tensile force F'_B which may lead to the bursting of the cable.

A small proportion of the tensile force is independant of the length of lay and may be absorbed by the stranding of the cores. This may lead to a widening of the stranding and thus to a tensile stress in the direction of the cable axis. Conductors of differing materials, conductor shapes and types will respond in different ways. The effects are however of minor importance and can be neglected when calculating the short-circuit withstand of a cable.

Appart from the areas of pressure and the tensile stress in surrounding covers the cable construction is a criterion for the determination of the mechanical short-circuit withstand of a cable.

Tensile Force F'_B

The tensile force in a cover surrounding the cores is

$$F'_B = \beta \, \frac{\mu_0}{2\pi} \frac{I_s^2}{a} \qquad (19.54)$$

and the permissible tensile force depends upon the tensile stress σ and the thickness δ_S of the covering

$$F'_{Bz} = \sigma \, \delta_S. \qquad (19.55)$$

The tensile force produced by the short-circuit F'_B must not exceed the permissible tensile force F'_{Bz}

$$F'_B \le F'_{Bz}.$$

From this the mechanical short-circuit capacity I_{sz} is

$$I_{sz} = \sqrt{\frac{2\pi a}{\beta \, \mu_0} F'_{Bz}} = \sqrt{\frac{2\pi a}{\beta \, \mu_0} \sigma \, \delta_s}. \qquad (19.56)$$

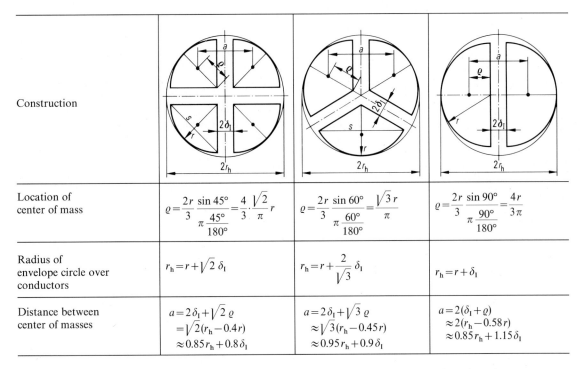

Construction			
Location of center of mass	$\varrho = \dfrac{2r}{3}\dfrac{\sin 45°}{\pi\dfrac{45°}{180°}} = \dfrac{4}{3}\cdot\dfrac{\sqrt{2}}{\pi}\,r$	$\varrho = \dfrac{2r}{3}\dfrac{\sin 60°}{\pi\dfrac{60°}{180°}} = \dfrac{\sqrt{3}\,r}{\pi}$	$\varrho = \dfrac{2r}{3}\dfrac{\sin 90°}{\pi\dfrac{90°}{180°}} = \dfrac{4r}{3\pi}$
Radius of envelope circle over conductors	$r_h = r + \sqrt{2}\,\delta_1$	$r_h = r + \dfrac{2}{\sqrt{3}}\,\delta_1$	$r_h = r + \delta_1$
Distance between center of masses	$a = 2\delta_1 + \sqrt{2}\,\varrho$ $= \sqrt{2}(r_h - 0.4r)$ $\approx 0.85\,r_h + 0.8\,\delta_1$	$a = 2\delta_1 + \sqrt{3}\,\varrho$ $\approx \sqrt{3}(r_h - 0.45r)$ $\approx 0.95\,r_h + 0.9\,\delta_1$	$a = 2(\delta_1 + \varrho)$ $\approx 2(r_h - 0.58r)$ $\approx 0.85\,r_h + 1.15\,\delta_1$

Fig. 19.26 Axial distances for the calculation of short-circuit forces in cables with sector-shaped conductors

For multi-core cable subjected to line-to-line or balanced three-phase short circuit as in Fig. 19.25 the value $\beta = 0.5$ apply. For sector-shaped conductors the axial distance a for axial centres of force points given in Fig. 19.26 may be used.

The permissible tensile stress σ for materials commonly used as covering (not given in all cases) are indicated below:

▷ outer protective covers (outer sheath) of PVC and PE,

▷ inner protective covers of extruded filler or lapped tapes,

▷ lead sheath ($\sigma \approx 13$ to 26 N/mm², see Table 3.2),

▷ aluminium sheath ($\sigma \approx 55$ to 65 N/mm², see Table 3.2),

▷ belted insulation in mass-impregnated cable ($\sigma \approx 56$ N/mm²) [19.18],

▷ spiral binder of steel-wire armour ($\sigma \approx 400$ N/mm²),

▷ transverse helical tape of copper wire screen ($\sigma \approx 120$ N/mm²),

▷ polymer or textile tapes as addition reinforcement.

It must be expected that the withstand capability for impact force as imposed by a short circuit is less than for continuous loading. It is recommended to use only two thirds of these quantities in calculation. The quantity of 56 N/mm² for belted insulation was determined for use under short circuit and can be used unmodified [19.18]. From the same source the quantity for cable lead of 14 N/mm² is applicable. The withstand of polymer coverings requires further consideration in respect of temperature (temperatures below 0 °C may also be critical, e.g. a short circuit on a cold cable; tested data are not available).

From the results of short-circuit tests on cables at operating temperature [19.19]

NYY 4×185 SM 0.6/1 kV

a quantity of 2 N/mm² was established for the PVC outer sheath. For the outer sheath at 20 °C a quantity of 8 N/mm² is selected; this is approximately two thirds of the test quantity of 12.5 N/mm² in DIN VDE 0207 Part 5.

For PE sheaths the same quantities may be used for calculation.

Multi-core cables can be reinforced, e.g. by the use of suitable polymer or textile tapes. These tapes are arranged in one or several layers (n_L) each comprising one or several overlapping tapes (n_{BL}) wound over the laid-up cores. The tensile strength $\sigma*$ of these tapes is generally quoted in N/m – in relation to the tape width b. If w' indicates the number of turns of the binder per unit length then the permissible tensile withstand F'_{Bz} of the binder is:

$$F'_{Bz} = w' \, \sigma* \, b. \tag{19.57}$$

With the condition

$$F'_{Bz} \geq F'_B$$

the mechanical short-circuit capacity of the binder is

$$I_{sz} = \sqrt{\frac{2\pi a}{\beta \mu_0} F'_{Bz}} = \sqrt{\frac{2\pi a}{\beta \mu_0} w' \, \sigma* \, b}. \tag{19.58}$$

The number of turns w' per unit length is

$$w' = \frac{n_L \, n_{BL}}{B \sqrt{1 + \dfrac{1}{\left(\dfrac{\pi \, d_{Bm}}{B}\right)^2 - 1}}} \tag{19.59}$$

with the mean diameter d_{Bm} of the binder and the factor B

$$B = n_{BL}(b + \varepsilon). \tag{19.60}$$

If the tapes are overlapped by an amount $\ddot{u}$ then $\varepsilon = -\ddot{u}$, if wound with a gap l then $\varepsilon = l$ must be inserted.

If the strength of the sheath and binder is considered then the sum of all permissible tensile forces to equations 19.55 and 19.57 becomes.

$$\sum F'_{Bz} = \sigma \, \delta_S + w' \, \sigma* \, b \tag{19.61}$$

and the mechanical short-circuit capacity I_{sz} is

$$I_{sz} = \sqrt{\frac{2\pi a}{\beta \mu_0} \sum F'_{Bz}}. \tag{19.62}$$

Surface Pressure F'_F

The surface pressure stresses are less in multi-core cables than in single-core cables because the forces are spread over the whole length of the cable instead of the clamping region only. In the tests, mentioned earlier, made on multi-core cables it was observed that deformation had occured at the core surface and this was attributed to the construction, e.g. the screens.

The permissible surface pressure F'_{Fz} of a conductor having a diameter d_L with an insulation thickness δ_I, which has a specific permissible pressure σ_F (see Section 19.4.3) can be found from

$$F'_{Fz} = d_L \, \delta_I \, \sigma_F. \tag{19.63}$$

This must always be greater or equal to the current related force F'_L to which the conductor is subjected

$$F'_{Fz} \geq F'_L.$$

Cable Construction

The best form of construction is one in which the cores, by the use of suitable fillers and sufficiently firm coverings, are secured in such a way that relative movement of one to another during short circuit is prevented.

In a cable with sector type conductors damage to the edges can arise when significant movement of the cores takes place. Filler material may move in between the cores and cause indentations. This must therefore be avoided particularly in the case of sector-shaped conductors.

Experience and Calculation Quantities

According to DIN VDE 0298 Part 2 no special strengthening measures are necessary for multi-core cables used on systems with low quantities of peak short-circuit currents. This applies to cables with rated voltages

0.6/1 kV for peak short-circuit currents of up to 40 kA,

above 0.6/1 kV for peak short-circuit currents of up to 63 kA.

It is assumed that short-circuit forces of these magnitudes in unarmoured PVC cables can be restrained by the outer sheath. Where short-circuit currents exceed these quantities special measures need to be taken.

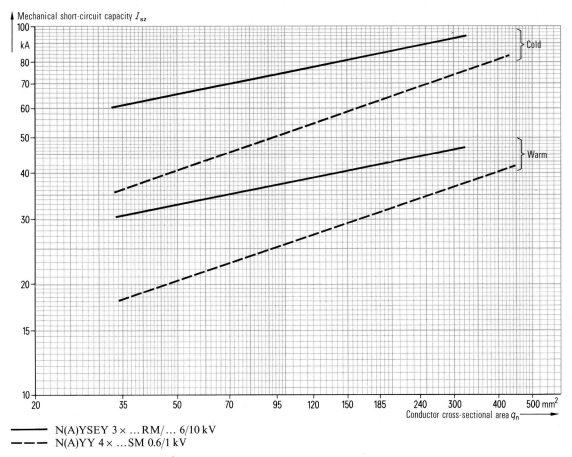

Fig. 19.27
Mechanical short-circuit capacity I_{sz} of multi-core cables with polymer insulation without special binding or strengthening

The mechanical short-circuit capacity for unarmoured PVC cables was calculated using the method described above and the results plotted in Fig. 19.27. The calculated values are supported by measured values from our own tests and also by [19.16] and [19.19], i.e. they are smaller than the measured values. Since the strength of the outer sheath of PVC alters with temperature by an unquantified amount these values are to be considered as guide values only. In this aspect they can also be applied to XLPE cables with PVC or PE outer sheaths.

Applying of a counter helix in a three-core cable with steel-wire armour is a suitable measure against the effect of short-circuit currents in excess of 100 kA peak. With unarmoured cables increased strength can be achieved by winding a suitable tape over the laid-up cores.

Mass-impregnated cables with lead sheaths are relatively well protected against impact forces. No failures are reported. The same comment applies to aluminium-sheathed cables.

For mass-impregnated cable without armour it is necessary to limit the short-circuit loading. As guide quantities for belted cables without armour for 0.6/1.0 kV and 6/10 kV the quantities given in Fig. 19.27 for cables at 20 °C may be used also.

Fixing Elements

With multi-core cables the effects of short-circuit forces are contained within the cable and do not stress the fixing elements. Fixing the cable by clamps is very much a matter of experience (see Section 29.5) where mainly the weight of the cable and a clear arrangement are major considerations. Cables supported on racks or similar structures arranged in a flat plane are fixed at points with due consideration of the thermo-mechanical forces.

Example 19.11

For the short-circuit condition of Example 19.10, but with a short-circuit duration of 0.6 s, a three-core cable with PVC insulation and copper conductor is to be selected.

For a cable in normal operation fully loaded to Table 19.5 the rated short-time current density is $J_{\text{thr}} = 115$ A/mm^2. With $t_{\text{kr}} = 1$ s

$$q_n \geq \frac{I_{\text{th}}}{J_{\text{thr}}} \sqrt{\frac{t_k}{t_{\text{kr}}}} = \frac{45.4 \times 10^3}{115} \sqrt{\frac{0.6}{1}} = 305.8 \text{ mm}^2.$$

$$(19.26\,\text{a})$$

Therefrom a cable of cross-sectional area 400 mm^2 must be selected. Since the cable is loaded only to a conductor temperature of 65 °C such that $J_{\text{thr}} = 119$ A/mm^2 a nominal cross-sectional area of

$$q_n \geq \frac{45.4 \times 10^3}{119} \sqrt{\frac{0.6}{1}} = 295.5 \text{ mm}^2 \qquad (19.26\,\text{a})$$

is calculated. Therefore a cross-sectional area of 300 mm^2 would suffice. The type reference of the cable selected is

NYSEY 3×300 RM/25   6/10 kV.

For further calculations the key data dimensions are:

Conductor diameter	$d_L = 20.6$ mm
Insulation thickness	$\delta_I = 4$ mm
Diameter over individual screens	$d_M = 32$ mm
Thickness of outer sheath	$\delta_S = 3.4$ mm
Overall diameter	$d = 79$ mm

The effective current generated force on the conductor is

$$F_L' = \alpha \frac{\mu_0}{2\pi} \frac{I_s^2}{d_M} = \frac{\sqrt{3}}{2} 0.2 \cdot 10^{-6} \frac{(110.2 \times 10^3)^2}{32 \times 10^{-3}}$$

$$= 65.7 \text{ kN/m}; \qquad (19.53)$$

The current generated force is therefore less than the force corresponding to the *permissible surface pressure*

$$F_{\text{Fz}}' = d_L \, \delta_I \, \sigma_F = 20.6 \times 10^{-3} \times 4 \times 10^{-3} \times 5 \times 10^9$$

$$= 412 \text{ kN/m}. \qquad (19.63)$$

In the outer sheath the peak short-circuit current causes a tensile force to develop of ($a = d_M = 32$ mm)

$$F_B' = \beta \frac{\mu_0}{2\pi} \frac{I_s^2}{a} = 0.5 \times 0.2 \times 10^{-6} \frac{(110.2 \times 10^3)^2}{32 \times 10^{-3}}$$

$$= 38.0 \text{ kN/m}. \qquad (19.54)$$

The permissible tensile force in the outer sheath of the cold cable at 20 °C with $\sigma = 8$ N/mm^2 is

$$F_{\text{Bz}}' = \sigma \delta_s = 8 \times 10^6 \times 3.4 \times 10^{-3} = 27.2 \text{ kN/m}$$

$$(19.55)$$

and the mechanical short-circuit capacity is

$$I_{\text{sz}} = \sqrt{\frac{2\pi a}{\beta \mu_0} F_{\text{Bz}}'}$$

$$= \sqrt{\frac{32 \times 10^{-3}}{0.5 \times 0.2 \times 10^{-6}} 27.2 \times 10^3} = 93.3 \text{ kA}.$$

$$(19.56)$$

For the cable at its operating temperature with $\sigma = 2$ N/mm^2 then

$$F_{\text{Bz}}' = 6.8 \text{ kN/m} \quad \text{and} \quad I_{\text{sz}} = 46.6 \text{ kA}.$$

The permissible tensile force F_{Bz}' and also the mechanical short-circuit capacity I_{sz} of the warm and cold cable are lower than the values required to satisfy this example. The cable must therefore be fitted with a binder to increase the mechanical short-circuit strength.

A binder tape is selected with a width $b = 60$ mm and a tensile strength $\sigma^* = 100$ N/m. The tape is applied in two layers ($N_L = 2$) with one tape per layer ($n_{\text{BL}} = 1$) and with 40% overlap. Thus

$$\varepsilon = -\ddot{u} = -b \frac{40}{100} = -60 \frac{40}{100} = -24 \text{ mm},$$

$$B = n_{\text{BL}}(b + \varepsilon) = 1(60 - 24) = 36 \text{ mm}, \qquad (19.60)$$

$$d_{\text{Bm}} \approx 2.15 \, d_M = 2.15 \times 32 = 68.8 \text{ mm},$$

$$w' = \frac{n_L \times n_{\text{BL}}}{B \sqrt{1 + \dfrac{1}{\left(\dfrac{\pi d_{\text{Bm}}}{B}\right)^2 - 1}}} \qquad (19.59)$$

$$= \frac{2 \times 1}{36 \cdot 10^{-3} \sqrt{1 + \dfrac{1}{\left(\dfrac{\pi \, 68.8 \times 10^{-3}}{36 \times 10^{-3}}\right)^2 - 1}}} = 54.8/\text{m}$$

and

$$F'_{Bz} = w' \sigma^* b = 54.8 \times 100 \times 10^2 \times 60 \times 10^{-3}$$
$$= 32.9 \text{ kN/m}. \qquad (19.57)$$

The permissible tensile force F'_{Bz} of the outer sheath of the cable at operating temperature is 6.8 kN/m and therefore the mechanical short-circuit capacity of the warm cable is

$$I_{sz} = \sqrt{\frac{2\pi a}{\beta \mu_0} \sum F'_{Bz}}$$

$$= \sqrt{\frac{32 \times 10^{-3}}{0.5 \times 0.2 \times 10^{-6}} (32.9 + 6.8) \, 10^3}$$

$$= 112.7 \text{ kA} \qquad (19.62)$$

19.4.3 Single-Core Cables and Fixing Methods

General

DIN VDE 0298 Part 2 stipulates, "single-core cables must be safely fixed to withstand the effects of peak short-circuit currents", which means they must withstand the stresses produced under short-circuit, remain in position such that neither the cable or fixing element is damaged. If in an installation high peak short-circuit currents are to be expected then after each short-circuit the fixing arrangement must be examined. Routine examination at regular intervals is recommended.

Normally cables are installed in trefoil formation. The following rules apply also for three single-core cables laid up. When installed in one plane side by side the load capacity may be increased but more fixings or clamping and space are necessary. If the load capacity of high-voltage cables of large cross-sectional areas is to be fully utilised, installation in one plane is appropriate with the consequental reduction of sheath/screen losses by single-point bonding or cross-bonding of sheaths/screens (see Section 21 page 322).

Cables installed side by side are fixed with cable clamps. The clamp must not form a magnetic circuit around any single cable otherwise hysterisis losses will develop thus feeding additional heat to the cable (Fig. 19.58).

If clamps are used in close proximity to one another on racks the longitudinal expansion forces of the cable, due to cyclic load currents or short-circuit cur-rents, especially where the wavy laying (see Section 19.3.3) cannot be fully accommodated, will be transmitted via the clamps to the cable racking.

For bunched cables either clamps or tape binding can be used (Fig. 19.29). The fixing and deflection of the racking is determined in Section 19.3.3 by consideration of the longitudinal expansion under temperature cycling and short circuit. For bunching cables by the use of binders or clamps the following applies.

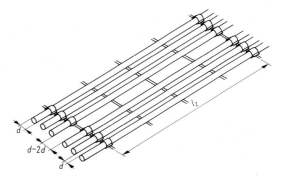

Fig. 19.28
Arrangement in one plane and fixing of single-core cables with non-magnetic clamps to a cable rack. On risers the maximum permissible fixing distance to DIN VDE 0298 part 1 is $l_z \leq 1.5$ m

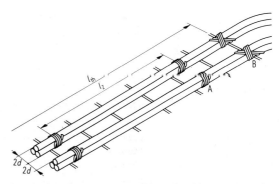

l_{th} Fixing distance on to cable rack taking account of expansion
A Tape binding (e.g. minimum 3 layers of self-adhesive tape) or strap clamps
B Cross tape binding (e.g. minimum 3 layers of self-adhesive tape) or fixing clamp

Fig. 19.29
Bunched arrangement and fixing of single-core cables on a cable rack. Greatest distance between fixings $l_z \leq 30d$ [19.31]

The distances between fixings, i.e. clamping, taping or binding distances, must be determined to three criteria (Fig. 19.30):

1. To avoid damage by kinking at the fixing points or by over bending half-way between the fixing points the permissible *deflection* Δ must not be exceeded. The cable is considered to be a rigid rod clamped at each end and uniformly loaded over the whole length by the force per unit length F'_L.

2. A further criterion is the *surface pressure* and the possible displacement of the conductor. This is related to the uniformly distributed force per unit length F'_L compressing the warm and therefore softend insulation in the areas of clamping or fixing tape width.

3. Lastly the *strength of the clamps* or bandages themselves must be considered as these are stressed in tension.

The smallest of the distances resulting from the above considerations is the one to be selected.

Bending Stress

It is common practice with cables to use the calculation method for a bar fixed at both ends and thus determine the transverse deflection Δ (Fig. 19.31) as a criterion. For cables with polymer insulation a permissible deflection of approximately 5% of the clamped length (l_D) has been established from tests. The quantities for cables with metal sheaths were calculated using the minimum bending radii to DIN VDE 0298 Part 1 Table 4 and these converted to linear measure. If it is required to avoid cables being bent more sharply than the specified permissible bending radii under short-circuit stress for all types of construction then clamping distances of six times the cable diameter should be used to Table 19.11.

Depending on the clamping distance a cable can be classified mechanically as falling between a bar and a rope. The oscillatory behaviour is of little consequence [19.28] and is therefore neglected.

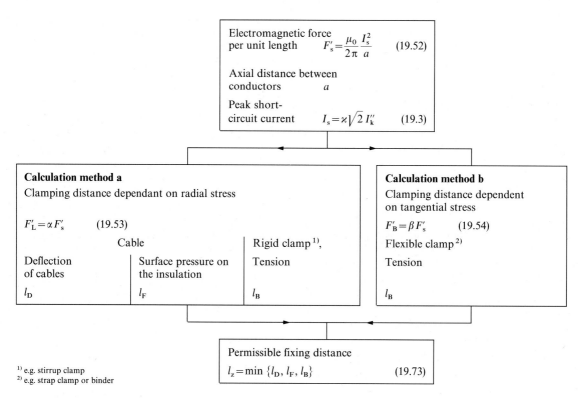

Fig. 19.30   Calculation of minimum fixing distance l_z for single-core cables.

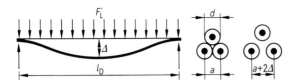

Fig. 19.31 Fixing distance related to deflection Δ

In order to consider the bending stiffness, which will differ depending on the type of construction, the cable is represented by the sum of the products $J_0 E$ of the major elements in the construction. The axial moments of inertia J_0 of the elements are illustrated in Fig. 19.32.

The E-modulus quantities for materials listed in Table 19.12 are estimated except for those of metals.

The influence of friction between the individual construction elements, likewise the spiral construction parts and compacting of stranded conductors, is not taken into consideration.

Table 19.11
Permissible deflection Δ and smallest fixing distance $l_{D\,min}$

Dimension	Type of cables		
	Cables with polymer insulation	Mass-impregnated cables	
		Lead sheath or corrugated aluminium sheath	Smooth aluminium sheath
Minimum bending radius to DIN VDE 0298, Part 1, Table 4	15 d	25 d	30 d
Permissible deflection Δ *as % of fixing distance*	5%	3%	2.5%
Minimum fixing distance $l_{D\,min}$	6 d		

Cylinder: solid conductor

$$J_0 = \frac{\pi d_L^4}{64}$$

Hollow cylinder:
Insulation, metal sheath, outer sheath

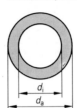

$$J_0 = \frac{\pi (d_a^4 - d_i^4)}{64}$$

Parallel cylinders:
Non-compacted stranded circular conductor

$$J_0 = \frac{\pi n_D d_D^4}{64}$$

n_D Number of wires

Compacted stranded circular conductor

$$J_0 = \frac{q_n^2}{4 \pi n_D}$$

n_D Number of wires
q_n Nominal cross-sectional area of conductor

Fig. 19.32
Calculation of axial moment of Inertia J_0 of conductor and coverings. (Friction between conductor wires and effect of laying-up is neglected)

Table 19.12
Elastic modulus of materials used in cable construction

Material	Elastic modulus E N/mm^2
Copper	115000
Aluminium	65000
Lead	17000
PVC insulation	10
PVC sheath	100
PE insulation	500
PE sheath	500
XLPE insulation	500

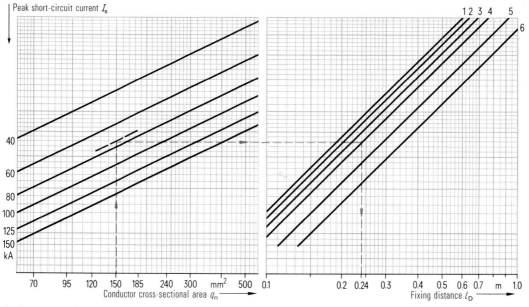

a) Copper conductor

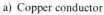

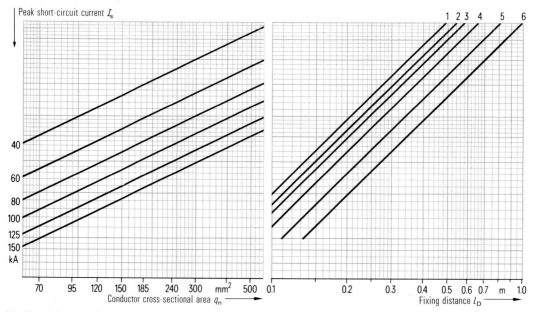

b) Aluminium conductor

Fig. 19.33
Fixing distance l_D, due to deflection Δ,
depending on construction and
on peak short-circuit current I_s
for single-core cables bunched in
trefoil formation

Curve	PVC cables	XLPE cables
1	0.6/1 kV	0.6/1 kV
2	3.6/6 kV	–
3	6/10 kV	3.6/6 kV
4	–	6/10 kV
5	–	12/20 kV
6	–	18/30 kV

If a force F'_L is equally distributed along a bar of length l which is fixed at both ends then the maximum deflection Δ at the centre of the bar (Fig. 19.31) is

$$\Delta = \frac{F'_L l^3}{384 \, \Sigma J_0 E}. \tag{19.64}$$

With the deflection k' in proportion to the length l

$$k' = \frac{\Delta}{l}$$

the permissible fixing distance dependant upon deflection is

$$l_0 = \sqrt[3]{\frac{384 (\Sigma J_0 E) \, k'}{F'_L}}. \tag{19.65}$$

The mean deflection measured over the length of the bar is

$$\Delta_m = 0.5\overline{3} \, \Delta. \tag{19.66}$$

The effective force can also be calculated with the corrected distance

$$a_k = a + 2 \times 0.5\overline{3} \, l_D \, k' \tag{19.67}$$

and one obtains the permissible fixing distance due to the deflection

$$l_D = \sqrt[3]{\frac{384 (\Sigma J_0 E) \, k'}{F'_L} \, \frac{a_k}{a}}. \tag{19.68}$$

The results of this cubic equation are

for $\quad q^2 + p^3 \geq 0 \quad l_D = \sqrt[3]{-q + \sqrt{q^2 + p^3}} +$
$$+ \sqrt[3]{-q - \sqrt{q^2 + p^3}},$$

for $\quad q^2 + p^3 < 0 \quad l_D = 2\sqrt{-p} \, \cos \frac{\varphi}{3},$

$$\varphi = \arccos \frac{q}{p\sqrt{-p}},$$

with $\quad q = \frac{l_0^3}{2} \quad$ and $\quad p = -l_0^3 \, 2 \times 0.5\overline{3} \, \frac{k'}{3a}.$

The results of the evaluation of equation 19.68 for the commonly used cables with polymer insulation are shown in Fig. 19.33.

Surface Pressure F_F

The hot pressure behaviour of cables with polymer insulation only permits a limited surface pressure without resulting in deformation. From short-circuit tests on 10 kV PVC cables of 240 mm² cross-sectional area having an insulation thickness of 4 mm where the conductor temperature increased from 70 °C to 160 °C a permissible specific area pressure σ_F of 5 N/mm³ was established. The reference area was determined from the projection of the conductor of diameter d_L to the insulation in the direction of force against the width b of the clamp. A further reference value is the insulation thickness δ_I.

To obtain the surface pressure rating the permissible specific surface pressure is multiplied by the insulation thickness so that – should deformation nevertheless occur – its magnitude is relative to insulation thickness.

Other types of insulation have not yet been investigated. For mass-impregnated cables the surface pressure should not be critical. For XLPE cables a value of 6 N/mm³ can be safely assumed.

At the area of fixing the force F'_L causes pressure on the insulation. The permissible specific pressure σ_F of the insulation material must not be exceeded. If F_{Fz} is the maximum permissible pressure it can be calculated from σ_F and area of the clamp width b (Fig. 19.34)

$$F_{Fz} = \sigma_F \, \delta_I \, d_L \, b, \tag{19.69}$$

from which can be found the fixing distance l_F relative to surface pressure from

$$l_F = \frac{F_{Fz}}{F'_L} = \frac{\sigma_F \, \delta_I \, d_L \, b}{F'_L}. \tag{19.70}$$

The results from equation 19.70 in respect of the cables with polymer insulation most commonly used are shown in Fig. 19.35.

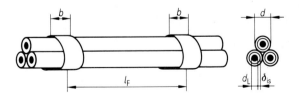

Fig. 19.34
Fixing distance l_F in respect of permissible surface pressure F_{Fz}

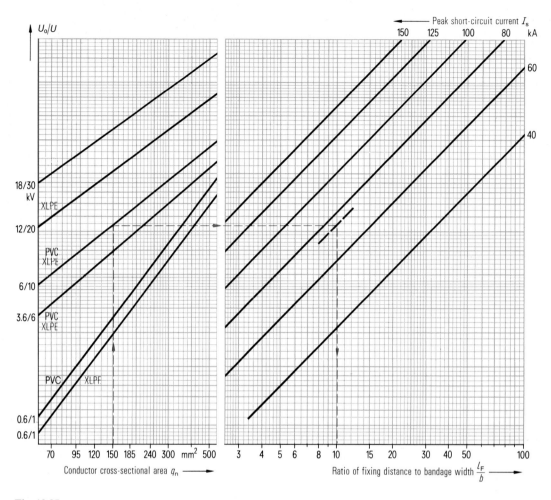

Fig. 19.35
Fixing distance l_F in relation to permissible surface pressure F_{Fz}, clamp or bandage width b and in consideration of type of cable construction and peak short-circuit current I_s of single-core cables bunched in trefoil formation

Stressing of Clamps and Binders

Clamps and binders are stressed under short-circuit current by radial forces (F'_L) and tangential forces (F'_B). In Fig. 19.36 the effects of these forces are illustrated for the instant of peak current in the upper cable. The upper cable is subjected to the radial force F'_L and each of the lower cables to the radial force $F'_L/2$ such that an equilibrium of forces results. The short-circuit forces tend to expand the clamp to a maximum circumference and this may lead to deformation or fracture.

The force $F'_L/2$ result in tension in the two fixing bolts shown in Fig. 19.36 to the same quantity.

In the top stirrup and the base plate tensile stresses are present due to the tangential force F'_B. The direction of forces within the stirrup are shown also in Fig. 19.36.

Stirrup clamps are normally stressed mainly by radial forces. Their strength or permissible stress F_{Lz} in respect of this force is indicated (see Table 19.13). The corresponding fixing distance is

$$l_B = \frac{F_{Lz}}{F'_L}. \tag{19.71}$$

Binders and strap clamps are stressed by the tangential force F'_B. This can be imagined if in Fig. 19.36 the stirrup and base plate is replaced by a binder.

The permissible stress F_{Bz} of some tapes and strap clamps is given in Table 19.13. The relevant fixing distance is

$$l_B = \frac{F_{Bz}}{F'_B}. \tag{19.72}$$

If l_z is the smallest fixing distance having taken all other criteria into account

$$l_z = \min\{l_D, l_F\}, \tag{19.73}$$

then the number of layers w of a binder or tape having a permissible tensile withstand force F_{Bz} is

$$w = l_z \frac{F'_B}{F_{Bz}}. \tag{19.79}$$

The fixing materials listed in Table 19.13 are not equally well suited for all areas of application. Glass fibre reinforced self adhesive polyester tape (e.g. scotch tape N° 45) are not always UV resistant or

Table 19.13
Permissible tensile forces F_{Lz} and F_{Bz} for fixing materials

Bunching and fixing material	Width b	Permissible Tensile force	
		F_{Lz}	F_{Bz}
	mm	kN	kN
Scotch-Band No. 45	25	–	0.6[1]
Permacell-Band P-162[2]	25	–	0.85[1]
Copper strap ECuF30 $\delta = 0.7$ mm	15	–	3.15
Copper strap ECuF30 with cyclops lock.			1.5
BICC steel strap clamp (three layers)	55	–	14.2[3]
BICC reinforced steel strap clamp (four layers)	55	–	29.6[3]
SIEMENS short-circuit proof stirrup clamp[4]			
Size 0 to 3	40	12[5]	–
size 4	50	12[5]	–
id-Technik, KS clamp	80	12.5[6]	–
id-Technik, KP clamp	80	25[6]	–

[1] at 100% safety with allowance for expansion
[2] in black which according to makers is largely weather-resistant
[3] makers figure
[4] Fixed to C rail 40 × 22
[5] 21 kN when using forged bolts
[6] makers liturature

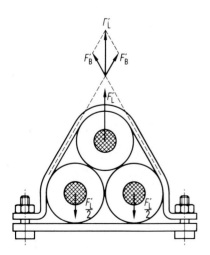

Fig. 19.36 Short-circuit forces in a stirrup clamp

weather-resistant and are thus not suited for use out-doors or in rooms with high humidity. The adhesive ages and the tapes lose their adhesive strength. Constant checking of the binders and refurbishment at regular intervals is therefore necessary.

Copper and steel straps are not suitable for the direct securing of cables to racks. If the cables are moved by the short-circuit forces in the axial direction the straps can cut into the outer sheath. However, these straps are suitable for use as bunching components.

The Siemens stirrup clamps are shaped to withstand the short-circuit forces and have radiused edges on the cable axis sides. They are resistant to weathering and suitable for use outdoors. Both bunching and fixing clamps are available.

Example 19.12

In this example the following main criteria are determined in respect of four ranges of system breaking power: the nominal conductor cross-sectional area q_n, fixing distance l_{th}, deflection for mounting h_1, as well for determining cable rack width h_2, the fixing distances l_D, l_F, l_B and from this l_z for bunching, also the fixing elements are selected.

It is assumed that rigid cable racking is to be used fitted with C profile rails 40×22 mm so that stirrup clamps can be installed safely with the use of hammer-head screws.

Working through Examples 19.1, 19.3, 19.9 and 19.10 (balanced three-phase short circuit) is recommended. To obtain solutions the relevant diagrams or curves are used.

The calculation for XLPE cable

N2XS2Y $1 \times \dots$ RM/...   6/10 kV

is made for a bunched installation on horizontal cable racks. The results are discussed for a breaking power of 250 MVA. The thermal load requires that a copper conductor is selected having a cross-sectional area of 95 mm^2.

In consideration of the thermal expansion of the cable Siemens stirrup clamps are required at distances l_{th} of 1.8 m on the racking. The deflection h_1 at installation is 27 mm which is approximately equal to the cable diameter. The racking width must allow for

a deflection of 48 mm which is equivalent to 1.7 times the cable diameter (see Fig. 19.22).

Interposed between the fixing clamps binders of copper tape are required with zyklop locks at pitches of 0.32 m. This calculated distance is determined by the withstand of the lock which is 1.5 kN. The fixing distance of 1.8 m is divisible by 6 and therefore conveniently gives a pitch for the binders of 0.3 m.

If Siemens stirrup clamps are used also for the bunching then $l_z = l_D = 0.34$ m. Both distances are measured between centre points of the clamps. For a clamp width of 40 mm the distance between clamps (pitch of clamps) could be selected to be $1.8/5 = 0.36$ m, since the free length on which the deflection is based is $0.36 - 0.04 = 0.32$ m which is still less than 0.34 m. It follows therefore that relative to a fixing distance of 1.8 m one bunching clamp can be saved.

The *fixing distances* l_b in the last lines of the example have been selected such that the clamps are distributed equally over the *distance in respect of thermal expansion* l_{th} without exceeding the *maximum permissible fixing distance* l_z.

		Units	Value			
Given						
U_n Rated voltage of switchgear		kV	10	10	10	10
S_a Rated breaking power of switchgear		MVA	250	500	750	1000
I_a Symmetrical breaking current						
$I_a = \dfrac{S_a}{\sqrt{3}\,U_n}$	Equ. (19.1)	kA	14.43	28.87	43.3	57.74
Far-from-generator short circuit						
$I_k'' = I_k = I_a$, $\varkappa = 1.8$						
t_k Short-circuit duration		s	0.5	0.5	0.5	0.5
Thermal short-circuit load capacity:						
From Fig. 19.7 one obtains $m = 0.1$, $n = 1.0$						
The thermal short-circuit duty						
$I_{th} = I_k'' \sqrt{m+n}$	Equ. (19.24)	kA	15.13	30.28	45.41	60.56
From Fig. 19.11a the required conductor cross-sectional area with the short-circuit capacity		mm^2	95	150	240	400
I_{thz} Thermal short-circuit capacity		kA	19	30	49	81
I_{kz}'' Permissible initial symmetrical shorth-circuit current		kA	18.5	29	47	78
d Outer diameter		mm	29	32	36	42
Mechanical short-circuit capacity:						
The short-circuit current is						
$I_s = \sqrt{2}\,\varkappa I_k''$	Equ. (19.3)	kA	36.73	73.49	110.22	146.98
from Fig. 19.25 one obtains $\alpha = \dfrac{\sqrt{3}}{2}$, $\beta = 1/2$						
The electromagnetic force per unit length is						
$F_s' = \dfrac{\mu_0}{2\pi}\dfrac{I_s^2}{a} = 0.2 \times 10^{-6}\dfrac{I_s^2}{a}$	Equ. (19.52)	kN/m	9.30	33.75	67.47	102.87
with $a = d$.						
The effective force on the conductor is						
$F_L' = \alpha F_s'$	Equ. (19.53)	kN/m	8.05	29.23	58.43	89.09
The effective force on the core covering is						
$F_B' = \beta F_s'$	Equ. (19.54)	kN/m	4.65	16.88	33.74	51.44
Fixing distances depending on the criteria						
① Fixing distance $l_z = 30d$ to Fig. 19.29		m	0.87	0.96	1.08	1.26
② Deflection l_D from Fig. 19.33		m	0.34	0.24	0.23	0.24
③ Surface pressure l_F/b from Fig. 19.35		–	28	10	6.7	5.2
④ Copper tape, $b = 15$ mm with Zyklop lock $F_{Bz} = 1.5$ kN						
$\quad l_B = \dfrac{F_{Bz}}{F_B}$	Equ. (19.72)	m	0.32	(0.09)	(0.04)	(0.03)

>

(Continued from page 313)

	Units	Value			
⑤ Siemens Stirrup clamp (Table 19.13)					
Size		0	0	1	3
Width b	mm	40	40	40	40
l_F with value $\dfrac{l_F}{b}$ to ③	m	1.12	0.40	0.27	0.21
with stud-bolt $F_{Lz}=12$ kN					
$l_B=\dfrac{F_{Lz}}{F'_L}$ Equ. (19.71)	m	1.49	0.41	(0.21)	(0.13)
with hammerhead screw $F_{Lz}=21$ kN					
$l_B=\dfrac{F_{Lz}}{F'_L}$ Equ. (19.71)	m	2.61	0.72	0.36	0.24
Maximum permissible fixing distance l_z					
$l_z=\min\{l_D, l_F, l_B\}$	m	0.31	0.24	0.23	0.21
to criteria		④	②	②	⑤ and ③

Deflection due to thermal expansion

Given:
Lowest ambient temperature $\vartheta_0 = 0\ ^\circ$C
Ambient temperature during installation $\vartheta_1 = 20\ ^\circ$C
Permissible operating temperature $\vartheta_2 = \vartheta_{Lr} = 90\ ^\circ$C

	Units	Value			
1) Selected is $l_{th}=50\ d$	m	1.45	1.6	1.8	2.1
with $h_0=0.5\ d$	mm	14.5	16	18	21
Deflection to Fig. 19.23 a					
– during installation $\dfrac{h_1}{d}=0.76$	–				
$h_1=0.76\ d$	mm	22	24	27	32
– for the sizing of width of cable rack $\dfrac{h_2}{d}=1.32$	–				
$h_2=1.32\ d$	mm	38	42	48	55
2) Correction of l_{th}, h_1 and h_2 in consideration of pitch of rungs of 300 mm, l_{thk}	m	1.8	1.8	1.8	2.4
e.g. $h_{1k}=22\ \dfrac{1.8}{1.45}=27$ mm h_{1k}	mm	27	27	27	37
e.g. $h_{2k}=38\ \dfrac{1.8}{1.45}=48$ mm h_{2k}	mm	47	47	48	63

Result

	Units	Value			
Fixing with Siemens stirrup clamps on cable rack on a pitch of l_{th}	m	1.8	1.8	1.8	2.4
Deflection when installed	mm	27	27	27	37
Deflection for sizing of cable rack	mm	47	47	48	63
Bunching by means of					
– Binders of copper strap at pitch of l_b	m	0.32	–	–	–
– Siemens stirrup clamps at pitch of l_b	m	0.36	0.225	0.225	0.20

19.4.4 Accessories

The short-circuit withstand of accessories is required to be proved by tests to DIN VDE 0278. Test are prescribed for peak short-circuit currents of 63, 80 and 125 kA. Accessories for use on circuits with peak short-circuit currents below 40 kA are not required to be tested. For installation the guidelines given in catalogues and installation manuals for the accessories, e.g. "Increasing the short-circuit strength" [19.32] must be observed:

In respect of the special measures for the increase of short-circuit withstand of an accessory the following are of note:

▷ Installation of cable lugs and making joints in the conductor by hard solder, welding or crimping.

▷ Securing of cable tails or core ends as they spread to the sealing ends by means of tie straps such that the outer cores are in V-formation in preference to a U-formation.

▷ Reinforcement of the dividing point of the cores in a multi-core cable by the use of a spreader clamp or by a tape serving of suitable material which if necessary is then cast into a resin block (to increase the mechanical strength a suitable mould can be used).

19.5 Symbols used in Formulae in Section 19

(Other symbols used are from Section 18)

All equations are written as quantitative equations. The symbols used in the formulae have a numerical value and also relate to a unit of measurement. The quantities are independant of the system of units, whichever compatable system of units is selected, e.g. (SI units) Système International d'Unités.

Longitudinally related dimensions "per unit value" such as electrical key data, thermal resistances and have symbols marked with an apostrophe stroke (DIN 1304). In the text only the electrical key data are marked as unit value. For thermal resistances and losses, to simplify the expression, this is not used.

Indices

b	Operating value
k	Short circuit
k1	Line-to-earth short circuit
k2	Line-to-line short circuit
k3	Balanced three-phase short circuit
n	Nominal value
r	Rated value
E	Earth fault, Earth

Symbols		*Unit*
a	Axial distance between conductors for calculation of short-circuit forces	m
b	Width of tape or band or clamp	m
b	Screen constant	$m/\sqrt{s}$
c	Specific heat	$J/(Km^3)$
c_i	Specific heat of covers under and over the screen	$J/(Km^3)$
d	Outer diameter of cable	m
d_{Mm}	Mean diameter of screen	m
d_D	Diameter of a single wire of either conductor or screen	m
d_L	Diameter of conductor	m
f_q	Factor for transverse load	–
h_o	Deflection due to thermal expansion of cable on no-load and at minimum ambient temperature	m
h_1	Deflection at temperature ϑ_1, during installation	m
h_2	Deflection due to thermal expansion at conductor temperature ϑ_2 for sizing of cable rack	m
k	Coefficient for calculation of thermal short-circuit capacity	$A\sqrt{s/m^2}$
k_1	Material coefficient for calculation of thermal short-circuit capacity	$A\sqrt{s/m^2}$
k_2	Temperature factor for calculation of thermal short-circuit capacity	–
k_3	Short-time rating factor	–
l	Length of cable or conductor	m
l_b	Fixing distance	m
l_{th}	Distance between clamp centres for fixing the cable to a structure in respect of thermal expansion	m
l_z	Maximum permissible clamping distance	m
l_B	Fixing distance relative to strength of clamp or bandage	m
l_D	Fixing distance relative to deflection	m
l_F	Fixing distance relative to surface pressure	m
m	Factor for the thermal effect of d.c. component	–
n	Factor for the thermal effect of a.c. component	–
n_D	Number of strands in a conductor	–

n_{BL}	Number of tapes per layer of a short-circuit restraining binder	–
n_L	Number of layers of tape in a short-circuit restraining binder	–
q_{el}	Electrical cross-sectional area	m^2
q_g	Cross-sectional area by weight	m^2
q_{geo}	Geometric cross-sectional area	m^2
q_n	Nominal cross-sectional area	m^2
r	Radius of the conductor sector (Fig. 19.26)	m
r_h	Radius of envelope circle over conductors	m
t_k	Duration of short-circuit	s
t_{kr}	Rated short-circuit duration (1 sec)	s
t_{kz}	Permissible short-circuit duration	s
w	Number of turns of a binder	–
w'	Number of turns of a binder relative to cable length	1/m
z	Exponent for calculation of k_3	–
$z+1$	Factor for longitudinal addition	–
A	Equivalent circumference of the cooling area of the screen	m
C'_E	Capacitance per unit length of cable, conductor-earth	F/m
C'_L	Capacitance per unit length of cable, conductor-conductor	F/m
C_{EN}	Conductor-earth-capacitance of network	F
C_{LN}	Conductor-conductor-capacitance of network	F
E	Modulus of elasticity	N/m^2
F	Conduction factor for heat transfer of the screen wires	–
F'_s	Electromagnetic force per unit length between two parallel conductors	N/m
F_{th}	Thermo-mechanical force	N
F'_B	Effective force (tensile) in a cover embracing all conductors of a system	N/m
F'_L	Effective force on a conductor	N/m
F_{Bz}	Permissible strength of a flexible clamp or binder in relation to $F_B = F'_B \cdot l_z$	N
F_{Fz}	Permissible surface pressure (force)	N
F_{Lz}	Permissible strength of a rigid clamp in relation to $F_L = F'_L \cdot l_z$	N
I_a	Symmetrical breaking current	A
I_{bb}	Ohmic component of operating current I_b	A
I_{bw}	Reactive component of operating current I_b	A
I_g	Geometric mean value of conductor currents under earth fault	A
I_{gM}	Geometric mean value of sheath currents under earth fault	A
I_k	Steady-state short-circuit current	A
I''_k	Initial symmetrical short-circuit current	A
I''_{kz}	Permissible initial symmetrical short-circuit current	A
I_s	Peak short-circuit current	A
I_{sz}	Mechanical short-circuit current carrying capacity	A
I_{th}	Thermally equivalent short-circuit current	A
I_{thr}	Rated short-time current	A
I_{thz}	Thermal short-circuit current carrying capacity	A
I_C	Capacitive line-to-earth fault current due to capacitance of network; in networks with insulated starpoint this value is the earth-fault current	A
I_{Cz}	Permissible capacitive line-to-earth fault current in consideration of temperature rise in cable	A
I_{CL}	Charging current in the conductor resulting from faulty part of network	A
I_F	Line-to-earth fault current, the current which flows at the point of fault, with only one earth fault, to earth or earthed parts	A
I_{Rest}	Earth-fault residual current which flows in networks with earth-fault compensation	A
J_o	Axial moment of inertia	m^4
J_{thr}	Rated short-time current density for $t_{kr} = 1$ s	A/m^2

P'_{vM}	Sum of sheath losses	W/m
R'_{L20}	d.c. resistance per unit length of conductor at 20 °C	Ω/m
R'_y	a.c. resistance per unit length of conductor	Ω/m
R'_{we}	Effective resistance per unit length of the cable under earth-fault condition	Ω/m
R'_A	d.c. resistance per unit length of armour	Ω/m
R'_M	d.c. resistance per unit length of metal sheath	Ω/m
R'_ϑ	d.c. resistance per unit length of conductor with conductor at temperature ϑ	Ω/m
S''_k	Initial symmetrical short-circuit power (short-circuit power)	VA
S_a	Breaking power of switch gear	VA
T'_{Kie}	Fictitious thermal resistance of cable under earth-fault condition	Km/W
U_e	Conductor-earth-voltage in normal operation $U_e = U_b/\sqrt{3}$	V
U_n	Rated voltage of network/system	V
X'_M	Reactance per unit length of metal sheath	Ω/m
α	Factor for mechanical stressing of conductor under short-circuit condition	–
α_{20}	Temperature coefficient for electrical resistance at 20 °C	1/K
α_{th}	Thermal expansion coefficient	1/K
β	Factor for the mechanical stressing of covers and layers under short circuit	–
β_i	Thermal penetration quantity	$W\sqrt{s}/Km^2$
δ	Wall thickness	m
δ_k	Effective screen wall thickness under short-circuit condition	m
δ_A	Thickness of one wire of flat steel-wire armour	m
δ_I	Thickness of insulation	m
δ_S	Thickness of outer sheath	m
λ_{1e}	Factor for calculation of sheath and screen losses under earth fault	–
λ_{2e}	Factor for calculation of armour losses under earth fault	–
ϱ	Thermal resistivity	Km/W
ϱ_i	Thermal resistivity of sheath over or under the screen	Km/W
ϱ_{20}	Electrical resistivity at 20 °C	Ω m
σ	Tensile strength	N/m^2
σ^*	Tensile strength of tape relative to tape width	N/m
σ_F	Permissible surface pressure related to insulation thickness	N/m^3
ϑ	Temperature	°C
ϑ_a	Conductor temperature at the commencement of short circuit	°C
ϑ_e	Permissible short-circuit temperature	°C
ϑ_o	Minimum ambient temperature	°C
ϑ_1	Ambient temperature during installation	°C
ϑ_2	Conductor temperature for sizing of cable rack	°C
μ	Factor for determination of symmetrical breaking current (to DIN VDE 0102)	–
$\varkappa$	Factor for determination of peak short-circuit current (to DIN VDE 0102)	–
Δ	Deflection	m
Λ	Factor for calculation of sheath or screen losses	–
Θ_0	Reciprocal value of temperature coefficient for electrical resistance α at 0 °C	K

19.6 Literature Referred to in Section 19

[19.1] Roeper, R.: Kurzschlußströme in Drehstromnetzen. 6. überarbeitete und erweiterte Auflage. Siemens AG, Berlin, München 1984

[19.2] Haubrich, H.-J.: Entwicklung der Kurzschlußströme in Energieübertragungs- und Energieverteilungsnetzen. etz-a Bd. 97 (1976), S. 286 bis 292

[19.3] Schreyer, L.: Kurzschlußströme in Hausinstallationen. Elektromeister Bd. 10 (1972), S. 593 bis 594

[19.4] Johannsen, A.: Einpoliger Fehler „Wischer"-kombinierte Sternpunktbehandlung in 10-kV-Mittelspannungsnetzen. etz Elektrotechn. Z. 102 (1981) H. 7, S. 367 bis 369

[19.5] Happoldt, H.; Oeding, D.: Elektrische Kraftwerke und Netze. Berlin: Springer, 1978

[19.6] Kahnt, R.; Lindner, H.: Untersuchungen über die Sternpunktbehandlung in Mittelspannungsnetzen der ländlichen Versorgung. Elektriz.-Wirtsch. 76 (1977) H. 13, S. 403 bis 409

[19.7] Winkler, F.: Strombelastbarkeit von Mittelspannungskabeln bei Erdschluß. etz Bd. 105 (1984), S. 178–182 und S. 518

[19.8] Rziha, V.E.: Starkstromtechnik Bd. 2, 8. Aufl., Berlin: Wilhelm Ernst und Sohn, 1960

[19.9] Vereinigung Deutscher Elektrizitätswerke VDEW: Die Kurzschlußleistung in Mittelspannungsnetzen und ihre Begrenzung. VWEW-Frankfurt am Main 1966

[19.10] Hecht, A.: Kurzschlußerwärmung von Kabeln. VDI-Forschungsheft Nr. 362, 1933

[19.11] Avramescu, A.: Über die Berechnung der Kurzschlußerwärmung. ETZ 59 (1938), S. 985 bis 988

[19.12] Kumlik, L.: Kurzschlußerwärmung von Kabeln. ETZ 56 (1935), S. 729 bis 731

[19.13] Goossens, R.F.: Versuche zur Bestimmung der zulässigen Kurzschlußerwärmung in 10-kV-Gürtelkabeln. Elektriz.-Wirtsch. 61 (1962), S. 393 bis 398

[19.14] Müller, H.Chr.: Zur Kurzschlußbelastbarkeit von 10-kV-Gürtelkabeln mit einer Aderisolation auf PVC-Basis. Dissertation, TH Karlsruhe, 1963.

[19.15] Brauch, A.; Steckel, R.-D.: Kurzschlußversuche an 10-kV-VPE-Kabeln mit drei verseilten und gemeinsam geschirmten Al-Massivsektorleitern. Elektriz.-Wirtsch. 74 (1975), S. 942 bis 945

[19.16] Kuhmann, H.; Rohde, N.: Dreiadrige 6- und 10-kV-PVC-Kabel mit Gießharz-Endverschlüssen für hohe dynamische Kurzschlußstrom-Beanspruchung. Elektriz.-Wirtsch. 73 (1974), S. 771 bis 775

[19.17] Winkler, R.: Erwärmung von Energiekabeln mit Kunststoffisolierung durch Kurzschlußströme. Elektriz.-Wirtsch. 76 (1977), S. 919 bis 921

[19.18] Gosland, L.; Parr, R.G.: A basis for short-circuit ratings for paper-insulated lead-sheathed cables up to 11 kV. ERA, Techn. Report F/T 195, 1960

[19.19] Gröber, R.; Stein, B.; Weitzel, H.G.: Versuche zur Ermittlung des Kurzerwärmungsfaktors und der Kurzschluß-Endtemperatur bei PVC-Kabeln mit Kupferleitern. Elektriz.-Wirtsch. 69 (1970), S. 578 bis 582

[19.20] IEC 20A Publication 949 (1988): Calculation of thermally permissible short-circuit currents taking into account non-adiabatic heating effects.

[19.21] Bott, J.; Schröder, G.: Zulässige Grenztemperaturen der Schirme von Starkstromkabeln bei Kurzschlußbelastung. ETZ B 21 (1969), S. 247 bis 250

[19.22] Bott, J.: Kurzschlußbelastbarkeit von Schirmen elektrischer Kabel. Bull. ASE 64 (1973), S. 918 bis 924

[19.23] Mildner, R.C.; Arends, C.B. and Woodland, P.C.: The short-circuit rating of thin metal tape cable shields. AIEE Trans., 87 (1968), S. 749 bis 758

[19.24] Arrighi, M.R.: Etude du comportement des ecrans metalliques a des courants de court-circuit. Bulletin Societe francaise des electriciens (1959), S. 649 bis 665

[19.25] EPRI EL-3014, Project 1286-2, Final Report, April 1983. "Optimization of the design of metallic shield-concentric conductors of extruded dielectric cables under fault conditions".

[19.26] Buchert, H.; Pays, M.; Pinet, A.: Short-circuit test results on HN 33-S-23 Cable screen. Electricite de France, Oct. 1984

[19.27] Eichhorn, K.Fr.: Messungen und Berechnung der transienten Temperaturen von Kabelschirm und Isolierung im Kurzschluß. IEE Second International Conference on Power Cables and Accessories 10 kV to 180 kV, Nov. 1986

[19.28] Foulsham, N.; Metcalfe, J.C.; Philbrick, S.E.: Proposals for installation practice of single-core cables. Proc. IEE, Vol. 121 (1974), S. 1168 bis 1174

[19.29] IEC Publication 724 (1984): Guide to the short-circuit temperature limits of electric cables with a rated voltage not exceeding 0.6/1.0 kV.

[19.30] Lehmann, W.: Elektrodynamische Beanspruchung paralleler Leiter. ETZ 76 (1955), S. 481 bis 488

[19.31] Siemens-Montagevorschrift E MA/MS 320-1.220 10/1983: Kabelverlegung auf Pritschen und Steigetrassen.

[19.32] Siemens-Montagevorschrift E MA/MS 319.220 9/1978: Erhöhte Kurzschlußfestigkeit an Endverschlüssen.

20 Resistance and Resistance per Unit Length of Conductor

20.1 Resistance per Unit Length on d.c.

The maximum permissible quantities for resistance per unit length R'_{20} on d.c. at 20 °C are given in DIN VDE 0295 (see Section 1).

The d.c. resistance varies with temperature ϑ. At the permissible operating temperature ϑ_{Lr} it becomes, according to equation 18.6,

$$R'_{\vartheta r} = R'_{20}[1 + \alpha_{20}(\vartheta_{Lr} - 20\ °C)]. \tag{20.1}$$

The temperature coefficients for the electrical resistance α_{20} at 20 °C for the conductor materials are:

$$\text{copper} \quad \alpha_{20} = 0.00393 \quad \text{and}$$

$$\text{aluminium} \ \alpha_{20} = 0.00403.$$

Rating factors for the term $1 + \alpha_{20}(\vartheta_{Lr} - 20\ °C)$ can be taken from Table 20.1.

Table 20.1
Rating factors for conductor resistance for temperatures other than 20 °C

Conductor temperature °C	$1 + \alpha_{20}(\vartheta_{Lr} - 20\ °C)$	
	copper	aluminium
20	1.0	1.0
25	1.0197	1.0202
30	1.0393	1.0403
35	1.0590	1.0604
40	1.0786	1.0806
45	1.0983	1.101
50	1.118	1.121
55	1.138	1.141
60	1.157	1.161
65	1.177	1.181
70	1.197	1.202
75	1.216	1.222
80	1.236	1.242
85	1.255	1.262
90	1.275	1.282
95	1.295	1.302
100	1.314	1.322

20.2 Resistance per Unit Length on a.c.

When operated on single-phase and three-phase alternating current, additional frequency dependent losses develop in the conductor and metallic coverings of the cable. These can be expressed as an effective resistance per unit length R'_w. From equation 18.5 for a temperature increase of the conductor to the permissible operating temperature ϑ_{Lr} is

$$R'_{wr} = R'_{\vartheta r}(1 + y_s + y_p)(1 + \lambda_1 + \lambda_2). \tag{20.2}$$

The factors for the resistance increase of the conductor due to skin effect y_s, proximity effect y_p, sheath loss factor λ_1 and the armour loss factor λ_2 must be calculated according to [18.2, 18.7 or 18.8].

The effective resistance per unit length R'_w is practically constant for the permissible operating temperature ϑ_{Lr}. Its quantity changes insignificantly with varying temperatures of screen, sheath or armour. For project design therefore the resistances per unit length R'_{20}, $R'_{\vartheta r}$ and R'_{wr} for 50 Hz may be taken from the tables in Part 2 of this work [18.17].

The additional losses increase with increasing cross-sectional area. They are very small for plastic cables without screen or armour. In multi-core cables with screen, metal sheath or armour higher additional losses occur.

In single-core cables additional losses develop due to induced currents in the metal sheath, screen or in the non-magnetic armour if these are bonded, as is common practice, at joints and are earthed at sealing ends (see Sections 21.2.1 and 21.2.2). Single-core cables with steel-wire armour enclosing the full circumference develop significant additional losses. If the armour wires are not touching and are equally spaced around the cable circumference (open armouring) the additional losses are markedly reduced.

Steel-tape armour which is commonly used on three-core cables must not be used on single-core cables for operation on a.c. circuits as this would result in unacceptably high additional losses and would increase the inductance per unit length significantly.

In unarmoured cables and also in single-core cables with non-magnetic armour, the additional losses can be kept to very low levels, providing the cables are not installed side by side with about 15 cm pitch but are bunched touching in treefoil.

20.3 Current Related Losses

The ohmic losses to equation 18.3 for the permissible operating temperature $\vartheta_{L,r}$ and the rated current I_r can be found from

$$P'_{ir} = nI_r^2 R'_{wr}.\tag{20.3}$$

Quantities for P'_{ir} can be taken from the tables [18.17]. The ohmic losses for a current I, other than I_r can be calculated using the following equation

$$P'_i = P'_{ir}\left(\frac{I}{I_r}\right)^2.\tag{20.4}$$

21 Inductance and Inductance per Unit Length

21.1 Inductance per Unit Length of a Conductor System

The inductance L of a solid circular conductors of infinite length against radius and distance of conductors arranged as shown in Fig. 21.1 [21.1, 21.2] is

$$L = \frac{\mu_0}{2\pi} l \ln \frac{a}{\varrho} \tag{21.1a}$$

$$= \frac{\mu_0}{2\pi} l \left(\frac{1}{4} + \ln \frac{a}{r_L} \right). \tag{21.1b}$$

Where

μ_0 magnetic space costant
   $(\mu_0 = 4\pi 10^{-7}\,\text{H/m})$
a Axial spacing between conductors
ϱ Equivalent radius of conductors
   $(\varrho = 0.779\, r_L)$
r_L Radius of conductors
l Length of conductors

The inductance per unit length L' per conductor in a conductor loop is

$$L' = \frac{\mu_0}{2\pi} \ln \frac{a}{\varrho} \tag{21.2a}$$

$$= \frac{\mu_0}{2\pi} \left(\frac{1}{4} + \ln \frac{a}{r_L} \right). \tag{21.2b}$$

With equation 21.2 the inductance per unit length of a conductor in a single-phase system can be calculated assuming that one conductor acts as flow and the other as return. Equations for the inductance per unit length of conductors in a symmetrical three-phase system are given in Table 21.1.

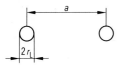

Fig. 21.1
Arrangement of two infinitely long solid conductors

In single-core cables with stranded conductors the conductor radius deviates only very little from the radius of a solid conductor of the same cross-sectional area, such that the equations can be applied directly for these.

Example 21.1

For a circuit loop comprising circular conductor of radius $r_L = 9\,\text{mm}$ located at an axial spacing of $a = 20\,\text{mm}$ the inductance per unit length is

$$L' = \frac{\mu_0}{2\pi} \ln \frac{a}{\varrho} = \frac{4\pi 10^{-7}}{2\pi} \ln \frac{20}{0.779 \times 9}$$

$$= 2.10 \times 10^{-7}\,\text{H/m} = 0.21\,\text{mH/km}. \tag{21.2a}$$

21.2 Single-Core Cables

21.2.1 Earthed at Both Ends

The inductance of single-core cables per unit length is dependent mainly on their axial spacing. For commonly used arrangements and spacings the inductance per unit length can be taken from tables in part 2 of this book. For conditions not provided for in the tables the quantities can be calculated by using the equations given in Table 21.1, providing that if the cables have metal sheaths or screens these are earthed only at one end. It is possible in principle to calculate the individual inductances per unit length of each of several systems of single-core cables installed in proximity. The rules for these calculations, however, are not included in this work. By suitable arrangement of cables and systems the inductance per unit length can be maintained approximately equal for each of them (see Section 21.2.2).

For cables containing metal sheaths or screens of copper wire or tape (here also referred to as a sheath) the calculation becomes more involved and in the following formulae mean values are determined. The conductor circuit and sheath circuit behave as the

Table 21.1
Inductance per unit length of single-core cables without sheaths or screens in a symmetrical three-phase system

$$\frac{\mu_0}{2\pi} = 0.2 \times 10^{-6} \ \text{H/m}$$

$$L' = \frac{\mu_0}{2\pi} \ln \frac{a}{\varrho} \tag{21.2a}$$

$$L'_R = \frac{\mu_0}{2\pi} \left(\ln \frac{\sqrt{a_{RS} \times a_{TR}}}{\varrho} + j\sqrt{3} \ln \sqrt{\frac{a_{RS}}{a_{RT}}} \right) \tag{21.3}$$

$$L'_S = \frac{\mu_0}{2\pi} \left(\ln \frac{\sqrt{a_{ST} \times a_{RS}}}{\varrho} + j\sqrt{3} \ln \sqrt{\frac{a_{ST}}{a_{RS}}} \right) \tag{21.4}$$

$$L'_T = \frac{\mu_0}{2\pi} \left(\ln \frac{\sqrt{a_{TR} \times a_{ST}}}{\varrho} + j\sqrt{3} \ln \sqrt{\frac{a_{TR}}{a_{ST}}} \right) \tag{21.5}$$

Mean inductance

$$L'_m = \frac{\mu_0}{2\pi} \ln \frac{\bar{a}}{\varrho} \tag{21.6}$$

Mean geometric distance in mm:

$$\bar{a} = \sqrt[3]{a_{RS} \times a_{ST} \times a_{TR}} \tag{21.7}$$

$$L'_R = L'_S = L'_T = L' \tag{21.8}$$

$$L' = \frac{\mu_0}{2\pi} \ln \frac{a}{\varrho} \tag{21.9}$$

$$a_{RS} = a_{ST} = a_{TR} = a$$

$$L'_R = \frac{\mu_0}{2\pi} \left(\ln \frac{\sqrt{2}\,a}{\rho} - j\sqrt{3} \ln \sqrt{2} \right) \tag{21.10}$$

$$= \frac{\mu_0}{2\pi} \ln \frac{a}{\varrho} + 0.0693 \times 10^{-6} - j\,0.12 \times 10^{-6} \tag{21.11}$$

$$L'_S = \frac{\mu_0}{2\pi} \ln \frac{a}{\varrho} \tag{21.12}$$

$$L'_T = \frac{\mu_0}{2\pi} \left(\ln \frac{\sqrt{2}\,a}{\varrho} + j\sqrt{3} \ln \sqrt{2} \right) \tag{21.13}$$

$$= \frac{\mu_0}{2\pi} \ln \frac{a}{\varrho} + 0.0693 \times 10^{-6} + j\,0.12 \times 10^{-6} \tag{21.14}$$

Mean inductance

$$L'_m = \frac{\mu_0}{2\pi} \ln \frac{\bar{a}}{\varrho} \tag{21.15}$$

Mean geometric distance in mm:

$$\bar{a} = \sqrt[3]{2}\,a = 1.26 \tag{21.16}$$

323

windings of a 1:1 transformer. A voltage to earth E_0' induced in the sheaths:

$$E_0' = I\omega M'. \qquad (21.17)$$

The mutual inductance per unit length M' per conductor is

$$M' = \frac{\mu_0}{2\pi} \ln \frac{2a}{d_{Mm}} \qquad (21.18)$$

with

$$d_{Mm} = d_M - \delta_M \qquad (21.19)$$

I Current in conductor
ω Angular frequency
   (at 50 Hz is $\omega = 2\pi f = 100\,\pi/s$)
d_{Mm} Mean diameter of sheath
d_M Outer diameter of sheath
δ_M Thickness of sheath
a Axial distance between conductors
   If the cables are laid side by side in flat formation the calculation must be made using the mean geometric distance $\bar{a}$

$$\bar{a} = \sqrt[3]{2}a \quad \text{(see Table 21.1)}.$$

In systems where joints are connected through and with the sheath earthed at one end only the voltage E_0' to earth will appear at the open end. The induced voltage can be measured between sheaths in a single-phase a.c. system as

$$E' = 2E_0' \qquad (21.20)$$

and in a three-phase a.c. system as

$$E' \approx \sqrt{3}\,E_0'. \qquad (21.20)$$

If the sheaths of the cables are bonded at both ends the induced voltage will cause a sheath current I_M to flow

$$I_M = \frac{E_0'}{\sqrt{R_M'^2 + X_M'^2}} \qquad (21.22)$$

with the a.c. reactance per unit length

$$X_M' = \omega M'. \qquad (21.23)$$

The sheath currents create a magnetic field which opposes that of the conductor currents. The inductance per unit length of each conductor is thereby reduced by the additional inductance per unit length $\Delta L'$,

$$\Delta L' = M' \frac{1}{\left(\dfrac{R_M'}{X_M'}\right)^2 + 1} \qquad (21.24)$$

and the resistance per unit length per conductor is increased by the additional resistance per unit length

$\Delta R'$ where

$$\Delta R' = R_M' \frac{1}{\left(\dfrac{R_M'}{X_M'}\right)^2 + 1}. \qquad (21.25)$$

At this the resistance per unit length of the sheath becomes

$$R_M' = \frac{1}{q_M \varkappa} \qquad (21.26)$$

with q_M as sheath cross-sectional area. For the conductivity of the sheath $\varkappa$ the value at the operating temperature must be used. For rough calculation the following values are approximate:

Metal sheath	Conductivity $\varkappa$ under load approx. $1/\Omega m$
Copper screen	50×10^6
Aluminium sheath	32×10^6
Lead sheath	4.2×10^6

Example 21.2

The characteristic quantities for a cable N2XS2Y 1×240 rm/25 12/20 kV installed in one plane are:

Conductor radius	$r_L = 9.3$ mm
Outer diameter of screen	$d_M = 34.0$ mm
Thickness of screen (including transverse helical tape)	$\delta_M = 0.7$ mm
Cross-sectional area of screen	$q_M = 25$ mm^2
Conductivity of screen at 20 °C	$\varkappa_{20} = 56 \times 10^6/\Omega m$
Longitudinal addition of screen	5%
Outer diameter of cable (reference diameter, see Part 2)	$d = 39.8$ mm

If the three cables are installed side by side with a gap of 70 mm (brick thickness) the axial distance between the cables becomes

$$a = d + 70 \approx 40 + 70 = 110 \text{ mm}.$$

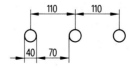

For $a/r = 110/9.3 = 11.83$ from Fig. 21.2 we get

$$L' - \frac{\mu_0}{2\pi} \ln \frac{a}{\varrho} \text{ H/m}$$

$$= 0.54 \text{ mH/km (per conductor)}. \qquad (21.2\,\text{a})$$

For the conductor inductances per unit length with only one end bonded and earthed, from Table 21.1 we get

$$L = \frac{\mu_0}{2\pi} \ln \frac{a}{\varrho} = 0.54 \text{ mH/km}$$

$$L'_R = \frac{\mu_0}{2\pi} \ln \frac{a}{\varrho} + 0.0693 \times 10^{-6} - j0.12 \times 10^{-6} \text{ H/m}$$

$$= L' + 0.0693 - j0.12$$
$$= 0.54 + 0.0693 - j0.12 \text{ mH/km}$$
$$= 0.61 - j0.12 \text{ mH/km}, \qquad (21.10)$$

$$L'_S = 0.54 \text{ mH/km} \qquad (21.11)$$

$$L'_T = \frac{\mu_0}{2\pi} \ln \frac{a}{\varrho} + 0.0693 \times 10^{-6} + j\,0.12 \times 10^{-6} \text{ H/m}$$

$$= 0.54 + 0.0693 + j\,0.12 \text{ mH/km}$$
$$= 0.61 + j\,0.12 \text{ mH/km}. \qquad (21.12)$$

The mean geometric distance is

$$\bar{a} = \sqrt[3]{2}\,a = \sqrt[3]{2} \times 110 = 138.6 \text{ mm}. \qquad (21.16)$$

The mean inductance per unit length obtained from Fig. 21.2 with

$$\frac{\bar{a}}{r} = \frac{138.6}{9.3} = 14.90$$

is

$$L'_m - 0.59 \text{ mH/km (per conductor)}.$$

To determine the induced voltage the mutual inductance per unit length must be calculated. The mean diameter of the aluminium sheath is

$$d_{Mm} = d_M - \delta_M = 34.0 - 0.7 = 33.3 \text{ mm};$$

with

$$\frac{\bar{a}}{d_{Mm}} = \frac{138.6}{33.3} = 4.16,$$

one obtains from Fig. 21.3

$$M' = 0.424 \text{ mH/km (per conductor)}.$$

With this for an induced voltage to earth for a cable length of 1 km and a short-circuit current of 30 kA.

$$E'_0 = 30 \times 10^3 \times 100\pi\,0.424 \times 10^{-3} = 3996 \text{ V/km}. \qquad (21.17)$$

The voltage between the open ends of the two screens is

$$E' = \sqrt{3} \times 3996 = 6921 \text{ V/km}. \qquad (21.21)$$

If the screens are bonded and earthed at both ends the inductance is reduced and the effective resistance of the conductor is increased.

With an a.c. reactance per unit length of

$$X'_M = 100\pi\,0.424 \times 10^{-3} = 0.133 \text{ Ω/km} \qquad (21.23)$$

and a screen resistance per unit length

$$R'_{M\,20} = \frac{1.05 \times 10^3}{25 \times 10^{-6} \times 56 \times 10^6} = 0.750 \text{ Ω/km} \qquad (21.26)$$

$$R'_M = 0.750(1 + 0.00393(80 - 20)) = 0.927 \text{ Ω/km}$$

the inductance per unit length of the conductors is reduced by the additional inductance per unit length

$$\Delta L' = 0.424 \frac{1}{\left(\frac{0.927}{0.133}\right)^2 + 1} = 0.009 \text{ mH/km}. \qquad (21.24)$$

The effective inductance per unit length is:

$$L' = L'_m - \Delta L' = 0.59 - 0.009$$
$$= 0.581 \text{ mH/km (per conductor)}.$$

The additional resistance per unit length of the conductor caused by the reaction of the induced currents in the screens is

$$\Delta R' = 0.927 \frac{1}{\left(\frac{0.927}{0.133}\right)^2 + 1}$$

$$= 0.0187 \text{ Ω/km (per conductor)}. \qquad (21.25)$$

For rough calculation with

$$\varkappa = 50 \times 10^6 \quad 1/\text{Ωm}$$

the relevant cable impedances become

$$R'_M = 0.989 \quad \text{Ω/km}$$
$$\Delta L' = 0.0075 \quad \text{mH/km}$$
$$L' = 0.582 \quad \text{mH/km}$$
$$\Delta R' = 0.0176 \text{ Ω/km}.$$

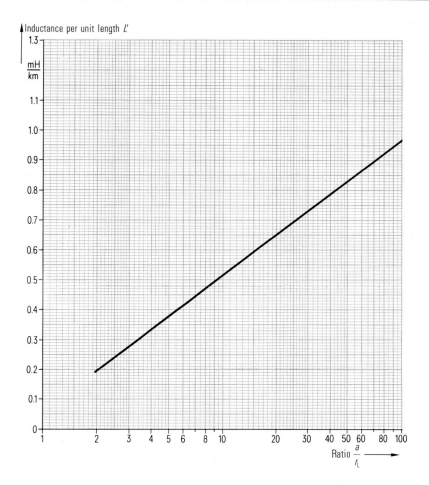

Inductance per unit length L'

$\dfrac{mH}{km}$

Ratio $\dfrac{a}{r_L}$

Fig. 21.2
Inductance L' per unit length of one conductor in a conductor loop to equation 21.2 b

21.2.2 Arrangement of Cables

If two busbar systems are coupled by a number of single-core cables in parallel, the inductance of each of the parallel cables should be equal as far as is possible to ensure equal current load sharing between the cables. This inductance is most unbalanced when cables of one phase are grouped and laid side by side next to one another. A preferred arrangement is for the cables of different phases to be grouped into systems such that the spacing of cables within one system is less than the spacing between systems. A fully symmetrical load sharing can be more readily achieved by the use of three-core cables because here, due to the uniformly laid-up cores, in normal operation there is no inductive reaction with neighbouring cables.

The clearance between two systems of single-core cables should be approximately twice the axial spacing of individual cables in a system. In addition the sequence of phases within a system is most important. Depending on the number of three-phase systems the following phase relationship is recommended:

RST TSR RST TSR etc.

With this arrangement the conductor inductances of the paralleled cables within a phase are approximately equal. The inductances of the phases R, S and T are however different. This is less of a disadvantage than unequal inductances between the parallel connected cables in any one phase. The following arrangement is most unfavourable

RST RST RST

since this results not only in unsymmetrical phase inductances between R, S and T but also unbalanced inductances of the paralleled cables in any phase.

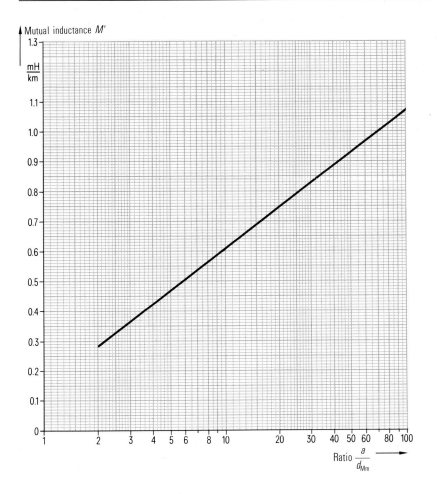

Mutual inductance M'

$\frac{mH}{km}$

Ratio $\dfrac{a}{d_{Mm}}$

Fig. 21.3
Mutual inductance per
unit length M' of a conductor due to inductance
of the sheath to equation
21.18

In installations on racks or cable trays the cables of the same phase must not be arranged side by side but on different platforms. Between the racks a distance of 300 mm or more should be maintained. On each platform room should be given to allow two different systems with opposing phase sequences to be installed.

RST   TSR

RST   TSR

RST   TSR etc.

With this arrangement the inductances of the paralleled cables is reasonably equal. The inductances of the phases, however, differ from one another but this is not so important as these busbar linking cable runs are normally short. Should only one system be necessary the triangular (trefoil) arrangement

  S
R T

will result in equal phase inductances. Where several systems are installed each in trefoil grouping it is of advantage to arrange the rotation

S   S   S   S
R T T R R T T R etc.

The trefoil arrangement of several systems above one another is not recommended since the inductances of the paralleled cables differ greatly one from another.

With single-core cables installed in air particular attention must be given to adequate fixing at short intervals so that the cables are not, in the event of a short circuit, because of the force developed by the high peak short-circuit current, hurled from their positions. The mechanical short-circuit capacity can be determined by reference to Section 19.4.

21.2.3 Earthing from Either One or Both Ends of Metal Sheath or Screen

When single-core cables are operated on single-phase or three-phase a.c. systems voltages are induced in the metal sheaths which are directly proportional to the current in the conductor, the frequency, the coefficient of mutual inductance between conductor and sheath and also the length of cable run.

In normal practical installations incorporating joint boxes the metal sheaths are bonded through and are bonded and earthed at the sealing ends completing a circuit for induced currents to flow in the metal sheaths. These create an additional loss and therefore a reduction in load capacity when on single-phase or three-phase operation as compared to their capacity on d.c. operation. The values shown in the tables of Part 2 take account of the reduction in load capacity caused by these additional sheath losses.

Non-magnetic armour or screens (e.g. in single-core polymer cables) also create similar additional losses when earthed at both ends and likewise this must be considered when determining load capacity.

If for economic reasons or where the load capacity of the cable is critical and these losses must be avoided, the metal sheaths or screens of the cables and their respective sealing ends must be earthed at one end only. At the other end of the cables the sealing ends must be fixed to a rack insulated from earth and from one another. This, however, allows the full induced voltages to appear at the non-earthed end between the metal sheaths/screens and earth. On a three-phase system the voltage between screens will be $\sqrt{3}$ times the induced voltage in each cable. In order to keep these voltages, which are proportional to cable length, within acceptable limits cables which are earthed at one end must be of short length (normally less than 500 m). In longer cable runs the individual voltages must be divided by installing cable joints which contain a built-in insulating barrier. The metal sheaths (screens) in the individual sections must be earthed only at one end.

Apart from the higher cost of installation for single-point earthing of sheath/screen further disadvantages exist. The earthing is less effective and if the cable is used to connect one station with another an earth connection cannot be made between them. This can result in more elaborate earthing measures being required at the stations. Since the induced current in the sheath is suppressed the reduction in conductor inductance is less. Special attention must be paid to the induced voltages which can occur at the free ends during short circuit and during switching (see also DIN VDE 0141).

21.2.4 Cross-Bonding of the Sheaths, Transposition of the Cables

There is another method of suppressing the induced current in the sheaths which leaves only a residual current: The metallic continuity of the sheaths is interrupted at the joints and sequentially cross-bonded as shown in Fig. 21.4a. This corresponds to a transposition of the sheaths, similar to the transposition of asymmetrically arranged overhead lines. In addition the cables themselves may be transposed (Fig. 21.4b). By these methods the earthing conditions are not impaired and the dangers of high induced voltages at the free ends are avoided. However the reduction factor for inductive interference is less.

Earthing at one end only, as well as cross-bonding of the sheaths or transposition are at present employed with extra-high-tension cables only, due to the high cost of installation and of specially constructed joint boxes as well as of the additional maintainance required. In medium-voltage installations the losses related to the current may be kept within reasonable limits by using the most favourable type of cable (see Fig. 21.5) [21.3 to 21.6].

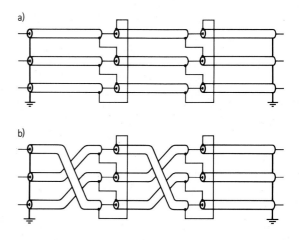

Fig. 21.4
Single-core cable system in flat formation with
(a) sheaths/screens sequentially cross-bonded and
(b) conductors sequentially transposed

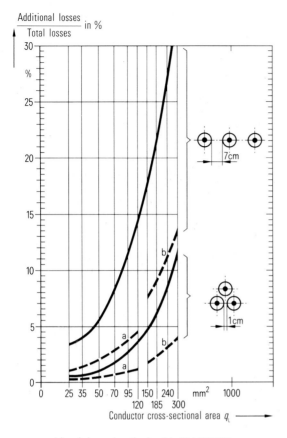

Additional losses / Total losses in %

— Aluminium-sheathed cable NAKLEY
— — — XLPE cable NA2XS2Y

a Screen cross-sectional area 16 mm²
b Screen cross-sectional area 25 mm²

Fig. 21.5
Additional losses as percentage of total losses
in single-core 20 kV cables

21.3 Multi-Core Cables

Basically the equations shown in Table 21.1 also apply to multi-core cables. The conductors are not normally circular and the cores are often arranged in a non-symmetrical formation as, e.g. in cables with four cores for 0.6/1 kV. The armour of the multi-core cables is normally of a ferromagnetic material which, because of the field concentrating effect increases the inductance. A correction of the calculation values using the results of measurements is therefore necessary. The inductance per unit length of the commonly used types is given in Part 2. The accuracy of these quantities is adequate for practical requirements.

21.4 Zero-Sequence Impedance and Zero-Sequence Impedance per Unit Length

Generally the zero-sequence impedance of cables is not a fixed quantity. It not only depends on the design of the cable, but also on environmental conditions. Other cables running in parallel, pipe lines, railway lines, etc., in addition to sheath, screen and armour of the cable itself, play their part here. The zero-sequence impedance of steel-armoured cables is furthermore dependent on the currents flowing in the conductor-earth loop (zero currents) due to the magnetization of the steel tapes or wires.

These are the reasons why for high-voltage cables it is not possible in praxis to calculate the zero-sequence impedance. Measurements on installed cables have to be carried out, in order to obtain usable values. It has been shown by numerous measurements, that the range of impedance values may be very wide, depending on the type of cable. The matter is different with multi-core cables (including cables with a concentric fourth conductor), which are normally used as low-voltage cables only, in so far as in this case the calculated values come very close to those obtained by measuring installed cables [21.7 to 21.11].

329

Table 21.2
Zero-sequence impedance. Measured quantities for cables installed in towns with test currents of 100 to 300 A

Rated voltage U_0/U kV	Type of construction	Type reference	Zero-sequence impedance Ω/km
6/10	Belted cables with paper insulation and lead sheath aluminium sheath	NAKBA $\quad 3 \times 120$ SM NAKLEY 3×120 SM	$2.11 + j\,0.62$ $0.83 + j\,0.31$
	PVC cables, three-single core cables in trefoil formation	NYSY $\quad 1 \times 120$ RM/10	$1.10 + j\,0.68$
12/20	S.L. cables with paper insulation and lead sheath	NAEKBA 3×150 RM	$1.34 + j\,0.66$
	PVC cables, three-single core cables in trefoil formation	2YHSY $\quad 1 \times 150$ RM/8	$0.72 + j\,0.59$
64/110	Single-core oil-filled cables, three cables in trefoil formation	NÖKUDY 1×240 RM/V 12/H	$0.54 + j\,0.34$
	External gas-pressure cables in steel pipe	NPKDvFStY 3×240 OM/V	$0.36 + j\,0.27$

21.5 Literature Referred to in Section 21

[21.1] Brüderlink, R.: Induktivität und Kapazität der Starkstrom-Freileitungen. Braun, Karlsruhe 1954

[21.2] Edison Electric Institute: Underground systems reference book. Edison Electric Institute, 1957

[21.3] Brookes, A.S.: The design of specially bonded cable systems. Electra No. 28, S. 55 to 81

[21.4] Skipper, D.J.: The design of specially bonded cable circuits (Part II). Electra No. 47, S. 61 to 86

[21.5] Eichhorn, K.F.: Erhöhung der Durchgangsleistung dreiphasiger Kabelsysteme mit Serienimpedanzen und Zusatzleitern. ETZ-Archiv (1979) 3 bis 8

[21.6] Dabringhaus, H.-G.: Übertragungsverhalten von Kabeln mit ausgekreuzten Mänteln. Elektrizitätswirtschaft 79 (1980) 809 bis 815

[21.7] Langrehr, H.: Rechnungsgrößen für Hochspannungsanlagen (AEG-Handbücher, 9). AEG-Telefunken, Berlin 1968

[21.8] Happoldt, H.; Oeding, D.: Elektrische Kraftwerke und Netze. Springer, Berlin, Heidelberg, New York 1978

[21.9] Balzer, G.; Gretsch, R.: Impedanzen und Kurzschlußströme in Niederspannungsnetzen. ETZ-A 95 (1974) 323 bis 332

[21.10] Balzer, G.: Nullimpedanzen von Kabel- und Freileitungsnetzen zur Einhaltung der Nullungsbedingungen. ETZ-B (1976) 175 to 180

[21.11] Hosemann, G.; Balzer, G.: Impedanzwerte zur Berechnung des Kurzschlußstromes nach DIN 57102 Blatt 2/VDE 0102 Teil 2. Elektronorm 29 (1975) 118 to 121

[21.12] Roeper, R.: Kurzschlußströme in Drehstromnetzen. 6., überarbeitete Auflage. Siemens AG, Berlin, München 1984

22 Capacitance and Capacitance per Unit Length

22.1 General

Table 22.1 summarises the equations for the calculation of capacitance per unit length, the capacitive and earth-fault current as well as the capacitive load. For the commonly used types of cable these quantities can be found directly from tables in Part 2 [18.17]. The quantities of capacitance per unit length are partly mean quantities derived from actual measurements. Due to manufacturing tolerances there will be a spread of quantities in practice.

22.2 Operating Capacitance per Unit Length C_b'

Due to inhomogenity of the insulation, the measured quantities deviate significantly from calculated quantities, especially because of interstices in non-radial field cables.

With *belted cables* the operating capacitance can be determined by using measurements of partial capacitances. A precondition for this is that the partial capacitances which are similar are also equal when compared to one another. The following partial capacitances can be measured:

1. From one conductor to the remaining conductors and common metal covering (metal sheath, screen, concentric conductor) – capacitance C_A'.

2. From all conductors linked together to the common metal covering – capacitance C_B'.

From these quantities the partial capacitances C_E' and C_L' and the operating capacitance per unit length C_b' can be calculated.

Cables with polymer insulation can be similarly treated, where they incorporate a *concentric conductor, metal screen or armour* over the laid-up cores. In cables with polymer insulation which do not contain a common metallic covering over the laid-up cores (e.g. NYY) the extent of the electrostatic field is indeterminable. Capacitance per unit length, especially to earth cannot be quoted.

In *radial field cables* it is sufficient to measure the capacitance to earth per unit length C_E' and this is both the operating and zero-sequence capacitance per unit length.

Where cables have *sector-shaped conductors and individual core screening* and the insulation is thin relative to conductor dimensions, e.g. cables for 0.6/1 kV, the calculation can be made approximately by assuming a circular conductor having the same circumference.

Quantities of the *relative permittivity* (permittivity of the dielectric) ε_r for insulation materials normally used at 50 Hz and 20 °C are given in Table 22.2.

The *relative permittivity* ε_r is temperature dependant. Over the range of normal operating temperatures for cables the variation is negligible for impregnated paper and polyethylene insulation. Significant variations occur, however, with PVC compounds. For PVC cables, guide values may be taken from the curves in Fig. 22.1 to correct the relative permittivity ε_r and thus enable the determination of the operating capacitance C_b', capacitive current I_C' and earth-fault current I_e'. In these curves a base value ε_{r20} which is the value at 20 °C is taken as unity.

Tabelle 22.1   Capacitance per unit length of cable

Arrangement		Eccentric arrangement	Single-core radial field cable	
Electrical system			*Single-phase a.c.*	
Operating voltage U_b between		Main conductor and metal cover		
Operating voltage to earth	$U_e =$	U_b	U_b	$U_b/2$
Operating capacitance per unit length of all conductors to earth	$C_b' =$	$\dfrac{2\pi\varepsilon_0\varepsilon_r}{\ln\dfrac{r_1^2 - c^2}{r_F r_1}}$	$\dfrac{2\pi\varepsilon_0\varepsilon_r}{\ln\dfrac{r_1}{r_F}}$	
Zero-phase sequence capacitance per unit length	$C_0' = C_E' =$	C_b'		
From measurement:				
One conductor to all other conductors and metal cover	$C_A' =$	C_E'		
All conductors to metal cover	$C_B' =$	C_E'		
Operating capacitance per unit length	$C_b' =$	C_E'		
Zero-phase sequence capacitance per unit length	$C_0' =$	C_E'		
Earth-fault current	$I_e' =$	$U_e\,\omega\,C_E'$		$2\,U_e\,\omega\,C_E'$
Capacitive current	$I_C' =$	$U_e\,\omega\,C_b'$		
Capacitive load of the cable (VA)	$P_C' =$	$n\,U_e^2\,\omega\,C_b'$		
Number of conductors, i.e. loaded conductors per cable	$n =$	1		

[1] Example: A concentric cable with earthed sheath as return conductor
[2] Two single-core cables in a single-phase a.c. system
[3] Also applies for multi-core cables with individually screened cores or metal sheathed cables in a three-phase system (construction type B and C)

Two-core belted cable	Three-core belted cable	Single-core radial field cable
		3)
	Three-phase a.c.	
	Main conductor	
$U_b/2$	$U_b/\sqrt{3}$	$U_b/\sqrt{3}$
$\dfrac{2\pi\,\varepsilon_0\,\varepsilon_r}{\ln\dfrac{2c(r_I^2-c^2)}{r_F(r_I^2+c^2)}}$	$\dfrac{2\pi\,\varepsilon_0\,\varepsilon_r}{\ln\sqrt{\dfrac{3c^2(r_I^2-c^2)^3}{r_F^2(r_I^6-c^6)}}}$	$\dfrac{2\pi\,\varepsilon_0\,\varepsilon_r}{\ln\dfrac{r_I}{r_F}}$
$\dfrac{2\pi\,\varepsilon_0\,\varepsilon_r}{\ln\dfrac{r_I^4-c^4}{2c\,r_F\,r_I^2}}$	$\dfrac{2\pi\,\varepsilon_0\,\varepsilon_r}{\ln\dfrac{r_I^6-c^6}{3c^2\,r_F\,r_I^3}}$	C_b'
$C_L'+C_E'$	$2C_L'+C_E'$	C_E'
$2C_E'$	$3C_E'$	C_E'
$C_E'+2C_L'$ $=2C_A'-C_B'/2$	$C_E'+3C_L'$ $=(9C_A'-C_B')/6$	C_E'
$C_B'/2$	$C_B'/3$	C_E'
$2U_e\,\omega\,C_E'$	$3U_e\,\omega\,C_E'$	
2	3	1

r_F Radius of the conductor or inner conducting layer (if present)
r_I Radius of insulation

Table 22.2
Electrical characteristics of the cable dielectric materials

Cable type	Electric space constant at operating temperature ε_r	Dielectric loss factor at operating temperature $\tan\delta$	Permissible operating temperature ϑ_{Lr} °C	Limiting value for U_0[1] kV	Insulation resistivity at 20 °C ϱ_I Ωm
Mass-impregnated cable	3.5	10×10^{-3}	65–80	38	5×10^{12}
Oil cable	3.6	3×10^{-3}	85	63.5	5×10^{12}
Gas-pressure cable	3.5	3×10^{-3}	85	63.5	5×10^{12}
PVC cable	8[2]	100×10^{-3}	70	6	7×10^{11}
PE cable	2.4	0.4×10^{-3}	70	127	10^{15}
XLPE cable	2.4	0.55×10^{-3}	90	127	10^{14}

[1] If the cable-rated voltage is equal to or greater than the given quantity U_0, the dielectric losses must be included in the calculation of load capacity
[2] at 20 °C $\varepsilon_r \approx 3$ to 4

333

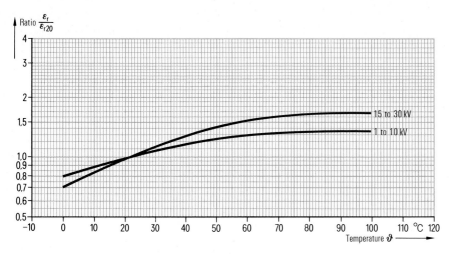

Fig. 22.1
Variation of relative permittivity ε_r of PVC compounds with temperature (also applies, for temperature dependance, to capacitance C'_b) affecting capacitive current I'_C and earth-fault current I'_e. Higher or lower quantities may result due to variations in the PVC compounds

22.3 Capacitive Current I'_C and Earth-Fault Current I'_e of a Cable

Fig. 22.2 illustrates the capacitive currents and earth-fault currents which flow in a cable operating in a three-phase system.

The capacitances per unit length are represented by concentrated capacitances shown between the conductors C'_L and between conductor and earth C'_E and are assumed to have symmetrical values. These result in capacitive currents I'_{CL} and I'_{CE} flowing. In arrangements A and B in normal operation capacitive currents flow only in the conductors. In the sheaths or screens the capacitive currents are out of phase by $2\pi/3$ and equate to zero such that the current to earth is zero. In the cables to arrangement C currents also flow in the sheaths and screens. At approximately the mid point of the cable run they will equate to zero, increasing towards each end until at the feeding end these currents achieve the value of capacitive current I'_C. At the bonding points (bonding points of sheaths or screens) the currents equate to zero.

In the cables of arrangements A and B the earth-fault current I'_e flows as shown in Fig. 22.2 b, e.g. from faulty conductor T to a sheath or a screen and from

there through the capacitances per unit length of conductor-sheath to the healthy conductors R and S which are energised at full line-to-line voltage $U_b = \sqrt{3}\,U_e$.

Because of the conductor-conductor capacitance per unit length, the currents produced do not influence the earth-fault current I'_e, they flow only in the conductors. They are created by the line-to-line voltage in normal operation as well as during the earth fault and are of the same quantity under both conditions.

In the cables of arrangement C currents flow in the sheaths or screens and are in phase with the conductor currents. The currents of the healthy phases are zero at the mid point of the cable run and reach at the two ends half the value of the conductor current at the feeding end. If the point of fault is situated in the immediate vicinity of the feeding end then the earth-fault current is approximately equal to the current in the faulted phase.

Fig. 19.3 illustrates the vector relationship of currents and voltage of the cable in arrangement A under earth-fault conditions. The capacitive currents I'_{CL} and I'_{CE} compliment one another and equate to zero in each ease. In arrangements B and C capacitive current I'_{CL} is not present and the diagram is simpler.

a) Normal operation

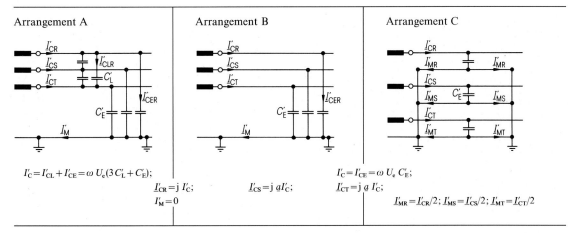

$$I'_C = I'_{CL} + I'_{CE} = \omega U_e (3 C'_L + C'_E);$$

$$I'_{CR} = j\, I'_C; \qquad I'_{CS} = j\, \underline{a} I'_C; \qquad I'_{CT} = j\, \underline{a}\, I'_C;$$

$$I'_M = 0$$

$$I'_C = I'_{CE} = \omega U_e\, C'_E;$$

$$\underline{I}_{MR} = \underline{I}'_{CR}/2;\ \underline{I}_{MS} = \underline{I}'_{CS}/2;\ \underline{I}_{MT} = I'_{CT}/2$$

b) Earth-fault operation

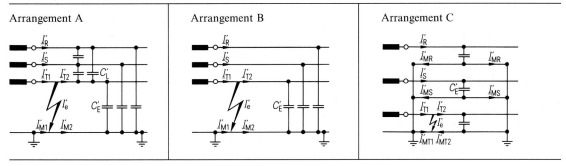

With insulated star point:

$$\underline{I}_R = j\, I'_{CL} + j\,[(1-\underline{a})/\sqrt{3}]\, I'_e/\sqrt{3};$$

$$\underline{I}_S = j\, \underline{a}^2\, I'_{CL} + I'_e/\sqrt{3};$$

$$\underline{I}_{T1} = j\, \underline{a}(I'_{CL} + I'_e);\ \underline{I}_{T2} = j\, \underline{a}\, I'_{CL};$$

$$\underline{I}_{M1} = 0;\ \underline{I}_{M2} = j\, \underline{a}\, I'_e;$$

$$\underline{I}_R = j\,[(1-\underline{a})/\sqrt{3}]\, I'_e/\sqrt{3};$$

$$\underline{I}_S = I'_e/\sqrt{3};$$

$$\underline{I}_{T1} = j\, \underline{a}\, I'_e;\ \underline{I}_{T2} = 0;$$

$$\underline{I}_{MR} = \underline{I}_R/2;\ \underline{I}_{MS} = \underline{I}_S/2;\ \underline{I}_{MT1} = \underline{I}_{MT2} = -\underline{I}_T/2$$

With ideal compensation ($X_D = 1/3\, \omega\, C_E$) (all other formulae above apply):

$$\underline{I}_{T1} = \underline{I}_{T2} = j\, \underline{a}\, I'_{CL}$$

$$\underline{I}_{M1} = \underline{I}_{M2} = j\, \underline{a}\, I'_e$$

$$\underline{I}_{T1} = \underline{I}_{T2} = 0$$

$$\underline{I}_{MT1} = \underline{I}_{MT2} = j\, \underline{a}\, I'_e/2$$

In general: $\underline{I}_e = j\, \underline{a}\, I'_e$ (at the fault point on conductor T); $\quad I'_e = 3\, U_e\, \omega\, C'_E; \quad \underline{a} = -\dfrac{1}{2} + j\dfrac{\sqrt{3}}{2}; \quad \underline{a}^2 = -\dfrac{1}{2} - j\dfrac{\sqrt{3}}{2}$

Arrangement A	"Belted cables" (e.g. NKBA), Cables with polymer insulation and common screen, sheath or armour (e.g. NYFGY) as well as overhead lines
Arrangement B	Three-core radial field cables with common screen, sheath or armour (e.g.: NYSEY)
Arrangement C	Single-core cables with screen (e.g.: N2XS2Y) or sheath and three-core cables with individually screened cores (e.g.: N2XSE2Y) as well as S.L. cables (e.g.: NEKBA)

The main conductors are marked with the subscripts R, S, T and sheath, screen and armour are identified M

Fig. 22.2 Capacitance currents in 3 phase a.c. cables

22.4 Dielectric Losses

Dielectric losses occur in cables operating on a.c. in a similar manner to those in a capacitor. As a result a small dielectric loss current flows such that the current I flowing through the dielectric deviates by a small loss angle δ from the 90° leading capacitive current I_C of an ideal loss free capacitor.

From Fig. 22.3

$$\tan \delta = \frac{I_w}{I_C}. \tag{22.1}$$

For a cable in a three-phase a.c. system the dielectric losses are

$$P_d' = n U_e^2 \, \omega \, C_b' \tan \delta \tag{22.2}$$

$$= n \left(\frac{U_b}{\sqrt{3}} \right)^2 \omega \, C_b' \tan \delta. \tag{22.3}$$

In general $U_b = U_n$.

The dielectric loss factor $\tan \delta$ of a cable is determined mainly by the material used for insulation but also depends on the construction of the cable (e.g. whether semi-conducting layers are included over the conductor and over the insulation). Discharges in gas-filled voids within the insulation represent a voltage-dependent increase in the dielectric loss factor.

The dielectric loss factor of most insulating materials is temperature dependent. The curves in Fig. 22.4 illustrate the characteristic variation of $\tan \delta$ in relation to temperature for various types of insulation. Deviations to a more or less degree can exist depending on the type and composition of the polymers respectively polymer compounds, insulating papers, impregnation fluids, etc. These deviations, however, do not provide a reliable basis for judging the quality of a cable.

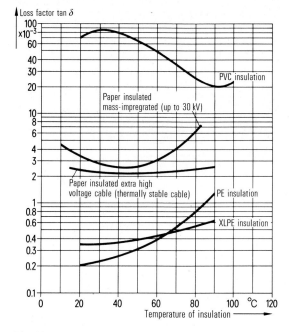

Fig. 22.4
Variation of dielectric loss factor $\tan \delta$ of cables with temperature (typical example)

The heat developed by the dielectric losses does not significantly increase the cable temperature except at higher voltages. When calculating the load capacity the dielectric losses are not considered for cables for rated voltages below the values of U_0 given in Table 22.2. For voltages above these the $\tan \delta$ values given in the table may be used.

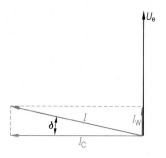

Fig. 22.3 Vector diagram of a capacitor

23 Insulation Resistance, Insulation Resistance per Unit Length and Leakage

The leakage is the reciprocal of the insulation resistance. For radial field cables with a diameter over the inner conducting layer of d_F and a diameter over the insulation of d_I, one obtains

$$G' = \frac{2\pi}{\varrho_1 \ln \dfrac{d_I}{d_F}}. \tag{23.1}$$

Quantities for the insulation resistivity ϱ_1 at 20 °C are given in Table 22.2. At operating temperature this quantities decrease by approximately two orders of magnitude in cables with polymer insulation and in mass-impregnated, oil- and gas-pressure cables by approximately one order of magnitude.

The quantity of the insulation resistance is not normally a measure for the operational safety of the cable; only when the low insulation resistance is attributable to ingress of moisture the operational safety is reduced. Because of the difficulties in laying down specific quantities for insulation resistance, the test regulations for power cables issued by individual countries do not include precise quantities. An indication of what quantities may be expected with new cables direct from the factory is given in Tables 23.1 and 23.2. It is normal for these quantities to be exceeded by several times.

The relationship of insulation resistance and temperature is shown in Fig. 23.1 for PVC cables and in Fig. 23.2 for cables insulated with impregnated paper. The curves in these figures are drawn with the base quantity at 20 °C equal to unity.

When the insulation resistance of cables is measured it must be noted that the insulation resistances of sealing ends and joint boxes are included in the measurement so that the results may be affected by, for example, a moist surface on a sealing end.

Table 23.1
Insulation resistance per unit length of PVC cables

Single-core cables with concentric conductor or screen and multi-core cables with individually screened cores (e.g. NYCY, NYSY, NYHSY, NYSEY)

Rated voltage U_0/U	Conductor cross-sectional area mm²	Insulation resistance per unit length, minimum quantity at 20 °C, conductor to screen or concentric conductor approx. MΩ·km
0.6/1 kV	up to 4	60
	up to 25	30
	up to 95	20
	above 95	15
3.6/6 kV	up to 50	80
	up to 150	50
	above 150	30
6/10 kV	up to 50	90
	up to 150	60
	above 150	40
12/20 kV	up to 70	500
	above 70	250
18/30 kV	all	500

Multi-core cable with common screen over laid-up cores (e.g. NYCY, NYCWY, NYKY, NYFGY)

Rated voltage U_0/U	Conductor cross-sectional area mm²	One conductor to the remaining conductors and screen approx. MΩ·km	All conductors to screen approx. MΩ·km
0.6/1 kV	up to 4	60	30
	up to 25	40	20
	above 25	30	15
3.6/6 kV	up to 50	100	80
	up to 150	60	50
	above 150	40	30

Table 23.2
Insulation resistance per unit length of paper insulated mass-impregnated cables

Belted cable

Rated voltage	Conductor cross-sectional area	Insulation resistance per unit length, minimum quantity at 20 °C	
		One conductor to all other conductors and metal sheath	All conductors to metal sheath
U_0/U	mm^2	approx. $M\Omega \cdot km$	approx. $M\Omega \cdot km$
0.6/1 kV	up to 4	400	200
	up to 16	250	150
	up to 50	150	100
	up to 150	120	80
	above 150	100	60
3.6/6 and 6/10 kV	up to 16	500	250
	up to 50	400	200
	up to 150	300	150
	above 150	200	100

Radial field cable
(single-core cables, S.L.-cables and H-cables)

Rated voltage	Conductor cross-sectional area	Insulation resistance per unit length, minimum quantity at 20 °C, conductor to metal sheath
U_0/U	mm^2	approx. $M\Omega \cdot km$
3.6/6 and 6/10 kV	up to 16	500
	up to 50	400
	up to 150	300
	above 150	200
8.7/15 and 12/20 kV	up to 50	700
	up to 150	500
	above 150	350
18/30 kV and above	up to 50	1000
	up to 150	750
	above 150	500

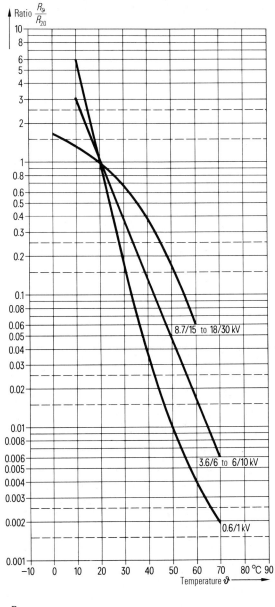

$$\frac{R_\vartheta}{R_{20}} = f(\vartheta)$$

R_ϑ Insulation resistance at the temperature ϑ
R_{20} Insulation resistance at 20 °C

Fig. 23.1
Variation of insulation resistance with temperature of PVC cables (average values). Depending on the composition of the PVC, higher or lower values may be found

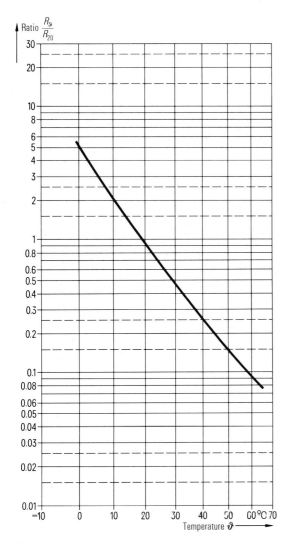

$$\frac{R_\vartheta}{R_{20}} = f(\vartheta)$$

R_ϑ Insulation resistance at the temperature ϑ
R_{20} Insulation resistance at 20°C

Fig. 23.2
Variation of insulation resistance with temperature of
mass-impregnated paper-insulated cables

24 Determination of Voltage Drop

24.1 General

In low-voltage networks it is especially important to check that the cross-sectional area selected on the basis of current load (see page 150 onwards) fulfils the requirements in respect of voltage drop.

For very long cable runs in medium-voltage networks an investigation is also recommended.

24.2 Short Cable Runs

In cable runs of short length on low voltage the capacitive current I_C can be neglected.

A cable[1] of length l, having an effective resistance per unit length R'_w and a reactance per unit length X'_L operating in a three-phase a.c. system and carrying current I_{be} will develop a voltage drop ΔU. If φ is the phase angle between the current I_{be} and the voltage U_{be} at the loaded end of the cable (Fig. 24.1):

$$\Delta U = \sqrt{3}\, I_{be}\, l(R'_{wr} \cos \varphi + X'_L \sin \varphi) \qquad (24.1)$$

(assuming a lagging current, i.e. inductive load, for a leading current the sign of X'_L must be reversed, in which case it is possible for ΔU to be negative).

[1] The term cable in this instance is used to represent a conductor which may be a cable, an insulated conductor or an overhead line

The percentage voltage drop, relative to the rated voltage U_n, is given by

$$\Delta u = \frac{\Delta U}{U_n}\, 100. \qquad (24.2)$$

The phase-angle difference between the supply end and loaded end voltages is:

$$\cos \delta = \frac{U_{be} + \Delta U}{U_{ba}} \qquad (24.3)$$

where U_{ba} and U_{be} are the voltages (line to line) at the supply end and loaded end of the cable.

The voltage drop becomes on d.c.

$$\Delta U = 2 I_{be}\, R'\, l \qquad (24.4)$$

and for single-phase a.c.

$$\Delta U = 2 I_{be}\, l(R'_{wr} \cos \varphi + X'_L \sin \varphi). \qquad (24.5)$$

When calculating the voltage drop in cables on d.c. systems and in cables of up to $16\,mm^2$ conductor cross-sectional area on a.c. systems, it is sufficient to take into account only the d.c. resistance at operating temperature. With conductors larger than $16\,mm^2$ cross-sectional area on a.c. systems, it is necessary to take into consideration both the effective resistance and the reactance. In the case of unarmoured cables, and especially for wiring cables and flexibles, the limit is considerably higher.

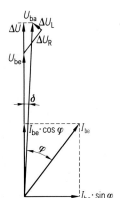

Fig. 24.1
Vector diagram showing the voltage drop of an inductive load

24.3 Long Cable Runs

If at operating voltages of $U > 10\,kV$ with a long cable run the calculated capacitive current (see Section 22) cannot be ignored in comparison with the load current, then a close approximation of the voltage drop and losses can be determined by reference to Figs. 24.2 and 24.3.

The operating capacitance is imagined to be represented by two capacitors of half quantity concentrated at each end of the cable as shown in Figs. 24.2.

To calculate the voltage drop due to the effective resistance per unit length R'_{wr} and the reactance per unit length X'_L a constant fictitious current I_f is assumed to flow, which is derived with the aid of the current flowing through the concentrated capacitance $\frac{C'_b l}{2}$ at the end of the cable

$$I_{ce} = \frac{U_{be}\, \omega C'_b\, l}{2\sqrt{3}}, \tag{24.6}$$

the fictitious current

$$I_f = \sqrt{(I_{be} \cdot \cos \varphi_e)^2 + (I_{be} \cdot \sin \varphi_e - I_{ce})^2} \tag{24.7}$$

and with

$$\cos \varphi' = \frac{I_{be} \cos \varphi_e}{I_f}. \tag{24.8}$$

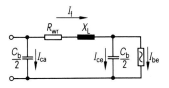

Fig. 24.2 Diagrammatic representation of the cable

The voltage drop is

$$\Delta U = \sqrt{3} \cdot I_f\, l (R'_{wr} \cos \varphi' + X'_L \sin \varphi') \tag{24.9}$$

and the cable power loss

$$\Delta P = 3 I_f^2\, R'_{wr}\, l. \tag{24.10}$$

The voltage at the supply end of the cable between the lines is

$$U_{ba} = \\ \sqrt{(U_{be}\cos\varphi' + \sqrt{3} I_f l R'_{wr})^2 + (U_{be}\sin\varphi' + \sqrt{3} I_f l X'_L)^2} \tag{24.11}$$

and the capacitive current of the concentrated capacitance at the supply end of the cable

$$I_{ca} = \frac{U_{ba}\, \omega C'_b\, l}{2\sqrt{3}} \tag{24.12}$$

From this can be calculated with

$$\cos \varphi'' = \frac{U_{be} \cos \varphi' + \sqrt{3} I_f l R_{wr}}{U_{ba}} \tag{24.13}$$

the current at the supply end of the cable

$$I_{ba} = \sqrt{(I_f \cos \varphi'')^2 + (I_f \sin \varphi'' - I_{ca})^2} \tag{24.14}$$

and the power input at the supply end of the cable

$$P_a = \sqrt{3}\, U_{ba}\, I_{ba} \cos \varphi_a. \tag{24.15}$$

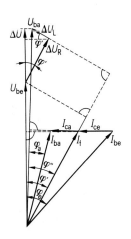

Fig. 24.3
Vector diagram of the voltage drop on an inductive load including the operating capacitance of the cable

Table 24.1 cos φ and relative values of sin $\varphi = \sqrt{1 - \cos^2 \varphi}$

cos φ	φ degrees	sin φ	cos φ	φ degrees	sin φ	cos φ	φ degrees	sin φ
1.00	0.0	0.000	0.76	40.5	0.649	0.52	58.7	0.854
0.99	8.1	0.141	0.75	41.4	0.661	0.51	59.3	0.860
0.98	11.5	0.199	0.74	42.3	0.673	0.50	60.0	0.866
0.97	14.1	0.244	0.73	43.1	0.683	0.49	60.7	0.872
0.96	16.3	0.281	0.72	43.9	0.693	0.48	61.3	0.877
0.95	18.2	0.312	0.71	44.8	0.705	0.47	62.0	0.883
0.94	20.0	0.342	0.70	45.6	0.714	0.46	62.6	0.888
0.93	21.6	0.368	0.69	46.4	0.724	0.45	63.3	0.893
0.92	23.1	0.392	0.68	47.2	0.734	0.44	63.9	0.898
0.91	24.5	0.415	0.67	47.9	0.742	0.43	64.5	0.903
0.90	25.8	0.435	0.66	48.7	0.751	0.42	65.2	0.908
0.89	27.1	0.456	0.65	49.5	0.760	0.41	65.8	0.912
0.88	28.4	0.476	0.64	50.2	0.768	0.40	66.4	0.916
0.87	29.5	0.492	0.63	50.9	0.776	0.39	67.0	0.921
0.86	30.7	0.511	0.62	51.7	0.785	0.38	67.7	0.925
0.85	31.8	0.527	0.61	52.4	0.792	0.37	68.3	0.929
0.84	32.9	0.543	0.60	53.1	0.800	0.36	68.9	0.933
0.83	33.9	0.558	0.59	53.8	0.807	0.35	69.5	0.937
0.82	34.9	0.572	0.58	54.5	0.814	0.34	70.1	0.940
0.81	35.9	0.586	0.57	55.2	0.821	0.33	70.7	0.944
0.80	36.9	0.600	0.56	55.9	0.828	0.32	71.3	0.947
0.79	37.8	0.613	0.55	56.6	0.835	0.31	71.9	0.951
0.78	38.7	0.625	0.54	57.3	0.842	0.30	72.5	0.954
0.77	39.6	0.637	0.53	58.0	0.848			

25 Economic Optimization of Cable Size

Losses occur with the transmission of electrical energy through cables. The cost of these losses K_v can be reduced by selecting a larger cross-sectional area. By increasing the area the prime cost K_a is increased. The economical solution is to select a cross-sectional area which has the least overall cost. This cross-sectional area is termed the economic cross-sectional area.

Detailed guidelines for the evaluation and reduction of losses in the network and therefore the overall cost are given, e.g., in the directives issued by VDEW [25.1, 25.2]. Here only the selection of the economic cross-sectional area of a cable system using the annual costs method is discussed.

The *annual cost* K of a cable installation is derived from the fixed annual costs that is from annual capital charges K_d and the annual costs of losses K_v,

$$K = K_d + K_v. \tag{25.1}$$

The capital charges for the cable installation are

$$K_d = \frac{K_a(T + T_R)}{100} \tag{25.2}$$

with T representing percentage amortization rate and T_R percentage additional charges.

The *prime cost* K_a includes the cost of the cable, accessories, installation and ground works. The cost of the accessories, installation and ground works are only to a small extent dependent on the cross-sectional area. These can therefore be neglected in determining an economic cross-sectional area.

The *amortization rate* T is calculated with

$$T = \frac{100\,q^t(q-1)}{q^t - 1}, \tag{25.3}$$

$$q = 1 + \frac{p}{100}, \tag{25.4}$$

or can be found from Table 25.1.

The *amortization period* t in public utility supply networks is normally 30 to 40 years.

Table 25.1
Amortization rates T in % of prime value

Annual rate of p interest	Amortization period t in years					
	10	15	20	25	30	35
%	Amortization rate T in %					
0.00	10.000	6.667	5.000	4.000	3.333	2.857
3.00	11.732	8.377	6.722	5.743	5.102	4.654
3.25	11.873	8.529	6.878	5.904	5.268	4.825
3.50	12.024	8.683	7.036	6.067	5.437	5.000
3.75	12.176	8.838	7.196	6.233	5.609	5.177
4.00	12.329	8.994	7.358	6.401	5.783	5.358
4.25	12.483	9.152	7.522	6.571	5.960	5.541
4.50	12.638	9.311	7.688	6.744	6.139	5.727
4.75	12.794	9.472	7.855	6.919	6.321	5.916
5.00	12.950	9.634	8.024	7.095	6.505	6.107
5.50	13.267	9.963	8.368	7.455	6.881	6.497
6.00	13.587	10.296	8.718	7.823	7.265	6.897
7.00	14.238	10.979	9.439	8.581	8.059	7.723
8.00	14.903	11.683	10.185	9.368	8.883	8.580
9.00	15.582	12.406	10.955	10.181	9.734	9.464
10.00	16.275	13.147	11.746	11.017	10.608	10.369

The *annual rate of interest* p is dependent on the situation of the capital market. Additional charges for maintenance and repair are accounted for by T_R with a value of 0.5 to 1% added to the amortization rate.

The *annual cost of losses* K_v in a feeder of N cables having a length of l and with an energy cost of k_a is

$$K_v = \varepsilon l N [k_a (T_v P'_i + T_B P'_d) + k_1 P'_i]. \tag{25.5}$$

For a feeder cable, that means a load only at the end of the cable having a *load distribution factor* $\varepsilon = 1$ [25.1].

The *ohmic losses* P'_i must be calculated for the maximum load (see Sections 18.2.1 page 152 and 18.2.4 page 180) of the operating current I_b using the effective resistance per unit length $R'_{w\vartheta}$ of the conductor at operating temperature ϑ (see Section 20 page 320).

An approximation can be made using the effective resistance per unit length R'_{wr} at the permissible operating temperature.

The *dielectric losses* P'_d are, from Section 22,

$$P'_d = n\omega C'_b \left(\frac{U_b}{\sqrt{3}}\right)^2 \tan\delta. \tag{22.3}$$

These losses are only of importance for PVC cables at 6/10 kV and high-voltage cables above 35/60 kV.

The *annual operating period* T_B represents the total number of operating hours over one year. The *utilization time of power losses* is

$$T_v = \mu T_B. \tag{25.7}$$

If the cable is live over the whole year for the calculation of dielectric losses $T_B = 8760$ hours is used. The duration of power loss, however, is calculated using the actual operating duration T_B of the installation.

The *loss load factor* μ is difficult to determine. If the load factor m is known (see Section 18.2.1) an approximate value for μ can be calculated. A simple equation is given, e.g. in the VDEW directives [25.1]:

$$\mu = 0.17m + 0.83m^2. \tag{25.8}$$

The annual utilization time of power losses T_v can also be determined from Fig. 25.1 providing the *annual utilization period* T_m is known:

$$T_m = m T_B. \tag{25.9}$$

The *load factor* m as shown in Section 18.2.3 is determined from the 24 hour load diagram.

The equation 18.51 which is used for the calculation of load capacity (see Section 18.4.3 page 200), gives quantities which differ only marginally from those of equation 25.8:

$$\mu = 0.3m + 0.7m^2. \tag{18.51}$$

With the load factor m the variation of load over 24 hours is taken into account. It must be borne in mind, however, that other variations, due to seasonal changes or on specific days, may be present. These can be determined only by a great deal of effort. To simplify, some guide values are given in Table 25.2 which may be used.

The additional capital costs to provide for power losses k_l, need only be considered under certain circumstances. A basis for this the incremental installation costs are to be taken. Details for this can be

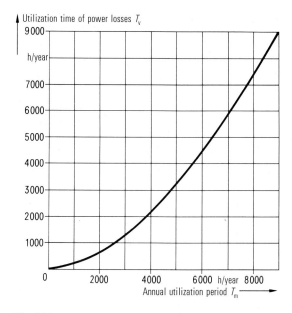

Fig. 25.1
Utilization time of power loss relative to annual utilisation period

Table 25.2
Guide values for the duration of power loss T_v taking into account seasonal and/or weekly load variations

Type of operation	Load factor m	Utilization time of power loss T_v h
Public utility load	0.7 0.85	3000 4300
Industrial load	0.85 1.0	4700 6300

found in the guidelines published by VDEW [25.1, 25.2]. The additional capital costs k_l result from the additional investment required in generation plant and network equipment. These investments are necessary in certain circumstances to generate and transmit the energy which is absorbed by transmission losses.

Example 25.1

For an operating voltage $U_b = 20$ kV with a cable run of length $l = 2.5$ km, a power load of 18 MVA (equivalent to 520 A) is to be transmitted for utility load operation. By reference to Section 18 with the specified operating conditions from Table 18.2 it is necessary for a bunched installation in the ground to use three cable systems of

   NA2XS2Y 1×120 RM/16   12/20 kV

each cable is therefore loaded with an operating current I_b of 173 A.

The prime cost K_a of the cable is DM 586 000 with copper at a current price of DM 300 per 100 kg and aluminium at DM 350 per 100 kg. With an annual interest rate $p = 6\%$ and an amortization period $t = 35$ years, from Table 25.1 the annual amortization rate is 6.897% of the prime cost. With an additional $T_R = 0.5\%$ for maintenance and repairs the annual capital charges are

$$K_d = \frac{586\,000(6.897 + 0.5)}{100} = 43\,346 \text{ DM/year}.$$

$$(25.2)$$

The utilization time of power losses is estimated to be $T_V = 3000$ h (see also Example 25.2). The ohmic losses are

$$P_i' = 1 \times 173^2 \times 0.328 \times 10^{-3} = 9.82 \text{ W/m}. \quad (18.3)$$

The dielectric losses are negligible and also the additional capital cost losses factor k_1 can be neglected.

With the cost of electrical energy $k_a = 0.1$ DM/kWh and three systems, i.e. $N = 9$ cables, the annual cost of losses

$$K_v = 1 \times 2.5 \times 10^3 \times 9(0.1 \times 10^{-3} \times 3000 \times 9.82)$$
$$= 66\,285 \text{ DM/year} \qquad (25.5)$$

and the annual cost becomes

$$K = 43\,346 + 66\,285 = 109\,631 \text{ DM/year}. \qquad (25.1)$$

The same method of calculation is used for the next larger cross-sectional area. The costs relative to cross-sectional area of cable are shown in Fig. 25.2. The economic cross-section of 240 mm² incurs the minimum of cost.

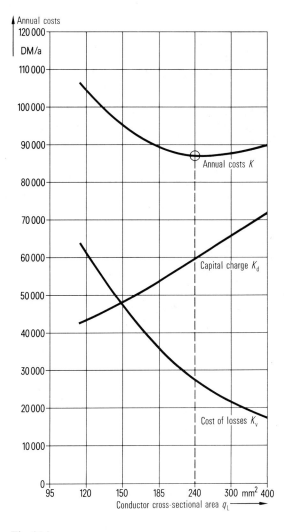

Fig. 25.2
Costs as a function of the cross-sectional area q_L. The economic cross-sectional area is 240 mm², see Example 25.1

Example 25.2

To illustrate the influence of the utilization time of the power loss, the cost of energy, the rate of interest and the amortization period, further examples have been worked through and the results are tabulated in Fig. 25.3. These examples were based in all other respects on Example 25.1. The broken line indicates the movement of the economic cross-sectional area to larger values.

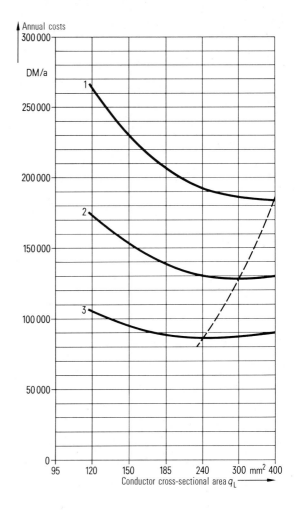

Curve		1	2	3
Utilization time of power loss T_v	h	6000	4500	3000
Cost of energy k_a	DM/kWh	0.15	0.125	0.10
Annual interest rate p	%	10	8	6
Amortization period t	a	15	25	35

Fig. 25.3
Determination of economic cross-sectional area (– – – –) from Example 25.2

25.1 Symbols Used in Formulae in Section 25

All equations are written as quantitive equations. The symbols used in the formulae have a numerical value and also relate to a unit of measurement. The quantities are independent of the system of units, whichever compatible system of units is selected, e.g. (SI units) Système International d'Unités.

Longitudinally related dimensions "per unit value" such as electrical key data, thermal resistances and losses have symbols marked with an apostrophe stroke (DIN 1304). In the text only the electrical key data are marked as unit value. For thermal resistances and losses, to simplify expression, this is not used.

k_a	Cost of electrical energy	DM/Wh
k_l	Cost for providing power losses	DM/Wh
l	Length of cable run	m
m	Load factor	–
n	Number of loaded conductors in the cable	–
p	Annual rate of interest in %	–
t	Amortization period	years
C'_b	Operating capacitance per unit length	F/m
I_b	Maximum operating current, e.g. during a 24 hour load cycle	A
K	Annual cost of cable run	DM/year
K_a	Prime cost of cable run	DM
K_d	Annual capital charges, fixed costs	DM/year
K_v	Annual cost of losses	DM/year
N	Number of cables in cable run	–
P'_i	Ohmic losses	W/m
P_d	Dielectric losses	W/m
$R'_{w\vartheta}$	Effective resistance per unit length of conductor at conductor temperature ϑ	Ω/m
R'_{wr}	Effective resistance per unit length of conductor at permissible operating temperature	Ω/m
T	Amortization rate %	–
T_m	Annual utilization period	h/year
T_v	Annual utilization time of power losses	h/year
T_B	Annual operating period,	h/year
T_R	Addition charges for repair and maintenance in %	–

U_b	Operating voltage line-to-line in three-phase a.c. system	V
ε	Load distribution factor	–
μ	Load loss factor	–
$\tan \delta$	Dielectric loss factor	–
ω	Angular frequency, $\omega = 2\pi f$	1/s

25.2 Literature Referred to in Section 25

[25.1] VDEW: Netzverluste. Eine Richtlinie für ihre Bewertung und ihre Verminderung.
2. Ausgabe, Verlags- und Wirtschaftsgesellschaft der Elektrizitätswerke mbH, Frankfurt (Main) 1968

[25.2] VDEW: Gestaltung und Betriebsweise von städtischen Mittelspannungsnetzen.
Verlags- und Wirtschaftsgesellschaft der Elektrizitätswerke mbH, Frankfurt (Main) 1977

Sheathed and non-sheathed cables are,
both during installation and
in operation, often subjected to high mechanical stress.
An important aspect of quality assurance
is the testing of insulating and sheathing materials
in respect of stretching and tensile strength

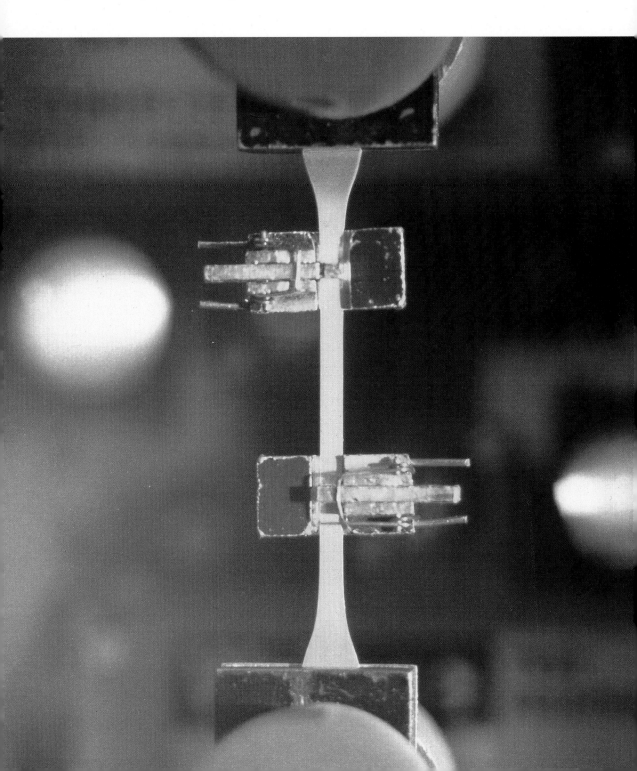

26 Interference of Power Cables with Control and Telecommunication Cables

With the further extension of power and communication networks it is more frequent that communication cables are run parallel to power cables or overhead lines or are fed into power or transformer stations. Especially in the event of earth fault on the power equipment, voltages may be induced in the communication cables as a result of inductive or resistive coupling and these may endanger personnel, equipment or the cables themselves or interfere with the transmission of communications.

Recommendations on permissible voltages, calculation and protection methods have been issued by CCITT[1] in its "Recommendations for the protection of telecommunication lines against adverse influences of power lines" and by the Association of German Electrical Engineers (DKE) in their guidelines DIN VDE 0228 Parts 1 to 4. According to these rules the longitudinal electromotive force EMF in the cables must not exceed the following values between conductor and earth:

---

[1] Comité Consultatif International Télégraphique et Téléphonique

---

Permissible values of longitudinal EMF in telephone cables

CCITT	VDE	
60 V	65 V	Long-time interference in communication circuits which are not terminated with isolating transformers
150 V	250 V	Long-time interference in communication circuits which are terminated with isolating transformers or other special conditions are provided.
300 V	300/500 V	Short-time interference (≤ 0.5 s) in communication circuits which are not terminated with protective elements (isolating transformers, surge arrestors or similar)

It is permitted to exceed the above values in exceptional circumstances. The rules issued by CCITT permit a longitudinal EMF of 650 V, where the cables have polymer insulation and the interference is from high-voltage equipment having increased operational safety.

The value of 300 V given in the VDE specifications applies only to telecommunication installations for German Post Administration and for any systems connected to these.

60%	60%	of the cable test voltage for short-time interference (≤ 0.5 s) of telecommunication circuits terminated with protective elements (isolating transformers, voltage arrestors or similar).

In normal practice communication cables are designed for a test voltage core to earth of 2 kV (rms value).

The permissible interference voltage is then 1200 V. It is the aim to design the cable installation such that the interference voltage of 1200 V is not exceeded, e.g. by cable with a favourable reduction factor. Higher voltages can be permitted if both the cable and all associated accessories are designed for a higher test voltage. If work is carried out on such installations additional protective measures are necessary (DIN VDE 0228 Part 1).

Where the cable is terminated with voltage arrestors or with an element connecting the conductors to earth, the calculated longitudinal EMF by induced interference between the protective elements may be accepted up to 120% of the cable test voltage. The cable installation is stressed under such conditions only up to 60% of the test voltage (partial interference i.e. short-circuit at the mid point of the appropriate cable run).

Interference voltages can be determined either by measurement or by calculation. The later method gives only approximate values because the parameters which also influence the magnitude of the interference voltage (e.g. specific ground resistance) are often not known exactly. In such cases values obtained by practical experience must be used.

In three-phase a.c. systems only an earth-fault current is of consequence since for normal operation the magnetic fields of the three-phase system cancel out. Only in telecommunication cables which run alongside overhead power lines for several kilometers in a distance of less than 50 m an interference during normal operation must be considered.

In systems with a solidly earthed star point a single earth fault may induce unacceptably high interference in neighbouring telecommunication lines. In systems with earth fault compensation (extinguishing coil) or insulated star point this can only occur if there are simultaneous earth faults on different points in two phases (double earth fault). As is discussed in CCITT and in DIN VDE 0228 dangerous interference is mainly attributed to high-voltage networks having solidly earthed star point. In networks with insulated star point or with earth-fault compensation to DIN VDE 0228 Part 2 the interference need only be considered where the extinguishing limit to Fig. 26.1 is exceeded. For railway signalling circuits and other communication circuits requiring a similar high degree of protection, these limits, in view of the required safety, are not applicable.

An interference by an earth-short circuit to DIN VDE 0228 Part 2 is not considered (see Fig. 26.1) in:

▷ three-phase cables with metal sheath or copper screen having equal conducting capacity, if the earth short circuit current is $\leq 2\,\text{kA}$ at system voltages of 20 kV and below;

▷ three-phase cables in steel pipe where the earth short circuit current does not exceed 15 kA;

▷ networks with star-point earthing through a non-permanent ohmic resistance.

The following provides a guide to the calculation for an estimation of dangerous interference in the event of earth fault in a power installation. For other cases, e.g. the interference during normal operation of an electrified railway track it is advisable to refer to the recommendations of CCITT and the VDE specifications. Apart from those guidelines in many situations

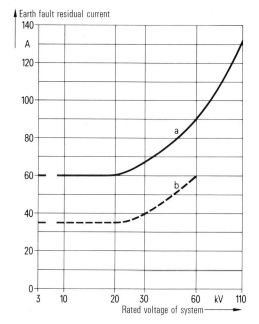

a In networks with earth-fault compensation
   (for cable systems up to 20 kV rated voltage with
   sections of overhead line the curve a also applies to
   insulated star point)
b In networks with insulated star point.

Fig. 26.1
Limit of extinguishing earth-fault residual current respectively earth-fault current to DIN VDE 0228 Part 2

it is necessary to study site conditions carefully to be able to judge the type of influence correctly.

In DIN VDE 0228 Part 4 the symbols used in formulae are shown with an apostrophe stroke where they are related to a reference frequency f'. However in deviation from this all, longitudinally related values, such as resistance, reactance and voltage per unit length, in line with DIN 1304 are also marked with an apostrophe stroke.

26.1 Inductive Interference

Whereas with overhead communication lines the influence of both the electrical and magnetic fields of nearby high-voltage power lines need to be investigated, with cables the magnetic field only is of significance in respect of interference. By the magnetic field formed by the circuit phase conductor-earth, an EMF is induced in the communication cable conductor to earth. The proportionality factor is the mutual inductance M_{Ee}. The value of mutual inductance depends upon the dimensions of the two circuits and their position relative to one another. If these cross at right angles, then the mutual inductance $M_{Ee} = 0$; if they are parallel to one another, then M_{Ee} is directly proportional to the length of the parallel run.

In the event of an earth-short circuit on a power line, an EMF E_i is induced in the conductors of the neighbouring communications cable, which is

$$E_i = \omega \cdot M'_{Ee} \cdot l \cdot I \cdot r \qquad (26.1)$$

E_i Induced EMF in volts
ω Frequency s^{-1}
M'_{Ee} Mutual inductance per unit length between the influenced and influencing cable (or line) in H/m
l Length of parallel run in m
I Interference inducing current (e.g. earth fault) in A for calculation of the longitudinal voltage (DIN VDE 0228 Part 2)
r Overall reduction factor (product of several reduction factors)

Such reduction factors are:

r_{Ki} Reduction factor for metallic covering of power cable
r_{Ku} Reduction factor for the metallic covering of telecommunications cable
r_S Reduction factor for rails
r_E Reduction factor for earth ropes braid
r_X Reduction factor for neighbouring pipes or other earthed metallic conductive parts

} Reduction factors for compensating conductors

The EMF E_i of a conductor to earth can be measured with a high-resistance voltmeter at one end of the affected cable when the other end is earthed. If both ends of the conductor are insulated from earth then, assuming equal earth-conductor capacitance on both sides of the centre point of the run, an EMF will appear at each end equal to half the value for the full run. Installations should always be dimensioned to accept the full EMF value in consideration of the case if one end only is insulated from earth, i.e. the full value of EMF can appear at the other end.

26.1.1 Mutual Inductance

The mutual inductance M_{Ee} is dependant upon the distance between the influenced and influencing conductors, the frequency and the specific earth resistance. In Fig. 26.2 the values of mutual inductance are shown for a frequency of 50 Hz with three values

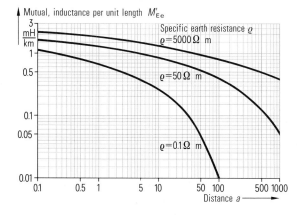

$\varrho = 5000\ \Omega m$ Rocky ground
$\varrho = 50\ \Omega m$ Loam, cultivated ground (central europe)
$\varrho = 0.1\ \Omega m$ Urban areas

Fig. 26.2
Mutual inductance between influenced and influencing conductors at $f = 50$ Hz

of specific ground resistance in relation to the distance between the influenced and influencing conductors. Intermediate values can be found in DIN VDE 0228 Part 1 Fig. 6.

Where a communication cable and a power line converge, the calculation must be made using a mean distance 'a' which is calculated from the values a_1 and a_2 at the beginning and end of the convergence

$$a = \sqrt{a_1 \cdot a_2}. \tag{26.2}$$

When the ratio $a_2/a_1 > 3$, then the calculation of smaller sections must be carried out and the part values for E_{ik} added linearly.

26.1.2 Inducing Currents

Accurate values for short-circuit currents (inducing currents) along the run can best be determined by measurement on a network model or by using data processing equipment. A rough calculation is possible if information is available on short-circuit or operation data at the input feed point of the power line (Busbars). Values can be taken from e.g. DIN VDE 0102 Part 1 and DIN VDE 0228 Part 2 or from the book "Short-circuit currents in three phase systems" by R. Roeper (see p. 319 [19.1]). According to the VDE specifications the short-circuit current may be taken into account with 0.7 of its value only. This so-called expectation factor takes into account that a coincidence of all unfavourable conditions (e.g. short-circuit current flowing along the whole of the interference length) is most unlikely. This factor, however, does not apply to cases of expected interference with railway signalling lines.

26.1.3 Current Reduction Factor of the Influencing Power Cable

The current reduction factor r_{Ki} is a measure of the effect of screening of a metallic cable covering. In the event of an earth fault the total effective short circuit current I_{k1} flows via the conductor to the point of fault and then divides into two parts one via the cable sheath and the other via the earth (part current I_{k1E}). The reduction factor in this circum-

stance is the ratio of the current returning through earth relative to the total earth short circuit current. The impedance of the current path to earth is formed from the self impedance of the metal coverings with current flow via earth and from the impedances to earth at the cable ends. The ohmic resistance of the earth circuit is negligible. The impedances to earth at the cable ends of power cables is generally very small. To simplify estimations of interference conditions therefore the earthing resistance values have not been included in the calculation. The current reduction factor r_{Ki} of an interference producing power cable is then:

$$r_{Ki} = \frac{R'_M}{\sqrt{R'^2_M + \omega^2 \cdot L'^2_E}} \tag{26.3}$$

r_{Ki} Current reduction factor
R'_M d.c. resistance per unit length of metallic covering in Ω/m
ω Frequency s^{-1}
L'_E Inductance per unit length of earth circuit (approx. 2×10^{-6}) in H/km

In order to achieve a favourable (smallest) reduction factor r_{Ki} a low resistance metallic covering is desirable.

Screened cables, e.g.

with PVC insulation   NYSY, NYSEY, NYFGY, NAYSY, NAYSEY, NAYFGY

or

with XLPE insulation   N2XS2Y, N2XSE2Y, NA2XS2Y, NA2XSE2Y

have only screens with relatively low conduction values. These are of copper wire having a geometric cross section of 16 to 35 mm^2 or of flat steel wire armour consisting of wires 0.8 to 1.4 mm thick (the influence of an open steel tape spiral binder is small). The reduction factor lies mainly between 0.87 and 0.98 and therefore is not of significant effect in most cases.

For cables with steel-tape armour (minimum two layers overlapping), the inductance of the steel-tape armour needs to be taken into account in equation 26.3 in the denominator of the equation. As the permeability of the steel-tape armour varies relative to the strength of the magnetic field the reduction factor for such cables varies depending on the magnitude of the influencing current. Because the calculation of reduction factors for steel-tape armoured cable with lead sheath is most complex, the values are represented by curves shown in Figs. 26.3 and 26.4. The sheath thicknesses and the steel-tape dimensions comply with the VDE specifications.

For unarmoured cables with lead or aluminium sheath the reduction factor r_{Ki} can be taken from Figs. 26.5 to 26.7. For three-core cable with 20% covering of flat-steel wire armour of 1.4 mm thickness (e.g. NÖKDEFOA) Fig. 26.8 applies. For cables in steel pipe (e.g. type ÖIGLuSt2Y, NIvFSt2Y and NPKDvFSt2Y) a reduction factor r_{Ki} of 0.3 can be used. The diameter of the steel pipes is approximately 120 mm and the wall thickness approximately 4 mm.

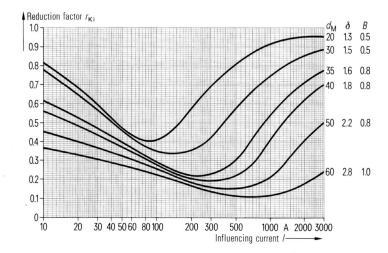

d_M Diameter over lead sheath in mm
δ Thickness of lead sheath in mm
B Thickness of steel tape in mm

Fig. 26.3
Reduction factor r_{Ki} relative to influencing current in cables with lead sheath and double steel-tape armour at $f = 50$ Hz

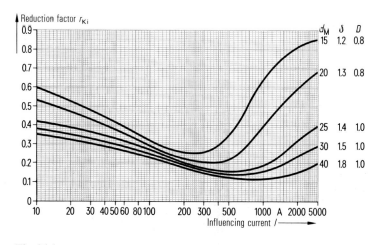

d_M Diameter over lead sheath in mm
δ Thickness of lead sheath in mm
B Thickness of steel tape in mm

Fig. 26.4
Reduction factor r_{Ki} relative to influencing current for S.L. cables with lead sheath and double steel-tape armour at $f = 50$ Hz

353

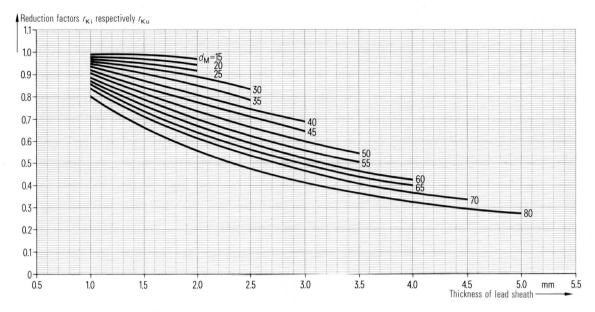

d_M Diameter over lead sheath in mm

Fig. 26.5
Reduction factors r_{Ki} and r_{Ku} for unarmoured cables with lead sheath at $f = 50\,\text{Hz}$

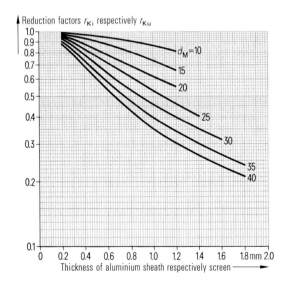

d_M Diameter over aluminium sheath respectively screen in mm

Fig. 26.6
Reduction factors r_{Ki} and r_{Ku} for non-armoured cables with smooth aluminium sheath respectively screen at $f = 50\,\text{Hz}$

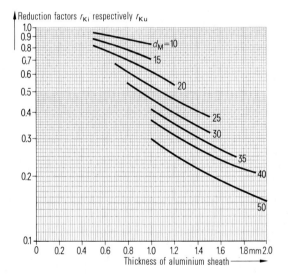

d_M Diameter over aluminium sheath in mm

Fig. 26.7
Reduction factors r_{Ki} and r_{Ku} for unarmoured cables with corrugated aluminium sheath at $f = 50\,\text{Hz}$

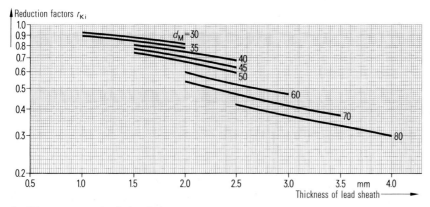

d_M Diameter over lead sheath in mm

Fig. 26.8
Reduction factor r_Ki for cables with lead sheath and 20% covering of flat steel-wire armour 1.4 mm thick at $f = 50$ Hz

26.1.4 Voltage Reduction Factor of the Influenced Telecommunication Cable

The voltage which is liable to endanger the telecommunication or control cables stresses the insulation between conductor and metallic covering. As the conductors do not carry current, the voltage in the circuit conductor – metal covering is equal to the voltage drop in the metal covering created by the current in the sheath longitudinal resistance. With sufficiently good earthing the voltage between conductor and metal sheath and between conductor and earth are equal. The voltage reduction factor r_Ku of an influenced cable is defined as the ratio of, induced EMF per unit length in a conductor of a cable having an earthed metallic sheath to, induced EMF per unit length in a conductor of a similar cable but without a metallic sheath.

The reduction factor of a telecommunication cable is only then effective when the metallic cable sheath is continuously earthed or at least earthed at the ends. A continuous earthing is also provided if the metallic cable sheath or the armour has a jute serving which is directly in contact with the ground. Metal sheaths of cables with an outer thermoplastic sheath, however, have no contact with earth. For the earthing of such cables at the ends of an influenced section the following is applicable:

The end points respectively the intermediate connection points of the telecommunication cable must be within the potential level of the earthing system (plate earth, ring main earth). The value of earth resistance is not critical as far as effective interference voltage at these points is concerned. The earthing system must be, apart from earthing the screen elements of the cable, also bonded to other existing earthing elements such as water pipes, lightning conductors and such like. Branch circuit interference is not critical providing these are connected to the influenced run through isolating transformers.

If these conditions are not fulfilled then the values of earthing resistances must be taken into consideration when the reduction factor is calculated. In the worst case in equation (26.3) the value of R'_M must be replaced with R'_f where:

$$R'_\text{f} = \frac{R'_\text{M} \cdot l + 2 \cdot R_\text{e}}{l} \qquad (26.4)$$

R'_f Fictitious value of d.c. resistance of metallic cable covering in Ω/m

R_e Mean value of earthing resistance at both ends in Ω

l Distance between earthing points in km

R'_M d.c. resistance per unit length of metallic covering in Ω/m

For the calculation of the reduction factor r_{Ku} (for influenced cables) in the same sense the same formula applies as used for the calculation of reduction factor r_{Ki} (for influencing cables). For non-steel tape armoured types

$$r_{Ku} = \frac{R'_M}{\sqrt{R'^2_M + \omega^2 \cdot L'^2_E}}. \qquad (26.5)$$

r_{Ku} Voltage reduction factor for influenced cable
R'_M d.c. resistance of metallic covering in Ω/m
L'_E Inductance per unit length of earth circuit (approx. 2×10^{-6}) in H/m
ω Frequency in s^{-1}

For r_{Ku} the curves to Figs. 26.5 to 26.7 are applicable. The values for cables with lead sheath respectively laminated aluminium sheath (thickness of aluminium 0.2 mm) can be taken from Fig. 26.9. The longitudinal resistance of a 0.2 mm thick aluminium screen is approximately equal to that of a lead sheath of the same diameter. For a cable with a copper screen of 0.12 mm thick the same values apply as for an aluminium screen of 0.2 mm thickness.

For influenced cables with steel-tape armour the reduction factor depends on the induced EMF E'_i at the location of the cable:

$$E'_i = \omega \cdot M'_{Ee} \cdot I \cdot r_{Ki} \cdot r_x \qquad (26.6)$$

with r_x representing a reduction factor of a compensating conductor (see page 357, for formula symbols see page 351 equation (26.1)).

The curves shown in Figs. 26.9 and 26.10 give relative values for reduction factor r_{Ku} for normal cable constructions with steel tape armour in respect of the induced EMF E'_i. Cable dimensions are based on DIN VDE 0816.

A corrugated steel sheath has similar protective properties to that of steel-tape armour. To achieve a favourable reduction factor the value of conductance must be increased, e.g. by adding a copper wire layer under the corrugated steel sheath.

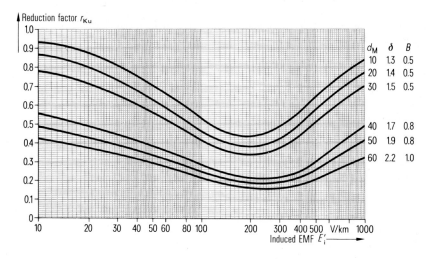

d_M Diameter over lead sheath in mm
δ Thickness of lead sheath in mm
B Thickness of steel tape in mm

Fig. 26.9
Reduction factor r_{Ku} for telecommunication cables with lead sheath respectively laminated aluminium sheath and two layers of steel-tape armour relative to the induced EMF at $f = 50\,Hz$

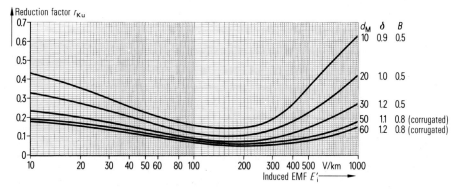

d_M Diameter over aluminium sheath in mm
δ Thickness of aluminium sheath in mm
B Thickness of steel tape in mm

Fig. 26.10
Reduction factor r_{Ku} for telecommunication cables with aluminium sheath and two layers of steel-tape armour relative to the induced EMF at $f = 50$ Hz

26.1.5 Reduction Factors of Compensating Conductors

If a power cable (1), a communication cable (2) and an earthed conductor (3) are run parallel to each other (Fig. 26.11) over the same distance, the calculation of interference is made with a reduction factor r_x of the compensating conductor to equation (26.7). Often the compensating conductor is a railway track or the earth braids of power cables (reduction factors r_s and r_E)

$$r_x = \sqrt{\frac{R_3'^2 + \omega^2 \left(L_3' - \dfrac{M_{13}' \cdot M_{23}'}{M_{12}'}\right)^2}{R_3'^2 + \omega^2 L_3'^2}} \qquad (26.7)$$

r_x Reduction factor of the compensating conductor
R_3' Resistance per unit length of the compensating conductor in Ω/m
L_3' Inductance per unit length of the circuit compensating conductor – earth (approx. 2×10^{-6}) H/m
M_{13}' Mutual inductance per unit length between (1) and (3) in H/m
M_{23}' Mutual inductance per unit length between (2) and (3) in H/m
M_{12}' Mutual inductance per unit length between (1) and (2) in H/m

The mutual inductances can be taken from the graph in Fig. 26.2.

In the Tables 26.1 and 26.2 a summary is given of the reduction factors r_s and r_E which apply to railway tracks and earthing ropes. In bad ground situations where conductivity is poor the reduction factors are more favourable, conversely where the conductivity is good the factors are less favourable. The rail reduction factors apply to influenced cables in the immediate proximity of the rail (distances up to about 5 m). For power cables which run in the same trench as the influenced cable but alongside the railway track the values are higher by approximately 20%.

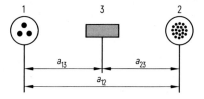

Fig. 26.11
Compensating conductor (3) between power cable (1) and telecommunication cable (2)

Table 26.1
Reduction factor r_S for railway track at a mean specific ground resistance of approximately $50\,\Omega\mathrm{m}$

Reduction factor for rails		r_S
Non electrified track	Single or multiple track	0.8 0.6
Electrified track	Single or double track Triple or multi-track Large railway stations	0.5 0.35 0.2

Table 26.2
Reduction factor r_E for earth ropes at a mean specific earth resistance of approximately $50\,\Omega\mathrm{m}$

Type of earth braid	Cross section mm^2	d.c. resistance Ω/km	r_E
Steel	70	2.31	0.98
Steel/copper	30/20	0.813	0.84
Steel/copper	42/28	0.586	0.78
Bronze	50	0.473	0.75
Bronze	70	0.345	0.69
Steel/aluminium	32/44	0.518	0.77
Steel/aluminium	32/185	0.145	0.62
Steel/aluminium	50/300	0.093	0.61

26.2 Noise Voltage in Symmetrical Circuits

The impedance values of the earth circuit loop of each conductor of a telecommunication or control cable differ to some extent, due to unavoidable manufacturing tolerances. Whilst the difference is very small in telecommunication cables, where conductors are laid up in either pairs or quads, it is considerably larger with the conductors of control cables as they are stranded in layers. This asymmetry may be caused, for example, by the capacity difference between conductors of a pair and the sheath – the so called earth coupling. Depending on the EMF, voltages appear between the conductors, which are termed, external voltages (unweighted noise voltage), which may cause distortion of the signalling circuits.

If, for example, for remote control purposes a circuit comprises one conductor in the outermost and one on the innermost layer of a cable, a certain percentage of the interference voltage is likely to appear between these conductors. The current induced by this transverse voltage may lead to maloperation of relays under certain conditions.

This transverse voltage is, of course, not only due to imbalance to earth within the cable, but also to that of the equipment directly connected to the conductors.

Only cables having conductors laid up in pairs and equipment with adequate balance to earth and dielectric strength should therefore be used in locations where high levels of interference voltage are expected.

26.3 Ohmic Interference

Apart from the interference with telecommunication cables caused by induction from power cables installed parallel, in the event of earth fault in the power cable they may be seriously influenced by ohmic coupling also. This type of influence is likely mainly with cables entering a power or transformer station. In the event of a shortcircuit to earth the voltage drop which develops on the earthing equipment, due to earth impedance Z_E, appears between the conductors of the telecommunication cable and earth reduced by the reduction factor r_{Ku}. The ohmic interference voltage is therefore

$$E_{\Omega k} = I_E \cdot Z_E \cdot r_{Ku}. \tag{26.8}$$

$E_{\Omega k}$ Interference voltage due to ohmic coupling in V

I_E Part of the earth current which flows from earthing network to earth in A. This current is never greater than the product of the earth-fault current and the largest reduction factor of one of the power conductors of the faulty network

Z_E Impedance to earth of earthing equipment in Ω

r_{Ku} Reduction factor of the telecommunication cables

The steel-tape armour on the telecommunication cable has little effect in reducing the interference due to ohmic coupling as the armour becomes magnetically saturated. The armour may again become effective, however, if the metallic covering is installed insulated within the interference zone and beyond. At the end of the insulated run the metal covering must again be earthed. The ohmic longitudinal voltage per unit length for the determination of the effective reduction factor can be calculated, in this case, with the following formula:

$$E'_i = \frac{I_E \cdot Z_E}{l_{is}}. \tag{26.9}$$

E'_i Ohmic longitudinal voltage per unit length in V/m

I_E Part of earth current which flows through the earthing arrangement to ground, in A

Z_E Impedance to earth of the earthing equipment in Ω

l_{is} Insulated cable length in m.

By suitably selecting the length of the insulated section the ohmic longitudinal voltage per unit length can be brought to a favourable value in respect of the magnetising of the steel-tape armour.

26.4 Inductive and Ohmic Interference

Inductive and ohmic interference may occur at the same time in a telecommunication cable if it is run into a power station or transformer station together with a power line or cable. The induced longitudinal EMF E_i, which should be calculated separately, and the interference voltage due to ohmic coupling $E_{\Omega k}$ are displaced by approximately a 90° phase relationship to each other. The total interference voltage therefore is:

$$E = \sqrt{E_i^2 + E_{\Omega k}^2}. \tag{26.10}$$

E Total interference voltage in V

E_i Induced longitudinal EMF in V

$E_{\Omega k}$ Ohmic coupled interference voltage in V

26.5 Details Required for Planning

For each project a detailed investigation of the protective measures to be adopted must be undertaken in order to achieve the best results both technically and commercially. Voltage arrestors or isolating transformers may be installed in existing telecommunication installations to protect them from interference voltages. Telecommunication cables for new installations should be selected however with coverings designed to reduce outside interference to an acceptable level.

Influencing Cables or Overhead Lines

1. Operating voltage and frequency

2. Type of star-point earthing to DIN VDE 0141 (solidly earthed, insulated star point, earthed via limiting coil or resistance)

3. Detail of three-phase overhead lines
   a) Cross-sectional area and material of main conductors and earth braid.
   b) Pictorial arrangement of overhead line supports giving dimensions and, in the case where several systems are supported by the same structure, the phase marking.
   These details are necessary only if overhead telecommunication cables are installed either on the same structures as the high-voltage lines or are laid in the ground parallel over a distance of several kilometers at a distance of ≤ 50 m from the three-phase power lines.
   c) If the power phase conductors are transposed, or if this is proposed for the future, then distances of the transposition points from the feeding power station or transformer station are required. (This data is also required under b.)

4. Arrangement of three-phase cables
   Type designation, diameter, material and thickness of metal sheath and armour (preferably drawing of cross-section with dimensions); number of parallel cables.

5. Earth short-circuit currents
   For this the sum of the zero phase sequence currents $3I_0$ is the determining factor which is the initial short-circuit a.c. current value. The inducing currents are relative to the distance between the earth short-circuit point and the feed point and this should be shown in diagrammatic form. The minimum requirements are:
   a) Short-circuit current available at the system input busbars.
   b) Highest possible value of inducing short-circuit current (sum of the zero-phase sequence currents $3I_0$) over the whole influenced length; in networks extinguishing coils or with insulated star point the double-earth fault short-circuit current over the whole run is the decisive factor.

6. Disconnect time of the protective circuit breaker under short-circuit conditions (single/double-earth fault)

359

7. Maximum earth-fault current with earth fault on one phase in networks with insulated or compensated star point (earth-fault residual current)

8. Maximum load capacity of every three-phase system

9. Earth-circuit resistances of the internal earthing systems of the feeder power station or transformer station.

Influenced Telecommunication Cable

1. Total length of influenced run (parallel with power line)

2. Distance between telecommunication cable and high-voltage line (where the distance varies over the length an exact route plan is necessary)

3. Distance between the input feed of the power line and the influenced cable run

4. Specific ground resistance along the influenced run.

5. Details of existing metallic earthed conductors close to the cable run, e.g. water pipe lines, gas pipes or cables (diameter, material, wall thickness); layout (possibly with indication of points where these earthed conductors are discontinuous), the type of overall insulation, earth contact resistance (e.g. per 1 m^2 of surface), cathodic protection.

6. Cable type and year of manufacture (where telecommunication cable already exists); construction, with indication of diameter, type and thickness of sheath and the armour (preferably sketch of cross-section)

7. Details of the transmission system used in the telecommunication cable (d.c. or a.c., operating voltage, communication or signal transmission, dialling by earth pulsing, network protection, railway signal conductors, data transmission...)

8. Type of cable termination (termination of all or part of the conductors with isolating transformer or voltage arrestors).

26.6 Calculated Example

Example 26.1

A telecommunication cable is to be installed close to a 220 kV overhead line. The following details are known:

220 kV overhead line, solidly earthed, $f = 50$ Hz

earth braid steel/aluminium 32/185 ($r_E = 0.62$ to Table 26.2 see page 358);

earth short-circuit current over the total influenced run $3I_0 = 8.5$ kA;

allowing for an expectancy factor $w = 0.7$ to DIN VDE 0228 in equation 26.1 for $I = 8.5 \times 0.7 = 5.95$ kA must be inserted;

length of parallel run $l = 5$ km, distance between telecommunication cable and overhead line $a = 100$ m (Fig. 26.2 $M'_{Ee} = 0.4$ mH/km at $\varrho = 50$ Ωm).

If the telecommunication cable is of simple construction with laminated sheath (aluminium screen 0.2 mm thick) is this construction adequate? The reduction factor for 25 mm screen diameter is only 0.95 approximately (Fig. 26.6), the test voltage between core and screen is 2000 V.

With equation (26.1) we get:

$$E_i = \omega \cdot M'_{Ee} \cdot l \cdot I \cdot r_E \cdot r_{Ku}$$
$$= 100\pi \times 0.4 \times 10^{-6} \times 5 \times 10^3 \times 5.95 \times 10^3$$
$$\times 0.62 \times 0.95$$
$$= 2202 \text{ V}.$$

For the telecommunication cable, therefore, the following possibilities exist:

a) All conductors can be equipped at the ends and at intermediate feed points, if present, with voltage surge arrestors connected to local earth points. The value of $E_i = 2202$ V is below the permissible value of 120% (when using voltage arrestors) of the cable test voltage.

b) If the cable, for transmission system reasons, cannot be connected to earth through arrestors in the event of interference, then it is only permissible for $E_i \leq 1200$ V (see page 349). All transmission circuits could however be terminated with isolating transformers. The required reduction factor must then not exceed the value:

$$r_{Ku} = \frac{1200 \times 0.95}{2202} = 0.52.$$

Therefore it is possible, e.g. to use an aluminium sheathed cable having a sheath thickness of 1.1 mm minimum at a diameter of 25 mm (Fig. 26.6).

c) as b) however a proportion of the transmission circuits cannot be terminated in isolating transformers because the systems are d.c. operated. Only $E_i \leq 500$ V is permissible in this condition (see page 349). The reduction factor of the cable must not exceed, therefore, the value:

$$r_{Ku} = \frac{500 \times 0.95}{2202} = 0.22.$$

If one attempted to achieve this value only by improving the screen conductivity value, the cost would prove uneconomical. In such cases a two layer steel-tape armour is arranged above a screen having sufficient conductivity.

The induced EMF is then, from equation 26.6:

$$E_i' = \omega \cdot M_{Ee}' \cdot I \cdot r_E$$
$$= 100\pi \times 0.4 \times 10^{-6} \times 5.95 \times 10^3 \times 0.62$$
$$= 463 \text{ V/km}.$$

From Fig. 26.10 it can be seen that at this field strength for an aluminium-sheathed cable with steel-tape armour and a 25 mm diameter sheath of thickness 1.1 mm a value of $r_{Ku} = 0.18$ is achieved. The value therefore is safety below $E_{ik} = 500$ V.

Example 26.2

A 20 kV cable is laid in close proximity to a cable of a railway signalling system (block system) alongside an electrified railway track. The German Railway Authority specify that the permissible interference voltage must not exceed 1200 V.

The following details are known:

    20 kV-cable, extinguishing coil earthed system;

    Power cable: S.L. cable NEKEBA 3×50 rm 11.6/20 kV (diameter over lead sheath 22 mm, thickness of lead sheath 1.3 mm);

    Double earth-fault current 3 kA (a probability factor is not permissible in respect of interference with signalling systems);

    Length of parallel run 5 km;

    Distance power cable – telecommunication cable 1 m (mutual inductance to Fig. 26.2 for $\varrho = 50 \, \Omega m$ is 1.3 mH/m).

Is the existing telecommunication cable with lead sheath and steel-tape armour satisfactory? Diameter of the lead sheath 30 mm, thickness of the lead sheath 1.5 mm, steel-taper armouring 2×0.5 mm.

From Fig. 26.4 the reduction factor r_{Ki} for the power cable, under an earth short-circuit current of 3 kA, is approximately 0.7. The track reduction factor r_s is 0.5. To determine the effective reduction factor for the telecommunication cable, we must first calculate the induced EMF to equation 26.6

$$E_i' = \omega \cdot M_{Ee}' \cdot I \cdot r_{Ki} \cdot r_s$$
$$= 100\pi \times 1.3 \times 10^{-6} \times 3 \times 10^3 \times 0.7 \times 0.5$$
$$= 429 \text{ V/km}.$$

By reference to Fig. 26.9 a reduction factor for the telecommunication cable of 0.45 is found. The total longitudinally induced EMF is to equation 26.6

$$E_i = E_i' \cdot l \cdot r_{Ku} = 429 \times 10^{-3} \times 5 \times 10^3 \times 0.45$$
$$= 965 \text{ V}.$$

This value is below the permissible limit of 1200 V.

27 Design and Calculation of Distribution Systems

27.1 Introduction

The supply of electrical energy entails:

Generation in a power station,

Transmission at high-voltage levels to load centres,

Distribution through medium-voltage and low-voltage systems to individual consumers.

A significant proportion of the overall investment cost for the supply of electrical energy from the power-station to the consumer is – due to the obligation to make a supply available over a wide spread area – attributable to the network at distribution level [27.1] (Fig. 27.1).

Medium- and low-voltage distribution systems in comparison with high-voltage systems with transmission function require a high proportion of cable. It is typical for these distribution systems, that they have a large number of components invariably of the same type (cable, transformers etc.). When determining these components certain minimum requirements must be fulfilled, e.g. in respect of their electrical and mechanical characteristics. Since each consumer must

be connected, irrespective of power requirement, an optimised loading of the system is, therefore, not always achievable.

The majority of consumers requiring energy for lighting, power and heating are supplied from low voltage. Low-voltage systems therefore require the greatest proportion of costs of electrical energy distribution. In the public supply, the low-voltage system feeds consumers with low or small power requirements such as households, small industrial units, shops and workshops.

For special tariff consumers, e.g. large scale consumers such as industry or department stores, frequently consumer-owned supply terminals exist for taking power direct from the medium-voltage system. These special consumers may also have their own medium-voltage system as well as an extensive low-voltage system (house installation system).

The intensive and comprehensive use of electrical energy preconditions a suitable "quality of supply" which means adherence to a set of agreed or standard values (voltage, frequency) and a high degree of reliability of energy supply. Quality of supply and cost

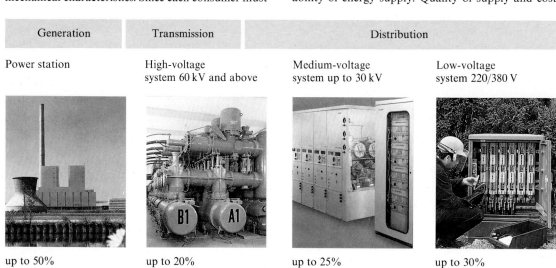

Generation	Transmission	Distribution	
Power station	High-voltage system 60 kV and above	Medium-voltage system up to 30 kV	Low-voltage system 220/380 V
up to 50%	up to 20%	up to 25%	up to 30%

Fig. 27.1 Proportion of investment cost in the supply of electrical energy

must have an acceptable relationship to one another and must take account of the requirements of the majority of consumers. Where particularly high-grade requirements are demanded it may be necessary for the consumer to meet the resulting additional costs [27.2].

The crucial point of the technical-economical optimisation lies, for the cost-intensive distribution systems, in the selection of

▷ uniform component parts

▷ economic voltage steps

▷ suitable forms of systems.

The use of a minimum number of types and sizes of, e.g. cables and transformers as uniform component parts, allows cost effective construction of systems and also simplifies project design work. The selection should be such that a wide range of system and consumer structure can be accommodated.

A significant contribution to avoid unnecessary losses in transmission and transformation is provided by the selection of suitable voltage steps. Often in the course of the development of a network, existing voltage steps can become uneconomical and new voltage levels will need to be established.

Over a period of time, systems may become complex and require reorientating by the introduction of simple systems, which is a precondition for the optimum operation of distribution systems.

The realisation of these complex conditions makes it necessary to carefully plan the future extensions of existing distribution systems as well as of new installations.

27.2 Determination of Power Requirement as a Basis for Planning

Distribution systems are closely associated with the consumer. Selection of component parts and their arrangement are therefore influenced directly by local conditions and the power requirements of consumers. To prepare for the design work for a distribution system, therefore the existing load and its local distribution must be established and the future development estimated.

A complex area of supply, depending on its geographical structure and natural limitations (wide roads, railway tracks, courses of rivers) needs to be subdivided into smaller areas. The power requirement for several stages of construction has to be determined, based on plans for building development and consumer structure, e.g. housing estates, trade or industrial development.

27.2.1 Load Requirement of Dwellings

The predominant proportion of the load in low-voltage systems of the public supply is attributable to the requirements of residential properties. Depending on the equipment included, the sum of the rated power of all electrical appliances within one dwelling (WE) can be up to 30 kW or more. The real peak demand for one dwelling will, however, be significantly less than this value because not all the apparatus will be in use at the same time. Fig. 27.2 shows the peak demand H_0 for a dwelling.

As the number of dwellings to be supplied by a part of a system (rising mains, main cables, substations, medium-voltage cables) increases, the proportional peak demand per dwelling relative to the total load on this part of the system decreases (simultaneity factor).

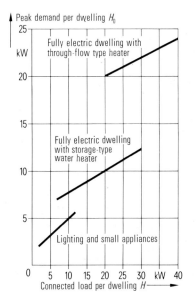

Fig. 27.2
Peak demand H_0, relative to connected load H of a dwelling

The system-load portion P to be expected results from the peak demand H_0 of a dwelling multiplied by the simultaneity factor g_n. Previously this was calculated taking the number of dwellings n to be supplied using an equation developed by Rusk [27.3]:

$$P = g_n \cdot H_0$$

$$g_n = g_\infty + \frac{1 - g_\infty}{n}.$$

From later investigations, especially in connection with supply systems for large towns, the following formula was evolved [21.4]:

$$g_n = g_\infty + \frac{1 - g_\infty}{n^{3/4}}.$$

This formula was used as a basis for the curves shown in Fig. 27.3. This figure shows values gained from practical experience for the system load dependent on the number of dwellings each equipped differently with electrical appliances. The simultaneity factor g_∞

for an infinite number of consumers is given for each curve. The curves are based on average connected demand per dwelling. For a housing estate with, e.g. 50 dwellings with an average connected demand of 20 kW, one can expect a system-load portion of 2 kW per dwelling.

With a smaller number of dwellings the full peak demand can occur. Moreover where special consumer load patterns arise, e.g. a works housing estate with shift working, a higher simultaneity factor can be expected. In this, through-flow type heaters, because of their larger power demand (approximately 18 to 24 kW) are more noticeable than storage-type heaters (approximately 6 kW). This must be considered when sizing the rising main, the service fuse and the service cable.

In the systems with a large number of consumers the difference is equalised since the operating times of the through-flow heaters are comparatively short.

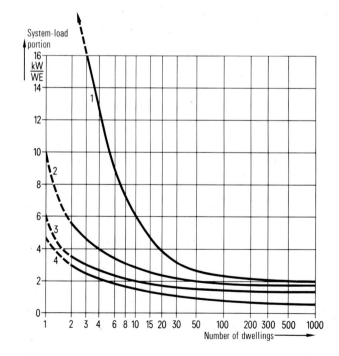

1 Fully electric household with through-flow heater ($g = 0.06$)
2 Fully electric household with storage-type water heater ($g = 0.2$)
3 Household with only electric cooker and appliances
4 Household with only electric lighting and other appliances ($g = 0.2$)

Fig. 27.3
System-load portion per dwelling (WE) relative to number of dwellings

For rough calculations for the planning of systems supplying approximately 50 dwellings or more, the following values may be used:

Type of dwelling	System-load portion P per dwelling (mean values for approx. 50 WE dwellings or more)
Without electric cooker or water heater (partly electric dwelling)	1 kW per WE
With electric cooker and water heater (fully electric dwelling)	2 to 3 kW per WE
With electric heating (all electric dwelling), with either direct or night-storage heating	5 to 15 kW and higher per WE

Load Increase in Existing Supply Systems

In existing systems load increase can occur, e.g. due to extensions and modernisation of dwellings or commercial premises.

The expected load is calculated using an annual rate of increase based on the present load and preceding development. Depending on the type of building development in the area of supply and the probable infilling of vacant spaces, growth rates of between 2 to 10% per year can arise.

Furthermore, without a general load increase due to alterations or change of consumers, local deviations of the load must be expected requiring a suitable extension to the system.

In order to fully estimate the future load, it is necessary to study building plans and area utilization plans of the relevant area.

27.2.2 Load Requirements of Special Consumers

When estimating the expected load for an area, the requirements of special consumers connected to the medium-voltage system must be considered (commercial enterprises, office blocks, hospitals, industrial premises etc.). Changes in the development structure of an area have an influence on the type of consumer and their size and proportional power requirement. If necessary therefore the load prognosis must be established separately for the requirements of low-voltage and medium-voltage consumers.

As far as this is done for a larger area of supply (e.g. a trading estate) mostly annual incremental rates and general estimation, e.g. of load density, for the forecast of load requirement will normally suffice. For the planning of internal supply systems for special consumers or for a more accurate forecast of load requirement for the local supply utility, an internal load analysis is necessary. As a basis for this, values obtained from practical experience or measured values from similar projects can be used.

Supply Systems in Buildings

For the estimation of loads for large building complexes the physical arrangement (vertical or horizontal) of the individual consumers and from this the distribution of load centre within the building must be taken into consideration.

Apart from consumer equipments spread across an area, e.g. light fittings and small appliances, mostly also concentrated loads (lifts, air conditioning equipment, sprinkler systems or large kitchens) must be supplied. For the consumer devices over an area, often specific values per unit area (Watts/m^2) are used. In the following some typical values are given which in real applications need to be verified because of the specific building or consumer situation [27.5]:

Lighting	10 to 25 W/m^2	
air conditioning	1 to 3 kW/equipment	
office buildings	100 W/m^2	2 kVA/work place
lifts	10 to 50 kVA/lift	
hotels	3 to 4 kVA/room.	

In as much as installed power is indicated, the real load consumer values need to be estimated using simultaneity factors e.g.

ventilation and air conditioning	$g = 0.9$ to 1
lighting	$g = 0.7$ to 0.95
data processing	$g = 1$.

Industrial and Trading Premises

In industrial and trading premises for the estimation of load the methods of manufacture and type of installed machinery are the determining factors. Depending on the type of industry it must be considered whether the loads are spread over the area with nearly equal load requirements or whether there are consumer groups with great differences in connected load value and with load peaks which can occur at differing times.

Expecially in heavy industry the connection of large individual loads must be considered.

From the installed power of the individual machines of an industrial plant, the relevant simultaneity factor g and the appropriate power factor $\cos \varphi$ real system load can be determined. As guide values for the determination of total load, the following demand per unit area may be used [27.6]:

Repair workshops,
automatic lathes,
weaving and spinning mills   50 to 100 kW/m^2

machine tool manufacture
mechanical workshops
and welding plant   70 to 300 kW/m^2

press shops, hardening,
steel smelting and rolling mills   200 to 500 kW/m^2

The simultaneity factor will differ dependent upon the type of apparatus connected (e.g. for machine tool manufacture $g = 0.4$, for crane installations $g = 0.2$). For lighting a simultaneity factor of $g = 1$ is used. The power factor is normally between $\cos \varphi = 0.5$ to 0.7 where no power factor correction is installed.

System Standby Supply Plant

In supply systems for large buildings and industrial plants, because of either regulations (DIN VDE 0108) or for operational or economic reasons, a very high reliability of supply is required. Therefore standby power supplies are installed (diesel generators, static converters with battery back-up). Depending on the power demand which must be met by this plant this may influence the choice for type of plant and system arrangement. For the estimation of the load this means separation of important consumers, which must be supplied in the event of main supply failure, from consumers of non-essential loads for which supply from the general system will suffice.

27.2.3 Total Load

To plan the incoming supply of the system under consideration from a higher level of voltage or from a power station requires knowledge of the total load to be expected. The time related differing load peaks of individual system parts are taken into account

when determining the total power requirement by the use of a simultaneity factor. As an example for the estimated peak-load of individual substations supplying an urban area, the simultaneity factor g of 0.6 to 0.7 for the total peak load is used.

It is recommended that the estimated load values are compared with measured real values from time to time and deviations considered when planning extensions to the systems. This is of particular importance for long term development of public utility distribution systems.

27.3 Planning of Distribution Systems

27.3.1 General

Simple system configuration, clearly arranged operation conditions and flexibility in respect of extensions of the distribution system must be objectives for the planning. Only careful planning provides the foundation to be able to supply electrical energy reliably and economically.

When extending existing systems, firstly an analysis of the actual capacity and operational reliability of the system is required. Load-flow and short-circuit calculations will emphasise weak points in the system and give an initial indication where improvements must be made. In this evaluation not only improvements based on minimum requirements for voltage stability, load capacity and short-circuit safety, but future demand must also be considered.

Also in new installations or extensions of supply systems the functional arrangement of systems is of particular importance. The selection of system configuration is dependent on the load and also structural make-up of the area or building to be supplied. Systems which have grown over a long period have not always the optimum configuration, especially where the load situation has changed in the meantime. The simplification of system configuration as a basis for the economical development and reliable operation is, next to the system calculation, an important part of the system planning (system architecture). This requires the creative selection of several alternative solutions. Numerous local conditions together with individual experience and planning philosophy thereby influence the decisions of the planning engineer. Because of this set of examples which follow cannot claim to be complete but can provide suggested methods for planning work on real projects.

The planning of distribution systems close to the consumers must be derived from the low-voltage system to which the majority of the consumers are connected. Position and power requirement of the consumers determine the construction of the low-voltage system. The location of feed points (substations) from the super imposed medium-voltage system should be positioned as close as possible to the load centre on the low-voltage side. The local position of substations and consumer stations and their power requirement are the load points of the medium-voltage system and influence the location of runs of medium-voltage cables. The positions available for feeder, transformer and system stations as well as the prospective routes for cables and overhead lines restrict optimised planning and often a compromise solution is necessary. It is the task of the planner to highlight, at an early stage, the consequences of the compromise indicating the additional costs and higher losses involved.

To achieve further improvement of the network, a change to a higher voltage level or application of new technology often means the development of several planning variants followed by a detailed comparison of these from both technical and economical aspects. The precondition for this is the fulfilment of performance requirements in respect of voltage stability, load capacity and short-circuit capability. These must be verified by suitable system calculations, e.g. load-flow calculations for normal operating conditions and fault conditions as well as short-circuit calculations.

An economical evaluation of the technically comparable variants must embrace the cost of investment, of operation and maintenance of the systems as well as the cost of losses.

All evaluations must be conducted for several extension stages according to the time-related progression of the load prognosis and the effect of the various improvement measures at differing times on the total cost (dynamic system planning). The evaluation of the cost situation can be carried out by using various methods of cost calculation (annuity method, cash-assets method).

Criteria of variants which cannot be expressed as costs, e.g. updating with prospective technologies, clearly arranged system configurations etc. can best be determined from efficiency analysis.

27.3.2 Selection of Distribution Voltage

In general the voltage levels of the low-voltage and medium-voltage systems are fixed for the utility supply authorities.

For the low-voltage system of the public supply, according to IEC [27.7] a uniform standard value of 230/400 V is recommended. In industry where a high proportion of load is in motors then in new installations 660 V is also employed.

The medium-voltage systems lying above the low-voltage systems must fulfil two main functions: It must be sufficiently powerful to transmit the high incoming power from the main substations (feeding from the high-voltage systems) and its component parts on the other hand transmit energy economically to numerous system substations and consumer stations. The optimum value for medium-voltage systems therefore lies within the voltage levels of 11 to 22 kV.

In many countries with high-load density, the simplification of voltage stepping in the public supply area is more or less well established. In countries were there is a significant rising demand for power the selection of voltage levels in the medium-voltage systems forms a particularly important part of the network planning. Frequently because of the hitherto development, numerous voltage steps are found and because of several transformation steps additional costs for investment and losses are incurred. It must, however, be checked whether these voltages levels are adequate in the future for increasing demand or whether a higher voltage system should be superimposed. In this aspect it must be assessed whether an existing intermediate voltage can be omitted partially or even completely or should be revised. For industrial systems in particular, an in-plant generation may exist and this, together with the possibility of installing high voltage motors, must be considered.

Fig. 27.4 indicates the main stages and variants of solutions when selecting and reviewing distribution voltages.

Where the voltage stepping exists at 20/6 kV, the lower level of 6 kV should be releaved by extending the 20 kV system so that in the long term the 6 kV system is dismantled.

With the voltage stepping of 30/6 kV an economically viable solution is to change the 6 kV system to 10 kV and feed this directly from the high-voltage system (110 kV).

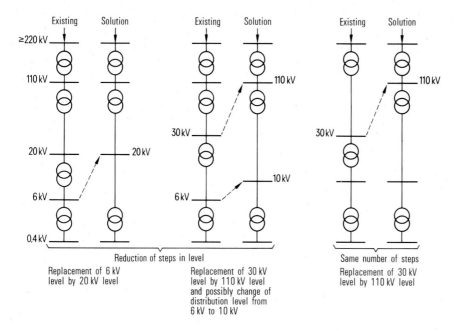

Fig. 27.4 Typical solutions in voltage selection

A great variation between the high-voltage and the existing 30 kV level may lead to the introduction of, for example, 110 kV instead of 30 kV as an economical solution. Generally between the feeding high-voltage level and the low-voltage systems the minimum number of steps should be incorporated to avoid high investment cost and additional losses at the transformation steps.

27.3.3 Low-Voltage Systems

System Configuration and Types of Operation in the Public Supply

Whether an area is supplied via cables or overhead lines has to be decided in consideration of local conditions and the economics of these alternatives. In areas with low-load density an overhead line may be a cost effective solution. However, today also in rural areas load values are reached for which, together with an architectural point of view the establishment of a cable network or the changeover from overhead line to cable is economically justifiable.

Depending on load density and type of structural arrangement of buildings, differing system configurations result for the low-voltage cable system.

In a conventional low-voltage system of the public supply the cable runs (mains cable) follow the route of highways (Fig. 27.5). At road junctions the cables are joined in cable distribution cabinets (nodes). Substations should wherever possible feed into the load centre and have a sufficient number of branches.

Dwellings are normally connected by means of a spur (service cable) from a branch or T-joint box or from a through-type joint box on the main supply cable. Service boxes have the advantage that no terminal points are required on the main cable. The through-type joint boxes offer possibilities for changing connections in the event of a fault but also have disadvantages because of their numerous terminal points.

Whether cables in a road are installed on one side or both sides depends upon the width of the road, the specific cost of installation and on load density. Where buildings are openly spaced (great distances between houses, narrow roads) and hence low-load density, the installation of cable on one side of the road may suffice. In close spacing of houses and hence high-load density, installation of main cables on both sides is generally more favourable. Streets with particularly high-power demand may be supplied with double cables on each side of the road with alternating connection of the houses where it may not be convenient to relieve the low-voltage system by the installation of separate substations (e.g. for special consumers).

Depending on the interlinking of the low-voltage connections in the distribution nodes, the low-voltage

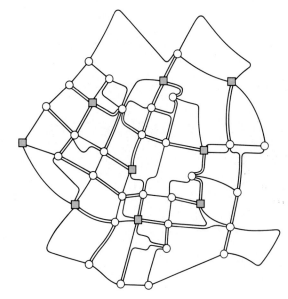

▨ Network station

○ Cable distribution cabinet

— Power cable with aluminium conductor NAYY

Fig. 27.5 Example of a conventional low-voltage network

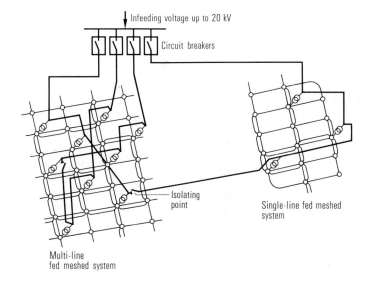

Infeeding voltage up to 20 kV

Circuit breakers

Isolating point

Single-line fed meshed system

Multi-line fed meshed system

Ϙ Transformer substation

— Meshed system

○ Cable distribution cabinet

Fig. 27.6
Examples of multi line fed and single-line fed meshed systems

369

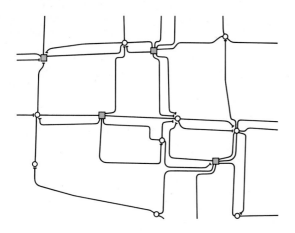

Fig. 27.7
Conventional L.V. system operated as radial network

▣ Network station
○ Cable distribution cabinet

system may be operated as a radial or as a meshed system. Security of supply, voltage stability and load equalization are all fulfilled by the use of a meshed system especially where numerous consumers of differing size with differing types of load are connected.

In single-line fed meshed system, all substations connected via the low-voltage network are supplied from a single medium-voltage line (Fig. 27.6).

In multi-line fed meshed systems, the substations of several medium-voltage lines can be connected in parallel through the low-voltage system. The meshing of the low-voltage system therefore does influence the development of the medium-voltage system. The connection of the substations to the medium-voltage cables must be coordinated with the operational requirements of the low-voltage system to avoid overloading if an individual medium-voltage line is opened (strategic location of stations). Because of these restrictions encountered in respect of flexible extension and operation of the network, meshed systems have lost prominence in the public supply sector, in spite of their other advantages.

Low-voltage systems because of the simple and clearly operation are today mainly operated as radial systems (Fig. 27.7). Each substation has its own area

of supply. Changeover possibilities to other substation areas are mostly provided for in the distribution cabinets, which allow for full or part reserve in the rare event of a substation failure.

Extension of a Low-Voltage System

The adaption of an existing low-voltage system to an increasing power demand becomes economically by introducing additional substations into nodes which currently are not directly fed (Fig. 27.8). In this way the load of the cable can be reduced in order not to exceed during the total period a maximum value, even if the load demand of the consumers is multiplied. This presumes that a sufficiently sized cross-sectional area of cable had been installed uniformly at the outset.

With increasing load density, the distance between substations is therefore reduced. The low-voltage system with high-load density becomes practically a pure connecting system (Fig. 27.9). The substations are operated without interlinking or possibility of reconnection. This type of system has advantages in very high-load density areas, e.g. in housing estates with electric storage heating.

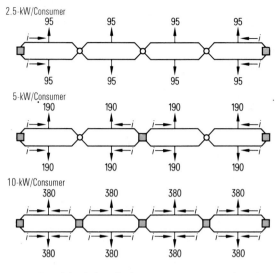

2.5-kW/Consumer

5-kW/Consumer

10-kW/Consumer

→95   Load in A (total of consumers on one shared cable)

▨   Substation (in feeding)

○   Cable distribution cabinet

i→   Cable circuit (max. 190 A at all stages of extension)

Fig. 27.8
System extension, due to increasing load density, by additional substations

Systems of Buildings

Large buildings with a high-load requirement are mostly supplied as special consumers from the medium-voltage system and have their own transformers as infeeding into an internal low-voltage system (installation system). Depending on the extent of the building complex, this low-voltage system may be widely branched or for a high rise building will consist only of rising mains with floor distribution systems.

For economical arrangement and operation, the system should be of simple configuration (radial system with reconnecting facilities) with short distances between infeed and load centre points being the planning objective. Feeder cables or conductors of great length should be avoided in favour of a decentralised installation with transformers immediately adjacent to the load centre points (e.g. a transformer for lifts and ventilation in the roof area of high-rise buildings or a transformer in the air conditioning centre). With such an installation a medium-voltage supply is then also necessary (Fig. 27.10).

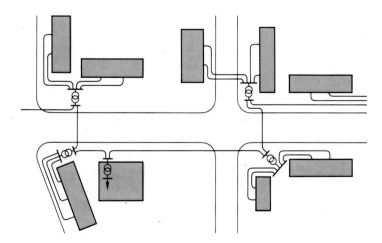

Fig. 27.9
Low-voltage connection network for a housing estate with high-load density

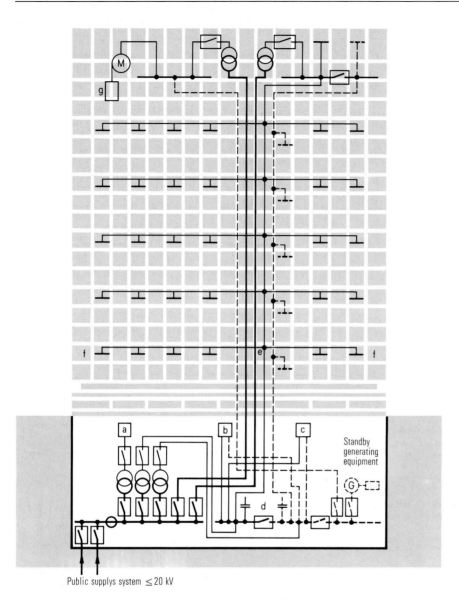

Public supplys system ≤ 20 kV

a   Cooling machinery
b   Heating, ventilating, pressurised water
c   Smoke stack, fire pumps
d   Central power factor correction (automatic)
e   Floor distribution system
f   Sub-distribution
g   Lifts

Fig. 27.10   Installation system in a high-rise building

Industrial Supply Systems

To feed the high loads in industrial supply systems and maintain short current paths on the low-voltage side, the transformers should be installed within the workroom close to the load centre points. The equipment and machines (loads) necessary for the production are mostly connected to a sub-distribution board.

These sub-distribution boards are the load points of the low-voltage system, e.g. within a workroom. When selecting the component parts for an industrial installation, the frequently used connecting systems such as Siemens 'Usystem' or the busbar trunking system (L-system) should be considered.

In industry radial systems are often employed (Fig. 27.11), depending on the requirement of security of supply with appropriate reconnection facilities in the sub-distribution boards (dublicate system).

For equipment with heavy load fluctuations, e.g. welding equipment, meshed systems (Fig. 27.12) may offer advantages in respect of voltage stability and load balancing.

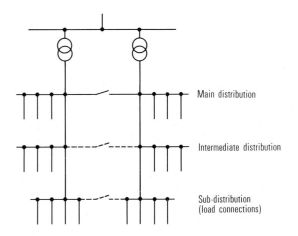

Main distribution

Intermediate distribution

Sub-distribution
(load connections)

Fig. 27.11
Example of low-voltage system (radial system) with reconnection facilities

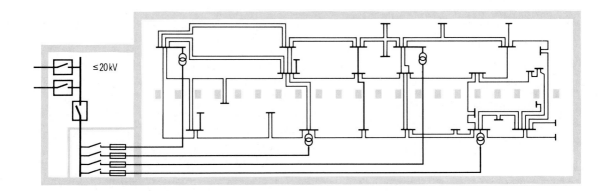

≤ 20 kV

Fig. 27.12 Example of meshed system in a workshop

Location of Substations

An important precondition for the economical performance improvement of low-voltage systems is the selection of optimum locations for the infeed points. The transformers should be situated as near as possible to the load centre to avoid transmission losses. Fig. 27.13 illustrates an example of three different variants for the connection of a consumer with a power requirement of up to 400 kVA to a 10 kV supply:

In variant 1 the substation is arranged in the vicinity of the 10 kV cable. The power is transmitted on the low-voltage side by cables.

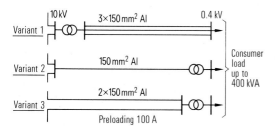

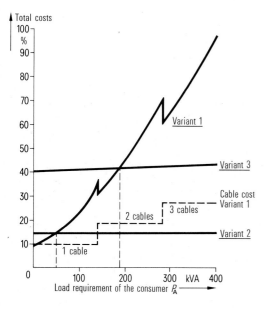

Fig. 27.13
Cost comparison of variants for siting and connection of a transformer station

Variant 2 shows a transformer arranged close to the consumer connected with a branch cable on the 10 kV side.

In variant 3 the substation is arranged close to the consumer and the 10 kV cable is looped-in.

For rough comparison of the variants, only the investment costs of cables and their installation together with the costs for losses are considered.

The cost curves show that for a consumer power requirement P_a of approximately 50 kVA variant 2 and likewise for P_a of approximately 200 kVA variant 3 shows advantage over variant 1. The location of the transformer station therefore deserves particular attention. From these cost relationships it can also be deduced that the capacity of individual transformers should be prudently selected to avoid larger transmission capacity and hence larger transmission losses in the low-voltage system [27.8].

Component Parts of the Low-Voltage System

Increasing load density requires more substations in the low-voltage system. The low-voltage cables therefore must be sized adequately from the outset and the siting of stations must be provided in advance to be able to adapt the low-voltage system optimal to the future development of load.

To achieve a reasonable improvement of the low-voltage system, uniform component parts should be employed. The range of economic conductor cross-section of the cable of copper conductors is generally between 70 and 120 mm^2 (correspondingly with conductors of aluminium 95 to 150 mm^2). Although the earth-works account for a significant proportion of the total cost, the cost of losses is also significant and therefore when in doubt a larger cross section should be selected.

Transformers used in substations of the public supply have rated capacities of up to 630 kVA. Since failures in transformers are very rare, it is sufficient to provide one transformer to each station. Only for heavy unit loads or very high-load density, e.g. in industrial plant, larger rated capacities are justified. Here also it may be necessary to consider, as reserve, a second or even several transformers.

27.3.4 Medium-Voltage Systems

Public Supply

The medium-voltage system should have a distribution function only, i.e. the medium-voltage cable should directly supply the substations and consumer stations which feed the low-voltage systems. Feeder cables (cables which serve only for the transportation of energy without any consumer or substation connected) should only be planned where long distances without load points are to be bridged and where it is not economic at the time to use a higher voltage level. This assumes that the high-voltage substations (main substations) have to be positioned directly at the centre of load of the medium-voltage system and that the high-voltage cables are brought into the load centre of the distribution areas.

The medium-voltage cables are normally connected by load break switches without protective function and looped in to the substations (cable-line link with stations). The cable-line links begin and end at switching stations (transformer station, remote station) equipped with circuit breakers including short-circuit protection. The number of stations looped into one cable-line link is limited by the load capacity of the cable (having reserve capacity also for other cable lines) and by operational considerations (number of stations effected in the event of a cable failure).

With a fault on one cable, the faulty part between two stations is isolated at both ends and the remaining cable-line links are switched to other cables. Depending on the location of the fault (e.g. at the beginning of a line) it may be necessary to supply the total load of the connected substations from another cable line. A corresponding reserve of capacity must therefore be built-in at the planning stage.

The extension of the medium-voltage distribution system is dependent to a great extent on the building development within the area. The systems having grown over a long period of time often are not optimised as far as their layout is concerned. Numerous cross connections between the individual cable-line links and a large number of substations with switch gear equipments may make operation of such systems less obvious. The connection of new stations and the installation of new cable runs open the possibility to disentangle and respectively simplify the arrangement.

For the simple and clear arrangement of distribution systems in the main two basic configurations are

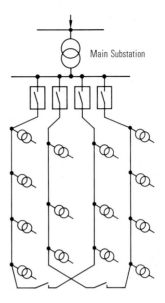

Main substation:
Circuit breaker with time-over current protection

Fig. 27.14 Medium Voltage open ring system

used, namely the ring system or a spur system with remote stations which may be varied if need be.

The ring system consists of cable-line links which are connected at both ends to the same feed point (Fig. 27.14). These are mostly operated in a substation close to the load centre in open form (open rings). With this method of operation, as opposed to closed ring operation, in the event of a fault the breakdown affects only one half of the ring.

As short-circuit protection of the cable branch circuits time-overcurrent protection in the feeder station is sufficient. Additional short-circuit indicators in the substations allow the faulty section to be quickly found and isolated in a short time. When suitable short-circuit indicators and auxiliary cables are used the fault indication can be transmitted to the system control center so that isolation of the faulty section can be carried out by remote control. The load connected to the cable is to be selected such that also in an unfavourable situation of fault interruption the total section can be fed from one side. The maximum loading of a ring cable is therefore 50% for each half ring to allow for full reserve capacity. This has the advantage that in operation, even with uncontrolled switching, overloading is not possible.

In a system with remote stations, the outgoing line cables from the feed point are connected at their remote ends at a switching station (remote station) (Fig. 27.15). If such a network is to be operated closed, the minimum protection requirement is for directional overcurrent protection in the remote station together with overcurrent protection at the feeder station or better distance protection is used on both ends.

Because of the differing lengths and load of the individual cable-line links, it is preferable to investigate

	Feed station:
1	Circuit breaker with time-delayed overcurrent protection

	Remote station:
2	Circuit breaker with directional overcurrent protection

Fig. 27.15
Medium-voltage system with remote station (closed loop operation)

the load distribution during normal operation and in conditions of fault and accordingly allocate a maximum permissible loading to the cables.

Whether a ring system or a system with remote station is selected is also affected by the extent of the area to be supplied, the position of the feed point (transformer station) and the routing of cables. For areas of supply with high-load density and a central position of the transformer station, a ring system is favoured. Extensively stretched areas which have the transformers situated at the periphery together with cable routes all approximately in the same direction lead to the selection of the remote station system.

Naturally numerous other variants of system configurations are possible or may also be present due to development of the system. In addition variations result due to local conditions (rural area, housing estates, urban areas) and the system resulting from them (overhead lines, cables). Independent of these, a largely simplified arrangement of the system to the aforementioned basic configurations is desirable.

When the subordinate low-voltage system is operated as a meshed sytem at the planning stage of the medium-voltage network, it must be observed that the associated medium-voltage lines start at one feed point and can be linked together for the commissioning of the system.

If several transformer stations are present within one urban area, each transformer station supplies its own section of medium-voltage system. Parallel operation of the transformer stations via the medium-voltage system is not recommended because of possible circulating currents and the increased level of short-circuit current.

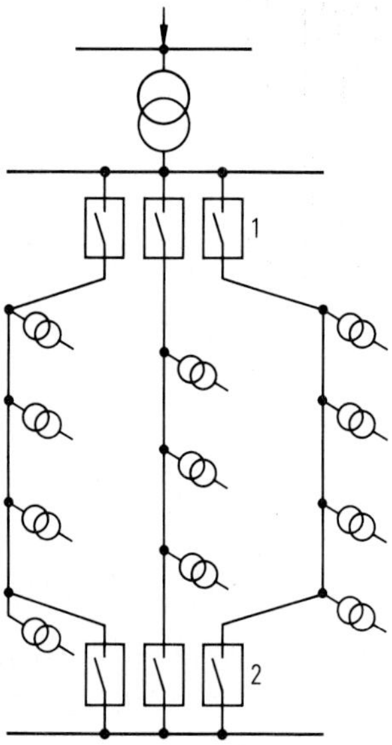

Expansion of the Medium-Voltage System

With increase of load or physical expansion of a medium-voltage system, firstly one would attempt to enlarge the existing installation and reinforce the present system. For small increases in power requirement, the cost of extension will lead towards this solution.

For large alterations in power demand or extension to the system, however, because of significant transmission losses and because of the need to incorporate new means of transmission, costs will quickly rise.

Extension of an existing system and transformer station

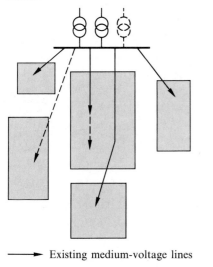

Extension by new transformer station with system splitting

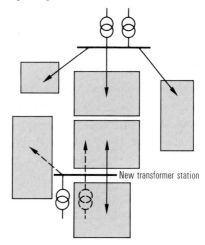

New transformer station

——▶ Existing medium-voltage lines

----▶ Planned medium-voltage lines

Areas of supply of medium-voltage systems

Fig. 27.16
Basic variants for extension of a medium-voltage system (M.V. lines shown simplified)

In this case new feed points need to be constructed, that means with increasing load new transformer stations must be incorporated into the existing system. The existing supply areas need to be split and the load of existing transformer stations relieved by the new stations (Fig. 27.16). The resulting load curve for the transformer over the total planning period is shown in Fig. 27.17.

The rated power of the transformers, with certain limitations, remains constant over the total time period and hence also the short-circuit power within the range of a transformer station remains unchanged. For the design planning of the medium-voltage switching station, definite base values are therefore fixed.

The inclusion of a new transformer station (feed point) in a medium-voltage system invariably leeds to a significant rearrangement of the system. At such an occasion therefore the existing system should be investigated and where possible the configuration updated. For the necessary planning work, the requirements for new substations of the associated low-voltage network should also be considered.

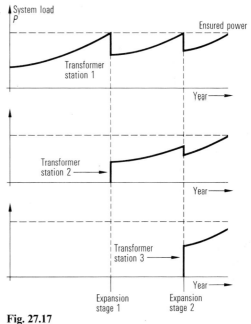

Fig. 27.17
Load curves of transformer stations during system extension stages

Distribution Systems in Large Buildings

Distribution systems in buildings differ from systems in the public supply because of the short distances between the decentralised substations. The part of the cost for switchgear relative to the total cost is therefore higher than the part of cost attributable to cables especially if, e.g., a ring system with circuit breaker and protection is selected to supply the substations (Fig. 27.18 a).

A significantly more favourable cost relation for providing a comparable degree of reliability of supply is achievable for such supply systems by a radial arrangement (distribution-centre feed system) (Fig. 27.18 b.)

The transformer branches are connected directly to the main substation via load-break switches and high-voltage fuses. Furthermore the cross-sectional area of branch cables can be considerably less than that required for a ring system. The reliability of supply for this system configuration, due to the reduced number of component parts, is higher than that of a ring system.

The cost advantage is limited by the number of transformers per substation and by the length of the branch cables. In extensive building complexes therefore several centre distribution systems are arranged for individual buildings or groups of buildings to reduce the number and length of branch cables. The individual main substations are operated, e.g. via a closed feeder system with selective fault detection (Fig. 27.19).

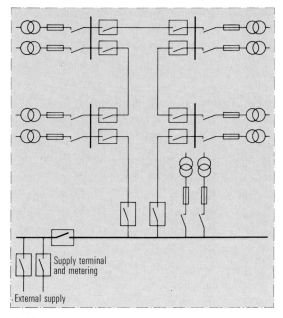

a) Ring system

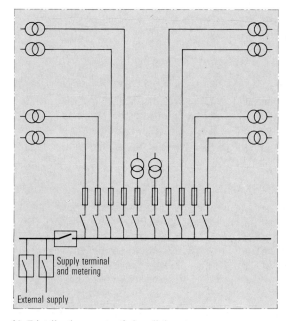

b) Distribution centre fed radial system

Fig. 27.18
Variants for medium-voltage systems in buildings

Industrial Supply Systems

In industrial plant the medium-voltage system is designed similar to that of a building. If in-plant generation is included, the generally higher short-circuit powers must be considered. For reasons of cost two separate part systems with different fault levels may be favoured.

A system with high short-circuit capacity (feeder system) supplies the different centre points as, e.g. the feed from the superposed public supply, in-plant generation or high-voltage motors for large drives.

The common distribution system (connected via short-circuit limiting reactors) supplies the majority of consumers (Fig. 27.20).

Often because of the particularly clear arrangement and simplified mode of operation, a purely radial system is selected for industrial plant which, however, is by suitable system separation in the switching stations designed as a "double system". With the possibility of reconnection (often automatically) a high operational reliability may be achieved (Fig. 27.21).

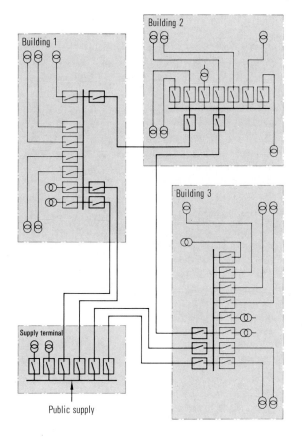

Fig. 27.19
Supply to an extensive building complex with feed system and several centre point feed systems

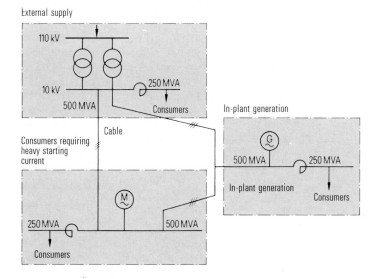

—⌒— Short-circuit limiting reactor

Fig. 27.20
Industrial medium-voltage system
(extension with two different levels)

379

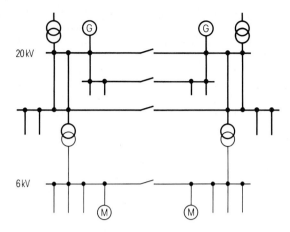

Fig. 27.21
Example of medium-voltage radial system with differing voltage levels

Standby Power Supply

With consumers of high power demand the necessary standby generation supply may reach power values which for economic reasons make it necessary to employ the standby power supply generators on the medium-voltage level. The medium-voltage system must be arranged such that on failure of a system supply and run-up of the standby plant either load shedding is possible or better an individual part of the system requiring standby power has to be isolated (Fig. 27.10).

Component Parts of the Medium-Voltage System

For the dimensioning of the medium-voltage cables mostly the maximum short-circuit power is the determining factor which is established in the planning phase (see Section 19.1). For this reason and the costs of switching equipment in the medium-voltage system in mind, the short-circuit power must be limited. This can be achieved by sub-division of systems, selection of a suitable short-circuit voltage of the infeeding transformers or by separate operation of the transformers.

In general one would not exceed, in 10 kV systems, a capacity of 250 MVA (in some cases also 350 MVA) and in 20 kV systems, 500 MVA.

With a disconnection time corresponding to the second time step of the selective protection of 0.6 seconds for medium-voltage cables this results, for the above short-circuit values, in a conductor cross-sectional area of 95 mm² for copper or 150 mm² for aluminium.

The selected cross-section should be installed uniformly throughout the system and also in the system branch lines. High ambient temperatures, unfavourable ground conditions or cable installations running parallel may make it necessary to use larger cross-sections. For special cases, e.g. for feeder cables, larger cross-sections should invariably be provided.

Switching equipment (circuit breakers and short-circuit protection) in the medium, voltage network serve also the function to to allow the necessary operational switching (reconnection) or fault recognition (short circuit disconnection). However, the provision of these must not impair the clarity of the system arrangement by being too great in number. Switch stations should therefore only be installed at preselected points (transformer stations, remote stations, former mainstations). In all other stations (substations) load break switches will suffice.

For both types of plant, in respect of protection of personnel, ease of operation and maintenance, the common instructions and regulations for the erection of switchgear must be observed.

Charge Current Compensation and Star-Point Treatment

Cables have, due to the way they are constructed, a higher capacitance per unit length than overhead lines which means that charging currents and earth capacitance currents are higher. The charging currents in medium-voltage networks of the public supply are mostly compensated by the reactive power demand of the system transformers and the consumers. Compensation equipment is therefore generally not required. In systems of buildings and industrial plant, however, due to the length of cable run being shorter, the inductive wattless power of the system is not fully compensated. In order to operate the system within the parameters for power factor laid down by the supply utility, additional power factor correction equipment (capacitors) is normally necessary.

With a single-pole fault (connecting conductor – earth) the voltage within the system is displaced. Depending on the treatment of the star point (Fig. 27.27)

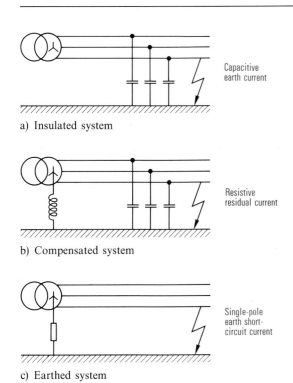

a) Insulated system

Capacitive
earth current

b) Compensated system

Resistive
residual current

c) Earthed system

Single-pole
earth short-
circuit current

Fig. 27.22
Various arrangements of star-point treatment

this results in a capacitive earth current (insulated system), a mostly resistive residual current (compensated system) or a single-pole short-circuit earth current (earthed system).

With insulated star-point, systems should be designed for operation with an earth current of only a few amperes. The earthing of the star point through a Petersen coil provides operating conditions most suitable for overhead line systems because many faults are cleared by self-extinguishing and do not lead to shut-down of the whole supply. In cable systems, however, non-localised transient earth faults create weak points in the system which may later lead to a double earth fault. Nevertheless many cable systems are operated in the compensated mode especially where there is a galvanic connection to an overhead system of the regional supply.

Where the sub-system is electrically separated from the regional supply, e.g. by the introduction of a new transformer station, then star-point earthing should be considered. By connecting an ohmic resistance or

reactance coil in the star-point earth connection, the single-pole earth fault current can be limited to a value of some 2000 A maximum. To provide for selective disconnection, the short-circuit protection should be of triple-pole type. Star-point earthing facilitates with clear and simple arrangement of systems a quick and, depending on the investment, also automatic fault location and isolation [27.9, 27.10].

The Superimposed High-Voltage System

The regional transmission of electrical energy is mostly carried out with voltages up to 132 kV, in Germany at 110 kV. There is an increasing trend for this voltage level, especially in areas of high-load density, to also be used for energy distribution replacing the already existing levels of 30 and 60 kV. The high-voltage system takes on the function of a distribution system and at the same time the proportion of cable used on this voltage level is also increasing.

Transformer stations, i.e. the input points from the high-voltage to the medium-voltage systems have to be, due to the load situation, situated at the load centre of urban areas and for reasons of space are to be constructed from indoor type. The SF_6 switchgear technique offers a compact solution for this. High-voltage distribution systems are mostly operated in a closed systems employing the relevant protective measures (distance protection, differential protection).

27.4 System Calculation

27.4.1 Basics

A number of influences such as building and electrical load structure or operating conditions affect the layout, i.e. the architecture of the system. To ensure that the technical conditions and relevant regulations are complied with, the necessary calculations must be made to verify this, i.e. load-flow and short-circuit investigations [27.11, 27.12].

With the load-flow investigations the voltages and currents of the individual components of operating equipment for different conditions of operation (peak load, light load, fault conditions) must be determined. An evaluation of the results of these calculations will highlight any excess or shortfall in respect of operational values, such as overloading of equipment or insufficient voltage at individual points of the system, providing also an on-load performance rating for the system including the losses.

Investigations into short-circuit currents will indicate the maximum levels of current which can occur in the system under fault conditions. These values can then be used to evaluate the dynamic and thermal stresses which the equipment must withstand and also the required interrupting capacities of the circuit breakers. Moreover to the short-circuit calculations in low-load operating conditions or for various special system switching combinations must be considered to verify the response conditions for the selective protection.

For these tasks data processing software programs are employed which make it possible to calculate expansive and complex systems. As a supplement to this, the voltage drop and short-circuit calculations which can be carried out with the aid of pocket calculators are discussed in various publications [27.13, 27.14].

In the following some common aspects for the low-voltage system are given.

The lengths of cable runs in the low-voltage system are limited by three criteria:

In systems with low-load density the minimum short-circuit current necessary to operate the protective de-vice to disconnect the fault, with high-load density the permissible voltage drop determines the maximum cable length for a given cross-sectional area. The maximum permissible load of the cable in respect of temperature rise according to VDE is the only determining factor in areas with high-load density such as in urban systems and industrial plant. Fig. 27.23 indicates the qualitative influence which the minimum short-circuit current, the voltage drop and the load capacity relative to load density have on the maximum permissible cable length.

In the low-voltage system a voltage drop of 3 to 5% of rated voltage between the output terminals of the transformer and the incoming supply fuse of a dwelling should not be exceeded since voltage sensitive apparatus such as radio and T.V. receivers may largely depend on the maintained level of voltage. However, it may not be possible to maintain this value at all parts of the system. In exceptional cases therefore where the high investment cost for connection is not economical, e.g. for a consumer at a peripheral point, a voltage drop of 5 to 7% may be acceptable. For an installation within a house an addition voltage drop of some 2% is to be considered.

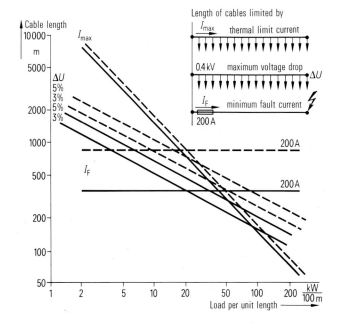

--- Cable 150 mm^2 Al
— Overhead lines 70 mm^2 Al

Fig. 27.23
Influence of load density on planning criteria of a low-voltage system

In medium-voltage cable systems the conductor cross-section selected is often sufficiently sized for the load condition after taking account of the requirements for short-circuit withstand. Problems in maintaining voltage levels can hardly be expected but it is recommended to investigate the load capacity because of possible unfavourable installation conditions (see Section 18.2).

27.4.2 Calculation of a Low-Voltage System

In the following an example is taken of a housing estate supplied from a substation by three low-voltage cable circuits showing the calculation method of voltage values and the selection of cross-sectional areas (Fig. 27.24 and Table 27.1).

The voltage drop ΔU (% of operating voltage U) for the transmitted power P is:

$$\Delta U = \frac{P \cdot l \cdot 100 (R_{\mathrm{W}} \cdot \cos\varphi + X_{\mathrm{L}} \sin\varphi)}{\Delta U^2 \cos\varphi}$$

where ΔU = Voltage drop in %
P = Transmitted power in kW
l = Length of cable run in m
R_{W} = Resistance in Ω/km
X_{L} = Reactance in Ω/km
U = Operating voltage in V.

The voltage drop is directly proportional to the load moment $P \cdot l$. For rough determination of the conductor cross section Fig. 27.25 may be used for the oper-

ating voltage $U_{\mathrm{r}}/U = 220/380$ V [1]. These values apply to a cable where the load is at the remote end from the supply. Where the loading is equally spread along the cable, half the cable length is used for the calculation.

The cables selected on the basis of maximum permissible voltage drop of 3%, where the three routes have different cross-sections must be investigated regarding the load capacity for each route separately. Each circuit here takes approximately 100 A (60 kW). This value lies below the permissible load capacity. All cables are dimensioned adequately.

If for local reasons the feed point must be located at S' then, because of the additional cable run S to S', an additional voltage drop occurs. Depending on the type of consumer and its simultaneity factor, the power carried by the run S to S' will be less than the sum of the peak loads of the three lines.

If one assumes a simultaneity factor of 0.8 then the load moment of the cable run $S-S'$ becomes 21,600 kWm. In order that a voltage drop of 1.2% is not exceeded, 2 cables each 185 mm² with copper conductors should be installed in parallel. If it is required to maintain a voltage drop in the system not exceeding 3% with the station repositioned then the cross-sectional areas of circuits 1, 2 and 3 must be increased to 150, 120 and 70 mm² respectively.

Apart from the increased investment for the reinforcement of the cable lines plus the feeder cables S to S' also additional losses occur which again have a negative influence on the cost evaluation [27.15].

This example emphasises the importance of locating a substation as close as possible to the load centre to avoid losses and extra voltage drop which all lead to additional cost. The most suitable location of the substation will be near the 60 kW-consumer together with short connections to the two other branches. The 250 m line may be avoided and the supply system becomes a ring solution.

In practical applications because of the advantages during installation and in operation uniform conductor cross-section is installed throughout. Since the installation costs are relatively high when determining cross-sectional areas the projected load of future years should be allowed for. When calculating a system therefore it is not the first intention to accept the currently applied cross-section but to answer the question whether the cross-section, used as a norm by the supply authority, is adequate.

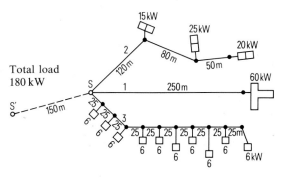

S = feed point (substation)
S' = alternative siting

Fig. 27.24
Low-voltage radial system with three circuits and one feed point

---
[1] New standard $U_{\mathrm{r}}/U = 230/410$ V superseding the previous voltage range

Table 27.1   Calculation data or example to Fig. 27.24

Circuit	Consumers		Length of circuit		Load moment	From Fig. 27.25 this gives	
	Number	Load P kW	Part length	Total length l m	$P \cdot l$ kW m	Conductor cross-section (copper)	ΔU
1	1	60	250	250	15 000	$q_1 = 95\ \text{mm}^2$	2.7%
2	1	15	120	120	1 800		
	1	25	120 + 80	200	5 000		
	1	20	200 + 50	250	5 000		
					Σ 11 800	$q_2 = 70\ \text{mm}^2$	2.8%
3	10	6	25	25	150		
			25 + 25	50	300		
			50 + 25	75	450		
			75 + 25	100	600		
			100 + 25	125	750		
			125 + 25	150	900		
			150 + 25	175	1 050		
			175 + 25	200	1 200		
			200 + 25	225	1 350		
			225 + 25	250	1 500		
					Σ 8 250	$q_3 = 50\ \text{mm}^2$	2.7%

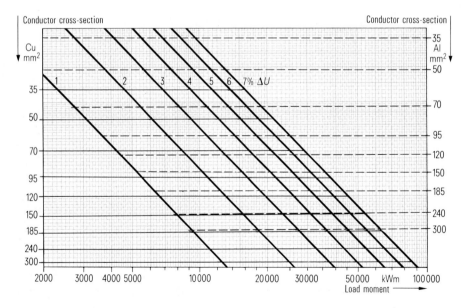

Fig. 27.25
Cable cross-sectional area of copper (Cu) and aluminium (Al) conductors with % voltage drop (on 220/380 V supply, cos $\varphi = 0.9$, conductor temperature 50 °C)

27.4.3 Investigations of Protective Measures Against Excessive Touch Voltage

The protective measures necessary are laid down in DIN VDE 0100 Part 410. This gives disconnection conditions for overcurrent protection devices used in TN system (multiple-earthed system). In earlier regulations the minimum disconnecting current was given as a multiple of the rated current of the protective device (k factor). Maximum disconnecting times are now given for fuses which are

▷ 0.2 seconds for circuits up to 35 A with socket outlets or circuits of mobile operating equipment of protection class I and

▷ 5 seconds for all other circuits in consumer installations.

Exceptions are made for the public supply distribution systems and other distribution systems with overhead lines or with cables buried in the ground as well as for totally insulated supply systems. For these systems it is sufficient if at the beginning of the cable section to be protected, an overcurrent protective device is installed and if the minimum fault current will exceed the maximum test current which is determined for the overload range applicable for the apparatus connected.

With N.H. fuses as are normally used in this type of system, the maximum test current corresponds to approximately 1.6 times rated current (this relates to a k factor of 1.6 as opposed to the value, used previously of 2.5). The disconnect time in the event of a fault in an unfavourable point in the system can under such conditions extend to 3 hours.

In TN systems therefore when selecting fuses not only the transmitted load but also the minimum fault current must be considered. In general this is the single-pole earth short-circuit current in the network. For the calculation of this value the recommendations of DIN VDE 0102 Part 2 apply. Standard formulae are included applicable to various types of system as well as a survey of impedances of operating equipment including zero-sequence impedances (expressed as a ratio of impedances).

For a cable run, fed from one side the minimum short-circuit current can be calculated, assuming the impedances of transformers and the superimposed system can be ignored, from the following formula:

$$I_{k\,min} = \frac{\sqrt{3} \cdot c \cdot U_{NT}}{\sqrt{(2 + R_{OL}/R_L)^2 \cdot R_L{}^2 + (2 + X_{OL}/X_L)^2 \cdot X_L{}^2}}$$

where

$I_{k\,min}$ = Minimum short-circuit current in A
R_L = Effective resistance of the line conductor at 80 °C in Ω
R_{OL}/R_L = Ratio of effective resistances zero-phase sequence to positive phase sequence
X_L = Reactance of the line conductor in Ω
X_{OL}/X_L = Ratio zero-phase sequence reactance to positive phase sequence reactance
U_{NT} = Rated voltage of secondary of the feed transformer involved (in a 380 V system, e.g. 400 V)
c = Factor 0.95

In Figs. 27.26 to 27.32 the fault currents derived using the above formula can be read off directly. The curves are valid only where the conductor cross-section is uniform over the total cable length and where the reactance values comply with DIN VDE 0102.

If differing conductor cross-sections have to be considered, e.g. a system cable feeding a long service cable then the impedances of the two cable sections must be added vectorally and the resultant sum values of the factors and impedances applied in the above formula.

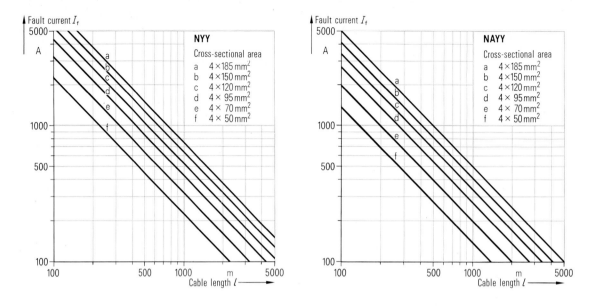

Fig. 27.26
Fault currents of NYY and NAYY (4 conductors, return path through fourth conductor) (DIN VDE 0102 Table 12)

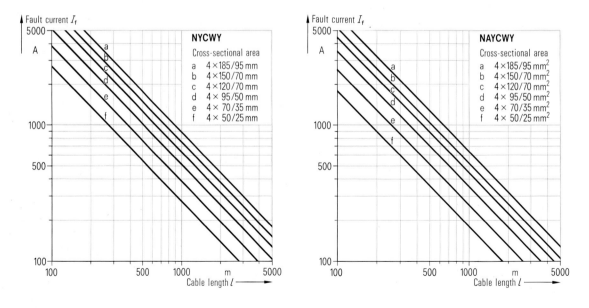

Fig. 27.27
Fault currents of NYCWY and NAYCWY, also NYCY and NAYCY (4 conductors + screen, return path through fourth conductor and screen) (DIN VDE 0102 Table 13)

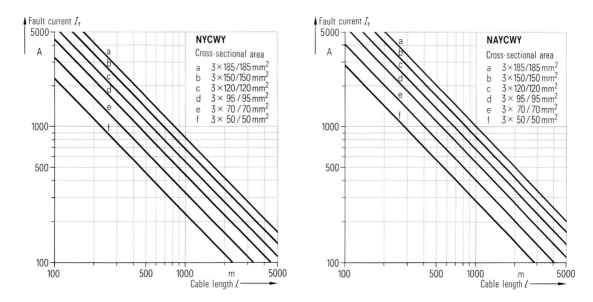

Fig. 27.28
Fault currents of NYCWY and NAYCWY, also NYCY and NAYCY (3 conductors + screen of the same cross-sectional area, return path through screen) (DIN VDE 0102 Table 15)

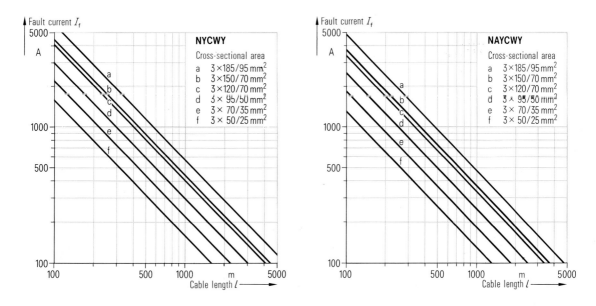

Fig. 27.29
Fault currents of NYCWY and NAYCWY, also NYCY and NAYCY (3 conductor + reduced size screen cross-section, return path through screen) (DIN VDE 0102 Table 16)

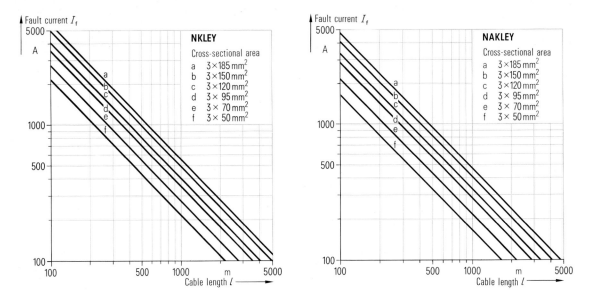

Fig. 27.30
Fault currents of NKLEY and NAKLEY (3 conductors, return path through sheath) (DIN VDE 0102 Table 14)

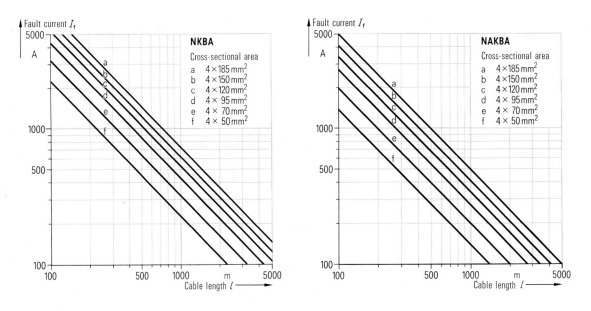

Fig. 27.31
Fault currents of NKBA and NAKBA (4 conductors, return path through the fourth conductor)
(DIN VDE 0102 Table 17)

388

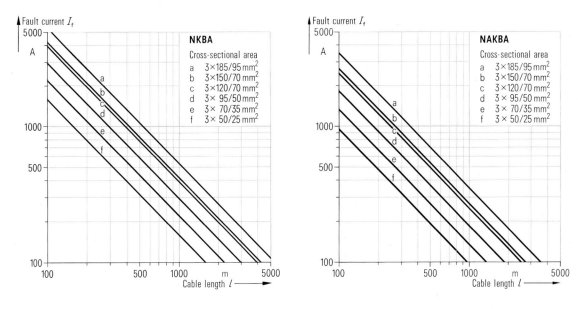

Fig. 27.32
Fault currents of NKBA and NAKBA ($3\frac{1}{2}$ conductors, return path through conductor with reduced cross-section) (DIN VDE 0102 Table 18)

Contact resistance and arc voltage drop may have a marked effect on reducing a fault current. In addition the effect of circuit impedance is related to the differing return circuit configuration for the cable whether via screen or earth. The values given in the figures ignore impedance of transformers and also of the superposed system. Particularly transformers with low nominal capacity will increase the total impedance and can further reduce the short-circuit current. Because of these factors and because of the influence of the ambient surroundings on the zero-impedance of the cable (e.g. water pipe work) in case of doubt measurements should be made to verify whether disconnect conditions are satisfied. In view of the extremely long disconnect time at minimum short-circuit current in a public system (1.6 times rated current of the fuses) this should be avoided wherever possible.

27.4.4 Investigation of Short-Circuit Protection and Discrimination

DIN VDE 0100 Part 41 contains measures necessary for the protection of cables against thermal stress due to overload or short circuit. The public supply distribution systems are not embraced by these regulations. However, these regulations are generally adhered to also in these systems to ensure adequate protection of the cables.

An important contribution for the reliable supply of energy is the selective application of protective devices. In radial systems where part sections also need protection the fuse ratings are tapered according to the reducing load to achieve discrimination of disconnection. In the event of short-circuit only, the last protective device next to the fault will respond. The cable run from the supply to this device will remain operational.

In industrial plant and in systems of buildings, fuses and circuit breakers are connected in series. Discrimination of the protective devices must be investigated by comparing their current time characteristics. Together with this, the superposed medium-voltage system may require to be considered. In order to check the correct and discriminative responses of protection

389

devices especially in expansive systems, the maximum and minimum short-circuit currents must be evaluated.

For the coordination of the protection equipments within a system with several voltage and (or) distribution levels, corresponding current time diagrams including all data of the protection devices and also system configuration must initially be established.

For large systems these tasks require comprehensive manual activity coupled with a great deal of experience, however, today much of this work can be done by data processing equipment. Fig. 27.33 illustrates a typical example of a data processed system plan and functional protection grading diagram [27.16, 27.17].

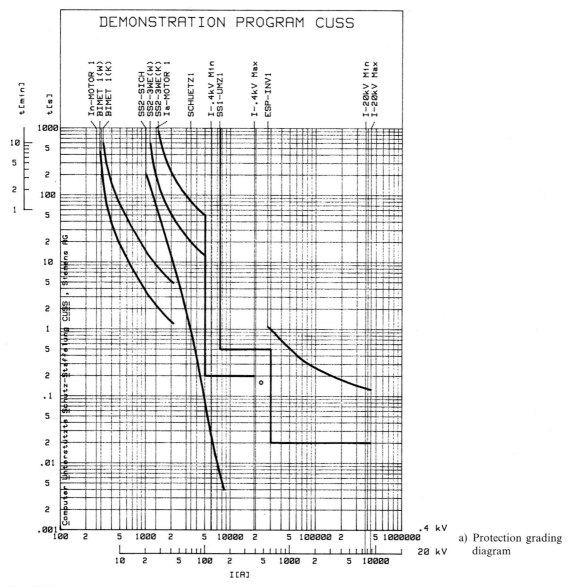

a) Protection grading diagram

Fig. 27.33
Computer-drawn current time curves (protection grading diagram) of protective devices and system plan

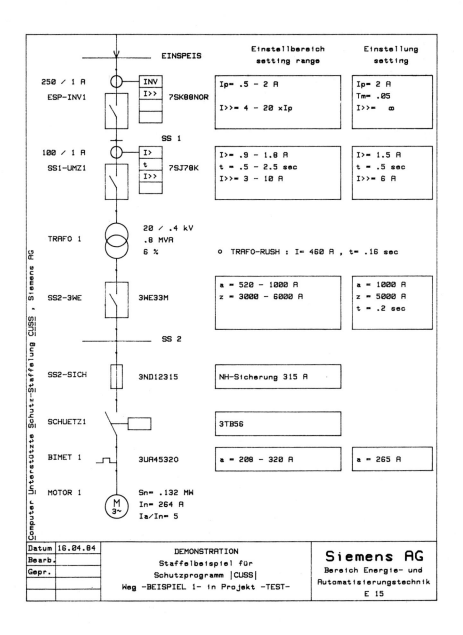

b) System plan

In a meshed system fuses of identical rated current and identical characteristics must be used to achieve discriminative disconnection. Fig. 27.34 illustrates a section of a meshed system and as an example shows current flow at point b with a short circuit in the direction of point e. Discriminative disconnection is ensured if, depending on the type of fuse, the proportion of short-circuit current is not larger than 70/80% of the total short-circuit current. The faulted cable between the two mesh nodes e and b is disconnected by the operation of fuse 2; the fuses 3, 4 and 5 remain intact such that the remaining system is not affected.

With four or more outgoing lines from a mesh node then normally it is rare that problems arise in respect of discriminative operation. Where only three outgoing lines are connected it is possible that a part current is greater than 80% of the total current and thus discriminative operation can no longer be ensured. To achieve discrimination, technical alterations are necessary in the system, e.g. the installation of a double cable system with equal cross-sectional area for the line having the largest part short-circuit current. With this method the mesh node point will have in effect four outgoing lines. Under unfavourable conditions, discriminative disconnection may be discarded where the necessary investment becomes excessive.

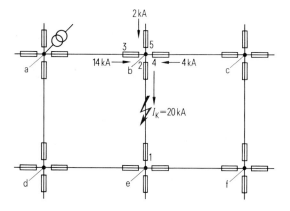

Fig. 27.34
Discriminative disconnection of a short circuit in a meshed system (see text)

27.4.5 Computer-Aided System Calculation

For simple system configurations the operating values such as voltage drop at the nodes, load current or short-circuit current can be calculated in a short time by means of a pocket calculator. Complex networks, however, even for simple layouts, require considerable time for full calculation especially for the preparation of a clear presentation of the results. For these reasons in distribution systems, with their multitude of component parts and the large number of variants possible solutions to be investigated, support by data processing equipment is a decisive contribution for rational processing of system calculations. Load-flow programs and programs for the calculation of single-, double- and three-pole short circuits are, today, standard equipment for a system planner. Of particular importance for the system planner is the flexibility of feed-in of the individual element data and a suitable evaluation of the results.

In addition to these standard calculations, programs are used today or are currently undergoing further development to provide a largely data processing technical support of planning work including optimisation of costs, e.g.

> determination of optimum routing of cables and locations of stations from predetermined installation possibilities,

> determination of system isolation points due to minimum losses,

> dimensioning of protection devices in low-voltage systems relevant to short-circuit and load conditions,

> optimisation of cable cross-sectional area.

Of particular importance in the planning of distribution systems is the graphical presentation. For the systems documentation, system plans which correspond to map-related positions (cartographically correct) and the associated plans can be made available in digital form for later recal by a plotter. In addition all data for the operating equipment, together with the graphical record are stored in a data bank. They are then available for documentation and the calculation of the systems.

Information systems of this kind are today an important aid for documentation, planning and the operation of distribution systems. Apart from electrical systems, also piped systems (water, gas) are recorded in a similar manner.

27.5 Literature Referred to in Section 27

[27.1] Schaller, F.: Verteilung elektrischer Energie (Diareihe). Siemens AG, Berlin, München 1974

[27.2] Erche, M.; Schaller, F.: Versorgungsqualität in elektrischen Verteilungsnetzen. Kommunalwirtschaft (1984) Heft 2, S. 57 bis 61

[27.3] Rusk, S.: The Simultaneous Demand in Distribution Network Supplying Domestic Consumers. ASEA Journal 29 (1956) Heft 5, S. 59 bis 61

[27.4] VDEW (Hrsg.): Planung und Betrieb städtischer Niederspannungsnetze. VDEW, Frankfurt 1984

[27.5] Seip, G.G. (Hrsg.): Elektrische Installationstechnik. Siemens AG, Berlin, München 1985

[27.6] Planung von Niederspannungsversorgungsnetzen in der metallverarbeitenden Industrie (Druckschrift). Siemens AG, Berlin, München 1981

[27.7] International Electrotechnical Commission Standard. IEC Standard Voltages. Publication 38, Sixth edition 1983

[27.8] Schaller, F.: Energieverteilungsnetze müssen billiger werden. Betriebstechnik (1976) Heft 7/8, S. 78 bis 80

[27.9] Kahnt, R.; Körner, H.: Niederohmige Sternpunkterdung in Mittelspannungs-Kabelnetzen. Elektriz.-Wirtschaft 67 (1968) Heft 13, S. 336 bis 342

[27.10] Körner, H.; Vehling, W.; Kahnt, R.: Ein Netzleistungssystem für die Betriebsführung des Hoch- und Mittelspannungsnetzes – Planung und Vorbereitung. Elektriz.-Wirtschaft 84 (1985) Heft 4, S. 113 bis 117

[27.11] Belkhofer, H.; Erche, M.: DV-Systeme für die Netzplanung. Elektrotechn. Z. 100 (1979) Heft 2, S. 70 bis 74

[27.12] Belkhofer, H.; Hofbauer, P.; Linker, K.W.; Oberländer, J.; Planung elektrischer Energieversorgungsnetze. Elektrotechn. Z. 103 (1982) Heft 6, S. 303 bis 308

[27.13] Siemens AG (Hrsg.): Handbuch der Elektrotechnik. Siemens AG, Berlin, München 1971

[27.14] Roeper, R.: Kurzschlußströme in Drehstromnetzen. 6. überarbeitete und erweiterte Auflage. Siemens AG, Berlin, München 1984

[27.15] VDEW (Hrsg.): Netzverluste. Eine Richtlinie für ihre Bewertung und ihre Verminderung. VWEW, Frankfurt 1978

[27.16] Schaller, F.: Schutz in Verteilungsnetzen von Industrie und Gebäuden. ETG-Fachberichte 12 – Selektivschutz. VDE-Verlag (1983), S. 75 bis 89

[27.17] Sachs, U.; Schaller, F.; Stöber, K.-J.: DV-unterstützte Selektivitätskoordinierung für Schutzgeräte mit Überstrom-Zeit-Verhalten. Siemens Energie & Automation 7 (1985) Heft 1, S. 24 bis 28

The preparation of cable ends and the installation of cable accessories require great accuracy and care.
The sealing ends form the end connection of the cable can also serve at the same time as termination to the electrical apparatus as in this example showing the triple encapsulated entries into an SF_6 insulated high voltage installation

Laying and Installation

28 Cable Identification Marking

28.1 Manufacturers Marking, VDE-Marking

To facilitate identification of the manufacturer of a cable, the sheath of cables with polymer insulation bears a manufacturers mark. For polymer insulated cables with lead sheaths and for ships cables, an identification thread is incorporated which is trade mark protected (mark of origin). For polymer insulated cables to VDE the outer sheath must bear, at distances not exceeding 50 cm, the year of manufacture and the VDE mark. For cables with a diameter of ≥ 10 mm, in addition, a longitudinal marking is required. Instead of the VDE marking, cables to DIN VDE 0265 and ships cables to DIN VDE 0261 may have the black/red VDE identification thread.

Examples of the marking of polymer insulated cables of Siemens manufacture:

> 1987 PROTODUR ◁VDE▷ 0271
> as well as continuous marking in meters.

> 1987 PROTOTHEN-X ◁VDE▷ 0273
> as well as continuous marking in meters.

Colours of the Siemens identification thread.

> red/white/green/white.

In paper insulated cables, below the top layer of insulation the VDE identification strip is applied in to signify the origin e.g.,

> 1987 Siemens AG VDE 0255.

The lettering is repeated at distances of approximately 30 cm.

28.2 Colours of Outer Sheaths and Protective Coverings

The colours to be used for the outer sheaths of cables and flexibles are given in VDE 0260 "Leitsätze für die Farbe von Außenmänteln und Außenhüllen aus Kunststoff oder Gummi für Kabel und isolierte Leitungen", outer sheaths or protective coverings of polyethylene (PE) always contain a carbon additive to achieve an improved durability and are therefore coloured black.

For sheath and covering colours to DIN ICE 304 "Standard colours of insulation of low frequency cables and wires", refer to Table 28.1. It must be noted that coloured PVC sheaths under the influence of sulphur compounds especially hydrogen sulphide change in colour to black. The sulphur compounds occasionally found in the soil result from bacterial decomposition of organic matter under exclusion from air, from sewage or faecal substances and also from some types of town gas.

Table 28.1   Colour of outer sheaths

Item	Type of Cable	black	yellow	grey	blue	red	white	ivory
1	Power cables for rated voltages up to 0.6/1 kV,	×						
	but for underground mines		×					
	Intrinsically safe installations in locations with explosion hazard				×			
2	Power cables for rated voltage over 0.6/1 kV with PVC-sheath					×		
	with PE-sheath	×						
3	Power cables for rated voltages up to 1 kV	×					×	
	but for cables used in damp locations			×				
	Rubber-sheathed cables (NSSHÖU) for use in underground mines and in industry		×					
	Intrinsically safe installations in locations with explosion hazard				×			
4	Power cables for rated voltage exceeding 1 kV[1]					×		
	but cables for fluorescent lamps		×					
5	Heavy-duty telephone cables	×						
	but for power supply and industrial plants (incl. underground mines)			×				
	Intrinsically safe installations in locations with explosion hazard				×			
6	Telephone switching cables			×				
7	Telephone installation cable	×		×				×
8	Telephone-sheathed cables for underground mines			×				
	but Intrinsically safe installations in locations with explosion hazard				×			

[1] Trailing cables for rated voltages exceeding 1 kV and cables with PE sheath may also be black if they are not intended for use in underground mines

28.3 Core Identification for Power Cables[1] up to $U_0/U = 0.6/1$ kV

In multicore cables the cores are identified to DIN VDE 0293 as follows

Number of cores	With protective conductor (suffix "J")	Without protective conductor (suffix "O")	With concentric conductor
2	green/yellow, black	brown, blue	black, blue
3	green/yellow, black, blue	black, blue, brown	black, blue, brown
4	green/yellow, black, blue, brown	black, blue brown, black	black, blue brown, black
5	green/yellow, black, blue, brown, black	black, blue brown, black, black	black with printed numbers
6 and more	green/yellow, further cores black with printed number	black with printed numbers	black with printed numbers

For cables with mass impregnated paper insulation the following applies

▷ natural colour is brown

▷ green/natural colour is green/yellow.

The core colour of single-core cables is always black; apart from this also green/yellow is permitted. The numerical designation with 6 or more cores must be clearly identifiable and the colour must be in contrast to the colouring of the core. The numbering of the cores commences at the centre with 1. Two markings following one another must always be situated inversely in opposing positions. A line which underlines these markings signifies the base of the marking.

Identification markings used previously were as follows:

gnge green/yellow marked protective (PE) or combined neutral and protective conductor (PEN)

gn-nat green/natural colour marking denotes protective or combined neutral and protective conductor in cables with paper insulation

[1] For flexible cables and wires see page 82

sw core marked black
hbl core marked light blue
br core marked brown
nat core marked natural colour (only in cables with paper insulation)

Apart from colour identification of cores in multicore cables with paper insulation it was also permissible to mark the cores which were not used as PE or PEN conductors with printed numerals.

Cores to be used as follows:

As protective (PE) or combined neutral and protective (PEN) conductor	Exclusively the core marked green/yellow or green/natural	The core must not be used for any other purpose. An exception applies only to supply authority networks in the Federal Republic of Germany. Here for cross sectional areas greater than 6 mm² this is also permissible for earthed neutral conductor
As neutral conductor (N)	The blue core	The core may also be used as a phase conductor

The concentric conductor, the aluminium sheath or corrugated copper sheath of a cable may also serve as PE or PEN conductor or as neutral conductor but must not be used as a main conductor.

Previously Used Colour Identification of Cores

Cables with colour identification to previous standards exist in many installations.

N° of cores	Polymer insulated cable	Paper insulated cable
2	light grey, black	natural, blue
3	light grey, black, red	natural, blue, red
4	light grey, black, red, blue	natural, blue, red, blue/natural
5	light grey, black, red, blue, yellow	–

The following were used:

As neutral (N) or combined neutral and protective conductor (PEN)

> in paper-lead cables: the core coloured natural
> in aluminium-sheathed cables: the aluminium sheath
> in polymer insulated cables without concentric conductor: the grey core.

As protective conductor

> in all cables without concentric conductor: the red core.

It must be observed that the cores coloured red, grey or natural were also permitted to be used as main conductors.

In multi-core cables having more than 5 cores the cores were coloured grey, the counting core blue with a directional core marked yellow. The counting always commences from the centre core (first core) to the adjacent blue core and continues in the direction of the yellow core. In Fig. 28.1, for example, the blue cores have the numbers 2, 8 and 20. The yellow cores have the numbers 3, 9 and 21.

At the "A" end the yellow core is situated in a clockwise direction next to the blue whereas at the E the cores are in mirror image (Fig. 28.1).

Sequence of Core Colours

It is common practice for the ends of the cable to be designated E (start) and A (finish). The colour se-

quences mentioned on page 397 appear at the A end in a clockwise direction and at the E end in a anti-clockwise direction. In cases where jointing of cores of the same colour is required in a cable joint, the individual lengths must be arranged in the required direction such that the A end of one corresponds with the E end of the other cable. It is necessary therefore to check each drum of cable to establish which end is outer most.

28.4 Core Identification for Cables for Rated Voltages Exceeding $U_0/U = 0.6/1$ kV

For rated voltages exceeding 0.6/1 kV the cable cores are no longer marked differently, i.e. the insulation of polymer insulated cables does not contain a colour additive and also paper insulation does not include coloured paper layers or printed numerals.

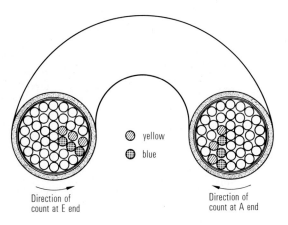

⊘ yellow
⊕ blue

Direction of count at E end   Direction of count at A end

Fig. 28.1
Counting direction of cores at E end and A end of polymer insulated control cables

29 Laying the Cables

29.1 Transporting

Cables are generally supplied wound on wooden drums. Short lengths of cable having a weight of up to approximately 80 kg can also be supplied in rings which are protected by a wrapping of crepe paper or similar.

The size of the drum is dependant upon the length, the weight and outer diameter of the cable. The maximum carrying capacity of the drum must not be exceeded. Dependant upon cable type, rated voltage and the conductor cross-sectional area, the drum core must have a diameter not less than a given minimum which is a multiple of the cable outer diameter d. Cable drums with smaller core diameters than specified by the cable maker are not permitted.

Subject to diameter and weight limitations up to 2 500 m of cable can be supplied on one drum. When decoiling long lengths it must be noted that the inner end moves and it is therefore necessary to interrupt the paying out from time to time to re-secure the inner end which becomes longer and longer. If this is not done then at the points of transition to the next layer kinking can occur. Decoiling at too high a rate may cause the layers to become entangled on the drum.

Transportation over land is normally carried out by rail or by lorry. In order to prevent drums moving during transportation these must be securely wedged to prevent movement. Before unloading it is necessary to verify that the drum is received in proper condition. In the event of damage a damage report must be initiated by the carrier in order that damages may be claimed where appropriate.

For the unloading of cable drums usually a crane or a ramp is necessary. If such aids are not available an auxiliary ramp must be constructed, consisting of wooden planks or girders, having an inclination not greater than 1 in 4. When errecting the planks or girders the weight of the drum and contents must be taken into account. When rolling the drum on the ramp this must be controlled by ropes and with the aid of winches or block and tackle. Furthermore

it is good practice to place a layer of sand some 20 cm deep at the base of the ramp to act as a cushion and brake.

Whether the cable is in coils or on drums, even if small and not heavy, these must not be thrown to the ground when unloading. Even if these fall on to soft padding the cables would be damaged by the fall.

The arrow painted on the side of the drum indicates the direction of rolling. If the drum is rolled in the opposite direction then the layers of cable would become loose.

For transportation of the drum of cable to the location of installation it is good practice to use a cable trailer equipped with loading and unloading facilities. Where the trailer is equipped with bearings for the drum axle the cable can be either decoiled with the waggon stationary (Fig. 29.1) or paid out from the waggon while it is moved forward at low speed.

Fig. 29.1
Paying out of cable from a cable trailer

If cables are not to be laid directly from the trailer into the trench, the drums should be unloaded at a point from where the cable is to be decoiled later. Rolling of the drum should be avoided.

Cable rings should wherever possible be stored flat. Rings not laid flat should be properly supported.

29.2 Preparation for Laying the Cable

The cable should, wherever possible, be decoiled from the top of the drum. For this the drum must be positioned such that the arrow on the drum points opposite to the direction of rotation for decoiling. The drum is lifted on its axle by the aid of winches such that the plank used for braking cannot become wedged beneath.

The cable must be inspected for external damage which, e.g. may have been incurred by incorrect roll-

ing of the drum. Since the laying of cables is often carried out by unskilled labour it is necessary to emphasise that the cable is a high-value commodity and is very sensitive to damage and must be handled with the necessary care. In order to avoid damage to the corrosion protection and the insulation, the cables must not be dragged over sharp objects and must not be bent too sharply (see Table 29.1).

It must be possible to brake the drum at any time in order to avoid, in the event of a sudden stoppage, continuation of decoiling which would result in sharp bending of the cable (Fig. 29.2). Avoidance of kinking is especially critical under all circumstances. As a brake, if need be, a simple plank can be used. When decoiling the cable drum is turned by hand in order to avoid unduly high tensile stress which, in particular, is applicable to small diameter unarmoured cables.

Small size cable rings may be uncoiled by hand on laying. Larger rings are best pulled off from a rotary table with the coil laying horizontally. Under no circumstances must the turns be lifted off coils or drums laying flat as this results in twisting and will cause damage to the cable.

Cables must be heated prior to laying, where the cable temperature is below 5 °C for mass impregnated cable or −5 °C for polymer insulated cable, otherwise

Table 29.1

Minimum permissible bending radii r when laying cables

	Paper insulated cables		Polymer insulated cables
	with lead sheath or corrugated aluminium sheath	with smooth aluminium sheath up to 50 mm diameter	
Multi-core cables up to $U_0/U = 0.6/1$ kV over $U_0/U = 0.6/1$ kV	$15 \times d$	$25 \times d$	$12 \times d$ $15 \times d$
All single core cables	$25 \times d$	$30 \times d$	$15 \times d$

d Outer diameter of the cable

Where a bend is to be made once only as for example immediately before a sealing end then, providing proper procedures are carried out (heating to 30 °C and bending over a form tool), the values of r can be reduced to 50% of those above.

Fig. 29.2 Pulling cable off a drum

the insulation and corrosion protection will be damaged during bending. These values apply to the cables themselves and not to the ambient temperature. Either the drums must be stored for several days in a heated building or heaters or hot air blowers applied at a sufficient distance. During this warming process the drums should be rotated at intervals. It must be ensured that the heating penetrates to the drum core evenly. During transportation the cable must be covered with tarpaulin sheeting to minimise heat loss. The laying work must be carefully planned and carried out quickly without delay before the cable cools excessively.

To avoid ingress of moisture it must be observed that the end capping of the cable is not damaged. Cut points of cables must be immediately (before laying!) capped (see page 434).

29.3 Differences in Level of the Cable Route

Paper insulated cables with standard mass impregnation must not be used where the permissible differences in level of the route are exceeded.

Vertical Installation

The guide values apply only for the ends of a cable (e.g. at a pole) where a suitable sealing end is used with facility for topping-up the impregnation fluid.

Cable type	Rated voltage U_0/U kV	Maximum difference in level m
Belted cable	up to 3.6/6 6/10	50 20
Triple sheathed cable	6/10 to 8.7/15 12/20 to 18/30	30 15

Steep Gradients

For all cables with $U_0/U = 0.6/1$ kV up to 18/30 kV the following gradients are permissible

up to 4% downward gradient without restriction of cable length

up to 10% downward gradient maximum 500 m cable length

Where these values are to be exceeded the polymer insulated cable or paper insulated cable with special mass impregnation (sv) must be used (see pages 35 and 100).

29.4 Laying of Cables in the Ground

29.4.1 Cable Route

For the progression of a cable route in built-up areas it is most suitable to use a paved pedestrian area or in overland routes to follow a foot path. The depth of trench is dependant on the number of cables to be laid above one another in the same trench, furthermore in urban areas and on industrial sites it also depends on any gas or water pipes which exist or may be laid in the future. The cable laying uppermost below a paved pedestrian area or foot path should be at a depth not less than 0.6 m and below roads at a depth of not less than 0.8 m. Where cables are laid at shallower depths they must be protected by e.g. concrete slabs of sufficient thickness.

Fig. 29.3 Laying out cable in a trench

401

rmally covered with a layer of sand ved) soil 10 cm thick and then, to rface damage during subsequent workings, are covered with bricks, plastic plates or similar devices. If covers are not provided warning tapes of plastic are normally used to mark the cable route.

Where high-voltage and low-voltage cables are laid in the same trench, it is the practice to lay the high-voltage cable in the lowest position. The high-voltage cables are then embedded in sand and covered by protective slabs. Above these on an additional layer of sand the low-voltage cables are laid. In such a cable arrangement the current load capacity, because of the mutual heating effect and drying out of the soil, is reduced (see page 201).

Where cables are required to be identified from the outside they can be marked after laying by fitting cable identification strips at a pitch of approximately 3 m.

Where the cable route runs in close proximity to or actually crosses the path of post office, railway or waterway installations the appropriate official regulations must be observed.

In areas of massing of energy cables, in view of current carrying capacity and drying out of the soil in many instances greater separation is necessary.

If control cables and high-voltage cables run on parallel routes for any great distance the magnitude of interference must be investigated. The same applies where the route is in close proximity to or crosses railway installations or communication networks of the post office (see Section 26).

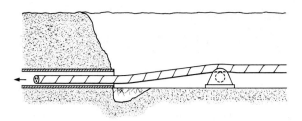

Fig. 29.4 Depression at pipe entry

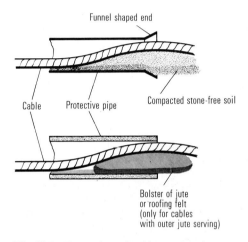

Fig. 29.5 Protection of cables at the pipe entry

Crossing of Roads

When the route crosses a roadway the cables must be drawn into pipes or cable duct blocks which extend beyond and under the pavements.

It is advisable always to provide reserve space in pipes or pipe ducts to avoid the necessity to reexcavate when adding cables at a later date. Pipe cavities which are not occupied immediately should be sealed off. Cables should not be allowed to rest on the sharp ends of the tubes or tube blocks. Where possible steel pipes should be fitted with funnel-shaped ends.

If the entrance of a pipe or pipe duct is level with the base of the trench, before the cable is fed in, a depression should be made in the trench floor so that as the cable is drawn in stones and other objects are not taken into the pipe causing jamming (see Fig. 29.4). After drawing in the cables these should be bolstered at each end such that they are touching the upper surface of the entry (Fig. 29.5).

The pipe bore should have a diameter of at least 1.5 times the outer diameter d of the cable. Pipe bends should, in respect of pulling the cable through, have a minimum radius corresponding to those shown on page 400 where for the value of d the pipe outer diameter should be used.

29.4.2 Laying of the Cables

The following methods may be employed for laying cables:

> Paying out from a cable trailer (Fig. 29.1),
> Laying by hand (Fig. 29.3),
> Laying by motor driven rollers (Fig. 29.11),
> Pulling off by winches,
> Ploughing in (Figs. 29.16 and 29.17).

Fig. 29.6   Cable roller

Fig. 29.7   Corner rollers

Paying out from a Cable Trailer

Providing there are no obstructions in the trench or its vicinity, cables may be payed out direct from the cable trailer. However it must be ensured that during the paying out the drum is manually rotated and braked in accordance with the laying speed to avoid high-tensile force or sharp bending of the cable.

Laying by Hand

Cable rollers (Fig. 29.6) placed at distances of between 3 to 4 m make laying easier. Corner rollers or similar devices should be provided at any bend in the route, always maintaining the minimum bending radii of the cables (Fig. 29.7).

If the cable is not guided by rollers it must be guided by hand. The men supporting the cable should be spaced at between 4 to 6 m along the cable.

If there is not sufficient labour to lay a long length of cable in one operation the cable drum can be jacked up at a mid point on the route and the required length pulled-off from the top in the direction 'A' (Fig. 29.8). The drum is then further rotated and a vertical loop formed approximately 3 to 4 m long which is pulled off in the direction B. The cable is lifted sideways over the drum flange towards the trench and the remaining cable pulled off from below in a loop. The twist in the cable must be spread over a length of 4 to 6 m. The cable can then be continued to be pulled off in a loop and placed directly in the cable trench.

If obstructions in the trench prevent laying of the second half of the cable from above, the cable should be laid out in a figure eight on the side of the original pull. Here it must be noted that with the cable laid

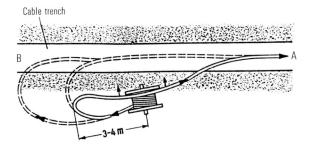

Fig. 29.8   Pulling off the cable in a loop

out in a figure eight arrangement the cable can only be pulled off again in that same direction from which it was first looped into the configuration (Fig. 29.9).

Laying by Motor Driven Rollers

Motorised rollers are used to pull the cable off the jacked up drum. It is advantageous to use rollers driven by electric motors (Fig. 29.10) installed in the cable trench at distances of 20 to 30 m. Where sharp

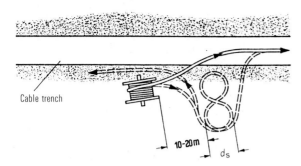

Cable trench

10–20 m

d_s

$d_s = 60 \times$ cable outer diameter with a minimum of 2 m

Fig. 29.9 Pulling off a cable in a figure eight

bends occur it may be necessary to place such a roller at both the commencement and the end of the bend. All rollers are supplied from a main distribution board via small sub-distribution centres and controlled from a central point (Fig. 29.11). The pulling force is distributed over the total length of the cable such that mechanical stresses are minimised hence cables with low tensile strengths can be installed in this way.

At the cable leading end only a small pulling force is necessary to guide the cable. For this duty 3 or 4 men are sufficient or a small electric winch may be used, synchronised with the cable rollers.

To provide power for this installation equipment a power feed of some 15 kVA is required, alternatively a portable generating set is necessary.

Fig. 29.10
Laying a cable by motor driven rollers

Pulling off by Winches

Pulling off by winch is possible only if there are very few bends or other obstructions in the route.

After releasing the cable end from the drum a pulling stocking (Fig. 29.12) is placed over the end and tied in position. A rope is secured to the eye of the pulling stocking. The soldered sealing (lead cap) cap of paper insulated cables must never be used for pulling.

When laying unarmoured cables or steel-tape armoured cables with the aid of a winch, the rope can be secured to the cable via a pulling head (Fig. 29.13) which grips directly on to the conductors. With this method it must be ensured that the pulling head

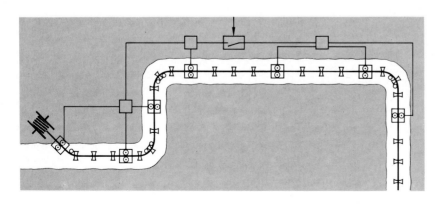

Fig. 29.11
Arrangement
of motor-driven rollers
in a cable route

Fig. 29.12   Cable stocking

Fig. 29.13   Pulling head

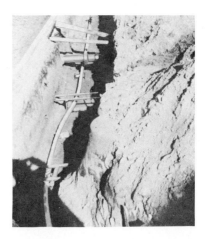

Fig. 29.14   Cable support rollers in a trench

clamps all conductor wires equally and that a suitable seal prevents the ingress of water into the conductors and insulation at the cable end.

To ensure cables are not overstressed or damaged when being pulled off by a winch (see Table 29.2), the following precautionary measures are necessary:

▷ Continuous monitoring of the tensile force. Where a rope tension measuring device is not available, the winch is placed on rollers (e.g. cable rollers) so that it is moveable and then it is anchored through a dynamometer.

▷ On the winch (e.g. at the drive of the wind drum) a shear pin or other suitable device is incorporated, calibrated to ensure the maximum stress is not exceeded. If for unforseen circumstance excessive forces occur the shear pin breaks and the pulling operation is interrupted.

▷ The cables and the pulling rope must, especially at bands, be guided over rollers (Figs. 29.14 and 29.15).

When pulling into steel or plastic pipes, even when the pipes are lubricated, large pulling forces may be required and where bends total 300° this may equal 100% of the cable weight. As a lubricating agent for steel or plastic pipe boiled alkaline free soap or a mineral free lubricant, e.g. SIKAGLEIT should be used.

All cables, in particular single-core cables should not be straightened after laying, but left slightly meandering, to allow for longitudinal expansion and contraction during thermal cycling (changes in current loading).

Fig. 29.15
Corner and guide rollers on a curve in the trench

Table 29.2
Permissible pulling force in the laying of low-voltage and medium-voltage cables (guide values)
(DIN VDE 0298 Part 1)

Means of pulling	Type of cable	Formula	Factor
With pulling head attached to conductors	All types of cable	$P = \sigma \cdot A$	$\sigma = 50\ \mathrm{N/mm^2}$ (Cu-conductor) $\sigma = 30\ \mathrm{N/mm^2}$ (Al-conductor)
With pulling stocking	Elastomer insulated cable[1] without metal sheath and without armour (e.g. NYY, NYSY, NYSEY, NYCWY etc.)	$P = \sigma \cdot A$	$\sigma = 50\ \mathrm{N/mm^2}$ (Cu-conductor) $\sigma = 30\ \mathrm{N/mm^2}$ (Al-conductor)
	All wire armoured cables (e.g. NYFGY, NAYFGY etc.)	$P = K \cdot d^2$	$K = 9\ \mathrm{N/mm^2}$
	Cable with metal sheath without pull resistant armouring: Single-sheathed cable (e.g. NYKY, NKLEY etc.)	$P = K \cdot d^2$	$K = 3\ \mathrm{N/mm^2}$
	S.L. cable (e.g. NEKEBA, NAEKEBA etc.)	$P = K \cdot d^2$	$K = 1\ \mathrm{N/mm^2}$

[1] When laying 3 single-core cables simultaneously with a common pulling stocking, the same maximum pulling force applies, whereas the pulling force for 3 laid-up single-core cables is 3 times that of a single-core and for 3 non-laid-up single-core cables is 2 times that of a single core

P Pull in N
A Total cross sectional area in mm^2 of all conductors (but not screen or concentric conductor)
d Outside diameter of cable in mm
σ Permissible tensile stress of conductor in N/mm^2
K Empirically derived factor in N/mm^2

Example

Permissible tensile force.

a) When pulling a cable using a pulling head clamped to the conductors
Cable type NYCWY 3 × 120 SM/70 0,6/1 kV
Permissible pulling force:
$p = \sigma \cdot A$
$p = 50 \cdot 3 \cdot 120 = 18\,000\ \mathrm{N}$
(The cross sectional areas of concentric conductor and screen are not included)

b) When pulling a cable using a pulling stocking
Cable type NYSEY 3 × 120 RM/16 6/10 kV
Permissible pulling force:
$p = \sigma = A$
$p = 50 \cdot 3 \cdot 120 = 18\,000\ \mathrm{N}$

Cable type NYFGY 3 × 120 RM 3.6/6 kV
$d = 47\ \mathrm{mm}$

Permissible pulling force:
$p = k \cdot d^2$
$p = 9 \cdot 47^2 = 19\,881\ \mathrm{N} \approx 19\,900\ \mathrm{N}.$

The data in Table 29.2 apply also for armoured cables where the pulling rope is not attached to the cable with the aid of a pulling stocking but is coupled direct to the wire armouring. For cables with pull resistant special armouring, e.g. round steel-wire armouring the permissible pulling force relative to the armour withstand must be determined such that an extension of 0.2% is not exceeded.

On horizontal routes and with well constructed cable trenches with easy running support, corner and box rollers and pulling by winch the following levels of pulling force can be anticipated:

Type of route	Pulling force expressed as % of weight of the cable being pulled in
Without serious bending	15 to 20%
With 2 bends each of 90°	20 to 40%
With 3 bends each of 90°	40 to 60%

Plough-Laying of Cables

In open terrain the cables may be plough-laid directly into the ground where circumstances permit, no obstructions, e.g. pipe runs which cross the route, and where protection of the cable with plastic plates or similar is not acceptable. This type of cable laying is particularly cost effective (Figs. 29.16, 29.17). Cables with PE sheath are particularly suitable for this form of laying.

Fig. 29.16
Plough laying of three bunched single-core XLPE cables for 12/20 kV with a heavy-duty cable plough

Laying of Single-Core Cables

For the laying of single-core cables under practical conditions, depending on local circumstances several methods have proved effective:

▷ Pulling-off and laying individual lengths in sequence,

▷ simultaneously pulling-off three lengths from three cable drums,

▷ laying of three pre-laid-up cables,

▷ Plough laying of three bunched cables (Figs. 29.16 and 29.17).

If the three lengths are laid in sequence care must be taken that the cable already laid is not damaged by the cables following (e.g. by chaffing or abrasive action). This applies particularly for the crossings under roadways or in other ducting situations. In these cases, also if the cables on the remainder of the route are bunched in triangular formation, a separate duct must be provided for each cable. These ducts must be of non-magnetic material (e.g. not of steel pipe). If, when drawn through, the cables are to be rebunched in the cable trench suitable non-magnetic distance pieces are available. Bunching with the aid of plastic tape at pitches of 5 to 7 m is normal practice.

Fig. 29.17
Plough laying of three bunched single-core XLPE cables for 12/20 kV with a light-duty cable plough

If on site sufficient space is available the simultaneous pulling off of three single-core cables from three individual drums has advantages. In a bunching bench the three cables are brought together for bunching in triangular formation for bunching and can then be laid as a single cable. The system is also proven where the three drums are carried on a suitably adapted flat back lorry.

Pre-laid-up cables, when laying in the ground, need not be bunched. They can readily be pulled from the drum and laid as a single cable. For this system where winch assistance is applied it is satisfactory if the rope is attached to a pulling stocking fitted over the laid-up cables and secured in the proper manner. The pulling forces are then equally distributed over the laid-up arrangement.

29.5 Laying of Cables Indoors

For laying indoors and in cable tunnels, cables must be used which have a flame-retardent outer sheath, e.g. cables with PVC sheath or armoured paper insulated cables without jute serving. In special cases for reasons of fire protection special constructional measures or additional protective arrangements, e.g. painting with flame resistant materials or the use of special cable[1] may be demanded.

The cables are fixed to walls or ceilings with cable clips or laid on cable trays or racks.

The distance between cable clips or supporting points when laid on cable trays should in a horizontal arrangement not exceed the following values:

for non-armoured cables:
20 times outside diameter

for armoured cables:
30 to 35 times outside diameter

max. distance of:
80 cm

On vertical routes the distance between cable clips may, depending on the type of cable and type of cable clip, be increased. Distances of 1.5 m should not be exceeded.

If simple cable clips are used for securing polymer insulation and cables with polymer insulation and outer sheath, a curved metal strip (inverted through) should be inserted between the base and the cable.

29.5.1 Cables on Walls, Ceilings or Racks

Indoors the cables are either fixed to walls or ceilings or laid out on racks (Fig. 29.18). During the planning stage the space requirement (taking account of current-load capacity, cable massing and permissible bending radii) as well as the strength of supporting structures and other component parts are to be considered.

For cable racking sufficient space must be provided. These are either fixed to the wall or are installed

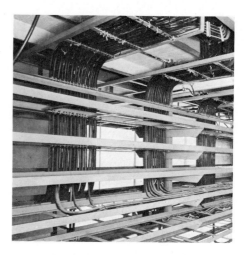

Fig. 29.18 Cables laid on racking

free standing such that the cables can be laid from the side and are readily accessible for replacement or rearrangement. The racking and the cables must be arranged so that the heat dissipation from the cables by convection is not hindered.

29.5.2 Cable Tunnels and Ducts

The main advantage of using cable tunnels or ducts is the ease of access for replacement or extension of the installation without extensive workings.

Especially, where there is great cable massing, the cables installed in ducts can normally be subjected to higher loading than when laid in the ground. A disadvantage is the high cost of supply and installation of the ducts. For this reason installation in ducts is normally restricted to buildings and around outdoor switchgear plant.

In walk through ducts (cable tunnels) the cables are for practical reasons laid on cable trays above one another. Multi-core cables which are laid horizontally on the duct floor or on the trays do not require fixing.

The cables should be laid in the duct with a space between each approximately equal to the cable diameter. The load capacity of the cable may be determined by reference to the information given on page 230. Since this is dependant on ambient air temperature a good natural ventilation should be pro-

---

[1] Halogen free cables (Section 14.3) and halogen free flexible cables (Section 8.4) with improved performance in the event of fire

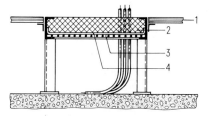

1   Cover plate
2   Fire protective mortar
3   Steel mesh 75 mm × 75 mm galvanised
4   Aluminium foil 0.4 mm thick

a) barrier between floors

b) barrier using a system which is commercially
    available
    (MCT-feed through)

c) Bulkhead in cable tunnel with fire-protective mor-
    tar sealing and fire-resistant steel door

Fig. 29.19
Examples of barriers in walls and ceiling through
feeds as well as in cable tunnels or ducts

vided in the duct (openings for ingoing and outgoing
air). Where necessary forced ventilation must be provided.

With due consideration of the possibility of spreading
of fire the cable ducts must be fitted with barriers
at least at the point of entry into operation rooms,
switching stations, etc. (Fig. 29.19).

When cables from a trench join into a cable duct,
the junction must be treated as when feeding through
walls or under road crossings where they are drawn
into pipes or pipe blocks. The inner diameter of the
pipes shall be a minimum of 1.5 times the outside
diameter of the cable. These pipes should be laid
slightly inclined towards the trench and be suitably
sealed at the ends to avoid ingress of water
(Fig. 29.20). For this sealing shrink on sleeving may
be used alternatively a hessian or similar material
saturated with epoxy resin SP supplemented by a
final sealing with the cast resin. When carrying out
this operation the casting temperature must not exceed 100°C. Cement or phenol containing mastic, e.g.
tar or bolsters saturated with tar must not be used
for this purpose. Cables having a PVC outer sheath
should be bandaged with polyethelene tape, when this
form of sealing is used, in the area of the feed through.
A further suitable method of sealing is to use a non-
solidifying elastic silicone mastic.

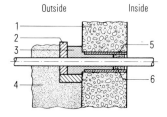 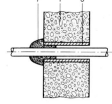

a) with cast resin   b) with silicone
    sealing compound   mastic

1 Brick wall
2 Brick or clay former
3 Filling compound
4 Sand backfill
5 Plastic, concrete or steel pipe
6 Seal of unsaturated jute rope or plastic tape
7 Elastic silicone compound (non-solidifying)

Fig. 29.20   Examples of wall feed-through

29.6 Cable Clamps

When fixing cables to steel structures or to steel supports (e.g. C rail to DIN EN 50024, angle iron) which are in turn secured to walls or ceilings, especially if the cable has a thermoplastic outer sheath it must be ensured that any clamps used have protective shoes or suitable shields. Without such protection the cables can be damaged by localised pressure from the clamps (Figs. 29.21 to 29.25).

Fig. 29.23
Ordinary cable clamps with protective shoes

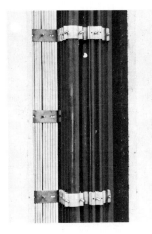

Fig. 29.21
Fixing of control cables stacked above one another in a vertical shaft

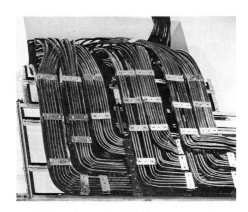

Fig. 29.24
Cables installed with Siemens cable clamps

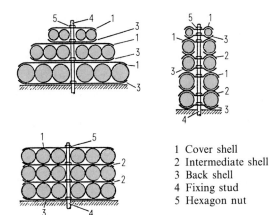

1 Cover shell
2 Intermediate shell
3 Back shell
4 Fixing stud
5 Hexagon nut

Fig. 29.22
Examples of cable installations using Siemens cable clamps

Fig. 29.25
Cable installation showing subsequent additions

29.6.1 Types of Clamps

Siemens have developed cable clamps where because of their design the installer is compelled to use the protective shoes provided. The clamps consist of simple coordinated building elements which can be readily mounted even in situations of great massing or in difficult cable routing positions. With these it is possible to extend the system especially when this entails adding layers above one another up to the limit of the carrying capacity of the steel supports.

All Siemens cable clamps consist of protective shoes, back shells, cover shells and fixing elements. These are designed for fixing to

▷ C rail and slotted strip by square-headed stud

▷ angle iron with hook bolt

▷ sheet metal with hook stud

▷ flat bar with screw clamp

The cable clamps can be attached to wooden structures by using double-ended bolts having a wood screw thread at one end or with the aid of plugs these can be attached to concrete or masonry walls.

The protective cover shoes are made of steel or aluminium manganese (AlMg3). All steel parts are zinc plated and chrome passivated thus providing corrosion resistance suitable for tropical conditions. AlMg3-clamps are preferred where severe conditions (e.g. aggressive or salt-laden atmosphere with high humidity) prevail. Because these are non-magnetic they are also recommended for the fixing of single-core cables.

The following types of clamp are available:

▷ Single-cable clamps with single or double fixing points for use with single-core cables.

▷ Double-cable clamps for two cables of approximately equal outside diameter.

▷ Multi-cable clamps for a number of cables of the same outside diameter (in the range 10 to 20 mm), especially for the common fastening of control, measuring and signaling cables.

To extend the installation, extension studs are screwed to the existing elements and the installation can then be built up as before.

If the current load capacity of the cables and the strength of the support structure permits with the aid of the appropriate intermediate shells, several layers of cables may mounted above one another.

29.6.2 Arrangements and Dimensions

Single-Cable Clamps

Single-cable clamp with single fixing point for cables up to 40 mm outside diameter

Cable outside diameter mm	Dimensions in mm				Weight per 100[1]
	(Length of clamp) 45 mm)	A	B	C max.	net kg
13 to 16		26	13.5	21	3.2 (1.1)
17 to 20		30	15.5	22.5	3.3 (1.1)
21 to 24		34.5	18	26.5	3.5 (1.2)
25 to 32		42.5	21.5	35	4.6
33 to 40		50	25	43	5.2

Double-Cable Clamps

Cable outside diameter mm	Dimensions in mm				Weight per 100[1]
	(Length of clamp) 45 mm)	A	B	C max.	net kg
10 to 16		40	24	18	3.4 (1.2)
17 to 20		47	31.5	22	4.2 (1.4)
21 to 24		53.5	35	26	4.8 (1.7)
25 to 32		66	42	35	8.5 (2.9)
33 to 40		79	48	43	10 (3.4)

Single-cable clamps with two fixing points for cables up to 70 mm, with top shell of AlMg3 up to 90 mm outside diameter

Cable outside diameter mm	Dimensions in mm				Weight per 100[1]
	(Length of clamp) 45 mm)	A	B	C max.	net kg
20 to 24		47	37	28	5.2 (1.8)
25 to 32		54	42	36.5	6.5 (2.2)
33 to 40		64	51	44.5	9.8 (3.4)
41 to 48		68.5	56	52	12.5 (4.3)
49 to 58		76	63	62	19.5 (6.7)
59 to 70		90	76.5	74	20.5 (7.1)
71 to 90		128	110	97	– (13.0)

Intermediate shells

10 to 16		40	24	7.5	1.1 (0.4)
17 to 20		47	31.5	11	1.4 (0.5)
21 to 24		53.5	35	12.5	1.7 (0.6)
25 to 32		66	42	15	3.4 (1.2)
33 to 40		79	48	18.5	4.5 (1.6)

Intermediate shells for single-cable clamps with two fixing points

25 to 32		54	42	21.5	4.1 (1.4)
33 to 40		64	51	22	6.5 (2.3)
41 to 48		68.5	56	28	9
49 to 58		76	63	30	11
59 to 70		90	76.5	34	12.5

[1] Without fixing stud; values in ( ) are for clamps of AlMg3

Multi-Cable Clamps
(for fixing to angle iron spacers are required)

Clamping width up to mm	Dimensions in mm				Weight per 100[1] net kg
	(Length of shells 45 mm)	A	B	C max.	
35		46	–	depending on cable diameter	3.3
50		62.5	–		4.2 (1.4)
75		90	–		5.8
100		121	72		12.6

		Intermediate clamps			
35		46	–	10	1.6
50		62.5	–	11	2.1 (0.7)
75		90	–	12.5	2.8
100		121	72	13.5	6.2 (2.1)

Foam Packers

Foam packers to be used with all multi-cable clamps to reduce pressure points on cables and compensate for differences in diameter (up to 5 mm)

Clamping width mm	Supplied	Dimensions approx. mm
35 50	in strips	40 × 500 × 8
75 100	cut with holes	40 × 80 × 8 40 × 120 × 8

Spacers

Spacers for fixing with double-cable clamps for cables from 10 to 16 mm outside diameter and all multi-cable clamps on angle iron with hook bolts

No. of cables that can be clamped side by side

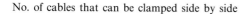

Clamping width in mm	Cable diameter in mm									
	8	9	10	11	12	13	14	15 to 16	17	18 to 20
35	4	4	3	3	3	2	2	2	2	–
50	6	5	5	4	4	4	3	3	3	–
75	9	8	7	6	6	5	5	4	4	4
100	12	11	10	9	8	7	7	6	6	5

[1] Without fixing stud; values in ( ) are for clamps of AlMg3

Fixing Elements for Cable Clamps:
Bolts, hook bolts, screw clamps

Fixing element	For fastening to	Size mm	Illustration
Square head stud	C Rail	Profile: 28/12, 30/15, 40/20, 40/22, 40/25, 48/26, 50/30	
Clamp stud	Slotted strip	30 × 4, slot size 70 × 8.5	
Hook bolt	Angle iron	40 × 40 × 5 45 × 45 × 5 50 × 50 × 5	
Hook stud	Plate	Minimum thickness 2	
Screw clamp with M 6 screw	Flat bar	Thickness up to max. 8	
Double ended screw with wood screw thread	Wood	Minimum thickness 30	
	Masonary with wall plugs		
Extension bolt	For adding to existing clamps		

Application and dimensions of fixing elements (bolts and hook bolts with M 6 thread)

Cable Suitable for clamps	Cable Outer diameter mm	Number of cable tiers	Headed studs for C rail	Headed studs for slotted strip	Hook bolt	Hook stud
			Length of thread in mm			
Single-cable clamp with one fixing point	13 to 16	1	20	36	14	14
	17 to 24		28	36	20	20
	25 to 32		36	36	28	28
	33 to 40		42	66	36	36
Single-cable clamp with two fixing points	20 to 24	1	28	36	20	20
	25 to 32	1	36	36	28	28
		2	66	66	57	57
		3	120	–	–	–
	33 to 40	1	42	66	36	36
		2	94	–	85	–
	41 to 48	1	51	66	42	42
		2	94	–	–	–
	49 to 58	1	66	66	57	57
		2	120	–	–	–
	59 to 70	1	66	66	57	57
		2	120	–	–	–
	71 to 90	1	75[1]	–	60[1]	–
Double-cable clamp	10 to 16	1	20	36	28	14
		2	42	66	42	42
		3	66	66	57	57
		4	94	–	85	–
	17 to 20	1	28	36	20	20
		2	42	66	36	36
		3	66	66	57	57
		4	94	–	85	–
	21 to 24	1	28	36	20	20
		2	51	66	42	42
		3	94	–	85	–
		4	120	–	–	–
	25 to 32	1	36	36	28	28
		2	66	66	57	57
		3	120	–	–	–
	33 to 40	1	42	66	36	36
		2	94	–	85	–
		3	120	–	–	–
Multi-cable clamp	up to 75	1	28	36	28	28
		2	42	66	57	57
		3	66	66	85	–
		4	94	–	85	–
	up to 100	1	36	36	36	36
		2	51	66	57	57
		3	94	–	85	–
		4	94	–	85	–

[1] With M 8 thread

30 Installation Guide

30.1 Preparation of Cable Ends

Paper-Insulated Cables with Lead Sheath

The ends of paper-insulated cable must be protected against both leakage of impregnating compound and the ingress of moisture. Where a cable is cut and is not installed immediately the ends must be sealed in a suitable manner, e.g. with a soldered-on lead cap, with sealing compound and metal cover or with a shrink-on plastic cap incorporating a suitable adhesive. It is not sufficient to bandage the end with insulating tape only.

When commencing installation the cable must be subjected to a "crackle test" to ensure moisture has not penetrated the cable. This applies especially for repair work also. Where moisture is detected the cable must then be cut to a point where the test does not indicate moisture. To carry out the crackle test cable oil or transformer oil is heated in a suitable vessel to a temperature of 140 °C while being stirred continually (beware of oil coming into contact with flames and also hot oil splashing). Then strips of paper taken from the belt insulation, from the core insulation close to the conductor as well as from the outer layers are immersed in the hot oil. If the oil begins to bubble moisture is present in the insulation.

When making the test protective gloves should be worn and furthermore the samples of paper taken should be of sufficient length to avoid danger from splashes. The samples must not be touched by hand or be brought into contact with moist objects as this would invalidate the test. The oil must not be heated to above 150 °C since in some synthetic oils depolymerisation could set in causing frothing without moisture being present.

The ends of the cable prepared for termination must be protected by sealing ends and all joint points protected by joint boxes. With high-voltage cables the sealing end also serves to control the decrement of the electric field.

Paper-Insulated Cables with Aluminium Sheath

The preparation of the ends of aluminium-sheathed cables is – with the exception that the aluminium sheath needs to be pretinned with rub-on solder – the same as for cables with lead sheath. The same accessories, as for lead-sheathed cables, are used. Since there is no steel-wire armouring then installation is eased. Further detailed guidance should be obtained from the manufacturers instructions.

Cables with Polymer Insulation for 0.6/1 kV

Cables with polymer insulation do not contain impregnating mass and the insulation is not moisture sensitive. Therefore for installation indoors in dry rooms sealing ends are not necessary (Fig. 30.1). The preparation of the ends of these cables is particularly simple. The thermoplastic outer sheath is cut by the aid of a sheath culting tool or knife with a controlled depth of cut and is removed. A longitudinal cut and

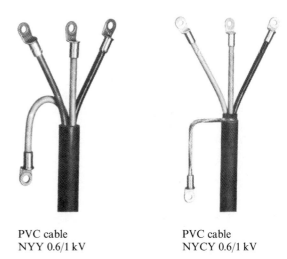

PVC cable   PVC cable
NYY 0.6/1 kV   NYCY 0.6/1 kV

Fig. 30.1
Prepared cable ends ready for connection
(indoor type)

Fig. 30.2
Slitting the outer sheath

Fig. 30.3
Circumferential cutting of the outer sheath

circumferential cut is made and the outer sheath is opened to the depth of the inner sheath or tape layer (Figs. 30.2 and 30.3). Care must be taken not to cut into the core insulation. The sheath is then easily removed. Where the cores have XLPE insulation these must be protected from solar radiation or ultra violet rays by shrink-on tubing or equivalent.

For outdoor use PVC and XLPE cables with $U_0/U = 0.6/1\,\text{kV}$ are fitted with sealing ends of cast resin or a shrink-on cap (see page 434). For the jointing of these cables, cast resin joint boxes are available which are suitable and safe in operation in unfavourable ambient conditions, e.g. aggressive soils or in water (see page 429).

Cable with Polymer Insulation for Rated Voltages
$U_0/U = 3.6/6\,\text{kV}$ to $18/30\,\text{kV}$

The ends of the cables must, as for paper insulated cables, be protected from the ingress of moisture. Cut points where not installed immediately must be reliably sealed at the ends. For this purpose suitable

shrink-on caps or plastic caps with suitable moisture-proof bonding to the sheath are available. A bandage alone of e.g. insulating tape is not sufficient.

In contrast to low-voltage cables all cables with rated voltage $U_0/U = 3.6/6\,\text{kV}$ and above must be fitted with sealing ends suitable for the connection (see page 424).

For cables with PE sheath it has been found that, due to the excellent insulating properties of this material, under certain conditions static charges can build up on the screen. To avoid danger to personnel by an accident – secondary to a discharge electric shock – from touching the screen or from cutting such cable, e.g. after unreeling from a cable drum or after laying, it is recommended that either the screen or all cutting tools are earthed. This can be done by connecting the screen or tools by means of a flexible cable to an earthed part of the installation.

30.2 Earthing of Metallic Sheaths and Coverings

Multi-Core Cables with Lead Sheath and Armouring

As indicated in Fig. 30.4 the lead sheaths are connected to the armour which at cable joints is connected through with a copper braid or lead inner casing and is earthed at the cable sealing ends. For S.L. cables this applies similarly, however, for the connection with the lead sheath and the armour different systems apply, depending on the design of the accessory selected.

Multi-Core Cables with Aluminium Sheath

These cables do not normally have armouring. The aluminium sheath is connected through at the joints and earthed at the ends. Where the aluminium sheath serves as PE or PEN conductor then an earthing braid is required having conductivity properties equal to that of the aluminium sheath. In house service boxes the aluminium sheath can be cut in a spiral form, to expose the cores, using the spiral strip as a connection and avoid breaking the sheath.

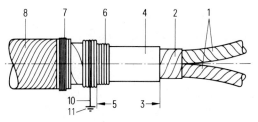

a) Belted cable

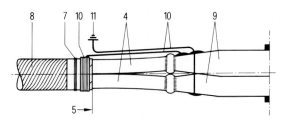

b) S.L. cable (joint box)

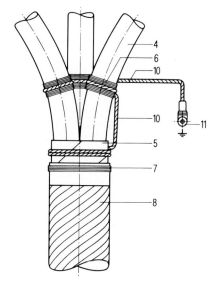

c) S.L. cable (sealing end)

1 Core
2 Belt insulation
3 End of lead sheath
4 Lead sheath
5 End of armouring. Armour to be cleaned with a file, coated with flux and tinned by using a soldering iron
6 Binding of tinned copper wire (diameter 2 mm) approx. 5 to 8 turns on to armour and lead sheath, for sealing ends for S.L. cables soldered to lead sheath only

7 Binding of tinned steel wire (dia. 1 mm) on to two or three layers of insulation tape
8 Jute serving
9 Inner lead casing
10 Earthing braids (soldered to armour and inner lead casing or to the lead casings of the three cores)
11 Connection to cast iron protective joint box or to the station earth

Fig. 30.4   Connection of lead sheaths and armouring (Typical arrangement)

Table 30.1
Sizes of copper connecting braids for aluminium-sheathed cables

Conductor cross-sectional area mm²	Cross-sectional area of copper braid mm²	Conductor cross-sectional area mm²	Cross-sectional area of copper braid mm²
3 × 16	25	3 × 120	70
3 × 25	35	3 × 150	95
3 × 35	35	3 × 185	95
3 × 50	50	3 × 240	120
3 × 70	50	3 × 300	150
3 × 95	70		

Single-Core Cables with Lead or Aluminium Sheath

In general the same guide lines apply as for multi-core cables. At the sealing ends earthing is carried out as normal, such that the metallic coverings of the individual single-core cables are cross connected.

If it is necessary to avoid the additional losses in the lead or aluminium sheath and if need be in the armour then sheathes and armour must be cross-linked and earthed at one end only (see page 328).

Cables with Polymer Insulation

The concentric conductor, the screen or the armour of PVC or XLPE cables are to be connected through

417

at cable joints and also earthed at the sealing ends in the same manner as the metal sheaths of paper insulated cables.

For cables with concentric conductor or a screen of copper wires, the copper wire layer is twisted together and earthed. For cables with a screen of copper tapes or steel wires, this screen must be earthed at the cable ends with a soldered on or crimped copper braid.

For screened single-core cables with polymer insulation, the same guide lines apply as for the relevant paper insulated cable with metal sheath.

30.3 Conductor Jointing[1]

Copper Conductors

Copper conductors are normally soldered into cable lugs or connection bolts using the soft solder L-Pb Sn 40 (Sb) or are through-connected using soft soldered copper sleeves. For spliced joints hard solder must be used and also in situations where increased mechanical stress on the conductor or where short-circuit temperatures exceeding 160°C are to be expected.

In recent years compression and crimp-on connectors are used in increasing numbers and these are suitable for short-circuit temperatures of up to 250°C.

Aluminium Conductors

Connection and jointing systems include soldering, screw clamping, welding or crimping. Because of the high temperatures involved, welding of connections is not generally used for cables with polymer insulation. Where conductor connections are to be welded the powder metal welding process is recommended. In cable joints as well as in sealing ends crimped or screw connections are nowadays preferred.

Solder connections are used only in exceptional circumstances where these can be subsequently protected against corrosion. In these joints the individual

Fig. 30.5
Tinning of conductors using rub-on solder L 211

strands of the conductor must be carefully cleaned and by using a tool or knife freed of all traces of oxidised skin. The strands must then each be tinned over their whole circumference using rub on solder L 211 (Fig. 30.5). These aluminium strands can then be reformed into a conductor and soldered into cable lugs or connection bolts using solder L-Pb Sn 40 (Sb).

Various Crimping Forms

Because of the ease of handling this type of jointing has gained great preference. There are a number of forms of crimping, e.g. hexagonal compression, deep impression crimping. Tools and accessories are available from specialist manufacturers.

[1] Additional information can be found in handbook "Abschluß- und Verbindungstechnik bei Starkstromkabeln" by H. Klockhaus and G. Wanser.
(Published by Verlags- und Wirtschaftsgesellschaft der Elektrizitätswerke mbH VWEW, Stresemannallee 23, 6000 Frankfurt (Main) 70)

Spade lug   Ring lug

Copper cable lug DIN 46 235   Aluminium cable lug

Copper and aluminium
jointing sleeve

a) Connection by hexagonal compression crimping

Cu cable lug   Cu cable lug DIN 46 235
DIN 46 234

Cu-jointing sleeve

b) Connection by impression crimping

c) Connection by deep impression crimping

Fig. 30.6
Examples of permanent jointing

Screw Jointing

In a large number of applications and particularly for branch connections of cables, e.g. the so called terminal rings for low voltage joints for domestic feeders, screw connectors have retained popularity (Figs. 30.7 and 30.8).

a) Terminal ring

b) Compact terminal

Fig. 30.7
Example of insulated branch connection terminals

Fig. 30.8
Example of screw-clamp connector

419

31 Repair of Damage to Outer Sheath

31.1 Outer Sheath of Polyvinylchloride (PVC) and Polyethylene (PE)

Dependant upon the degree of damage, the stressing of the cable sheath during installation in the ground on racks in ducts as well as the required voltage withstand of the sheath during corrosion protection testing, a suitable method of repair should be selected from Table 31.1.

If the material of the outer sheath cannot readily be determined and neither can the cable type be established from the markings and relevant VDE specifications then a flame test may be necessary to establish whether the sheath material is of PVC or of PE.

▷ PVC burns, giving off dense black soot, and pungent fumes develop as long as it is held in a flame.

▷ PE burns unaided once ignited and the material liquifies and drops away. The fumes have a smell of burning candle wax.

All repairs should be carried out, with a few exceptions, after the cable has been paid out. Damage observed before the installation must firstly be bandaged with self adhesive PVC tape and sealed in such a manner that, particularly for medium-voltage cables, with XLPE insulation water is prevented from penetrating under any circumstance. Points of damage which after installation are no longer accessible, e.g. when installed in a pipe or conduit, must however be properly repaired immediately.

Table 31.1
Methods of repair of PVC and PE outer sheaths relative to the degree of damage and type of installation

Method of repair (suitable for a test voltage up to 5 kV)	Type of damage							
	Abrasions, small tears not extending around the whole circumference (depth up to half-sheath thickness)				Larger tears, damage extending over the whole circumference (depth greater than half-sheath thickness)			
Type of installation	In ground or outdoor		Rack or cable tunnel		In ground or outdoor		Rack or cable tunnel	
Sheath material	PVC	PE	PVC	PE	PVC	PE	PVC	PE
Shrink-on repair sleeve	×	×	×	×	×	×	×	×
Lapping of F-CO 31 (self welding)	−	−	×	−	−	−	×	×
Cast resin sealing	×	−	×	−	×	−	×	−
Lapping of self adhesive PVC tape	−	−	×	−	−	−	−	−
Anealing[1]	×	×	×	×	×	×	×	×

[1] Only to be carried out by specialists

420

Fig. 31.1
For severe damage to the cable sheath the damaged section of sheath is cut away and replaced with bandaging with self-welding tape to the original sheath thickness.

Fig. 31.2
Repair of damage to a cable sheath by double layer bandaging

If the cable sheath is damaged such that the whole circumference is affected then the damaged section of sheath must be cut away and replaced by bandaging with self-welding tape (e.g. F-Co 31) up to at least the original outer diameter of the sheath (Fig. 31.1).

Repair by Self-Welding Tape

Before bandaging the damaged part and the cable sheath must be cleaned, extending to each side 100 mm beyond the damage, with solvent. The self welding tape F-Co 31 is then applied by winding over the damaged part or, for larger damaged areas, over the bandage (Fig. 31.1) and extending over the sheath 100 mm to each side.

As mechanical protection two layers of either Scotchrap tape or self-adhesive PVC tape (Fig. 31.2) should be applied.

Repair with Shrink-On Sleeve

A shrink-on sleeve, which is particularly easy to instal, comprises a longitudinally split shrinkable tube and this is recommended as a universal method of repair. This type of sleeve must be cut sufficiently long such that after being shrunk on to the cable it extends beyond the damaged part to each side by a distance of $l = 3 \times$ cable outer diameter with a minimum of 100 mm (Fig. 31.3a). A longitudinal shrinking of some 10% must be allowed for.

The damaged area of the cable sheath must be cleaned with solvent, corresponding to the length of the sleeve and roughened with emery cloth. The repair sleeve is then laid around the damaged portion and the press stud snap locking engaged (Fig. 31.3b).

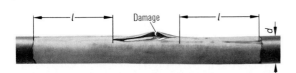

a) Repair sleeve to extend length l on either side beyond extent of damage (where $l = 3d \geq 100$ mm)

c) Shrinking on of repair sleeve using yellow flame propane torch

b) Laying on of a shrink-on sleeve containing thermochrome colour marks

d) Damaged sheath repaired using a repair sleeve

Fig. 31.3 Repair of damaged sheath by use of a shrink-on repair sleeve

The repair sleeve is then shrunk on to the cable by heating from the centre outwards using a yellow propane gas flame or hot air blower (Fig. 31.3c) until the thermochrome colour markings become invisible. This indicates the correct shrink temperature is reached. The melting adhesive on the inner surface of the repair sleeve must soften and be seen to squeeze out at both ends (Fig. 31.3d).

Cast-Resin Sealing for Cables with PVC Outer Sheath

The damaged part of the cable sheath is cleaned with solvent and roughened with emery cloth extending beyond the damage by 100 mm in both directions. Where the damage extends to a depth such that the inner laid-up cores are exposed, a layer of self adhesive PVC tape must be wound over the damaged area, with half lap, to prevent the liquid resin penetrating the cable (Fig. 31.4a).

A PVC tube serves as a mould for casting in the resin, the inner diameter of the tube being 10 to 20 mm greater than the cable diameter. The mould must also be 160 mm longer than the damaged section. The tube is cut longitudinally such that it can be fitted around the cable. For small areas of damage the mould normally used for cable joints (e.g. PV) will also suffice.

To seal the mould at each end a bandage of PVC tape is applied (Fig. 31.4b). This end bandage is located at least 50 mm beyond the damage extremity, or for cables with d of 35 mm or larger three times the cable diameter, to each side. The split PVC tube (joint mould) is placed over the end seal bandages and both ends finally sealed with PVC tape (Fig. 31.4c).

Resin (Protolin 80) can then be poured in. The PVC mould is normally left on the cable. After casting the repaired cable must not be bent at this point.

a) Lapping of cable cores with self adhesive PVC tape

$a \geq 50$ mm, with $d \geq 35$ mm $a = 3d$

b) Sealing with PVC tape extending distance a to both sides beyond damaged area

Lapping with Self Adhesive PVC Tape

This method of repair must be used only in dry rooms, indoors or in ducts for control, SIMATIC and power cables with PVC sheaths up to a diameter of 25 mm.

In cases of abrasions or small tears the sheath must be cleaned with solvent over a distance extending 100 mm to each side beyond the damaged part. Only after this procedure is the PVC tape wound over the damaged part and to both sides over the undamaged sheath for a distance of 100 mm.

This part of the sheath must only be subjected to light mechanical stress. The cable may be laid using cable laying rollers. The repaired portion must again be checked after laying is completed.

c) PVC tube fitted ready for casting

Fig. 31.4
Preparation of PVC sheath for casting in resin

Multi-Layer Corrosion Protection of Cables with Aluminium Sheath

Before undertaking a repair of PVC sheath of a aluminium-sheathed cable, it must be checked whether the plastic tape wound over the aluminium sheath has been damaged. If damage is found then the outer

PVC sheath must be completely removed extending 20 to 30 mm beyond the damaged area of tape on both sides. If the PVC outer sheath is then repaired by bandaging it must be ensured that the damaged portion of tape on the aluminium sheath is covered to an extent of at least 20 mm beyond the limit of damage.

31.2 Jute Servings on Cables with Lead Sheath

Damaged jute servings on cable with lead sheath are repaired by taping with Evo-Elt bandage.

32 Cable Accessories

32.1 Fundamental Objectives

Sealing ends form the termination points of a cable and serve as a connection to electric apparatus or machines or switchgear. Depending on the system rated voltage and the cable construction, the following objectives must be fulfilled:

▷ connection of the conductor (see Section 30.3),

▷ sealing of the cable against ambient influences (e.g. ingress of water),

▷ protection of core insulation (e.g. against U.V. radiation),

▷ controlled decrement of the electric field strenght on medium- and high-voltage cables (see Section 32.3),

▷ insulation from earthed parts (see Section 32.4).

Cable joints connect lengths of cables in long transmission routes or at points of repair and must fulfill the following functions:

▷ connection of the conductors (see Section 30.3),

▷ insulation of the conductors and, especially in medium-voltage cables, reestablishment of the elements of the cable (see Sections 32.3 to 32.7),

▷ protection against all ambient conditions of the ground,

▷ establishment of branch points for service cables in low-voltage cables.

Materials used in the manufacture of accessories must be compatible with materials in the cable and must have the following characteristics:

▷ resistance to thermal deformation,

▷ resistance to thermal oxidisation,

▷ permanently elastic,

▷ hydroscopically stable,

▷ outer protective covers must be resistant to ambient influences especially when used in the ground,

▷ easy to apply.

32.2 Requirements

The cable accessories must not only comply with electrical requirements but must also satisfy requirements in respect of ambient which can differ significantly depending on location, ground, indoors or outdoor.

Relevant test procedures for type testing, batch testing and routine testing of accessories in order to prove suitability are laid down in DIN VDE 0278. The test procedures are orientated to the operational requirements and are sub-divided for the individual product range:

Part 1 General,

Part 2 Cable joints $U_0/U > 0.6/1 \text{ kV}$,

Part 3 Cable joints $U_0/U = 0.6/1 \text{ kV}$,

Part 4 Sealing ends for indoor installations $U_0/U > 0.6/1 \text{ kV}$,

Part 5 Sealing ends for outdoor installations $U_0/U > 0.6/1 \text{ kV}$.

Table 32.1 provides a simplified summary of the criteria and the associated testing [32.1].

The tests are laid down in a coordinated sequence to provide stressing which realistically relates to that which may occur in operational conditions (Fig. 32.1).

Table 32.1   Operation stresses imposed on accessories and associated proof testing

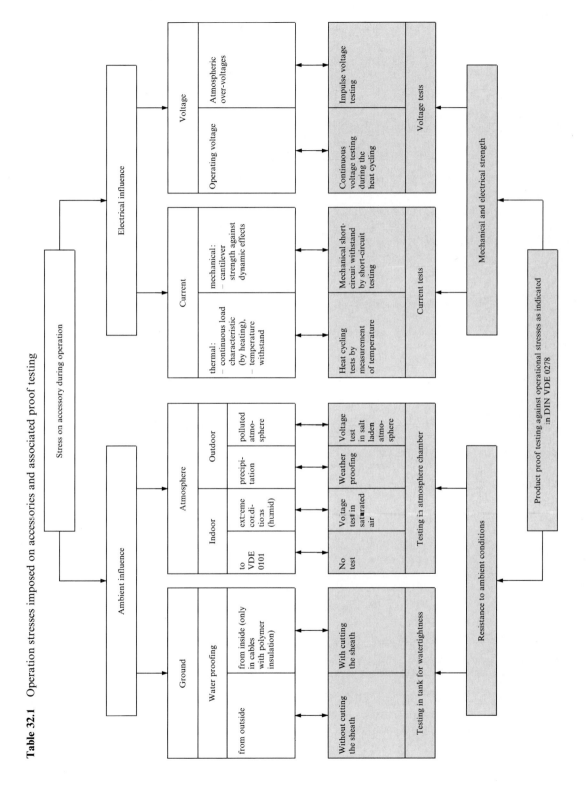

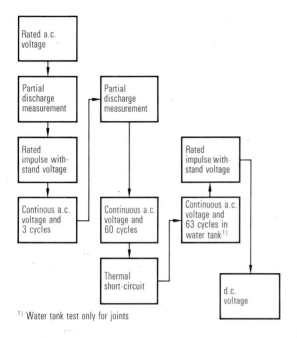

¹⁾ Water tank test only for joints

Fig. 32.1
Testing cycle for joints and sealing ends for cables with polymer insulation

32.3 Stress Control

The electric field is disturbed at the stripped end of medium-voltage and high-voltage cables since the inner electrode (conductor) continues on whereas the outer electrode (the conducting layer with sheath respectively screen) is cut back and removed in a defined area. This results in a high concentration of field in this area unless special measures are taken to control this.

In order to construct a control of the field two capacitances must be taken into consideration (Fig. 32.2):

▷ the capacitance between the individual surface elements and the removed conducting layer (C_1),

▷ the capacitance between the surface elements and the conductor (C_0).

Proportional to the relationship of the inverse of these capacitance values, potentials develop betweeen the surface elements and the end of the outer conduct-

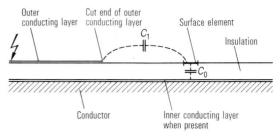

Fig. 32.2
Effective capacitances at cable ends [32.2]

ing layer. In this the initial voltage U_a at which surface discharge commences is derived from the following function:

$$U_a \sim E_d \sqrt{\frac{C_1}{C_0}}.$$

E_d Breakdown field strength at the periphery of the stripped end of the conducting layer

If the surroundings of the core consists of e.g. air, then C_1 is relatively small in relation to C_0, such that the larger proportion of the voltage appears between the surface element and the conducting layer which leads to discharge if the breakdown strength of air is reached.

To control the field (Fig. 32.3) it is arranged that either C_1 is made larger or C_0 is made smaller:

▷ In the commonly applied capacitance field control C_0 is reduced by forming, such that the radius at the cut end of the outer conducting layer is artificially increased by a stress cone.

▷ In the refractive control method C_1 can be enlarged by, for a sealing end, adding a sleeve of insulation material which has a very high dielectric constant.

An alternative method is to use resistive stress control. Here the capacitance C_1 is bypassed by a resistance per unit length. The capacitive current due to capacitance C_0 produces a voltage drop in this resistance per unit length which causes the existing potential to decay in a controlled manner. This method is frequency dependant and therefore is not effective on d.c. For this reason d.c. voltage testing cannot be applied to installations where sealing ends have resistive field control.

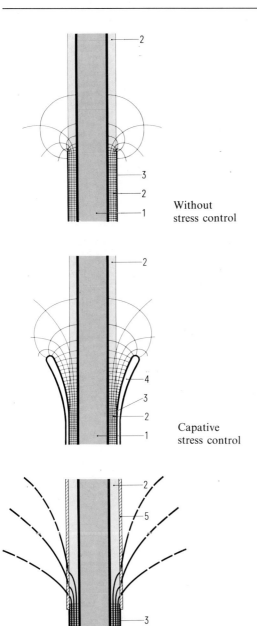

Without
stress control

Capative
stress control

Refractive
stress control

1 Conductor   4 Stress cone
2 Insulation   5 Stress control tube
3 Outer conducting layer

Fig. 32.3
Capacitive and refractive stress control

32.4 Fundamental Principles for the Construction and Installation of Accessories

Depending on the type of accessory (joint or sealing end) and voltage range in respect of the provision of quick and safe installation several techniques have been developed:

▷ compound filling technique,

▷ cast-resin technique,

▷ shrink-on technique,

▷ lapping technique,

▷ push-on technique,

▷ plug technique.

High-dynamic forces under short-circuit conditions may demand additional measures in both the construction and mounting of sealing ends. Detailed guidance is given in Section 19.4.4 "Mechanical short-circuit withstand".

32.4.1 Compound Filling Technique

The compounds normally used for filling accessories are bitumen based. These meltable compounds must be poured hot and are generally used for mass impregnated cable accessories. It must be born in mind that these compounds have a considerable degree of shrinkage on cooling approximating some 8 to 10%.

Cold pouring compounds are insulating filler compounds comprising two components namely oil and bitumen powder (asphalt). These are mixed in the cold state and solidify by expansion to a compact mass.

Table 32.2 provides a summary of hot poured filling compounds to DIN VDE 0291 Parts 1 and 1a. The types of cable and examples of application are also shown for each individual compound.

The most important property of all filler compounds is goodness of adhesion to paper insulation polymers and metals. They must also have good flow characteristics during handling and also an adequate voltage withstand. Compounds based on bitumen have these properties inherent in the basic material.

The voltage withstand of the compounds is approximately 60 kV/mm such that they are suitable for me-

Table 32.2
Hot poured compounds to DIN VDE 0291 Parts 1/2.72 for power cable accessories [32.3]

	Compound type	Cable type	Examples of applications	Main function
1	SN	Paper-insulated cables	For joints and sealing ends up to 10 kV (knock-off: allows for quick access to cables)	Insulation function
			For protective joints which provide only mechanical and corrosion protection	Corrosion protection function
2	SE	Paper-insulated cables	For pole sealing ends 1 kV which are subjected to strong solar radiation	Insulation function
3	SP	Paper-insulated and PVC cables	For joints and sealing ends up to 10 kV subjected to high humidity and for high-mechanical stress	Insulation function
			For protective joints which provide only mechanical and corrosion protection	Corrosion protection function
4	SY	PVC cables	For jointing and branch joints up to 1 kV	Insulation function
5	SZ	Paper-insulated and PVC cables	For joints and sealing ends up to 1 kV	Insulation function

dium as well as low voltage. Because these compounds do not offer high-mechanical strength and furthermore can absorb moisture after casting, in comparison with cast resin, they need to be contained in a leak-proof housing, e.g. of cast iron. The high temperatures for pouring (between 120 to 180 °C) demand great care in handling during installation.

The cable cores, especially to avoid thermal ageing, must not be heated excessively. For cables with polymer insulation, filling compounds based on bitumen have little importance and in many instances are unsuitable since the change effect between bitumen and the construction elements of the cable, especially in XLPE medium-voltage cables may lead to problems (e.g. change of conductivity of the extruded conducting layer due to expansion). For XLPE insulated cables outdoor type sealing ends, with porcelain insulators (support function), are used filled with an insulating compound based on polybutene, which does not react with XLPE (compounds M 24, M 40, M 43; see Table 32.3). Also for the sealing ends, e.g. with porcelain insulators, for mass impregnated cable a compound based on polybutene is used which is also compatible with older cables having an impregnation of oil resin compounds.

Table 32.3
Coordination of filling compounds and cable accessories

Type and identity mark of accessory	Filling compound used for:	
	normal ambient temperatures	higher ambient temperatures
Sealing ends		
for indoor installation		
EoD	M 24	M 40
for outdoor installation		
FEL	M 24	M 40
FEL-2Y, FEP	M 40	–
Joints		
up to 10 kV	M 11 (Sp)[1]	M 12[1]
above 10 kV		
protective joints	M 11 (Sp)[1]	M 12[1]
lead inner casings	M 24	M 43
spreader boxes		
AS	M 12[1]	
SS	M 11 (Sp)[1]	

[1] Compound based on bitumen, abbreviated marking in ( ) indicates classification to DIN VDE 0291

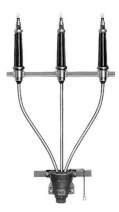 

Sealing end EoD
with transparent
cast-resin
insulators

Sealing end FEL
with porcelain
insulators

Fig. 32.4
Sealing ends for mass impregnated cables
$U_0/U = 12/20$ kV and $18/30$ kV

32.4.2 Cast-Resin Techniques

In the design of accessories epoxy (EP) and polyurethane (PUR) resins are used. Both resin groups harden by poly addition. Fig. 32.5 indicates the relatively long stretched molecules of resin and the short molecules of hardener which lie firstly in a random formation. Hardening occurs when the hardener molecules adhere to certain points of the resin molecules (the black dots in the figure indicate chemical groups capable of reaction) and which then form a three-dimensional network.

The advantages of the polyurethane cast-resin compounds are primarily in their ease of installation:

▷ The heat generated by the reaction is low for PUR cast resins, therefore quantities up to 5 liters can be cast at one time without affecting the insulation by excessive thermal stress.

▷ PUR materials adhere to PVC insulation without the use of a catalyst.

▷ The PUR material has a high resistance to impact, high thermal stability and excels also in low brittleness.

▷ With PUR resins only few internal stresses develop during hardening which allows uniform hardening.

The sealing tightness of a cast-resin accessory requires perfect bonding between the cast-resin material and the conductor insulation, which is dependent on both the mechanical and the specific adhesion.

The *mechanical adhesion* results from the anchoring of the cast resin in the pores and indents in the surface of a firm shape or body.

The *specific adhesion* is the most important force in the adhesion of two bodies and is attributed to the valency forces inside the material.

The cast-resin products PROTOLIN, manufactured by Siemens were developed for a great variety of applications in cable engineering. These cast-resin products have

▷ good electrical and mechanical properties,

▷ high resistance to ageing and weather as well as

▷ high resistance to attack by chemicals in the ground.

The most important characteristics and applications of these resins have been summarised in Table 32.4.

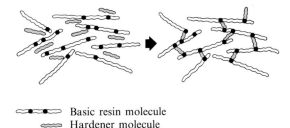

⬤▬⬤▬ Basic resin molecule
〰️ Hardener molecule

Fig. 32.5 Reaction process in cast resin hardening

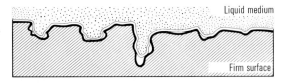

Liquid medium

Firm surface

Fig. 32.6
Mechanical adhesion, penetration and anchoring of the liquid medium (adhesive) in surface indentations (greatly magnified)

Table 32.4 Summary of the characteristics of cast resins important for their application

Cast resin	PROTOLIN 51	PROTOLIN 72	PROTOLIN 80	PROTOLIN 84
Regulation	VDE 0291 part 2 PUR, Typ GMH	VDE 0291 part 2 PUR, Typ GMW	VDE 0291 part 2 PUR, Typ GNW	VDE 0291 part 2 PUR, Typ GNW
Base	Polyurethane	Polyurethane elastified	Polyurethane elastified	Polyurethane
Packaging	Resin and hardener in two-part tins or in two-part bag (Fig. 32.7)	Resin and hardener in two-part tins (Fig. 32.8)	Resin and hardener in two-part tins (Fig. 32.9) or in two-part bag	Resin and hardener in two-part tins
Hardening time	1 to 3 hrs	1 to 3 hrs	$\frac{1}{4}$ to 2 hrs	$\frac{1}{4}$ to 2 hrs
Shelf life	24 months	24 months	24 months	24 months
Application	Joints for 3.6/6 kV PVC cables Transition joints for 6/10 kV cables	Indoor sealing ends for – medium-voltage PVC cables – rubber and thermoplastic-sheathed cables	Low-voltage accessories for – PVC-XLPE cables – mass impregnated – rubber and thermoplastic-sheathed cables Transition joints 20 kV Mechanically stressed accessories and resin filler for joint tubes	Low-voltage accessories for – PVC cables – mass impregnated – rubber and thermoplastic-sheathed cables Mechanically stressed accessories and resin filler for joint tubes
Special notes	For abnormally high-ambient temperatures or for very large filling volumes PROTOLIN 51 H must be used as it has longer hardening time	For PVC insulation good adhesion to insulation is achieved; XLPE cables require a bedding of tapes	Very good adhesion and therefore good tightness of sealing to XLPE cables	PROTOLIN 84 is less elastic but is less costly

Fig. 32.7
PROTOLIN 51 in two-part tins or in two-part bag
for cable joints (3.6/6 kV)

Fig. 32.8
PROTOLIN 72 in two-part tins for indoor sealing
ends for PVC cables, rubber and thermoplastic-
sheathed cables

Fig. 32.9
PROTOLIN 80 in two-part tins or in two-part bags
for low-voltage accessories for PVC and XLPE cables,
mass-impregnated cables, rubber and thermoplastic-
sheathed cables as well as for accessory parts subject
to mechanical stress

Application of Cast Resin

Both cast resin and accessory parts must be protected from moisture and must be completely dry when installed. For the cleaning (degreasing) of parts ready for casting in, either a solvent or emery cloth is used.

Packaged in Two-part Tins (Fig. 32.10)

The base resin is to be thoroughly stirred using a suitable, dry, grease-free implement. After adding the hardener both are mixed until the mixture is homogeneous. Avoid the inclusion of air bubbles by slow stirring. The resin mixture can then be poured into the prepared former which normally is of plastic.

Packaged in Two-part Bags (Fig. 32.11)

Resin and hardener are sealed in a two compartment bag separated by a clamp and seal (two-part bag). This bag is further protected by an outer bag which is lined with aluminium foil. This prevents water penetration into the cast resin. To prepare the resin the outer protective bag is removed together with the clamp and bag divider seal. The base resin and hardener can then be mixed by kneading the bag by hand. This mixing operation must be continued for at least three minutes. One corner of the bag is then cut away and the resin poured into the pre-formed mould.

a) Tearing open the seal b) Stirring resin, stirring hardener c) Adding the hardener

d) Mixing by stirring e) Forming a spout f) Pouring into accessory

Fig. 32.10
Using cast resin packed in two-part tin

a) Opening outer protective bag b) Removal of the seal and clamp c) Mixing by kneading

d) Cutting the bag e) Pouring into accessory

Fig. 32.11
Using cast resin packed in two-part bag

32.4.3 Shrink-On Technique

The heat-shrink system is used nowadays for a great variety of applications in cable accessories, especially for 1 kV straight joints.

As a basic material thermoplastic polyethelene is selected which, after being moulded into shape, is cross-linked and pre-stretched. This pre-stretched form (e.g. a tube) when heated will recover to its original dimensions. In this operation the tube "shrinks" on to the subjacent parts. A well adhesive-tight bonding with the insulation or the cable sheath is achieved by lining the inner surface of the moulded part or the tube with a melting adhesive. The adhesive melts with the heat applied in the shrinking process so that any unevenness in the cable surface, such as notches or grooves on the part to be surrounded are sealed.

There are basic differences between moulded and extruded parts.

With moulded parts the thermoplastic processing and cross-linking are achieved chemically by peroxides. Typical moulded parts are, e.g. end caps, as used for the sealing of cable ends.

Extruded parts (e.g. tubes) are produced in two stages:

▷ In the first stage the material is only thermoplastically shaped to the required form.

▷ At the second stage the cross linking is realized by electron radiation.

The moulded or extruded parts are then expanded while heated and then cooled while stretched. With the cooling the internal tension is frozen which, because of the rigidity of the material, is then locked in. When subsequently reheated (i.e. by flame or hot air) the molecular mobility of the cross-linked and stretched material increases with the material softening, allowing the frozen-in tension to contract the material to its original dimensions.

The shrinking temperature is between 110 and 150°C. Certain materials (low-temperature materials) will commence shrinking at approximately 70°C. As a heat source, a hot air blower or soft flame may be used. Shrink-on products are easy to instal, simple, quick and independent of ambient temperature. They can also be used, because of their flexibility, as additional protection against mechanical stress, e.g. as protection against kinking in flexible connections.

Table 32.5
Characteristic data for shrink-on products

Tensile strength	12 to 20 N/mm^2
Elongation at break	200 to 600%
Breakdown strength	80 to 200 kV/cm
Temperature resistance	-50 to 105 °C
Peel strength depending on the coating	40 to 120 N/25 mm
Storage life	practically unlimited

Shrink-On Tubes

In the accessory range shrink-on tubes are used for a variety of purposes:

▷ protection of cores of XLPE insulated cables against UV radiation (especially in outdoor installations),

▷ insulation and outer protection for straight joints (see Fig. 32.12),

▷ sealing of transition joints against leakage of the paper impregnating compound,

▷ outer mechanical protection for joints or joint boxes.

Fig. 32.12
Use of shrink-on tube over a straight joint in PVC cable for $U_0/U = 0.6/1$ kV

Fig. 32.13
End separation of a 4-core
PVC cable for $U_0/U = 0.6/1$ kV

Fig. 32.15
Shrinking-on of an end cap on a medium-voltage
cable with polymer insulation

Spreader Caps

With indoor installations the cable ends of cables
with polymer insulation can be sealed by using a
shrink-on spreader cap (Fig. 32.13) to prevent ingress
of dust and dirt. In outdoor installations cables (including the conductors) must be sealed to prevent
ingress of water.

Repair Sleeves

Shrink-on repair sleeves are used for the repair of
damaged sheaths (PE and PVC), these being coated
on the inner surface with a layer of heat melting adhesive (see Section 31). These sleeves incorporate an
easy to use snap-locking arrangement (Fig. 32.14).
Yellow, colour change thermochromatic dots indicate (yellow becomes black) when the correct shrink
temperature is reached. This ensures the adhesive has
melted and has reached the temperature required for
bonding.

End Caps

Shrink-on end caps serve to seal the ends of cables
with polymer insulation and mass impregnated
cables. They always have an inner layer of melting
adhesive (Fig. 32.15).

32.4.4 Lapping Technique

The lapping technique is used predominantly for
cables with paper insulation and for joints in cables
with polymer insulation for higher rated voltages.
The following are used:

▷ insulating tapes for the re-establishment of removed insulation such as
  – impregnated winding paper for mass impreg-
    nated paper cables,
  – high-voltage insulating tapes for cables with po-
    lymer insulation.

▷ conducting tapes for limiting the electric field as
  – partly metalised winding paper for mass im-
    pregnated paper cables,
  – contact tapes (conducting) for cables with po-
    lymer insulation.

**High-Voltage Insulating Tape for Joints
and Sealing Ends of Cables with Polymer Insulation
up to $U_0/U = 64/110$ kV**

The high-voltage insulating tape is manufactured
from an EPR base and is suitable for hand or machine wound insulation bandaging. Apart from me-

Fig. 32.14   Repair sleeve

chanical and electrical characteristics this tape offers resistance to ozone, chemicals, water immersion, humid atmospheres as well as suited to all operating and ambient temperatures.

After bandaging the tape vulcanises into a void-free insulation with a rubber like elasticity remaining constant at operating temperatures. The operating temperature aids in the rate of curing.

Contact Tape for Stress Control and Limitation of the Electric Field for Sealing Ends and Joints in Cables with Polymer Insulation

The contact tape is a low conductivity self-welding polymer tape. It is used:

▷ to smooth out uneven surfaces on conductors and connections,

▷ in connection with the outer conducting layer it is used for field control by bandaging to form a stress-cone.

Bandaging Paper for Mass-Impregnated Cable

Impregnated insulating paper in small rolls is used on mass-impregnated paper insulated cables to re-establish insulation in 20 kV to 30 kV joints. The joint boxes are filled with compound M 40.

The *partly metallized bandaging paper* is used for stress control in mass-impregnated paper cable sealing ends up to 30 kV. The bandaging paper is 50 mm wide and is metallized on both sides to a width of 25 mm. When these papers are wound on the core a stress-cone is formed.

32.4.5 Push-On Technique

Push-on accessories dominate the field of medium-voltage sealing ends but are also commonly used in medium-voltage joint boxes and high-voltage sealing ends.

In this technique, pre-manufactured work-tested parts are used. The quality of such an accessory is largely dependent on the manufacturer at the works, however this does not preclude the need for high-quality requirements also at the installation stage.

In Germany, for XLPE cables and because of there advantages, push-on sealing ends and joints of silicone rubber have been adopted (Figs. 32.16 and 32.17).

JAE 20   JAES 10   JAES 20

a) Push-on sealing ends for indoor installations

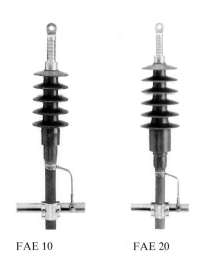

FAE 10   FAE 20

b) Push-on sealing ends for outdoor installations

Fig. 32.16
Silicone rubber-push-on sealing ends for single-core PROTODUR and PROTOTHEN cables
$U_0/U = 6/10$ kV, $12/20$ kV and $18/30$ kV

Fig. 32.17
Silicone push-on joint ASM for single-core XLPE
cables $U_0/U = 6/10$ and $12/20\,\text{kV}$

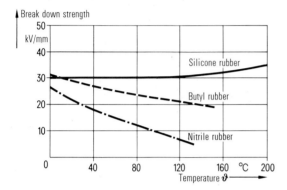

Fig. 32.18
Relationship of break down to temperature for
silicone rubber and synthetic rubber

Silicone rubber material described in Section 2.1.2
page 26 has characteristics which are of particular
advantage in the manufacture of accessories, namely:

▷ The good flow characteristics and low volumetric
shrinkage during cross-linking provides preconditions for close tolerance production of moulded
parts.

▷ Silicone maintains almost constant physical characteristics over a wide range of temperature (-50
to $+250°\text{C}$).

▷ The term $Si-O$ bonding (Fig. 2.12) effects the required temperature withstand and therefore resistance to aging.

▷ Silicone rubber has good elasticity independent
of temperature. Together with the thermal volumetric coefficient of expansion, which corresponds
in magnitude closely to that of cable insulating
materials, a split-free accessory is ensured when
subjected to heat cycling.

▷ The resistance to external influences (ozone, UV,
oxidisation, hydrolysis) provide preconditions for
good resistance to weathering, such that silicone
rubber is suitable as a material for outdoor accessories.

▷ The resistance to weak acid and alkalines, saline
solutions, minerals oils and microbes make it suitable for the production of cable joint boxes in
the ground.

▷ The break-down strength of silicone rubber is
found to be near constant over a wide range of
temperatures (Fig. 32.18). The $\tan \delta$ is less than
10^{-3}, the dielectric constant is < 3.

▷ In contrast to the elastomers based on hydrocarbons, in silicone under the condition of arcing
no conductive soot is produced but an insulating
silicic acid framework is formed. This provides
a high resistance to creepage.

▷ In contact with the cable materials no negative
interaction (contamination) occurs.

▷ By the addition of carbon, silicone can be made
conductive such that the specific volume resis-
tance can be made as low as $4\,\Omega\,\text{cm}$. This conductive silicone rubber is used for elements for field
control in push-on and plug accessories.

with
metal
housing

without
metal
housing

Fig. 32.19
Angled cable plug for single-core cables with polymer
insulation

32.4.6 Plug Technique

The plug technique offers the most compact form
of connection. The proven push-on technique with
silicone rubber insulation is here combined with a
screw-fixed disconnectable conductor connection.
Two basic types are available, the angled type and
T type. Typical areas of application are the connections to switchgear, transformers and electrical machines providing these have standardised cast resin
feed-through connectors to DIN 47636 which mate
with the plug.

32.5 Literature Referred to in Section 32

[32.1] Nickling, G.: Berücksichtigung der betrieblichen
Anforderungen und Beanspruchungen der Ka-
belgarnituren. Technische Mitteilungen, 72
(1979) Heft 7/8, Seite 479 bis 483

[32.2] Newi, G.: Garniturentechnik für Kunststoffkabel. Technische Mitteilungen, 72 (1979) Heft 7/8,
Seite 487 bis 495

[32.3] Waligora, H.J.: Heiß zu vergießende Füllmassen,
Kaltfüllmassen und kalthärtende Gießharze.
Technische Mitteilungen, 72 (1979) Heft 7/8, Seite
483 bis 487

[32.4] Klockhaus, H.; Wanser, G.: Abschluß- und Verbindungstechnik bei Starkstromkabeln. Verlags-
und Wirtschaftsgesellschaft

[32.5] VDEW-Kabelhandbuch. 4. Auflage, Verlags-
und Wirtschaftsgesellschaft der Elektrizitätswerke m.b.H., 1986

33 Cable Plan

When the installation of cables has been completed and all sealing ends and cable joints are secured in positions, a plan of the cables should be prepared to a preferred scale of 1:100 or 1:500 containing all data relative to the operation of the installation.

1. Cable type, cross sectional area of conductors, rated voltage.

2. Year of manufacture of the cable with year and month of installation.

3. Real laid length (length supplied less waste) between the centres of joints and between joint and sealing end bolt.

4. Location of cables and joints relative to fixed points, e.g. buildings, hydrants or boundary markers. Objects which may vary in position with time, such as natural water courses, trees, shrubs and such like are unsuitable for this purpose. The usefulness of kerbstones or milestones is limited as these may be moved in position due to road workings.

5. Variations in altitude along the cable route should be recorded on the plan by a 10- to 200 fold super-elevation. The altitude should be given in reference to sea level. Such a record is absolutely essential for oil-filled cables.

All subsequent alterations in the cable installation must be entered on this plan. This is important so that at a later date damage to cables may be avoided during construction work due to uncertainty of cable positions and also any faults may be quickly and safely located and corrected.

Measuring and Testing of Power Installations

34 Electrical Measurements in the Cable Installation, as Installed

For reasons of operational safety it is recommended to measure the electrical values in the cable installation, as installed, and record these in a cable log in order to have them available for comparison purposes and to assist in fault location.

Insulation Resistance

With *non-radial field cables* which have a metal sheath, concentric conductor, screen or armour over the laid-up cores the insulation resistance R_i (see page 337) of each individual core is measured to all other cores connected to the earthed metal sheath.

With *radial field cables* (single-core cables and multi-core cables with individually screened cores) the insulation resistance R_i is measured between each individual conductor and the metal sheath. The insulation resistance per kilometer R_i' is calculated from

$$R_i' = R_i \times l \qquad 34.1$$

R_i' Insulation resistance in $M\Omega \cdot km$
R_i Insulation resistance in $M\Omega$
l Cable length in km

When comparing these insulation resistance values with values measured at the factory, it must be borne in mind that the insulation resistance of a cable when installed complete with sealing ends and joints is always less than the factory-tested values of individually supplied cable lengths without accessories.

Conductor Resistance

For the measurement of this the conductors are short-circuited at one end by means of a connection which has at least the same conductivity. The additional resistance of this connection must be kept as low as possible by using substantial crimp or screw connections. By connecting a suitable measuring bridge to the other end of the cable the resistance of the conductor loop can be measured.

With three-core cables, for example, the three loops are measured

Conductor $1+2$
Conductor $2+3$
Conductor $3+1$

Providing the cables have been manufactured and installed in line with the standards, the measured values should be comparable one with another. Small differences, however, may be attributed to manufacturing tolerances of the conductors.

The resistance of one conductor is one half of the measured loop resistance R_s. The average conductor temperature at the time of the test must also be recorded.

Values which may be compared with those given in a resistance table based on a conductor temperature of 20°C can be determined from the following as resistance per kilometer:

$$R_{20}' = \frac{R_s}{l_s[1 + \alpha_{20}(\vartheta - 20)]} \qquad (34.2)$$

R_{20}' Conductor resistance at 20°C in Ω/km
R_s Measured loop resistance in Ω
l_s Loop length $\cong 2 \times$ cable length in km
α_{20} Temperature coefficient at 20°C,
     for copper $3.93 \times 10^{-3}/K$,
     for aluminium $4.03 \times 10^{-3}/K$
ϑ Temperature of conductor when measured in °C

Capacitance

The capacitance (see page 331) is measured in a similar manner to insulation resistance, for cables with *non-radial field*: each individual conductor to the other conductors connected to the earthed metal sheath, for *radial field* cables each individual conductor to the metal sheath.

A capacitance bridge is used for this measurement. The values obtained can be converted to µF/km.

35 Voltage Tests

35.1 General

Cable installations can be tested after completion using either d.c. or a.c. voltage. The level of test voltage and duration of application for low-voltage and medium-voltage cables is laid down in DIN VDE 0298 Part 1 and for high-voltage cables is laid down in DIN VDE 0256, 0257, 0258 and 0263. Tables 35.1 and 35.2 give a summary of the permissible test voltages and duration times extracted from the VDE and relevant IEC standards.

For cables with PE or XLPE insulation $U_0/U = 0.6/1$ kV to $18/30$ kV to DIN VDE 0298 Part 1 these values apply only for newly connected installations. For repeated testing and also for older installations containing such cables, a lower value of test voltage is recommended:

"In cables with PE or XLPE insulation which have been in service for some time, as well as for repeated testing it can not be precluded that by testing with high d.c. voltages faults may be initiated, whereas such damage would not have occured due to stresses under normal operation at the expected d.c. voltages. Until investigations in this field are concluded, testing should be conducted with a markedly reduced d.c. voltage, the level of which to be laid down by the user at his own responsibility."

It is recommended that values no higher than $2U_0$ should be applied.

For all voltage tests all conductive extraneous parts of the installation and associated apparatus (switches, current transformers busbars etc.) which are in range of the sealing ends or cables must be earthed. The minimum clearances in air as given in DIN VDE 101 for the relevant insulation rating are sufficient to avoid flashover to other parts of the installation during testing.

Where, to obtain the clearances, it would appear necessary to disconnect sealing ends, cable lugs or cable studs and where this may impose mechanical strain

on the cable, then these should remain connected and the isolating switches opened so that the test voltage is applied to them to one side only. In all cases the relevant safety codes of practice must be observed.

35.2 Testing with d.c. Voltage

Direct-current voltage tests are generally more convenient to conduct than a.c. voltage tests since, especially in long cable runs, only a low power is required. The financial outlay for a test transformer for d.c. testing is much less than for a variable transformer which for a.c. voltage tests must be suitably sized for the relatively high wattless power. Other advantages of the d.c. voltage test is the ease of transportation and the low value of collected load of the test equipment. Also the current (leakage current, fault current) flowing in the installation and in the sealing ends during d.c. testing can be more readily measured. This assumes that after the charging current has decayed, the steady state value will be less than 1 mA. In tests at voltages above 50 kV the current being measured may also include loss currents due to corona, depending on voltage, humidity of the air and other influences resulting in higher values of leakage.

In cables with paper insulation when tested with d.c. voltage, faults which are imminent – especially where there is ingress of moisture – can be recognised easily at the initial stage. For this reason alone d.c. is preferred to a.c. for the voltage testing.

For cable with polymer insulation (especially PE or XLPE insulation), a d.c. voltage test, because of the special characteristics of this dielectric has no such significance in respect of operational safety. ·

During testing flashover should be avoided. To prevent impulse voltage spikes occuring which exceed the rated level of impulse voltage, it is recommended in cable installations $U_0/U \geq 6/10$ kV to fit protective spark gaps on the sealing ends, during testing, calibrated to the rated peak-level impulse voltage (Table 17.2). When selecting the d.c. test voltage, the response voltage of the protective device together with its response tolerances must be considered.

If flashover occurs there is always a danger that the insulation may be damaged by secondary effects. For these reasons, especially for medium-voltage cables with PE or XLPE insulation, the d.c. test voltage should be restricted to the absolute minimum essential value and in all cases excessive test voltages should be avoided (see Section 35.1).

After carrying out d.c. voltage tests, the cable cores must be discharged slowly through a discharge resistor (with medium-voltage cable approximately 100 kΩ 1000 W), to avoid overstressing the insulation with pulse waves. After discharging the cable all cores should be earthed and left in this condition until immediately prior to making live. This action is necessary as after discharge, charge carriers may collect at the conductor and under certain circumstances significant levels of voltage, conductor to earth, may build up.

Testing of Cables with Non-Radial Field

For belted cable and cables with polymer insulation having non-screened cores the following test series may be used:

Conductor 1 to conductor $2+3+E$[1]
Conductor 2 to conductor $1+3+E$[1]
Conductor 3 to conductor $1+2+E$[1]

with full time duration for each test.

---

[1] E = Earth to which also must be connected all metallic coverings of the cable

Table 35.1
Test voltages for low and medium-voltage cables with rated voltages $U_0/U \leq 18/30$ kV ($U_m \leq 36$ kV)

Cable rated voltage U_0/U	Highest voltage for the equipment U_m	DIN VDE 0298 part 1		IEC 502	IEC 55-1	
		a.c. test voltage	d.c. test voltage [3]	d.c. test voltage [2] for cables with		
				Polymer insulation [3]	Paper insulation	
					Radial field cables	Belted cables
kV	kV	kV	kV	kV	kV	kV
0.6/1	1.2	–	5.6 to 8	6	6	6.5
3.6/6	7.2	7	20 to 29	18.5	18.5	23.5
6/10	12	12	34 to 48	25	25	34
8.7/15	17.5	–	–	37	37	–
12/20	24	24	67 to 96	50	50	–
18/30	36	36	76 to 108	76	76	–
Duration of each test		15 min	15 to 30 min	15 min	5 min	5 min

[2] IEC also includes tests on a.c. in single-phase and three-phase connection but they have not been included in the table
[3] For repetitive tests and for the testing of older installations for cables with PE and XLPE insulation, the guidelines in Section 35.1 must be observed

Testing of Radial-Field Cables

In systems of single-core cables with metal sheath, S.L. cables, H cables and cables with polymer insulation with individually screened cores, all conductors can be strapped together in parallel and tested against the earthed metallic covering.

35.3 Testing with a.c. Voltage

Where, in exceptional cases an a.c. voltage test is required then, in a three-phase circuit, one conductor is earthed while the other two are tested. The test being repeated on each conductor sequentially.

Table 35.2
d.c. test voltages and test duration for high-voltage cable installations with rated voltages $U_0/U \geq 26/45$ kV ($U_m \geq 52$ kV)

Cable rated voltage U_0/U	Highest voltage for the equipment U_m	Rated level of lightening impulse voltage U_{rB} (peak value)	d.c. test voltage			
			Oil-filled cable DIN VDE 0256 IEC 141-1 IEC 141-4	Gas-pressure cable DIN VDE 0257 DIN VDE 0258	Gas-pressure cable IEC 141-2 IEC 141-3	Cable with polymer insulation to DIN VDE 0263 IEC 840
kV	kV	kV	kV	kV	kV	kV
26/45	52	250	117	117	104	78
36/60	72.5	325	162	162	144	108
64/110	123	450 550	225 275	288	256	192
76/132	145	450 550 650	225 275 305	342	304	228
87/150	170	550 650 750	275 325 350	392	348	261
127/220	245	750 850 950 1050	375 425 475 520	508	508	–
220/380	420	1050 1175 1300 1425	525 590 650 715	–	–	–
			Duration of each test			
			DIN VDE 30 min, IEC 15 min	60 min	15 min	15 min

36 Locating Faults

In this book only a rough summary of possible methods of locating a fault can be described. The methods of measurement required to precisely locate a fault in a cable installation, differ depending on the type of fault and also on local conditions.[1]

A wide variety of measuring equipments and locating equipments are available to suit different methods of measurement and these can be divided into two groups:

a) conventional measuring methods using bridges,

b) fault location techniques using impulse reflection measuring equipment for roughly locating, supported by special equipment for precise locating.

The conventional measuring method is successful only under certain preconditions and a fault may need several measurement attempts to locate it. With impulse measuring equipment however the fault can be located, without the need for calculations, in a comparatively short time. There is no single piece of apparatus available to measure and locate cable faults of any type with accuracy, therefore in difficult conditions a number of preparatory auxiliary arrangements must be made. One differentiates between preliminary measurements, locating measurements and location with detection gear.

36.1 Preliminary Measurements

Preliminary measurements or qualitative measurements allow a judgement to be made as to the nature of the fault. One can establish which cores of the cable have an earth fault or short-circuit fault or which cores are open circuit. In addition, the resistance of the fault can be measured to evaluate the transitional resistances at the fault between cores or between each core and the metallic covering. Based on these findings and measurements the method for fault detection can be selected.

Testing for Earthed and Short Circuited Cores

All cores of cables are disconnected at each end at terminals which are fixed and insulated from one another and from earth.

The insulation resistance to earth is tested as described on page 439 by connecting one terminal of an insulation tester to earth (metal sheath) and the other to each conductor in sequence.

In testing for short circuit between cores the insulation resistance between conductors is measured using an insulation tester. If an impulse reflection measuring system is to be used it is firstly recommended to measure the transitional resistance at the point of fault using a battery-lead tester or a low-voltage measuring bridge. An insulation tester should not be used for this purpose because the high voltage of the tester may initiate a spark-over causing indication of false values of transitional resistance. It is then not possible to say whether preliminary locating tests with impulse reflection equipment will provide any reliable result. The magnitude of the operating voltage (low or high voltage), the cable type and the length of the cable are decisive in determining the lowest permissible value of insulation resistance.

Testing for Damaged Conductor Using Resistance Measuring

The cores are short circuited at one end of the cable (see page 439). Using a resistance bridge (Thomson or Wheatstone) the resistances of the individual core loops are measured and these values compared with those recorded at installation or otherwise by calculation from length and cross-sectional area of conductor.

By comparing present and previous values of the individual conductors, a reduction in cross section of one conductor, by damage, can seldom be recognised. If however the measured resistance of one conductor loop is markedly higher than that of another (perhaps in the range of $k\Omega$ or $M\Omega$ values) then it can be concluded that conductor severence has occured.

---

[1] Additional guidance is given in the handbook "Fehlerortungstechnik in Energie- und Nachrichtenkabeln" (Fault locating techniques in energy and communication cables) by Sinema, E. Vol. 75. Kontakt und Studium, expert publishing

36.2 Location Measurements by the Conventional Method

A carefully conducted preliminary set of measurements will provide clues as to the type of fault and help in the selection of a suitable location system.

For the most commonly occuring types of fault, proven methods of location which have been successful in practice are listed in Table 36.1.

The bridge measurements to MURRAY and GRAF methods are conducted using a portable measuring bridge. Galvanometer and slide wire (resistance combinations) are separately connected to the end terminal screws such that the transitional resistance and feed resistances are in series with the slide wire. In this way the resistance has little influence on the accuracy of the test whilst in the branches of the bridge formed by the cable conductors they could lead to significant errors of measurement. The short-circuiting of conductors at the remote end is carried out in the same manner as for resistance measuring.

The cable measuring bridge is equipped also for measurements to the Thomson method (for $0.4\,m\Omega$ upwards) and to the Wheatstone method using battery supply for insulation resistance measurement (up to $500\,k\Omega$). All ohmic resistance values relevant to fault location (conductor, fault transitional, insulation resistances) can be measured. In addition the cable measuring bridge is equipped for capacitance measurements calibrated by either internal or external standard capacitors.

For cable networks which do not warrant the use of a cable measurement vehicle equipped with elaborate apparatus, the portable cable bridge serves as basic equipment for cable-fault location. The bridge sensitivity can be increased, when using a centre zero μA-meter as a balancing instrument, by carefully increasing the measuring voltage from an anode battery such that faults with resistances up $50\,k\Omega$ can be located. It must be remembered that if the fault resistance collapses (breaks down) there is a danger of the slide-wire being overloaded. (Slide-wire protection resistance should be used!)

Table 36.1 Summary of methods of locating

Type of fault, existance of a return or auxiliary conductor	Suitable locating method
1. Conductor not severed	
Core to earth or core to core short circuit exists through a small or medium resistance (0–$50\,k\Omega$). Existing are:	
a) two auxiliary conductors	Voltage related measure ment (Fig. 36.1) or three-point measurement to GRAF (Fig. 36.4)
b) return lead of same resistance	Bridge measurement to MURRAY (Fig. 36.2)
c) one auxiliary conductor exists (parallel cable or overhead line)	Bridge measurement to MURRAY with conversion calculation of auxiliary conductor (Fig. 36.2)
d) no return or auxiliary connection exists	Current direction method to WURMBACH[1] (Fig. 36.6).
Earth fault exists with a high resistance (greater than $50\,k\Omega$)	High-voltage measuring bridge (Fig. 36.3) or burning through (see page 449) and then measuring with low-voltage bridge
2. Conductor is broken at point of fault	
a) earth fault on one or several cores	Current direction method to WURMBACH[1] (Fig. 36.6)
b) cores have good values of insulation (no core to earth or core to core fault).	a.c. measuring bridge for loss free capacitances (Fig. 36.5) or for capacitance comparison measurement

[1] Only for fault resistance of $100\,\Omega$ or less

Voltage Ratio Measuring

Procedure: Readings of the voltmeter are taken in switch positions a and b with identical values of current flowing at each reading, adjusted by resistance R.

h1 and h2 are auxiliary conductors

$$x = \frac{l \cdot U_a}{U_a + U_b} = l \frac{1}{1 + U_b/U_a} \qquad (36.1)$$

x Distance to fault in m
U_a, U_b Readings of Voltmeter in V
l Length of cable in m

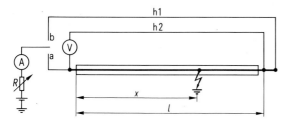

Fig. 36.1   Voltage ratio measurement

Bridge Measurement to MURRAY

Procedure: A balance is achieved by moving the slider along the slide wire S until galvanometer G indicates zero. If the return conductor R is different in length or cross-sectional area to conductor Q to be measured then the length of the return conductor

must be converted to a cross-sectionally equivalent length l_1 of the conductor to be measured. The value $l + l_1$ must then be substituted for $2l$ in the formula

$$x = 2l \frac{1}{1 + S_b/S_a} \qquad (36.2)$$

x Distance to fault in m
l Length of cable in m
S_b/S_a Ratio of proportional lengths of slide wire

Bridge Measurement with High Voltage

The measurement principle is the same as that for bridge measuring to Fig. 36.2. However, because high voltage is used a special model of measuring bridge is required

$$x = 2l \frac{1}{1 + S_b/S_a} \qquad (36.3)$$

x Distance to fault in m
l Length of cable in m
S_b/S_a Ratio of proportional lengths of slide wire

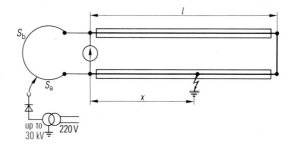

Fig. 36.3   Bridge measurements with high voltage

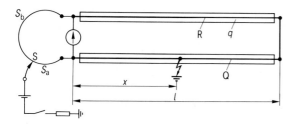

Fig. 36.2   Bridge measurements to MURRAY

Three-Point Measurement to GRAF

Procedure: Three measurements are made with switch in positions a, b and c.

By adjusting for balance on galvanometer G three points on the slide wire are found D_a, D_b and D_c.

h1 and h2 are auxiliary conductors

$$x = l \frac{S_b/S_a - 1}{S_c/S_a - 1} \qquad (36.4)$$

x Distance to fault in m
l Length of cable in m
S_a Distance on slide wire between D_a and A
S_b Distance on slide wire between D_b and A
S_c Distance on slide wire between D_c and A

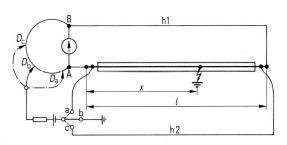

Fig. 36.4   Three-point measurement to GRAF

Bridge Measurement for Loss-Free Capacitances

Procedure: The bridge is balanced when there is no sound or minimum sound in the head phones.

Equipment required: Source of a.c. current at audible frequency (buzzer) and headphones (T) or sensitive instrument.

$$x = 2l \frac{1}{1 + S_b/S_a} \qquad (36.5)$$

x Distance to fault in m
l Cable length in m
S_a/S_b Ratio of proportional lengths of slide wire

As an alternative to this bridge measurement, by using a capacitance bridge, the capacitances of the severed core and a healthy core may be measured and the fault distance calculated

$$x = l \frac{C_x}{C} \qquad (36.6)$$

x Distance to fault in m
l Cable length in m
C_x Capacitance of the severed core in µF
C Capacitance of a healthy core in µF

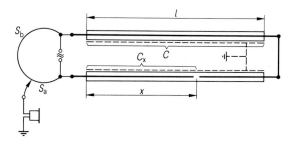

Fig. 36.5
Bridge measurements for loss free capacitances

Current Direction Measurement to WURMBACH

This method is suitable for low-resistance faults. In this method the clamps of a centre zero galvanometer are connected to the metallic covering of the cable at various points along the route (in the same direction in respect to one end). The distance between probes being 0.5 to 1.0 m.

The location of the fault is determined by noting the directional indication of the galvanometer in front of and behind the fault.

The fault lies at the point where the galvanometer indicates zero deflection. During testing the battery current must be switched on and off in order to eliminate possible errors of reading due to the effect of extraneous currents.

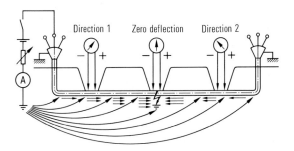

Fig. 36.6
Current direction measurement to WURMBACH

Voltage Step Method

For locating faults in sheaths of cables with aluminium sheath as well as earth faults in unscreened polymer insulated cables, the voltage step method based on the Wurmbach principle has gained in significance.

In this a d.c. voltage source (impulse generator) is connected to the sheath, screen or core which has the earth fault. The circuit is closed through the ground using an earthing rod. The current flow which forms in the ground along the cable run can be traced using galvanic probes in combination with a detector rod and differential transformer (Fig. 36.7).

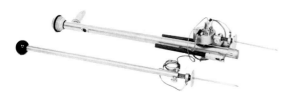

Fig. 36.7
Detector rod with galvanic probe and differential transformer, separate rod with galvanic probe

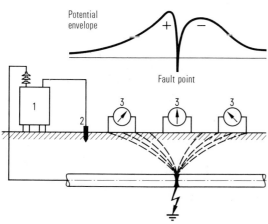

1 Impulse generator
2 Earthing rod
3 Galvanic probe with detection instrument

Fig. 36.8 Principle of the voltage step method

In the vicinity of the fault, where the current density is at the highest level, maximum voltage appears on the probe but at the point of fault it is zero, changing in sign on either side (Fig. 36.8). The easily measured polarity reversal at the point of fault leads to accurate location.

36.3 Locating of Faults by Pulse Reflection Method

This method of location is based on the principle of signal reflection. Impulses which are generated by the measuring apparatus are reflected at differing signal strengths from parts of the cable where the wave impedance changes, e.g. at the point of fault, joints, branch points etc. The reflected voltage signals are displayed on the oscilloscope monitor screen differing clearly in both sign and amplitude. The time which elapses between the initiation of the impulse and the receiving of the reflected signal is measured and halved. From the velocity of propagation, which can be calculated for cables to equation 36.7, together with travel time (t) of the impulses the distance to the reflection point and therefore to the fault can be determined. If the velocity of propagation of the impulses is denoted by v and the relative dielectric constant by ε_P

$$v = \frac{c}{\sqrt{\varepsilon_P}} \tag{36.7}$$

v   Velocity of impulses in m/s
ε_P Dielectric constant
c   Velocity of light in m/s

Conductor material and conductor cross section have no influence on the velocity. Since the dielectric constant is reasonably equal in cables of the same type these measurements can be conducted without knowledge of the conductor material or cross-sectional area. It is good practice however, prior to commencement of locating measurements, to measure the travel time on a healthy cable and from this result calculate the velocity. The distance to the fault can be calculated from the following formula:

$$l = \frac{v}{2} t \tag{36.8}$$

l   Distance to fault in m
v   Velocity of propagation in m/s
t   Measured time of travel in s

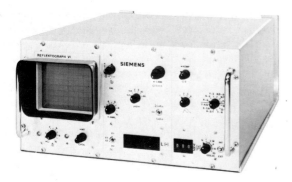

Fig. 36.9
Pulse reflection measuring device for fault location by
impulse reflection distance measurement

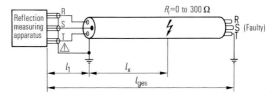

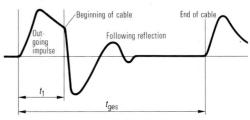

R   Healthy core with cable end open

An impulse generator produces measuring impulses
which are fed to the cable under investigation
through a flexible lead. The reflected return signal
is fed through an amplifier to the Y plates of a cathode ray tube. The deflection voltage for the time axis
is provided by an internal time circuit with selectable
scanning rates. On the monitor screen will be displayed the reflection image of the total length of
cable. The impulse travel time, respectively the distance to the fault location, can be read directly from
the apparatus.

Reflection measurement apparatus having a built-in
calculator has advantages. The reflected image of the
fault is followed by an adjustable light pen. The distance to the fault is then indicated as a digital read-
out in m.

The signal duration of the transmitted impulse is adjusted dependant upon the length of the cable and
the distance to the fault:

▷ short impulse for short cables or where the fault
  is close to the measuring point,

▷ long impulse for long cables or where the fault
  is remote from the measuring point.

If the fault resistance is low a strong negative reflection results (curve S for fault resistance $R_F \approx 0\,\Omega$);
where the fault resistance is high a weak negative
reflection results (curve S for fault resistance R_F
$= 100\,\Omega$). An open ended cable results in a positive
deflection and a short-circuited cable in a negative
deflection (Fig. 36.10).

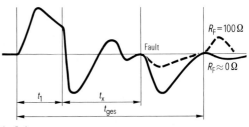

S   Faulty core

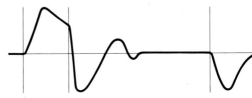

T   Healthy core with cable short circuited to earth

The length		Corresponding impulse travel time
l_1		t_1
l_x	Distance to fault	t_x
l_{ges}	Total length of cable under test	t_{ges}

Fig. 36.10
Images of a cable on the monitor screen of a
reflection measuring apparatus

Low Ohmic Faults

The application of the reflection measuring method is limited because the fault location to be measured must produce a visible deflection. For a power cable the transitional resistance of an earth fault or short circuit should not be more than approximately $500\,\Omega$ to ensure accurate measurement of the fault location. It is recommended therefore, prior to commencement of such measurement to verify the transitional resistance value by using a battery-operated cable tester or if necessary prepare the fault location as described in Section 36.4.

High Ohmic Faults

Impulse reflection measuring may also be used for the quick location of high ohmic faults or intermittent faults if the apparatus includes a synchronisation attachment.

With a high-tension generator (e.g. burn-through apparatus) high-voltage impulses are fed to the faulty core causing flashover at the point of fault.

The return travelling wave is detected by capacitive coupling to a high-voltage measuring device and is fed into the synchronisation attachment. The travel time of the impulse, respectively distance to fault, can be read off directly from the reflection measuring apparatus.

Application of Double Digital Memories

When using high voltage for the prelocation of faults the oscillograph trace cannot be repeated and therefore must be recorded photographically. Only in the outward swing method, where a more powerful high-voltage equipment is used to repeatedly charge and discharge the faulty cable increasing the number of flash overs, is a steady picture obtained which can be analysed. In addition to the difficulties in setting up photographic equipment and the subsequent time delay, addition problems can arise in analysing the photograph obtained. The principle of time measurement (displacement technique) used in the echo sounding system can not always be applied and measurements can only be made based on the calibrated time base within the impulse measuring apparatus.

Methods with quick repeat frequency facility are advantageous. These have the facility to electronically store a once only signal and therefore offer a noticeable saving in time and also more accurate measuring for locating intermittent faults. With this picture retention all the advantages of a modern echo sounding device can be applied. As memory stores magnetic tape, cathode ray oscilloscopes with memory tube and also digital stores are available.

For the prelocation of cable faults digital stores have particular advantages and it is expected therefore for this system to dominate in the future.

The memory offers the facility to hold a screen picture of faulty cables prior to and after treatment with burn-through or impulse apparatus. Even minute deviations in the two recalled, superimposed pictures can clearly indicate the point of fault.

The stored signal can also be recalled as often as necessary and be fed into a commercially available oscillograph or better still fed into the reflection measuring apparatus where it will be converted into a steady picture. With slow controlled data recall the output can be plotted on an XY plotter which can be calibrated with the relevant time scale to produce a picture for archive storage.

Time measurements using the stored picture information can be made by using an additional attachment to the memory as well as by using the impulse reflection device. When using the add-on time measuring attachment on the memory, measurements can be read from the oscillogram using the calibration time markers on the screen of 50 or 100 ns pitch. Alternatively the time can be measured in the normal manner using the proven displacement technique of the impulse apparatus viz. two periods or echo impulses are adjusted to coincide and the time differential for this moment is read off directly.

36.4 Preparation of Fault Point by Burn-Through

In order to measure and accurately locate a high-resistance fault using an impulse reflection device, the point of fault must be changed such that the transient resistance is of a suitably low value. To achieve this a cable test and burn-through device is required. In burning through the insulation material at the point of fault is carbonised such that a current carrying low ohmic transitional resistance is developed. The physical process in which heat is used to split the insulation material and form a firm carbon conductor path is termed "burn-through".

When using a powerful burn-through apparatus it is possible to burn the point of fault within a few minutes to produce a resistance value of 0.5 Ω, this however is not of great help since the point location often becomes difficult. The resistance at the fault is then too low being mostly a current carrying carbon bridge which is often non clearable.

With an impulse generator, flashover does not occur and therefore there is no audible noise. For this reason the burn-through system is today used only conditionally especially as technically sophisticated test equipment can achieve quick locating of a fault even of high resistance.

36.5 Locating Using Audio Frequency

In order to correct a cable fault, its exact location must be found. In most instances the routes of cable runs are recorded on the cable plan. If, however e.g. the route of an old cable is not known and the position and length of the cable require to be determined, ones first choice is to use inductive detection equipment.

Such apparatus consists of an audio frequency (tone) transmitter, a portable audio receiver with indicating instrument, headphones and two detector coils which are fixed to a rod so that they can be held close to the ground surface. For route tracing and for fault locating the cable is connected to the tone frequency transmitter (Figs. 36.11 and 36.12) and the route is walked carrying the tone frequency receiver. The intensity of the detection field can be determined with the headphones and the detection instrument.

For the location of faults with the tone frequency system it requires that a connection exists between two cores at the point of fault. This connection can be readily achieved using a modern burn-through device which allows earth potential free burn-through between two cores. In addition a low frequency should be used for detecting in order to reduce the effect of cross coupling with neighbouring cables to a minimum. The apparatus should also have a high degree of discrimination so that neighbouring power and control cables under voltage do not influence the detection process

Fig. 36.11
Fault locating with core-sheath field method

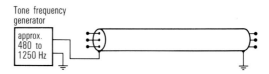

Fig. 36.12
Fault locating with sheath-earth field method

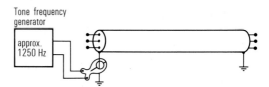

Fig. 36.13   Fault locating in a cable under voltage

By using a clamp-on current transformer to inductively inject the tone frequency into the cable sheath, the screen or the armour, the route of a cable under voltage can be located.

Route Locating of Cables by Sheath Field

In the core-sheath field method the tone frequency generator is connected between one core and the sheath while at the other end of the cable the core is connected to the sheath or screen and is earthed.

The current fed into the core is not equal to the sheath current since a proportion of this current returns through the ground. The differential field resulting from this can be picked up by the search coils and amplified through the audio receiver (Fig. 36.14).

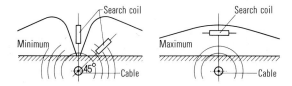

Fig. 36.14 Search coil application

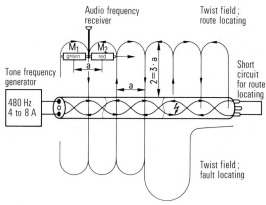

Fig. 36.15
Route and fault locating in cable with twist field
(lines of force are drawn 90° offset)

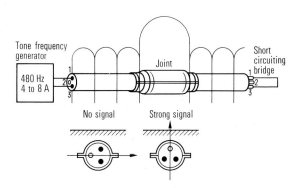

Fig. 36.16 Joint detection by twist field

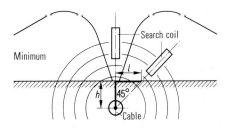

Fig. 36.17 Depth locating by twist field

The linking of the search coils by the lines of force of the core-sheath field is greatest when the coils are held horizontally and at right angles to the cable axis; the linking then generates maximum tone signal. If however the coil is moved horizontally to the cable a distinct and sharp tone minimum occurs.

In all route locating in which the sheath is used as either conductor or return path there is the possibility that neighbouring cables or pipes may cause signal distortion such that a false route is followed. In such cases the twist field method is more satisfactory.

Twist Field Method

If the tone frequency transmitter is connected to two cores which are short circuited at the remote end, the generated magnetic field axis follows the twist, or lay, of the cores. This field has two components which progress perpendicularly to the cable axis (cross and longitudinal field). Experience has shown it is advantageous to use the longitudinal field for locating in order to minimise the effect of extraneous fields which may arise from cables and pipes running parallel.

The coils are held horizontally and in the direction of the cable route. Voltages induced in the coils vary depending on the twist of the cores between minimum and maximum values (Fig. 36.15).

If the cable contains a prepared fault, up to the fault the so called twist field is received. Beyond the fault no current is flowing and hence no audio signal is detected.

At a joint the cable cores are spread. The twist distance is increased and the signal is stronger (Fig. 26.16).

The depth of lay of a cable can also be determined using twist field apparatus. In this method a vertically held coil is located over a minimum point and then tilted 45° until a minimum tranverse to the route is determined (Fig. 36.17). The distance between these two points is relative to the depth of lay ($l = h$).

Identification of Cables

With the aid of an impulse receiver connected to a clip-on current transformer any one cable in a bunch or group of cables can be identified or selected (e.g. for branching or other installation work. The conductor of the cable is fed with a pulsating d.c. voltage and the cable is detected by use of a clip-on current transformer.

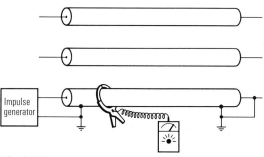

Fig. 36.18
Identification of cable by using clip-on current transformer

Locating by Sound Field Method

For accurate locating of faults in single-core cables and triple-sheathed cables the tone frequency methods are only successful in a few cases. For this reason today approximately 80% of cable faults are located using the sound field method. This method produces the quickest result and is very simple to apply. In this capacitors are charged by means of an impulse

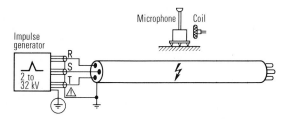

Fig. 36.19
Point locating of a fault using impulse voltage in a cable with metal sheath or screen

generator. The stored energy is then fed through a spark gap to the damaged cable. At the point of fault flashover occurs producing a loud bang. This bang can be detected by use of a geological microphone.

It is important that a safety impulse generator is used which ensures that if at the point of fault the cable is not discharged then this is done automatically and the cable disconnected. Otherwise at the point of fault a step voltage could be built up which would be dangerous to humans or animals. Cables which are laid very deeply cannot be detected, or only under certain conditions, by the sound field method.

36.6 Testing of Thermoplastic Sheaths

Voltage testing with the object of detecting faults in thermoplastic sheaths is only possible if metallic parts are situated immediately under the sheath, e.g. screen, armour or metal sheath.

Voltage Tests

To ascertain whether cables have become damaged during installation or during building operations while in operation, it is common practice to make a test using d.c. voltage up 5 kV (DIN VDE 0298 Part 1). To conduct this test the metal sheath, armour or screen must be isolated from earth at the sealing ends and if necessary at the joints. The insulating effectiveness to earth can then be checked by measuring leakage current to earth. Coverings which are undamaged can be expected to indicate a limiting value, e.g. for a cable 1 km long an insulation resistance of 2.5 MΩ resulting in a leakage current of 0.8 mA. In this way one can quickly determine whether the metal covering is in contact with earth.

Short circuiting of the screen to the earth clamp after charging it with a d.c. potential will provide positive indication. If the screen discharges with a visible and audible spark then the thermoplastic sheath is healthy; if a spark does not occur then it is evident the charge has leaked to earth through a damaged portion of the outer sheath. The cable so tested should then be subjected to further investigation or at least be monitored at more frequent intervals than in the normal routine test program.

Fault Locating

For the detecting and accurate locating of sheath damage faults, methods should be used which avoid secondary damage, e.g. by too high a voltage stress or excessive thermal stress at the point of damage. The previously described pre-locating of cable faults is difficult to apply to sheathing faults because the transitional resistance between the metal sheath and the ground is too large. A low ohmic burn-through at the fault point is also not practical because it would not be possible to form a carbon bridge to earth. This method is also most impractical as the energy released at the burn-through would not only damage the sheath but also the cable core insulation.

Possible suitable methods for the pre-locating are by the use of low-voltage or high-voltage measuring bridges or by locating using the differential voltage method. Because a cable may have several sheath faults at the same time these methods may not provide very satisfactory results.

Point Locating

For point accurate locating of the point of fault several methods exist. Mostly the audio frequency tone method and the impulse method are the preferred systems. In both a potential funnel develops at the fault location which can readily be detected with suitable measuring apparatus.

Tone Frequency Method

Locating with tone frequency requires a sufficiently good connection between the metal sheath or screen and earth. If an audible tone frequency applied to the sheath – the return path being formed through the ground – an electrical field radiates from the point of damage. Directly above the fault point the signal is zero. Using a suitable receiving apparatus this point can be located. Such a receiving apparatus consists of a tone frequency receiver and an associated detecting rod with capacitor probes. False measurements may occur due to the proximity of metal barriers, earthed lamp posts, metal fences or similar since these objects may cause distortion of the potential field.

Impulse Method

Using an impulse generator d.c. voltage impulses of between 2 and 4 kV are fed to the screen. These impulses create a potential funnel at the point of fault which can readily be detected. Sound detection by means of a geological microphone is not always possible since the transitional resistance is so large that no detectable acoustic noise is present. Point fault locating for practical reasons is therefore effected using a sensitive measuring instrument (measuring range approximately 10 to 300 mV centre-zero) in association with two detecting rods with insulated handles. These rods are connected by cables 10 to 20 m long to the balancing galvanometer. The cable route is walked using the rods at an initial pitch of 20 to 30 m. As the fault is neared the distance between the rods is reduced until a zero point is found and therefore the point of fault. It is important that only the injected impulses are registered and random deflections on the instrument due to stray earth currents are ignored.

In this locating method it is irrelevant whether the cable length contains several or only one sheath fault. All faults can be located without the need to repair the faults as found in sequence.

37 Construction and Resistance of Conductors

Table 37.1 Circular copper conductors

Rated cross-sectional area	Resistance at 20° C (max.) plain wires	metal-coated wires	Solid (class 1) Conductor diameter (max.)	Stranded (class 2) non-compacted Minimum number of wires	Conductor diameter (max.)	compacted Minimum number of wires	Conductor diameter (min.)	(max.)
mm²	Ω/km	Ω/km	mm		mm		mm	mm
0.5	36.0	36.7	0.9	7	1.1			
0.75	24.5	24.8	1.0	7	1.2			
1	18.1	18.2	1.2	7	1.4			
1.5	12.1	12.2	1.5	7	1.7			
2.5	7.41	7.56	1.9	7	2.2			
4	4.61	4.7	2.4	7	2.7			
6	3.08	3.11	2.9	7	3.3			
10	1.83	1.84	3.7	7	4.2			
16	1.15	1.16	4.6	7	5.3			
25	0.727	0.734		7	6.6	6	5.6	6.5
35	0.524	0.529		7	7.9	6	6.6	7.5
50	0.387	0.391		19	9.1	6	7.7	8.6
70	0.286	0.270		19	11.0	12	9.3	10.2
95	0.193	0.195		19	12.9	15	11.0	12.0
120	0.153	0.154		37	14.5	18	12.5	13.5
150	0.124	0.126		37	16.2	18	13.9	15.0
185	0.0991	0.100		37	18.0	30	15.5	16.8
240	0.0754	0.0762		61	20.6	34	17.8	19.2
300	0.0601	0.0607		61	23.1	34	20.0	21.6
400	0.0470	0.0475		61	26.1	53	22.9	24.6
500	0.0366	0.0369		61	29.2	53	25.7	27.6
630	0.0283	0.0286		91	33.2	53	29.3	32.5
800	0.0221	0.0224		91	37.6	–	–	–
1 000	0.0176	0.0177		91	42.2	–	–	–

Table 37.2 Shaped copper conductors

Rated cross-sectional area	Resistance at 20 °C (max.)	Stranded compacted (class 2)	Rated cross-sectional area	Resistance at 20 °C (max.)	Stranded compacted (class 2)
		Minimum number of wires			Minimum number of wires
mm^2	Ω/km		mm^2	Ω/km	
25	–	–	150	0.124	18
35	0.524	6	185	0.0991	30
50	0.387	6	240	0.0754	34
70	0.268	12	300	0.0601	34
95	0.193	15	400	0.0470	53
120	0.153	18	500	–	–

Table 37.3 Flexible copper conductors

Rated cross-sectional area	Resistance at 20 °C (max.)		Conductor diameter (max.)	Maximum diameter of strands	
	plain wires	Metal-coated wires	Classes 5 and 6	Stranded class 5	Stranded class 6
mm^2	Ω/km	Ω/km	mm	mm	mm
0.5	39.0	40.1	1.1	0.21	0.16
0.75	26.0	26.7	1.3	0.21	0.16
1	19.5	20.0	1.5	0.21	0.16
1.5	13.3	13.7	1.8	0.26	0.16
2.5	7.98	8.21	2.6	0.26	0.16
4	4.95	5.09	3.2	0.31	0.16
6	3.30	3.39	3.9	0.31	0.21
10	1.91	1.95	5.1	0.41	0.21
16	1.21	1.24	6.3	0.41	0.21
25	0.780	0.795	7.8	0,41	0.21
35	0.554	0.565	9.2	0.41	0.21
50	0.386	0.393	11.0	0.41	0.31
70	0.272	0.277	13.1	0.51	0.31
95	0.206	0.210	15.1	0.51	0.31
120	0.161	0.164	17.0	0.51	0.31
150	0.129	0.132	19.0	0.51	0.31
185	0.106	0.108	21.0	0.51	0.41
240	0.0801	0.0817	24.0	0.51	0.41
300	0.0641	0.0654	27.0	0.51	0.41
400	0.0486	0.0495	31.0	0.51	–
500	0.0384	0.0391	35.0	0.61	–
630	0.0287	0.0292	39.0	0.61	–

Table 37.4   Aluminium conductors

Rated cross-sectional area	Resistance at 20 °C (max.)	Circular conductor (class 2) stranded [1] compacted			Shaped conductor (class 2) stranded [2] compacted
		Minimum number of wires	Conductor diameter (min.)	(max.)	Minimum number of wires
mm^2	Ω/km		mm	mm	
25	1.20	6	5.6	6.5	6
35	0.868	6	6.6	7.5	6
50	0.641	6	7.7	8.6	6
70	0.443	12	9.3	10.2	12
95	0.320	15	11.0	12.0	15
120	0.253	15	12.5	13.5	15
150	0.206	15	13.9	15.0	15
185	0.164	30	15.5	16.8	30
240	0.125	30	17.8	19.2	30
300	0.100	30	20.0	21.6	30
400	0.0778	53	22.9	24.6	53
500	0.0605	53	25.7	27.6	–
630	0.0469	53	29.3	32.5	–

[1] Solid circular conductors (class 1) are permitted up to 300 mm^2. The diameters of non-compacted conductor 25 mm^2 to 63 mm^2 are found in DIN VDE 0295
[2] Solid shaped conductors (class 1) are permitted from 50 mm^2 to 240 mm^2

38 Conversion Table

Table 38.1 Comparison of metric, British and American standards of conductors of insulated cables

British Standards [1] BS			Metric range of cross-sectional areas [2] (complies with VDE)	American Wire Gauge (AWG [3])	
Conductor cross-sectional area sq. inch	Number and diameter of wires inch	Relative cross-sectional area in mm²	Cross-sectional area mm²	Relative cross-sectional area in mm²	AWG or MCM [3]
.001	3/.020 or 1/.036	0.65	0.75	0.653	19 AWG
				0.823	18
.0015	1/.044	0.97		1.04	17
.0020	3/.029	1.29	1.5	1.31	16
				1.65	15
.003	3/.036 or 1/.064	1.94		2.08	14
			2.5	2.62	13
.0045	7/.029	2.90			
.0050	1/.083	3.23		3.31	12
			4.0	4.17	11
.007	7/.036	4.52			
.008	1/.103	5.16		5.26	10
			6.0		
.01	7/.044	6.45		6.63	9
.013	1/.128	8.39		8.37	8
.0145	7/.052	9.35	10.0		
				10.55	7
.020	1/.160	12.90		13.30	6
.0225	7/.064	14.52	16.0	16.77	5
.03	19/.044 or 1/.192	19.35		21.15	4
.04	19/.052	25.81	25.0	26.67	3
			35.0	33.63	2
.06	19/.064	38.71		42.41	1
			50.0	53.48	1/0
.10	19/.083	64.52	70.0	67.43	2/0
				85.03	3/0
.15	37/.072	96.77	95.0	107.20	4/0
.2	37/.083	129.03	120.0	126.64	250 MCM
.25	37/.093	161.25	150.0	152.00	300
.3	37/.103	193.55	185.0	202.71	400
.4	61/.093	258.06	240.0	253.35	500
.5	61/.103	322.58	300.0	304.00	600
.6	91/.093	387.00		354.71	700
			400.0	405.35	800
.75	91/.103	483.87	500.0	506.71	1 000
1.0	127/.103	645.00	625.0		

[1] BS have adopted the metric range of cross-sectional areas to IEC 228

[2] To IEC 228

[3] The conductor cross-section is expressed as an AWG number (American Wire Gauge) and for large sizes in MCM (milli-circular mill)

Index